THE SUPERSYMMETRIC WORLD
The Beginnings of the Theory

SECOND EDITION

THE SUPERSYMMETRIC WORLD
The Beginnings of the Theory

SECOND EDITION

editors **G. Kane**
University of Michigan, USA

M. Shifman
University of Minnesota, USA

World Scientific

NEW JERSEY • LONDON • SINGAPORE • BEIJING • SHANGHAI • HONG KONG • TAIPEI • CHENNAI • TOKYO

Published by

World Scientific Publishing Co. Pte. Ltd.
5 Toh Tuck Link, Singapore 596224
USA office: 27 Warren Street, Suite 401-402, Hackensack, NJ 07601
UK office: 57 Shelton Street, Covent Garden, London WC2H 9HE

Library of Congress Control Number: 2024943399

British Library Cataloguing-in-Publication Data
A catalogue record for this book is available from the British Library.

Cover illustration: Victory over Eternity by Pavel Filonov © State Russian Museum, Petersburg

THE SUPERSYMMETRIC WORLD
The Beginnings of the Theory
Second Edition

ISBN 978-981-98-0060-5 (hardcover)
ISBN 978-981-98-0061-2 (ebook for institutions)
ISBN 978-981-98-0062-9 (ebook for individuals)

For any available supplementary material, please visit
https://www.worldscientific.com/worldscibooks/10.1142/14035#t=suppl

Desk Editor: Carmen Teo Bin Jie

Typeset by Stallion Press
Email: enquiries@stallionpress.com

FOREWORD TO THE SECOND EDITION OF "THE SUPERSYMMETRIC WORLD"

M. SHIFMAN

Theoretical Physics Institute, University of Minnesota, Minneapolis, MN 55455

shifman@umn.edu

The First Edition of this book was released in 2000, just before the symposium "Thirty Years of Supersymmetry" was held at the William I. Fine Theoretical Physics Institute (FTPI) of the University of Minnesota. Founders and trailblazers of supersymmetry descended on FTPI, as well as a large crowd of younger theorists deeply involved in research in this area. Remarkably, it was at this event that many of the early pioneers of the field met face-to-face for the first time. Table 1 presents the Contents of the historical part of the SUSY-30 Proceedings [1]. Since then 24 years have elapsed and significant changes happened in supersymmetry (SUSY). Its history definitely needs an update.

Below the reader will see a table which can be viewed as a starting point for the current edition. In 2000 supersymmetry explorations were on the rise and SUSY-based phenomenology attracted hundreds of young researchers, with around 2600 original papers published annually (see the graphs in Figs. 6.3, 6.4 on page 368). Expectations of an imminent experimental discovery of superpartners were prevailing in the community; the general mood was rosy.

The modern perspective on supersymmetry looks different. So far, all searches of SUSY in LHC experiments have had negative results, which is a great disappointment. The rosy mood of the early 2010s faded away.

On the other hand, the unique power of supersymmetry became obvious in mathematical physics, string mathematics, and such topics as black holes and critical (BPS) solitons, especially in strong coupling regime where SUSY hardly has any competitors. For the first time ever, the analytic proof of quark confinement was found in $\mathcal{N}=2$ super-Yang-Mills theory [2] slightly deformed by an $\mathcal{N}=1$ perturbation.

This is reflected in Fig. 1 which I designed especially for the current edition. (See also the graph in Fig. 6.5 on page 369). Figure 1 demonstrates

Table 1: The historic part in [1]. Table of contents

Part 1: Supersymmetry from the East	
N. Koretz-Golfand	Supersymmetry – 30
E.P. Likhtman	Around SUSY 1970
V. Akulov	Nonlinear way SUSY and $\mathcal{N}$ extended SUSY
S.I. Volkova, A.A. Zheltukhin	Glimpses of Dmitry Volkov life and work
V.A. Soroka	Supersymmetry and the odd Poisson bracket
Part 2: Supersymmetry from the West	
Pierre Ramond	Boson-fermion confusion: The string path to SUSY
J.H. Schwarz	String theory origins of supersymmetry
J.-L. Gervais	Symmetries: early days and nowadays
B. Sakita	Symmetries of fermions in the lowest Landau level
P. Fayet	About the origins of supersymmetric standard model
J. Iliopoulos	Non-renormalization theorems in global supersymmetry
L. O'Raifeartaigh	Chiral fermions on the lattice
P. West	Non-renormalization theorems in supersymmetric theories
M.F. Sonius	Recollections of a young contributor
S. Ferrara	Superconformal algebras and supergravities in higher dimension
A.H. Chamseddine, R. Arnowitt, P. Nath	Supergravity unification
Bryce de Witt	Electric-magnetic dualities in supergravity

that the "Pheno" branch of supersymmetry is drying out — its future is unclear at the moment. The "strong coupling" branch dealing with "imaginary" worlds and more mathematical applications of SUSY continues to grow.

Perhaps, it would be fair to add that the past searches for superpartners and the studies of very feebly-interacting goldstino/gravitino in a susy-breaking sector led to current searches for new feeble interactions in dark sectors. Life goes on, work continues, in particular, the hunt for axion-like particles (ALPs). One may hope that a new tree will sprout in Fig. 1.

"Supersymmetry is so beautiful and suggestive that most of us think it has got to show up sometime in nature, although so far it hasn't," said Steven Weinberg in his 1979 Nobel speech. I still believe in Weinberg's prophecy

Figure 1: The superworld tree in 2023. ©M. Shifman.

but it is reasonable to ask "when?" Unfortunately, I am not a prophet and the only answer I can give is "The future's not ours to see..."

> "We cannot guess the way
> Our words will echo through the ages...
> In real world, how it'll return..."[1]

In this volume we collected personal memories describing days long gone, with their excitement and even euphoria.[2] These testimonies were penned by the discoverers of supersymmetery, trailblazers, and pioneers themselves. Hopefully they will be used by future scholars, in particular, historians of science. Unfortunately, even in 2000 when the First Edition was being prepared it has been too late to obtain personal memoirs from Yuri Golfand (1922–1994) and Abdus Salam (1926–1996). We had to settle for their students

[1] A quote from a Russian poet, Fyodor Tyutchev (1803–1873). Translated by Lyudmila Purgina and Frank Jude.

[2] Or was it inebriation?..

and collaborators — Evgeny Likhtman in the first case, and John Strathdee and Peter West in the second.[3]

*****

Almost all of the early explorers of supersymmetry, its founding fathers, have left our world after 2000. This is a sad — but alas... — a natural process. Julius Wess and Bruno Zumino, who (among others) were the masterminds and discoverers of four-dimensional supersymmetry, passed away in 2007 and 2014 respectively. They were life-long friends and collaborators. I feel that Julius and Bruno, who gave so much to our community, deserve a special "In Memoriam" chapter in the current edition (see Chapter 1).

The available literature about the dawn of supersymmetry has significantly expanded since 2000. A wealth of information is presented in [1] and the accompanying volume [3]. In 2004, a fascinating "Concise Encyclopedia of Supersymmetry" was published [4] with more than 700 articles and a historical section written specifically for this Encyclopedia by E. Likhtman, Dmitry Volkov, V. Akulov, H. Miyazawa, G. Stavraki, V. Kac, V. Pakhomov, J. Łopuszanski, R. Haag, and D. Leites. Next, it is worth mentioning the collection [5] narrating the story of Felix Berezin, an outstanding mathematician who created a mathematical apparatus used in supersymmetric field theories. The integral over the anticommuting Grassmann variables that he introduced in the 1960s paved the way for the path integral formulation of quantum field theory with fermions, the heart of modern supersymmetric field theories and superstrings.[4] The Berezin integral is named after him, as is the closely related construction of the Berezinian, which may be regarded as the superanalogue of the determinant. In the same year the Dmitry Volkov Memorial Volume [6] was published in Ukraine in Russian. It contains both recollections of his colleagues and friends and Volkov's selected

[3]Here I have to make a few explanatory remarks. (i) In English literature Golfand's surname is written in two different ways — Golfand or Gol'fand. The apostrophe in the latter version indicates that a certain "mute" letter of the Russian alphabet is omitted. Phonetically this omitted letter makes the previous "l" soft, as in French; (ii) Evgeny Likhtman was Golfand's PhD student and a co-author of three crucial papers; (iii) John Strathdee was Salam's co-author for 27 years! He retired from ICTP in the late 1990s and hid somehwhere in New Zealand. Finding him there was not easy; I was told that he had left his mailing address only to a certain secretary at ICTP. I contacted her and eventually she entrusted me with Strathdee's address after I promised her to use it just once. Strathdee's reply to my letter contained seven lines, for which I am grateful, see Chapter 5, p. 248.

[4]See Marinov's article in Chapter 6, p. 356.

interviews spanning the years 1986–1991. Two detailed essays on Yuri Golfand appeared in Part 2 of the book [7].

In 2012 a profound collection *The Birth of String Theory* [8] was released. It contains a wealth of information on the early days of string theory some of which is relevant to our Chapter 2; see in particular Ramond's articles.

Approximately a year before his death, Stanley Deser published a book of recollections about his life and work [9]. The chapter entitled "A Big Year" will be especially interesting to readers since it narrates Deser's viewpoint on the story of the creation of Supergravity (SUGRA). Other creators of SUGRA published a number of reviews containing their recollections. The appropriate excerpts will be presented in Chapter 4 of the current Edition.

Finally, a remarkable connection between the late works of W. Heisenberg and the discovery of nonlinearly realized supersymmetry by Volkov and Akulov was revealed in [10].

Compared to the 2000 edition, the Second Edition is significantly expanded. First, as was mentioned, we added Chapter 1, *In Memoriam: Julius Wess and Bruno Zumino*, which presents recent recollections on Zumino and Wess. Chapter 4, *Local Supersymmetry (Supergravity)*, is completely revised, and now includes a number of articles which appeared after 2000. Chapter 7 containing a brief report on SUSY-50[5] was prepared specifically for the Second Edition. In the newly organized Appendix the reader can find English translations of some of the early papers by Golfand, Likhtman, Volkov, Akulov, and Soroka which were published in JETP Letters and therefore are not easily accessible. In a number of instances I added comments or explanatory remarks, see e.g. p. 483.

Other additions include an essay of John Iliopoulos on the early results, dreams and expectations of supersymmetry, as well as a brief review of the present situation (Chapter 5). In the same Chapter the reader will find recollections about Victor Ogievetsky (1928–1996), the inventor of the harmonic superspace for $\mathcal{N} = 2$ theories. Upon reflection, I decided that supersymmetry at *strong coupling* should be viewed as a "special branch" of SUSY (Fig. 1), and a few words about its pioneers are in order in the Second Edition. Correspondingly, I added an interview with Nati Seiberg in Chapter 5 devoted to the pioneers as well as my commentary.

Each article which was absent in the First Edition is marked by ⊎.

[5]The conference *50 Years of Supersymmetry* (SUSY 50) was organized by FTPI, University of Minnesota, on May 18–20, 2023.

I also correct two mistakes. The Foreword to the First Edition stated:

> Often students ask where the name "supersymmetry" came from. It seems that it was coined in the paper by Salam and Strathdee [11] where these authors constructed supersymmetric Yang-Mills theory [using superspace formalism]. This paper was received by the editorial office on June 6, 1974, exactly eight months after that of Wess and Zumino. Super-symmetry (with a hyphen) is in the title, while in the body of the paper Salam and Strathdee use both, the old version of Wess and Zumino, "super-gauge symmetry," and the new one.

When Bruno Zumino read this paragraph (it was long ago) he got upset and replied me in a letter which also contained a typewritten text of his talk [12]. He wrote:

> Dear Misha,
>
> [...] I did the same in my talk at 17th International Conference on High-Energy Physics. Please, read it.

I apologized and promised Bruno to correct my mistake. Now, many years later I fulfill the promise. The reader can find Bruno's 1974 talk added in Chapter 3, p. 140, just after Julius Wess' talk. If the reader looks through the above Bruno's talk he or she will note that in fact, apparently, Zumino and Salam had a pre-publication communication. The issue of Zumino's priority remains foggy. Nevertheless, the talk is worth reading.

Another unintentional omission was pointed out to me by Pierre Ramond. He suggested supplementing the narrative on Pierre Fayet's important contribution (MSSM),[6] by the reference to Dimopoulos and Georgi [14] who developed a supersymmetric model with softly broken SUSY in the context of Grand Unification in 1981.

As was mentioned on page vii, Abdus Salam was one of the SUSY pioneers whose role was absolutely crucial. He (and his life-long collaborator John Strathdee) invented superspace and superfields formalism [15] (this paper

---

[6]Namely, in the First Edition one reads:

> In the subsequent years, Pierre Fayet combined the Brout-Englert-Higgs mechanism with supersymmetry and introduced superpartners to the Standard Model particles [13]. This was the inception and birth of the minimal supersymmetric standard model (MSSM). Heavy quarks unknown in the 1970s, neutrino mixing angles and some details were introduced somewhat later [...],

Figure 2: One of the posters honoring Bruno Zumino's 90th anniversary.

was received by the Editorial Office on February 26, 1974). Ferrara-Wess-Zumino work on the same topic [16] was issued a little later (being received by the Editorial Office on May 3, 1974, it cites [15]). There are other important overlaps between Ferrara, Wess and Zumino, and Salam and Strathdee, e.g. the discovery of super-Yang-Mills theory (see [17], received by the Editorial Office on May 27, 1974),[7] and [18], received on June 6, 1974). In the First Edition brief recollections of Abdus Salam were presented only in the articles

[7]There is a curious typo in the published version of [17]. The editorial note reads "Received 27 May 1973." If correct, the arrival of the manuscript to the Nuclear Physics Editorial Office in May 1973 would be an acausal effect. As far as I know, this typo has never been corrected.

by Martin Sohnius and Peter West. For the Second Edition I solicited a more detailed essay from Peter West, and he kindly agreed (see page 250).

I have to add a remark on the 75-pages historical essay (p. 374) written by Rosanne Di Stefano 37 years ago which was published in the First Edition. A stream of relatively recent recollections authored by the Creators and Pioneers of SUSY and string theory (see p. ix) made some parts of this essay dated, especially Sections 8 and 9. Upon reflection, I decided to keep it intact in the current Second Edition because it contains a large number of details and names regarding the early days which are rapidly fading away in community's memory. More importantly, it gives an idea of the prevailing trends in the community in the 1980s — quite different from those we witness today almost 40 years later (Chapter 7, p. 461).

Preparing the Second Edition I noticed a change in the perception of SUSY history which is social rather than scientific. In the 1970s, 1980s, and 1990s, only a few early researchers in this area were aware of the contributions of Golfand, Likhtman, Volkov, Akulov and Soroka. Virtually none of the young theorists knew about works on four-dimensional supersymmetry and supergravity "before Wess-Zumino." This was one of the dark consequences of the 70-year Iron Curtain between the USSR and the rest of the world erected by the communist regime. The proceedings of SUSY-30 [1], the subsequent publications (the corresponding references can be found on page xiii) and a number of talks on the origins of SUSY I gave in the last two decades on various occasions led to a general recognition of the "contribution from the East." Now it has become common knowledge[8] which I view as my modest achievement in restoring "fair attribution" practice in our community.

Concluding this Foreword I have to mention that, unfortunately, Gordy Kane was unable to participate in preparation of the current edition. Therefore, I am solely responsible for all inaccuracies the reader may find in the added material.

Acknowledgments

With great pleasure I acknowledge many illuminating discussions with Pierre Fayet, Dan Freedman and Pierre Ramond. I am grateful to Ivo Sachs, Pierre Ramond, S. Moskaliuk, Hermann Nicolai, Mary K. Gaillard, John Iliopoulos, K. S. Stelle, Evgeny Ivanov, Emery Sokatchev, Natan Seiberg, Stanley Deser,

[8]See e.g. Chapter 2, Ramond's essay on p. 43.

Bayram Tekin, Peter van Nieuwenhuizen, Daniel Freedman, and S. Duplij for their contributions to the current version of this collection.

The cover of this book is based on a fragment from the painting *Victory over Eternity* by Pavel Filonov (1883–1941), a Russian artist, one of the leaders of the "Russian Avant-garde."[9] I use this opportunity to say thank you to Vera Kessenich who kindly helped me to obtain the reprint permission from the State Russian Museum in Petersburg.

I am very grateful to Lakshmi Narayanan — my World Scientific editor for many years — with whom I worked on around a dozen projects. Now, when she is retiring I would like to say thank you, Lakshmi.

This work is supported in part by DOE grant DE-SC0011842.

References

1. K. A. Olive, S. Rudaz and M. A. Shifman (Eds.), *SUSY-30. Proc. International Symposium celebrating 30 years of supersymmetry,* October 13–27, 2000, Nucl. Phys. **B** Proc. Suppl. **101** (2001).
2. *Electric-magnetic duality, monopole condensation, and confinement in $\mathcal{N} = 2$ supersymmetric Yang-Mills theory,* Nucl. Phys. B **426**, 19–52 (1994) [erratum: Nucl. Phys. B **430**, 485–486 (1994)].
3. *The Many Faces of the Superworld,* Yuri Golfand Memorial Volume, Ed. M. Shifman (World Scientific, Singapore, 2000).
4. *Concise Encyclopedia of Supersymmetry: And Noncommutative Structures in Mathematics and Physics,* Eds. Steven Duplij, Warren Siegel, Johnathan Bagger (Kluwer/Springer, 2004, 2nd Edition 2005).
5. *Felix Berezin: Life And Death of the Mastermind of Supermathematics,* Ed. M. Shifman, (World Scientific, Singapore, 2007); translations of this book in French and Russian are also available.
6. *Dmitry Vasielevich Volkov,* Eds. S.I. Volkova, A. Nurmagambetov, (Timchenko, Kharkov, 2007), in Russian.
7. *Physics in a Mad World,* Ed. M. Shifman, (World Scientific, Singapore, 2016).

---

[9] Filonov died of starvation during the Nazi Siege of Leningrad. Perhaps, I should mention here a curious "physics story" regarding Filonov's painting *Shostakovich's Symphony.* An inscription on the canvas read "19 January... eclipse 15-39 18 total eclipse." In 1986, a well-known nuclear physicist Yakov Smorodinsky learned with surprise [19] that none of the art experts could determine when exactly Filonov created this famous work. Smorodinsky looked up the dates and times of the eclipses which could be observed in Russia during Filonov's life; the lunar eclipse of 1935 fitted the data on the canvas. Thus, he managed to establish the exact date. Since then, Filonov's *Shostakovich's Symphony* is universally dated by 1935, with reference to Smorodinsky, see e.g. Irina Pronina, *We cannot guess the way our words will echo through the ages...,* in the Collection *Ya. A. Smorodinsky, Selected Works,* Classics of Science Series (URSS, Moscow, 2006, ISBN 978-5-8360-0537-5) page 537, in Russian.

8. *The Birth of String Theory*, Eds. Andrea Cappelli, Elena Castellani, Filippo Colomo, Paolo Di Vecchia, (Cambridge University Press, Cambridge, 2012).
9. Stanley Deser, *Forks in the Road: A Life in Physics*, (World Scientific, Singapore, 2022).
10. M. Shifman, *From Heisenberg to Supersymmetry,* Fortsch. Phys. **50**, 552–561 (2002); available at https://www.academia.edu/25243462/From_Heisenberg_to_Supersymmetry and https://web.archive.org/web/20170811214211id_/http://www.tpi.umn.edu/shifman/fulltext.pdf.
11. Salam and J. Strathdee, *Super-symmetry and Nonabelian Gauges*, Phys. Lett. B **51**, 353–355 (1974).
12. B. Zumino, *Bose-Fermi Supersymmetry*, Talk at the 1974 ICHEP Conference, Published in Proc. ICHEP '74, J. R. Smith and Geoffrey Manning (Eds.), Rutherford Appleton Laboratory. Chilton, 1974, pp. I.254–260.
13. P. Fayet, *Supergauge Invariant Extension of the Higgs Mechanism and a Model for the electron and Its Neutrino,* Nucl. Phys. B **90**, 104–124 (1975); *Supersymmetry and Weak, Electromagnetic and Strong Interactions,* Phys. Lett. B **64**, 159 (1976); *Spontaneously Broken Supersymmetric Theories of Weak, Electromagnetic and Strong Interactions,* Phys. Lett. B **69**, 489 (1977); *Mixing Between Gravitational and Weak Interactions Through the Massive Gravitino,* Phys. Lett. B **70**, 461 (1977); *Supersymmetric Theories of Particles*, Contribution to Orbis Scientiae, Coral Gables, FL, (Stud. Nat. Sci. **14**, 413–430 (1978)).
14. S. Dimopoulos and H. Georgi, *Softly Broken Supersymmetry and SU(5),* Nucl. Phys. **B 193**, 150–162 (1981).
15. A. Salam and J. A. Strathdee, *Supergauge Transformations*, Nucl. Phys. B **76**, 477–482 (1974).
16. S. Ferrara, J. Wess and B. Zumino, *Supergauge Multiplets and Superfields,* Phys. Lett. B **51**, 239 (1974).
17. S. Ferrara and B. Zumino, Supergauge Invariant Yang-Mills Theories, Nucl. Phys. B **79**, 413 (1974).
18. A. Salam and J. A. Strathdee, *Supersymmetry and Nonabelian Gauges,* Phys. Lett. B **51**, 353–355 (1974).
19. Noemi Smorodinskaya, *Yakov Abramovich Smorodinsky*, in M. Shifman (Ed.), *Under the Spell of Landau*, (World Scientific, Singapore, 2013), p. 163.

FOREWORD TO THE FIRST EDITION

G. L. KANE
Randall Laboratory, University of Michigan, Ann Arbor, MI 48109

M. SHIFMAN
Theoretical Physics Institute, University of Minnesota, Minneapolis, MN 55455

> "... One of the biggest adventures of all is the search for supersymmetry. Supersymmetry is the framework in which theoretical physicists have sought to answer some of the questions left open by the Standard Model of particle physics.
>
> Supersymmetry, if it holds in nature, is part of the quantum structure of space and time. In everyday life, we measure space and time by numbers, "It is now three o'clock, the elevation is two hundred meters above sea level," and so on. Numbers are classical concepts, known to humans since long before Quantum Mechanics was developed in the early twentieth century. The discovery of Quantum Mechanics changed our understanding of almost everything in physics, but our basic way of thinking about space and time has not yet been affected.
>
> Showing that nature is supersymmetric would change that, by revealing a quantum dimension of space and time, not measurable by ordinary numbers. ... Discovery of supersymmetry would be one of the real milestones in physics."
>
> E. Witten [1]

The history of supersymmetry is exceptional. All other major conceptual developments in physics and science have occurred because scientists were trying to understand or study some established aspect of nature, or to solve some puzzle arising from data. The discovery of supersymmetry in the early 1970's, an invariance of the theory under interchange of fermions and bosons, was a purely intellectual achievement, driven by the logic of theoretical development rather than by the pressure of existing data. Thirty years elapsed from the time of discovery, immense theoretical effort was invested in this field, over 30,000 papers published. However, none of them can claim to report the experimental discovery of supersymmetry (although there are some hints, of which we will say more later). In this respect the phenomenon is rather unprecedented in the history of physics. Einstein's general relativity, the closest possible analogy one can give, was experimentally confirmed within several years after its creation. Only in one or two occasions have theoretical predictions of a comparable magnitude had to wait for experimental confirmation that long. For example, the neutrino had a time lag of 27 years.

It would not be an exaggeration to say that today supersymmetry dominates theoretical high energy physics. Many believe that it will play the same revolutionary role in the physics of the 21st century as special and general relativity did in the physics of the 20th century. This belief is based on aesthetical appeal, on indirect evidence, and on the fact that no theoretical alternative is in sight.

The discovery of supersymmetry presents a dramatic story dating back to the late 1960's and early '70's. For young people who entered high energy physics in the 1990's this is ancient history. Memories fade away as live participants of these events approach the retirement age; some of them have already retired and some, unfortunately, left this world. Collecting live testimonies of the pioneers, and preserving them for the future, seems timely given the impact supersymmetry has already produced on the development of particle physics. Having said that, we note that this book did not appear as a result of a conscious project. Both editors had collected some materials for other activities [2, 3] and became aware of the other's interest and materials. Many people have been interested in how supersymmetry originated — the question often is asked in informal conversations — and how it can be such an active field even before direct experimental confirmation. We finally decided to combine materials, invite further ones, and edit this volume that makes available a significant amount of information about the origins of this intellectually exciting area. Most of it is in the words of the original participants.

In the historical explorations of scientific discoveries (especially, theoretical) it is always very difficult to draw a "red line" marking the true beginning, which would separate "before" and "after." Almost always there exists a chain of works which interpolates, more or less continuously, between the distant past and the present. Supersymmetry is no exception, the more so because it has multiple roots. It was observed as a world-sheet two-dimensional symmetry[10] in string theories around 1970; at approximately the same time Golfand and Likhtman found the superextension of the Poincaré algebra and constructed the first four-dimensional field theory with supersymmetry, (massive) quantum electrodynamics of spinors and scalars. Within a year Volkov and collaborators (independently) suggested nonlinear realizations of supersymmetry and started the foundations of supergravity. Using the terminology of the string practitioners one can say that the first supersymmetry revolution occurred in 1970–71 as the idea originated.[11] The second supersymmetry revolution came with the work of Wess and Zumino in 1973. Their discovery opened to the rest of the community the gates to the Superworld. The work on supersymmetry was tightly woven in the fabrique of contemporary theoretical physics. During the first few years of its development, there was essentially no interest in whether or how supersymmetry might be relevant to understanding nature and the traditional goals of physics. It was "a solution in search of a problem." Starting in the early 1980's, people began to realize that supersymmetry might indeed solve some basic problems of our world. This time may be characterized as the third supersymmetry revolution.

So, how far in the past one should go and where one should stop in the book devoted to the beginnings?

The above questions hardly have unambiguous answers. We decided to start from Ramond, Neveu, Schwarz, Gervais, and Sakita whose memoirs are collected in the chapter entitled *The Predecessors*. The work of these authors can be viewed as precursive to the discovery of supersymmetry in four dimensions. It paved the way to Wess and Zumino.

The central in the first part of the book is Chapter 2 presenting *The Discovery*. It contains recollections of Likhtman, Volkov, Akulov, Koretz-Golfand (Yuri Golfand's widow) and the 1999 Distinguished Technion Lecture of Prof. J. Wess, in which the basic stages of the theoretical

[10]The realization that the very same string theories gave rise to supersymmetry in four-dimensional space-time came later, see John Schwarz's remark on page 73. He notes: "The Ramond-Neveu-Schwarz string theory actually contains local space-time supersymmetry, but we were slow to realize that. The realization came after supersymmetry in four dimensions was thoroughly studied by others." The first evidence for space-time supersymmetry in superstring theory was published by Gliozzi, Olive, and Scherk (GOS for short) in September 1976 and January 1977 [4].

[11]In the Marxist terminology it would be more exact to say that this was a prerevolutionary situation. This nuance is too subtle, however, and cannot be adequately discussed in this article.

construction are, outlined.[12] Chapter 3 is devoted to the advent of supergravity. The fourth chapter is entitled *The Pioneers.* The definition of pioneers (i.e. those who made crucial contributions at the earliest stage) is quite ambiguous, as is the upper cut off in time which we set, *the summer of 1976.* By that time no more than a few dozen of original papers on supersymmetry had been published.

The selection of the contributors was a difficult task. We were unable to give floor to some theorists who were instrumental at the early stages (e.g. R. Arnowitt, L. Brink, R. Delbourgo, P.G.O. Freund, D.R.T. Jones, J.T. Łopuszański, P. Nath, Y. Ne'eman, V.I. Ogievetsky, A. Salam, E. Sokatchev, B. de Wit). Some are represented in other chapters (e.g. S. Ferrara whose 1994 Dirac Lecture is being published in Chap. 3.) Others are beyond reach. This refers to Abdus Salam and Victor Ogievetsky. The latter, by the way, wrote (together with L. Mezincescu) the first comprehensive review on supersymmetry which was published in 1975 [5]. Even now it remains an excellent introduction to the subject, in spite of the 25 years that have elapsed.

The question of where to draw the line tortured us, and we bring our apologies to all the pioneers who "fell through the cracks."

The second part of the book is an attempt to present a historical perspective on the development of the subject. This task obviously belongs to the professional historians of science; the most far-sighted of them will undoubtedly turn their attention to supersymmetry soon.[13] For the time being, however, to the best of our knowledge, there are no professional investigations on the issue. There was available a treatise written by Rosanne Di Stefano in 1988 for a conference proceedings which were never published. This is a very thorough and insightful review. On the factual side it goes far beyond any other material on the history of supersymmetry one can find in the literature. There are some omissions, mostly regarding the Soviet contributors, which are naturally explained by the isolation of the Soviet community before the demise of the USSR and relative inaccessibility of several key papers written in Russian. The Yuri Golfand Memorial Volume [3] which contains the English translation of an important paper by Golfand and Likhtman [6] as well as a wealth of other relevant materials, fills the gap. In addition, Springer-Verlag has recently published Memorial Volumes

[12]Unfortunately, our (probably, awkward) attempts to convince Prof. B. Zumino in the usefulness of this book failed — we were unable to obtain his contribution.

[13]So far (2024) this has not yet happened. –MS

in honor of Dmitry Volkov [7] and Victor Ogievetsky, [8] which acquaint the interested reader with their roles to a much fuller extent than previously.

The coverage of certain physics issues in Di Stefano's essay required comment; in a few cases we added explanatory footnotes. Di Stefano's essay is preceded by a relatively short article written by the late Prof. Marinov. It is entitled "Revealing the Path to the Superworld" and was originally intended for the Golfand Volume. This article presents "a bird's eye view" on the area. On the factual side it is much less comprehensive than Di Stefano's, but it carries a distinctive flavor of the testimony of an eye witness. Moreover, it reveals the mathematical roots of the discovery, an issue which is only marginally touched in Di Stefano's essay.

We are certainly not professional historians of science; still we undertook a little investigation of our own. Often students ask where the name "supersymmetry" came from? It seems that it was coined in the paper by Salam and Strathdee [9] where these authors constructed supersymmetric Yang–Mills theory.[14] This paper was received by the editorial office on June 6, 1974, exactly eight months after that of Wess and Zumino. Super-symmetry (with a hyphen) is in the title, while in the body of the paper Salam and Strathdee use both, the old version of Wess and Zumino, "super-gauge symmetry," and the new one. An earlier paper of Ferrara and Zumino [10] (received by the editorial office[15] on May 27, 1974) where the same problem of super-Yang–Mills was addressed, mentions only supergauge invariance and supergauge transformations.

Supersymmetry is nearly thirty years old. It seems that now we are approaching the fourth supersymmetry revolution which will demonstrate its relevance to nature. Although not numerous, we do have hints that this is the case. They are: (a) supersymmetry allows a stable hierarchy between the weak scale and the shorter distance scales such as the Planck scale or unification scale, (b) supersymmetry provides a way to understand how the electroweak $SU(2)\times U(1)$ symmetry is broken, so long as the top quark came out heavy (which it did), (c) gauge couplings unify rather accurately when superpartners are included in the loops,[16] (d) the Higgs boson is predicted to be light (LEP gives $M_H < 200$ GeV), and (e) the lack of any deviations

[14]See page x for a clarification.

[15]The editorial note says it was received on May 27, 1973. This is certainly a misprint; otherwise, the event would be acausal.

[16]An alternative way to say this is to say that the value of the weak mixing angle at the weak scale can be calculated accurately if one sets it to the value predicted by a unified theory at the unification scale.

from Standard Model predictions in the precision data at LEP and in other experiments is consistent with supersymmetry (it was anticipated that these deviations would be invisible).

Certainly, at the moment the indications are not conclusive. However inconclusive, they are the source of hope and enthusiasm for phenomenologically oriented theorists and experimentalists who would like to keep high-energy physics in the realm of empirical science.

Another aspect which came to limelight recently is the fact that supersymmetry became instrumental in the solution of highly nontrivial dynamical issues in strongly coupled non-supersymmetric theories, which defied solutions for decades. That of course does not imply that nature is supersymmetric, but it does add to the interest in supersymmetry.

Summarizing, in this book we bring together contributions from many of the key players of the early days of supersymmetry. We leave its relevance to our world to a future project.

28 August 2000

References

1. This quotation is from Witten's Foreword to G. Kane's book cited in Ref. 2.
2. G. Kane, *Supersymmetry: Unveiling the Ultimate Laws of Nature* (Perseus Books, 2000).
3. *The Many Faces of the Superworld*, Yuri Golfand Memorial Volume, Ed. M. Shifman (World Scientific, Singapore, 2000).
4. G. Gliozzi, J. Scherk and D. Olive, Phys. Lett. **65B**, 282 (1976); *Nucl. Phys.* **B122**, 253 (1977).
5. V.I. Ogievetsky and L. Mezincescu, *Usp. Fiz. Nauk* **111**, 637 (1975), [*Sov. Phys. — Uspekhi* **18**, 960 (1975)].
6. Yu.A. Golfand and E.P. Likhtman, *On the Extensions of the Algebra of the Generators of the Poincaré Group by the Bispinor Generators*, in I.E. Tamm Memorial Volume *Problems of Theoretical Physics*, Eds. V.L. Ginzburg *et al.* (Nauka, Moscow, 1972), page 37 [English translation: Ref. 3, page 45].
7. *Supersymmetry and Quantum Field Theory*, Proceedings of the D. Volkov Memorial Seminar, Kharkov, Ukraine, January 1997 [Lecture Notes in Physics, Vol. 509] (Springer-Verlag, 1998).
8. *Supersymmetries and Quantum Symmetries*, Proceedings of the International Seminar on Supersymmetries and Quantum Symmetries Dedicated to the Memory of V.I. Ogievetsky, Dubna, Russia, July 1997 [Lecture Notes in Physics, Vol. 524] (Springer-Verlag, 1999).
9. A. Salam and J. Strathdee, "Super-symmetry and non-Abelian gauges", *Phys. Lett.* **B51**, 353 (1974).
10. S. Ferrara and B. Zumino, "Supergauge invariant Yang–Mills theories", *Nucl. Phys.* **B79**, 413 (1974).

CONTENTS

List of Photographs

EPILOGUE TO THE FIRST EDITION

It is with deep sorrow that we have learned of the death of Professor Lochlainn O'Raifeartaigh on 18 November 2000. Just a few weeks before, he gave an engaging talk at the Symposium "30 Years of Supersymmetry" organized by the Theoretical Physics Institute in Minnesota. He was enthusiastic about supersymmetry on lattices, and outlined a program the final goal of which was solving supersymmetric gauge theories at strong coupling.

Prof. O'Raifeartaigh held closely to his heart the idea of this Volume. He presented several successive versions of his contribution (see p. 314) and polished them until he was fully satisfied. Lochlainn O'Raifeartaigh's work on mechanisms of supersymmetry breaking will forever be a part of the history of the development of supersymmetric theories.

The Editors

Figure 3: Lochlainn O'Raifeartaigh, 2000. Courtesy of Cormac O'Raifeartaigh.

Siddhartha Sen and Ivo Sachs: Lochlainn O'Raifeartaigh (1933–2000)[17]

Lochlainn O'Raifeartaigh was born in Clontarf, Dublin on March 11, 1933. He attended St Joseph's Primary School in Marino, Dublin and later Castleknock College. He then entered the University College at Dublin (UCD) where he had an outstanding undergraduate career, completing a joint mathematics and physics honors course as the best student in his year. One of his interests while a student at UCD, the Irish language, led him to join An Cumann Gaelach (the Gaelic Society). This interest never faded away — Lochlainn was fluent in Irish, French, and German.

After graduating from UCD in 1956, Lochlainn O'Raifeartaigh obtained a research fellowship from the Dublin Institute for Advanced Studies (DIAS) to work with the great Irish relativist J. L. Synge. After a year at DIAS, where he had the opportunity to attend lectures by Schrödinger, and after writing three papers in relativity, O'Raifeartaigh was awarded a traveling studentship to study under Walter Heitler, a former professor at DIAS, in Zurich. The University of Zurich had a distinguished record of excellence in theoretical physics. It was, after all, the place where Schrödinger discovered Quantum Mechanics. After completing his Ph.D. on "Non-Local Field Theories", O'Raifeartaigh returned to DIAS as an assistant professor. In 1968, O'Raifeartaigh was offered a senior professorship at DIAS which he accepted.

Lochlainn O'Raifeartaigh remained at DIAS for the rest of his life, building an internationally respected group in theoretical high energy physics, and always working at the forefront of the subject. His method of collaboration was to interact and discuss physics at the blackboard. These discussions were not polite, considerate exchanges, but rather exciting, heated arguments which continued until the issues being debated had been resolved.

Lochlainn O'Raifeartaigh's varied and deep contributions to physics were marked by his passion for clarity and his instinct for uncovering the fundamental features in a problem. His best known contributions are undoubtedly his "No-Go Theorem" in 1965 and his work on supersymmetry breaking in 1975. The "No-Go Theorem" brought to an end attempts to combine internal Lie group symmetries with Poincaré symmetry in a non-trivial way.

The O'Raifeartaigh mechanism for supersymmetry breaking was a result of a request from A. Jaffe at Harvard in 1974 to report on recent work by Wess and Zumino on supersymmetry in Aspen. This made him study

[17]From the DIAS website http://www.stp.dias.ie/Lochlainn/lochlainn.html, with kind permission of Professor Denjoe O'Connor and the authors.

the work of Wess and Zumino carefully. Later on, urged on by a remark from Pais that the problem of breaking of supersymmetry needed attention, O'Raifeartaigh quickly showed that supersymmetry could not be broken by a single chiral superfield. Initial attempts by O'Raifeartaigh to show that supersymmetry breaking by Higgs' mechanism was not possible failed, and eventually he was able to show that a minimum of three chiral superfields are needed for Higgs' mechanism to work. A delightful account of the way this work was done is contained in O'Raifeartaigh's contribution to the book "The Supersymmetric World" (Ed. D. Kane and M. Shifman). O'Raifeartaigh was a very productive and creative researcher. He published over 200 papers with over 60 collaborators! As new ideas came up, O'Raifeartaigh studied them, discussed them with passion, and made important contributions to them. A few examples of his many contributions can be given to illustrate the range of his interests.

Some of other O'Raifeartaigh's works worth mentioning are:

(i) Magnetic Monopoles

The surprising discovery by Polyakov and 't Hooft that magnetic monopoles could appear as solitonic excitations in certain gauge-Higgs field theories started a rigorous program of trying to understand what was involved in creating such monopoles. O'Raifeartaigh, in collaboration with P. Houston and S. Rouhani, studied axially symmetric monopoles in detail. They found that such configurations have intricate topological features. In particular, they showed that the Higgs field has to vanish on the axis connecting two monopoles of the same charge.

(ii) Toda Theory and W-Algebras

In a remarkable series of papers, Lochlainn O'Raifeartaigh, in collaboration with J. Balog, L. Feher, P. Forgacs and A. Wipf, tackled the problem of understanding the general properties of two-dimensional conformally invariant solvable field theories. It had been found that such theories are realizations of extensions of chiral Virasoro algebra including Kac-Moody algebras and the W-algebras which are polynomial extensions of the Virasoro algebra by higher spin fields. The study of such polynomial extensions was initiated by Zamolodchikov. O'Raifeartaigh and his co-workers established the exact relationship between Toda theories and WZNW models. Specifically, they showed that Toda theories could be regarded as WZNW models reduced by certain conformally invariant constraints. This result allowed them to elegantly show that the W-algebra of Toda theory followed from the algebra formed by gauge invariant polynomials of the constrained Kac-Moody currents and their densities. It also allowed them to obtain general solutions of the Toda field equations from simple WZNW solutions.

(iii) Seiberg-Witten Theory

The construction by Seiberg and Witten of an exact non-perturbative effective action for $\mathcal{N} = 2$ supersymmetric Yang-Mills theory created a tremendous amount of interest. Their effective action offered a way to understand quark confinement in this model. The result of Seiberg and Witten was based on the assumption of a certain duality relation. In collaboration with M. Magro, R. Flume, I. Sachs and O. Schnetz, O'Raifeartaigh showed that the Seiberg-Witten effective action was unique and that it could be obtained without using duality arguments simply by exploiting the supersymmetry of this model.

Lochlainn O'Raifeartaigh was a scholar in the history of high-energy physics and wrote a masterly book "The Dawning of Gauge Theory," a historical account of the quest for a unified theory of the fundamental forces of the physical world. O'Raifeartaigh received the von Humboldt Research Award in 1998 and the prestigious Wigner Medal in 2000 for his pioneering contributions to particle physics. He was elected a Member of the Royal Irish Academy at the age of 29, and was also a member of Academia Europea.

In all his work, Lochlainn O'Raifeartaigh's love of physics, his interest in understanding fundamental problems clearly, and his ability to present results with precision and clarity shone through.

1

In Memoriam: Julius Wess and Bruno Zumino

Figure 1.1: Julius Wess at CERN, 1960s.

Figure 1.2: Julius Wess delivering a lecture at Technion, Haifa, Israel, March 10, 1999. Courtesy of M. Marinov.

JULIUS WESS (1934–2006)[18]

STEPAN MOSKALIUK
Atominstitut, Technische Universität Wien, Austria
Email: moskaliukss@gmail.com

Julius Wess, one of the world's most prominent Austrian theoretical physicist, former Director at Werner-Heisenberg Institute in Munich, President of the Scientific Council of the Austro-Ukrainian Institute for Science and Technology in Vienna, died from a stroke on August 8, 2007 in Hamburg, Germany.

Julius Wess was born on December 5, 1934, in Oberwölz, Austria. He belonged to the generation of Austrian students who went through their secondary education (Gymnasium, Realschule) and started their university studies in the first decade after the Second World War. These were quite turbulent years immediately following the fall of the Nazi regime. Nevertheless Julius Wess became one of those physicists who gave back to Europe its role in science, when Europe was recovering from the enormous intellectual brain drain, especially in the German-speaking countries.

The scientific career of Julius Wess started in Austria. He received his Ph.D. in 1957 at Vienna University, where he was a student of Hans Thirring.[19] His Ph.D. examiner was one of the founding fathers of quantum mechanics, Erwin Schrödinger. Not too surprisingly, the start of his career coincided with the arrival of Walter Thirring at Vienna. Many Austrian physicists benefitted from Thirring's school. Julius Wess was no exception. It was there that he met his lifelong friend and collaborator Bruno Zumino.

Then his scientific trajectory brought him to CERN. In 1966 Wess became an Associate Professor at the Courant Institute at New York University, and in 1968 Full Professor at the Karlsruhe University. In 1990 he left Karlsruhe to become Director at the Max-Plank-Institute for Physics (Werner-Heisenberg Institute) in Munich and Professor at the Ludwig-Maximilians University. After his retirement he worked at DESY in Hamburg.

---

[18] From http://quark.itp.tuwien.ac.at/~diefaust/Scientists/Julius_Wess.html

[19] Hans Thirring (1888–1976) was an Austrian theoretical physicist, professor, and father of the physicist Walter Thirring. He won the Haitinger Prize of the Austrian Academy of Sciences in 1920.

In 1973, Julius Wess, together with Burno Zumino, discovered an extension of the notion of symmetry to a more general class called supersymmetry, which was later generalized in terms of the differential geometry of superspace. In this way supergravity emerged. The main novelty was that the symmetry generators which had been charcterized by commutators with the invention of supersymmetry were supplemented by anticommutators in the new extended supersymmetry algebra. The infinitesimal transformations generated by this algebra could change bosons into fermions and vice versa.

Supersymmetry has been independently discovered by three groups of authors: Yu. Golfand and E. Likhtman (1971); D. Volkov and V. Akulov (1972); and J. Wess and B. Zumino (1973). The motivations and starting points used by the three groups of authors were quite different:

- Golfand and Likhtman initially reported a construction of the super-Poincaré algebra and a version of massive super-QED. The formalism contained a massive photon and photino, a charged Dirac spinor and two charged scalars (spin-0 particles). Likhtman found algebraic representations that could be viewed as supersymmetric multiplets and observed the vanishing of the vacuum energy in supersymmetric theories at one loop.
- Contributions from Ukraine were made by Volkov and Akulov, who in 1972 tried to associate the massless fermion appearing due to spontaneous supersymmetry breaking with the neutrino. Within a year, Volkov and Soroka gauged the super-Poincaré group, which led to elements of supergravity. They suggested that a spin 3/2 graviton's superpartner becomes massive upon "eating" the Goldstino that Volkov and Akulov had discussed earlier. A mathematical basis for the work of Volkov and collaborators was provided by Berezin's and Katz's paper (published in 1970), where graded algebras were thoroughly studied. In his memoirs, Volkov also mentioned the impact of Heisenberg's attempts to obtain a fermion Goldstone particle.
- In the West, a breakthrough into the 4D superworld was achieved by Wess and Zumino in 1973. This work was done under the influence of the nascent two-dimensional supersymmetry on the string world sheet.

The Western researchers knew little (if at all) about the work done in the Soviet Union. Therefore, Wess and Zumino based their inspiration on common roots with string theory — another pillar of the modern theory — which in those days was referred to as the "dual model." Around 1969, the dual-resonance formula of strong interactions, found by Veneziano, was formulated in terms of four-dimensional harmonic oscillators. Nambu advanced the idea that these oscillators represented a relativistic string. After this, the scheme was reformulated as a field theory on the string world sheet.

The original theory was plagued by the fact that the spectrum contained a tachyon but no fermions and it was consistent only in 26 space-time dimensions. These problems motivated the search for a more realistic string theory which led to a super-string.

The fruits of Wess' world-famous collaboration with Bruno Zumino on supersymmetry have not yet found direct experimental verification. However, its basic idea became a fundamental building block in modern theoretical developments — from the unification of the basic interactions of elementary particles to the so far unique successful attempt of string theory to include the theory of gravity in the fundamental theory of Nature. It is very saddening that Julius Wess died before he could have learned about the experiments ultimately proving his revolutionary theory.

"They came up with quantities that you could measure and you could observe," said Roman Jackiw, a professor of physics at the Massachusetts Institute of Technology. "The understanding of anomalies, to be avoided in quantum versions of field theories, the basis of our interpretation of physical processes, will forever be linked to the name of Julius Wess too."

In his last years Julius Wess worked on non-commutative geometry and influenced, to a large degree, advances in this direction in which, like supersymmetry and supergravity, he opened new horizons of physical thinking.[20]

[20] Julius Wess' idea of the modern quantum theory based on his concept of symmetry — supersymmetry, quantum groups and gauge theories on non-commutative spaces — is presented in more detail in the book *Julius Wess: The Development of the Concepts of Symmetry in Physics*, Ed. S. S. Moskaliuk (TIMPANI, Kyiv, 2024), in the Series "Classics of World Science", vol. 5; available online at https://www.researchgate.net/profile/Stepan-Moskaliuk

REFLECTIONS ON SUPERSYMMETRY[21]

HERMANN NICOLAI
Max-Planck-Institut für Gravitationsphysik
Albert-Einstein-Institut
Mühlenberg 1, D-14476 Potsdam, Germany
Email: nicolai@aei.mpg.de

Abstract: Supersymmetry is a theme with many facets that has dominated much of high energy physics over the past decades. In this contribution I present a very personal perspective on these developments, which has also been shaped, in an important way, by my interactions with Julius Wess and Bruno Zumino.

1 Introduction

When Misha Shifman asked me whether I would be willing to contribute a chapter to this book with some memories of my encounters with Julius Wess, I did not hesitate for very long. For one, I am honored to have my name included in such an illustrious list of early contributors and pioneers of supersymmetry. Secondly, because I feel privileged to have been a witness when for a (world-historically speaking) fleeting moment the provincial town of Karlsruhe became an epicenter of theoretical physics, with Julius Wess and Bruno Zumino as the central and driving forces behind the development of supersymmetry. It does not happen often that a beginning graduate student is so lucky, as I was, to land in a hot spot of new theory developments, especially in a place that never makes it into the top one hundred of international university rankings. Of course, there were several other such hot spots, and the contributions to this volume vividly and from different angles tell the story how the idea of supersymmetry evolved in such diverse places. For this reason I will not dwell too much on topics that are already extensively dealt with in the book, but instead offer my personal perspective, covering some aspects that have received less attention in the other contributions, especially with regard to $\mathcal{N} = 8$ supergravity, as well as some thoughts concerning the current status of supersymmetry *vis-à-vis* theoretical high energy physics.

In writing this text I have decided not to include a list of references. The reason is simple: there would be either too few, or too many! At any rate, all references relevant to what I am saying can be easily located on

[21]Submitted January 8, 2024.

the electronic arXiv, or alternatively by consulting the companion articles in this book.

2 Early years in Karlsruhe

Shortly after enrolling at the University of Karlsruhe for my undergraduate studies in physics, it was clear to me that I would want to do my diploma thesis with Julius Wess who was at the forefront of exciting developments in theoretical physics. Even as students who knew hardly anything about the subject, we were keenly aware of this fact! So in 1974 I approached Julius with the request to become his diploma student. Yet before getting accepted I had to overcome a first hurdle, when he told me that before starting in earnest I would have to read the book *Introduction to the theory of quantized fields* by Bogoliubov-Shirkov, the then much used standard text on relativistic quantum field theory (QFT). This was clearly the device which he employed to scare away prospective students who were not 150% willing to commit. I remained undeterred even though I can now admit that I never read this formidable book to the end (and fortunately for me, Julius never checked). This was because I realized full well that if I was not accepted, the alternative would have been to settle for a boring diploma topic, or even (God forbid!) experimental work.

Thanks to the way the institute was run by Julius, life at the institute was very stimulating in many ways, not just with regard to physics. This included all kinds of social events and gatherings, perhaps most memorably the annual skiing excursions to Oberperfuss (Austria). The days in Oberperfuss were divided into physics sessions that took place in an old school building in the mornings and late afternoons, and ski outings in between. I was the only institute member who did not know how to ski, or even how to stand on skis properly. I still remember how on my first day I fell off the lift immediately on arriving at the top of the slope, feeling like the giant beetle in Kafka's novella "Die Verwandlung".[22] Nevertheless, after practicing for a couple of days I was proudly able to stand and slowly move on my skis, descending the slope by zigzagging across it horizontally (much to the annoyance of people racing down hill), which took me a good part of the afternoon.

At the institute in the Physikhochhaus the common room for postdocs and graduate students bore the label "Kinder" (children). Indeed we little

[22] The Metamorphosis.

ones had hardly any idea what the 'adults' (*alias* Julius and Bruno Zumino) were up to. But we greatly benefitted from the numerous prominent visitors. Lochlain O'Raifeartaigh, who discovered the first example of a Wess-Zumino type model that exhibits spontaneous breaking of supersymmetry, spent extended periods of time in Karlsruhe. Another highlight was a visit by Rudolf Haag, who started out his seminar with the question "Where is Martin Sohnius?". The main purpose of his visit was to work with Martin and Jan Łopuszański on their now famous classification of all supersymmetries of the S-matrix. Other visitors included André Martin and John Iliopoulos. John had been scheduled to give a supersymmetry seminar about Fayet-Iliopoulos breaking, but switched topics to talk about the c-quark which had just been discovered, and its proper interpretation in terms of the GIM mechanism. Pierre Fayet presented the first version of a supersymmetric extension of the Standard Model (SM) of particle physics, inventing a whole new vocabulary with neologisms like 'photino', 'gluino' and the like. Of course, the most prominent visitor was Bruno Zumino (which made us joke that there also had to be a superpartner by the name of 'Zumo'). Bruno came to Karlsruhe regularly, but when he was around we rarely saw him, with no idea of what they were doing because usually their office door was tightly closed or they worked at Julius' home.

In autumn 1975 I finished my diploma thesis on spontaneous breaking of global supersymmetry[23] which satisfied Julius, at least enough that he offered me the possibility of staying as a PhD student with him. For the topic of my doctoral thesis he suggested the problem of putting supersymmetry on the lattice. This looked like a win-win problem for an aspiring PhD student, because lattice gauge theory was the other freshly hot topic in particle physics at the time. But, as is common, the devil turned out to lie in the details. The idea did not really fly because of one main obstacle: the Leibniz rule is no longer valid for lattice derivatives, but nonetheless absolutely needed to re-assemble terms into total derivatives, as is required for supersymmetry invariance of the action. This effectively meant that there was no way of putting supersymmetry on the lattice without doing violence to it (and the problem gets worse if one wants to incorporate gauge symmetry). My attempts to get around the difficulty never led to fully satisfactory

[23] We did not publish any of our results because L. O'Raifeartaigh scooped us by getting there first! Instead, my first paper (in 1976) introduced a non-relativistic quantum mechanical version of supersymmetry, with a supersymmetric spin system built solely from Majorana fermions. This model has some similarity with the (supersymmetric) Sachdev-Ye-Kitaev (SYK) model that only became popular much later.

answers: for instance, the so-called SLAC derivative (the Fourier transform of the saw-tooth function $[k^\mu]$) which retains a rudimentary version of the Leibniz rule, is too non-local on the lattice to be of much use. Looking for a way out, I turned to constructive QFT, with the idea of proving rigorously the existence of Wess-Zumino model in the framework of constructive quantum field theory (recall that to this day we still have no example of a fully constructed interacting QFT in four dimensions satisfying all the Wightman axioms!). At first this looked promising because the model required only one divergent (wave-function) renormalization, unlike QED or ϕ^4 theory which needed three. Consequently, one needed to study only a one-dimensional renormalization map, a cutoff dependent map between bare and renormalized parameters considered in earlier work by R. Schrader. As one of the by-products this required reformulating supersymmetry in Euclidean signature, a challenge of the theory. While others concluded from this that $\mathcal{N} = 1$ supersymmetry cannot be Euclideanized, my proposal simply amounted to the statement that the correct Euclideanization procedure is the one that yields the proper Euclidean continuation of the Minkowskian supersymmetry Ward identities, in accordance with the Osterwalder-Schrader reconstruction theorem. These efforts earned me a seminar invitation to CERN, although I doubt that I was able to convince anyone, most importantly Bruno, of the relevance of what I was saying. In the end my 'constructive' attempts were also stymied because the joint UV and IR regulators required to rigorously define the functional (path) integral broke supersymmetry so badly that one could not really exploit the symmetry to its full worth. But at least I found a new way to characterize supersymmetry without using anticommuting variables, but instead just in terms of the bosonic functional measure that is obtained by integrating out fermions.

Meanwhile others at the institute were busy working on all kinds of different problems spawned by the discovery of supersymmetry, *e.g.* renormalization of supersymmetric gauge theories (O. Piguet and K. Sibold), or possible generalizations of supersymmetry to parasymmetry (N. Dragon). Considerable effort was spent on searching for off-shell superspace versions of $\mathcal{N} = 2$ gauge theories and supergravity in work involving R. Grimm and M. Sohnius, as well as Julius himself. As you can see, we had a whole new world in front of us to explore! Of course, Bruno and Julius continued their joint work, during the time I was there mostly concentrating on the superspace formulation of $\mathcal{N} = 1$ supergravity, building on the basic insight that even flat superspace requires torsion. I still recall one morning when Julius skipped the usual Austrian pleasantries, instead greeting me simply and directly with the statement "$\mathcal{L} = E$" — meaning that the supergravity Lagrangian in superspace is nothing but the supervielbein Berezinian. I could see how happy he was with this result, because what could be simpler than that for a gravitational Lagrangian?

A small fly in the ointment was the fact that one still had to take into account the superspace constraints, so the seemingly simple result could not be exploited directly to calculate loop corrections in supergravity.

Although we were all captivated by supersymmetry and in no doubt about its ultimate relevance, no one in Karlsruhe in those days thought that the available supersymmetric models could be directly relevant to real physics. Rather, the general attitude was that essential pieces were still missing from the picture — just like Yang-Mills theories are not viable without the crucial extra ingredients discovered only much later, namely the Brout-Englert-Higgs mechanism, and the subtle dynamical mechanism of confinement. The main advantage of supersymmetry emphasized in those days was not so much the absence of quadratic divergences (which only became the dominant narrative later, in connection with the hierarchy problem), but rather the fact that it overcame the Coleman-Mandula no-go theorem prohibiting the fusion of space-time and internal symmetries. The $\mathcal{N} = 1$ models that became fashionable for phenomenology in the 1980s precisely do *not* exploit this feature, which would require extended ($\mathcal{N} \geq 2$) supersymmetry. This is the basic reason for the (to me, esthetically unappealing) need to double up the known particle spectrum with unseen superpartners possessing the same internal quantum numbers.

3 Beyond simple supersymmetry

After finishing my PhD in October 1978 and an unsuccessful application for a CERN fellowship I moved to Heidelberg, where I spent most of my time learning QCD and perturbative quantization of gauge theories. I had already reconciled myself with the idea of spending the next few years there, when quite unexpectedly I was informed that I could finally join CERN because another fellowship candidate had declined the offer from CERN. So we moved to Geneva in November 1979. In the following years I continued to meet Julius regularly, rejoining the Karlsruhe institute as an associate professor for the two years between 1986 and 1988. After that our contacts became less frequent, as Julius accepted a joint offer from Ludwig-Maximilians-Universität München and the Max Planck Institute für

Physik, and finally moved there in 1990. Physics-wise, he had quit supersymmetry activities to start exploring new ground, turning his full attention to non-commutative geometry where, as he told me more than once, he saw a Nuevo El Dorado for theoretical physics.

At CERN my main focus switched to supergravity, an interest that had been ignited during a remarkable mathematical physics (M∩P) conference in Lausanne in 1979 that I was able to attend. More than anything, it was the brilliant performance of the French team represented by Bernard Julia and Joël Scherk that got me hooked. They both talked about their recent progress with maximal supergravity. Bernard reported on his discovery with Eugène Cremmer of the hidden exceptional $E_{7(7)}$ symmetry of maximal $\mathcal{N} = 8$ supergravity, while Joël gave a wildly funny presentation of their recent construction of $D = 11$ supergravity, drawing all kinds of parallels between the superworld and Nietzschean philosophy. This is when my 'extremist view' on supersymmetry slowly started shaping up: if it is to be supersymmetry, then either as much of it as possible, or none at all! For this reason I decided not to get involved in the activities on $\mathcal{N} = 1$ supersymmetry model building, both with regard to further theoretical developments such as developing the tensor calculus for supergravity matter couplings, as well as concurrent efforts to apply the theory to the phenomenology of the SM and its extensions. I rather concentrated my efforts on learning extended supergravity in the component (not superspace) formalism. In doing my first steps I was greatly helped by Paul Townsend, whose subtle sense of humor and very British understatement I have always admired, and Peter van Nieuwenhuizen, one of the fathers of supergravity, who generously accepted me as the ignoramus newcomer into the collaboration.

Around that time I started working with Bernard de Wit, aiming right away for maximal $\mathcal{N} = 8$ supergravity. Not only had Bernard done important preparatory work, but he also entrusted me with a big pile of handwritten notes. I spent the following weeks assiduously working my way through these notes, until I reached a stage where I could no longer see the forest for all the trees. Since our initial attempts at gauging the maximal $\mathcal{N} = 8$ theory had thus failed badly, we turned to $\mathcal{N} = 5$ supergravity as a kind of stopgap measure, a theory that by itself is only of limited interest (except perhaps as a testing ground for checking unexplained cancellations and UV finiteness properties of extended supergravities beyond four loops, as I learnt very recently from Zvi Bern). There we finally got things to work! Having convinced ourselves that $\mathcal{N} = 8$ would still refuse to cede, we decided to include an appendix in our $\mathcal{N} = 5$ paper explaining in all technical detail why $\mathcal{N} = 8$ supergravity could not be gauged. The paper was practically finished, so we decided to follow our usual habit of having another cup of coffee in the CERN cafeteria before releasing the paper. And there it suddenly struck us like a bolt of lightning out of the sky: we had made an elementary mistake in the calculation — so trivial that I cannot even remember now what it was! Needless to say that we rushed back to get to work right away on revising the paper. But then it still took several months to complete the construction of gauged $N = 8$ supergravity.

Optimism that $\mathcal{N} = 8$ supergravity might have something to do with the real world was still running high in 1982: on my first visit to DAMTP[24] none less than Stephen Hawking assured me that, in his opinion, this was the best candidate for a unified theory. He immediately put a graduate student (Nick Warner) on the problem of finding suitable 'realistic' symmetry breaking vacua of the $\mathcal{N} = 8$ potential. Although I kept coming to Cambridge as a visitor numerous times afterwards, I could never quite figure out whether Stephen continued to stand by this claim, or whether he had secretly switched allegiance to string theory at some point. Independently, at around the same time there were attempts to relate the $\mathcal{N} = 8$ supergravity to 'real physics', most notably by Ellis, Gaillard, and Zumino, who tried to exploit a conjecture by Cremmer and Julia, according to which the (chiral!) SU(8) R-symmetry of the theory could become dynamical. There was a short-lived burst of excitement when they found three $\bar{\mathbf{5}} \oplus \mathbf{10}$ fermion multiplets of SU(5), but those were accompanied by huge towers of 'junk particles' that were obviously useless for realistic applications. There was another proposal made by M. Gell-Mann, which I found even more intriguing, and will come back to it below. Furthermore, hopes for an 'easy' argument that $\mathcal{N} = 8$ supergravity could be finite to all orders were dashed when it was realized in a series of papers by Howe and Lindström, Kallosh and later Howe, Stelle and Townsend, that the theory admits (linearized) counterterms from three loops onward.[25] At that point it looked like all roads were blocked.

---

[24]The Department of Applied Mathematics and Theoretical Physics, the University of Cambridge, UK.

[25]To be sure, the question of finiteness (or not) of $\mathcal{N} = 8$ supergravity remains up in the air, in spite of stunning computational advances by Z. Bern, L. Dixon and collaborators, with results now reaching up to five loops. However, for all I know, the seven-loop calculation that might finally settle the question remains beyond reach. *Idem* for the question whether the known 'seed' counterterms can be made fully compatible with non-linear supersymmetry and non-linear $E_{7(7)}$ invariance.

On top of these failings there was another major development that pushed $\mathcal{N} = 8$ supergravity aside as a leading candidate theory for unification. This was the advent of heterotic string theory in 1984 which rolled over the CERN theory division like a tsunami, burying everything else underneath it. People were busily scurrying to jump on the bandwaggon, being certain that all problems would soon be solved, with an almost unique path from the heterotic theory to SM low-energy physics. The expectation that physics would be over within a few weeks resulted in a coherent state of collective inebriation that in retrospect looks rather absurd to me. Nevertheless, the general attitude from that moment on and for many years thereafter remained that anyone *not* working on string theory and the heterotic string was completely out of the loop. Too bad for our poor colleagues who had invested all their stock into non-string approaches to quantum gravity!

In spite of these developments I decided to stick with maximally supersymmetric theories, and to persist with my decision not to join the booming MSSM[26] and string model building industry (which at one point prompted John Ellis to exclaim "Hermann, we'll never turn you into a good phenomenologist!"). Excursions into string theory proper were only sporadic, perhaps most notably with our proposal (with A. Casher, F. Englert and A. Taormina) to derive all consistent superstring theories from the $D = 26$ bosonic string, including a mechanism to get space-time fermions out of the bosons, and thus explain the emergence of supersymmetry from a purely bosonic theory. Somewhat to my surprise, this proposal generated a huge amount of attention, getting me an invitation from M. Gell-Mann to present these ideas at his newly founded Santa Fe Institute. There, however, I failed to convince the fully assembled string establishment of the virtues of our insights. The main drawback was that despite all efforts our construction remained entirely kinematical — like most subsequent constructions of string vacua! — with no compelling dynamical explanation why the $D = 26$ string should do such a thing. Towards the end of my stay at CERN, I also wrote one paper with Sergio Fubini, showing how to take the 'square root' of the Yang-Mills action in eight dimensions, in an attempt to gain better access to the maximal ($\mathcal{N} = 4$) super-Yang-Mills theory. This led to the discovery of what we called the octonionic instanton. Sergio, one of the true pioneers of string theory, was a very kind person. He often told me what life was like at MIT when they did their ground-breaking work there, and hardly

[26]Minimal Supersymmetrc Standard Model.

anyone would bother to show up at their seminars. Talk about volatility in the public perception of what is good theory!

Over the following years I devoted much time to Kaluza-Klein supergravity and the question of consistent truncations beyond the linearized approximation, mostly in collaboration with B. de Wit (I think the importance of this problem is still not fully appreciated in works on the phenomenology of effective low energy actions obtained from higher dimensions). Further topics were all kinds of things exceptional, such as quasi-conformal realizations and minimal unitary representations of exceptional non-compact groups, and $E_{8(8)}$ in particular, where I learnt a lot from Murat Günaydin, and the construction of maximal gauged supergravity in $D = 3$ with Henning Samtleben, where, again, $E_{8(8)}$ played a pivotal role. However, my greatest fascination remained with B. Julia's early papers conjecturing the appearance of the *infinite-dimensional* exceptional symmetries E_9 and E_{10} in the further reduction to two and one dimensions. Of these, E_9 symmetry (or more precisely, $E_{9(9)}$) is now understood to be an extension of the Geroch group in general relativity, but E_{10}, the maximal rank hyperbolic Kac-Moody algebra, remains a total enigma. What struck me most there was the unexpected link between two seemingly disparate developments in physics and pure mathematics. Namely, in both cases the search for distinguished structures leads to unique answers. On the physics side, trying to make Einstein's theory as (super-)symmetric as possible, one arrives at a unique answer ($D = 11$ maximal supergravity), while on the other side mathematicians identified a uniquely distinguished Lie algebra, the maximally extended hyperbolic Kac-Moody algebra E_{10}, by way of classifying Dynkin diagrams. In fact, E_{10} has been shown to contain all simply laced hyperbolic Kac-Moody algebras as subalgebras, and is thus an all-encompassing mathematical entity — much like $D = 11$ supergravity, which keeps popping up in all discussions of M theory.

4 Supersymmetry and fundamental physics

The last time I saw Julius Wess, who had meanwhile moved to Hamburg, was in Israel on the occasion of Eliezer Rabinovici's and Shimon Yankielowicz' 60th birthday celebrations[27] in Jerusalem and Tel Aviv.[28] Julius seemed to

[27]Editor's note: The Rabinovici-Yankielowicz-Fest took place in 2006. -M.S.

[28]This was the first time I broke ranks with the supersymmetry community when reporting about my work with K. Meissner, where we proposed that it is *conformal symmetry* rather than supersymmetry that stabilizes the electroweak scale.

be in great shape, and we agreed that he should soon visit me in Potsdam. So I was absolutely shocked when I learnt that he had suffered a stroke shortly after; he died in 2007, and the planned visit never came to pass. Tragically, he thus also missed the chance to see the outcome of the LHC experiment which he had been eagerly awaiting: over the years he had become increasingly confident that LHC would find superparticles (whereas Bruno Zumino seemed more skeptical, as far as I could tell from his review talks that I was able to attend). Despite his optimism, and surely being aware that experimental confirmation of supersymmetry would earn them a very nice reward, Julius was always modest enough to admit the possibility that his expectations could turn out to be totally wrong. There was no question in his mind that experiment is the ultimate arbiter, and that either way one would have to accept the outcome of the experiment.

Half a century has now passed since the discovery of supersymmetry. During this time the subject has developed enormously, with stupendous advances on many fronts, some of which are also documented in this book. Supersymmetry has been a major driving force of developments in mathematical physics and pure mathematics. So it is definitely here to stay! Nevertheless, we now (in 2024) have to face up to the fact that supersymmetry, at least in the form championed over many years, is off the table as a realistic option for real life particle physics. 15 years of LHC searches have not produced a shred of evidence for superpartners of any kind. Quite to the contrary, the integrated results from LHC strongly indicate that the SM could happily live up to the Planck scale more or less *as is*, *without* supersymmetry or other major modifications.[29]

So where are we to go from here? Should we simply give up on the idea that supersymmetry is of any relevance to the real world, as demanded by a growing chorus of voices? Or should we pin our hopes on the next (100 TeV?) collider that, if it is ever built, will finally reveal that Nature *is* supersymmetric? If such hopes rely on further pursuing the (no longer 'low' energy) $N = 1$ supersymmetry option I am afraid they may well be disappointed, because the main motivating argument, namely solving the hierarchy problem, has largely evaporated. On the other hand, looking at the current disarray and confusion reigning in 'Beyond the SM' particle physics,

---

[29]One should also note the following point as possible evidence against supersymmetry. While supersymmetry strongly prefers BPS-type static configurations (such as AdS space), it is neither compatible with a positive cosmological constant nor, more generally, with a time-dependent cosmology — but this is what we see when we look out of the window!

I feel that simply abandoning the idea of supersymmetry for 'real' physics altogether would throw out the baby with the bath. Without *some* guiding principle, we may simply end up in an unnavigable multiverse of ideas (just look at the wildly diverse tableau of approaches to quantum gravity!), with no prospects of ever validating or falsifying any of them all the way.

To end this contribution on a more speculative note, let me outline where I see a possible way forward. This is to search for concepts *beyond* space-time supersymmetry, while taking the hints from what we have learnt from supersymmetry studies over the years. In other words, throw out the bath, but try to save the baby! And here I would first of all like to point out the remarkable fact that, with the possible exception of the asymptotic safety program, and despite their differences, *all* current approaches to quantum gravity are united in the belief that something dramatic happens to space or space-time near the Planck scale, in the sense that classical space-time must dissolve or 'de-emerge' there. This is analogous with continuum fluid dynamics which emerges from an underlying discrete structure of atoms and molecules. But if there is no space-time to begin with, the basic superalgebra

$$\{Q^i_\alpha, \bar{Q}_{\dot{\beta}j}\} = 2\delta^i_j \sigma^\mu_{\alpha\dot{\beta}} P_\mu$$

and possibly even the distinction between space-time bosons and fermions, cannot be fundamental. If consequently, space-time and concomitant concepts such as general covariance and gauge symmetries are to be emergent, the one million (or better: one billion?) dollar question is: *What* do they emerge *from*?

While there are no answers, there is also no lack of ideas and suggestions. One of them is a specific proposal that I have been involved in with Thibault Damour, Marc Henneaux and Axel Kleinschmidt, and which proceeds from the well-known BKL (short for: Belinski-Khalatnikov-Lifshitz) analysis of cosmological solutions of Einstein's equations near a space-like (cosmological) singularity. According to this analysis spatial points decouple near the singularity, effectively realizing a dimensional reduction to one (time) dimension at the singularity. But this is exactly the reduction for which one expects the hidden symmetry of maximal supergravity to get enlarged to the maximally extended hyperbolic Kac-Moody symmetry E_{10}! The hypothesis is therefore that the true symmetry reveals itself only in a 'near singularity limit', in perfect analogy with the full electroweak symmetry only manifesting in the high energy limit. So E_{10}, not supersymmetry, would be the key player! Towards a more concrete realization of this idea we have proposed a 'space-less' one-dimensional sigma model, mapping the supergravity dynamics in space-time to a null-geodesic motion on the infinite-dimensional coset manifold $E_{10}/K(E_{10})$. While capturing some essential features of maximal supergravity, this ansatz remains incomplete for many reasons. The outstanding challenge is to explain in full detail how E_{10} can replace space-time based QFT and the emergence of classical concepts such as space-time and gauge symmetries from a purely algebraic construct.[30] One would expect quantum theory to play an essential role in understanding how this happens. A further difficulty is that we have hardly any idea how to physically interpret the states associated with the imaginary roots of E_{10}, starting with 'dual graviton'. One encouraging feature, especially in connection with the incompatibility of supersymmetry with time dependent cosmological solutions, is that the $E_{10}/K(E_{10})$ sigma model does *not* admit static solutions.

What is clear, however, is that E_{10} goes beyond supersymmetry in the sense that it 'knows' everything that supersymmetry 'knows'. For instance, the information about the bosonic constituents of the maximal supergravity multiplets ($D = 11$, IIA, IIB,...) can be directly read off from the E_{10} Dynkin diagram, simply by 'slicing' it in different ways w.r.t. its various finite subdiagrams. Likewise, the fermionic sectors of the maximal supergravity multiplets are governed by the maximal compact (or 'involutory') subgroup $K(E_{10})$ of E_{10}, an infinite prolongation of the R-symmetry groups of maximal supergravities. For instance, the IIA and IIB fermion multiplets at a given spatial point are again obtained by decomposing the $K(E_{10})$ under its finite-dimensional R-symmetry subgroups. Moreover, one can expect that due to the unique properties of its root lattice (being the unique even self-dual Lorentzian lattice in ten dimensions), E_{10} will eventually play a key role in ensuring the (non-perturbative) quantum consistency of the theory, in the same way that the perturbative finiteness and consistency of string theory are ensured by *modular invariance*, whereas supersymmetry is mainly needed to get rid of tachyons. So there is some evidence that, at the most basic level, duality and modular symmetries are indeed more important than supersymmetry.

There is yet another, different, argument pointing in the same direction, though perhaps more obliquely. In 1983, M. Gell-Mann proposed what

[30]P. West has proposed a conceptually very different scheme based on the indefinite (but non-hyperbolic) Kac-Moody algebra E_{11}. His scheme does not invoke dimensional reduction.

he described as "a last ditch attempt to salvage the haplon interpretation of $\mathcal{N} = 8$ supergravity". The SO(8) gauge group of the theory contains a vector-like $SU(3) \times U(1)$, but it is immediately evident that this SU(3) cannot be identified with the color symmetry $SU(3)_c$. Noticing that after the removal of eight Goldstinos (to break all supersymmetries) one is left with the right number of spin-$\frac{1}{2}$ fermions ($48 = 3 \times 16$, corresponding to three generations of quarks and leptons, including right-chiral neutrinos), he had the very strange idea that the supergravity SU(3) should be identified with the diagonal subgroup of $SU(3)_c$ and a hypothetical family symmetry $SU(3)_f$. Putting the quarks and leptons into the appropriate representations of $SU(3)_c$ and $SU(3)_f$ (there is only one way to do this properly) he obtained exact agreement! In addition, the electric charge assignments almost work, in that the U(1) charges are systematically off by a 'spurion charge' shift of $\pm\frac{1}{6}$. So the proposal fails only by very little, and in a very systematic fashion. Not long after, Nick Warner and I were able to show that this scheme is *dynamically* realized at one of the stationary points of the $\mathcal{N} = 8$ potential. For a few days we were ecstatic, trying very hard to push the agreement further — *alas*, in vain. Since the proposal was greeted with understandable skepticism by many colleagues (who had mostly placed their bets with $\mathcal{N} = 1$ low-energy supersymmetry), I eventually decided to put it aside, since I did not want to get trapped in a quixotic chase after a mirage.

Nevertheless, the idea stuck in my mind. After all, if there are really only 48 fundamental spin-$\frac{1}{2}$ fermions, no more, no less, what other scheme could possibly explain this, with a *unique* answer as the outcome? Besides, the scheme can be immediately and easily falsified if even only one extra fundamental spin-$\frac{1}{2}$ fermion were to be detected. In 2015, when it began to dawn on everyone that LHC would not produce the predicted cornucopia of new particles, Krzysztof Meissner and I therefore decided to have another look, by asking ourselves what it would take to rectify the mismatch of electric charges. As it turned out, this could be achieved by a surprisingly simple deformation of the U(1) group by the generator

$$\mathcal{I} := \frac{1}{2}\big(T \wedge \mathbb{I} \wedge \mathbb{I} + \mathbb{I} \wedge T \wedge \mathbb{I} + \mathbb{I} \wedge \mathbb{I} \wedge T + T \wedge T \wedge T\big)$$

acting on the tri-spinor χ^{ijk} of $\mathcal{N} = 8$ supergravity which transforms in the $\mathbf{56} \equiv \mathbf{8} \wedge \mathbf{8} \wedge \mathbf{8}$ of SU(8). Here, the SO(8) matrix T represents the imaginary unit in the $SU(3) \times U(1)$ breaking, with $T^2 = -\mathbb{I}$, which in turn implies $\mathcal{I}^2 = -\mathbb{I}$. Because the triple wedge product $T \wedge T \wedge T$ is *not* an SU(8) element, there is no way of incorporating this deformation into $\mathcal{N}=8$

supergravity. The 'spurion shift' associated with $\mathcal{I}$ and needed for matching theory and observation is therefore incompatible with supersymmetry.

But now the remarkable fact (demonstrated jointly with Axel Kleinschmidt) is that, at least in the one-dimensional reduction, this deformation *can* be incorporated into $K(\mathrm{E}_{10})$. This could mean that matching the observed SM fermion spectrum with fundamental theory requires more than merely picking the right Calabi-Yau manifold, GUT[31] group or fermion multiplets, but may involve understanding in detail how such structures could emerge from a space-time-less pre-geometric context, and that we may have to go all the way to fancy infinite-dimensional symmetries such as E_{10} and $K(\mathrm{E}_{10})$ to make things work. Then the question is no longer whether $\mathcal{N} = 8$ supergravity is the right theory. Rather, it is whether and how E_{10} can replace space-time supersymmetry as a guiding principle, and what kind of pre-geometric theory it is that lies beyond $\mathcal{N} = 8$ supergravity and realizes these symmetries. And, most importantly, whether and how the partial matching with SM physics can be completed beyond merely kinematical considerations.

Since it could easily take another 50 years to figure all this out (assuming there is any truth in these admittedly unorthodox ideas), Krzysztof Meissner and I have recently started looking for possible experimental signatures that might provide observational tests of our hypotheses, in an attempt to search for the proverbial needle in the haystack. We propose to do this by concentrating on the massive gravitinos which are the only other fundamental fermions that would emerge from the $\mathcal{N} = 8$ supermultiplet. Unlike the MSSM gravitino, these gravitinos do carry SM charges and would thus participate in (non-chiral) SM interactions, which makes them directly detectable at least in principle. Taking into account the U(1) shift also for the gravitinos (which again requires $K(\mathrm{E}_{10})$) we have a new (stable) dark matter candidate here, but of a very unusual type. The gravitinos would have to be fractionally charged, extremely massive, and at the same time extremely rare (with an abundance of roughly one gravitino in a cube of $\mathcal{O}(10\,km)$ side length), so as to have escaped detection until now. We have argued that, if such gravitinos were really found, we would be able to address two outstanding open problems in astrophysics, namely (1) explaining the origin of supermassive primordial black holes in the very early universe (via

[31]Grand Unified Theory.

the condensation and gravitational collapse of lumps of supermassive gravitinos in the early radiation period), and (2) understanding the origin of the ultra-high energy cosmic ray events that have been observed over many years at the Pierre Auger Observatory in Argentina (via the mutual annihilation of supermassive gravitinos in the 'skin' of neutron stars). We are currently exploring ways to search for such particles in upcoming underground experiments, such as JUNO,[32] as such objects are evidently far beyond the reach of high energy colliders.

5 Conclusions

Independently of whether the ideas sketched above are on the right track or not, I remain attached to the Einsteinian point of view that we should try to understand and explain first of all *our* universe and *our* low-energy world, and that in the end there should emerge a more or less unique answer. I believe that 50 years of supersymmetry have brought us a wee bit closer to this goal, though not as close as many would have wished. Of course, this point of view runs counter to currently prevalent views according to which the only way out of the vacuum conundrum of string theory is the multiverse. But if Nature must pick the 'right' answer at random from a huge ($>10^{272\,000}$?) number of possibilities, I see no hope that we would ever be able to confirm or refute such a theory, nor to reach a truly fundamental understanding of the world that we see.

Already in 1929, and in connection with his first attempts at unification, Albert Einstein published an article in which he states with wonderful and characteristic lucidity what the criteria should be of a 'good' unified theory: (1) to describe as far as possible all phenomena and their inherent links, and (2) to do so on the basis of a minimal number of assumptions and logically independent basic concepts. The second of these goals, also known as the principle of Occam's razor, he refers to as "logical unity" ("logische Einheitlichkeit"), and goes on to say: "Roughly but truthfully, one might say: we not only want to understand *how* Nature works, but we are also after the perhaps Utopian and presumptuous goal of understanding *why* Nature is the way it is, and not otherwise."

To which I have nothing more to add!

[32]The Jiangmen Underground Neutrino Observatory (JUNO) is a medium baseline reactor neutrino experiment under construction at Kaiping, China.

Acknowledgments

I would like to thank Misha Shifman for giving me this opportunity to provide my personal perspective on the story of supersymmetry which owes so much to the legacy of Julius Wess and Bruno Zumino. I am also grateful to all my collaborators (including those, whose names do not appear in this text) for the collaborations and lasting friendships that I have been able to build over many years. My special thanks go to Axel Kleinschmidt and Stefan Theisen for their friendship and unwavering support over many years.

A FEW WORDS ABOUT JULIUS[33]

M. SHIFMAN
Theoretical Physics Institute, University of Minnesota, Minneapolis, MN 55455
shifman@umn.edu

I am proud that I was personally acquainted with Julius Wess. We first met in 1999 when I was working on the Yuri Golfand Memorial Volume. I invited him to contribute, and he accepted this invitation with enthusiasm. After that, we met many times, mostly at various conferences in Germany, Croatia and elsewhere. I was lucky to discuss with Julius various questions of theoretical physics, from supersymmetry to space-time noncommutativity, and listen to his recollections on how supersymmetry was born.

In physics Julius was a visionary who paved the way to generations of followers. In everyday life he was a kind and modest person, always ready to extend a helping hand to people who were in need. I remember him telling me how concerned he was about the fate of theoretical physicists in Eastern Europe after the demise of communism. Especially close were his ties with the Croatian theorists. He was instrumental in providing them with financial support through European grants. In particular, the "Wess" grants were used to organize summer schools for graduate students and postdocs.

Julius' ties with Israeli physicists also bore a special character. He was an unconditional supporter of Israel and realized its difficult situation — the very existence of this country was not for granted. Wess visited Israel on a regular basis. On page 3 you can see a photograph of Julius lecturing at Technion, Haifa, on March 10, 1999, during one of such visits.

To show his solidarity and support to Ukraine, Julius joined the International Advisory Board of the Ukrainian Journal of Physics and assumed the position of the Scientific Council President of the Austro-Ukrainian Institute for Science and Technology in Vienna.

*****

Oberwölz, a small medieval town, is situated at the foothills of the Austrian Alps. Its residents enthusiastically explain to the visitors their uniqueness — Oberwölz is the smallest town in Austria, with only one and

[33]Editor's note: Photographs below are mine. –MS

a half thousand inhabitants. That's where the University of Graz organizes conferences on promising areas in theoretical physics almost every year in late summer or early autumn.

One of the conferences I attended took place in 2008. Julius Wess died very early, at the age of 72 — what a terrible loss! — in 2007. Reading his obituary in *Physics Today*, I suddenly noticed that he was born in Oberwölz in 1934. I mentioned this to Willy Plessas, a chief organizer of the Oberwölz conference, and added that Wess was probably Oberwölz's most outstanding son, and it would be right to erect a memorial plaque at his birthplace.

The next time I came to the Oberwölz conference in 2010. A smiling, joyful Willy Plessas greeted me and blurted out with pride:

"Hooray! Tomorrow there will be a ceremony of unveiling the memorial plaque. Two years ago I spoke with the mayor of Oberwölz and he agreed."

Figure 1.3: Wess Memorial Plaque, Hauptplatz No. 5, Oberwölz.

Below is a quotation from a local bulletin:

> Municipality of Oberwölz International Physics Symposium took place in the Youth and Family Guest House on August 30th–September 4th, 2010. After it was announced at the last symposium in 2008 that the well-known physicist Julius Wess was born in Oberwölz in 1934, the memorial plaque was ceremoniously unveiled as part of this symposium.[34] It was erected at Julius Wess' birthplace, at Hauptplatz No. 5. Fifty physicists and scientists from 22 nations and 4 continents (North and South America, Asia, Australia and Europe) attended the unveiling of the memorial plaque for Prof. Julius Wess at his birthplace which currently hosts Stadtbäckerei Zechner (Zechner City Bakery). They met for a week on the topic of "QCD and Strings: Elements of a Universal Theory."
> In addition to the intensive scientific work, there was also time for some leisure activities. The guests were culturally impressed by a chamber music concert in the Spitalskirche with works

Figure 1.4: Paying homage to Julius Wess at the plaque opening ceremony.

[34]The guests were welcomed by Mayor Günther Bischof and Professor Willibaldf Pressas. In 2010 the symposium was again organized and carried out by the university professors Dr. Willibald Plessas from the Karl Franzens University of Graz and Dr. Harald Fritzsch from the Ludwig Maiximilians University in Munich.

> by Franz Schubert and Johann Sebastian Bach with Bertin Christelbauer on the cello and Werner Karlinger on the harp. A trip to the Eselsbergergraben and a visit to the district capital Murau were organized for them.

The memory of Julius Wess as a trailblazer in physics and an exceptionally bright and kind person should not fade away with time. He was a role model for many.

Figure 1.5: Bruno Zumino, LBL, 1985.

Figure 1.6: Bruno Zumino, 2000s. AIP Emilio Segre Visual Archives, Courtesy of Mary K. Gaillard.

BRUNO ZUMINO ⛨

(April 28, 1923–June 21, 2014)[35]

National Academy of Sciences

Bruno Zumino, an influential theoretical physicist, was known for his proof of the CPT theorem (the combined operations of charge conjugation, parity, and time reversal) with Gerhart Lüders; the elucidation of chiral Lagrangians with Julius Wess and others; the discovery of supersymmetry in four space-time dimensions with Wess; and the formulation of supergravity — the supersymmetric version of Einstein's theory of relativity — with Stanley Deser. Part of Zumino's rich legacy is that supersymmetry — a theory that doubles the number of particles present in the Standard Model of particle physics — became the subject of intense experimental searches using the Large Hadron Collider (LHC) at the European Organization for Nuclear Research (CERN) near Geneva, Switzerland.

Born in Rome, Italy, Zumino received his doctoral degree in mathematical science at the University of Rome in 1945. He held postdoctoral and teaching positions in Rome until 1949, when he accepted a research appointment at the Max Planck Institute for Physics in Göttingen, Germany, under the direction of Werner Heisenberg. There he met Richard Courant, who invited him to what is now the Courant Institute of Mathematical Sciences at New York University. In 1961, Zumino became head of NYU's Department of Physics, a position he held until moving to CERN in 1969.

He joined the physics faculty of the University of California, Berkeley, in 1981, where he spent the rest of his career. Among Zumino's achievements there, he made important contributions in the areas of differential geometry and quantum groups — an alternative approach to incorporating gravity into the quantum field theory that describes the strong, weak, and electromagnetic interactions — and his teaching was deemed as distinguished as it had been at New York University. During his first year at Berkeley Zumino taught a seminar course on supersymmetry and supergravity that was attended by advanced graduate students — and also by essentially all of the particle theory faculty and Lawrence Berkeley Laboratory staff.

Zumino was widely appreciated not only for his classroom teaching but also for his mentorship of graduate students and postdoctoral scholars. He referred to his research students as his "kids," and they were very devoted to him. His last student, Darren Shih, wrote, "I will always respect Bruno as one of the greatest persons I have had the honor to work with and to learn from in my life."

Similarly, upon learning of Zumino's passing many of his colleagues wrote that he was a "giant of physics," renowned for "the elegance of his thinking, lecturing, and writing," as well as a warm and generous person whom they remembered for his sense of humor and the "twinkle in his eye."

BRUNO ZUMINO: A BIOGRAPHICAL MEMOIR ⊎ [36]

MARY K. GAILLARD
Center for Theoretical Physics
Department of Physics
University of California (Berkeley), CA 94720-7300
Email: mkgaillard@lbl.gov

Abstract: Bruno Zumino, an influential theoretical physicist, was known for his proof of the CPT theorem (the combined operations of charge conjugation, parity, and time reversal) with Gerhart Lüders; the elucidation of chiral Lagrangians with Julius Wess and others; the discovery of supersymmetry in four space-time dimensions with Wess; and the formulation of supergravity — the supersymmetric version of Einstein's theory of relativity — with Stanley Deser. Part of his rich legacy is that supersymmetry — a theory that doubles the number of particles present in the Standard Model of particle physics — is currently the subject of intense experimental searches using the Large Hadron Collider at the European Organization for Nuclear Research (CERN) near Geneva, Switzerland. Bruno was elected to the U.S. National Academy of Sciences in 1985. He received many awards and honors, including the 1987 Dirac Medal and the 1988 Heineman Prize of the American Physical Society.

Bruno was born in Rome, Italy, on April 28, 1923, the son of Romeo Achille Zumino, an engineer from a family of artists in Majana, Italy, and Concetta Bruno of Rome. Growing up under Mussolini's reign and during the years of World War ll left painful memories that Bruno spoke of rarely, except to mention that his education was delayed while he was in hiding to avoid being drafted into the fascist army. He resisted pressure from his father to become an engineer. Instead, Bruno studied physics and mathematics. He received his doctoral degree in mathematical science, under the supervision of Bruno Ferreti, at the University of Rome in 1945.

Bruno held postdoctoral and teaching positions in Rome until 1949, when he accepted a research appointment at the Max Planck Institute for Physics in Göttingen, under the direction of Werner Heisenberg. There, Bruno met Richard Courant, who invited him to what is now the Courant Institute of Mathematical Sciences at New York University (NYU). Bruno held research and teaching positions at NYU until 1956, when he was fired, so he told me, because he did not want to work on the applied physics topic favored by his supervisor. Bruno moved to the Stevens Institute of Technology in Hoboken, New Jersey, but returned to the NYU faculty a year later, becoming head of the department of physics in 1961, a position he held until moving to CERN in 1969. Bruno became an American citizen on January 8, 1962 — a decision largely motivated by the congressional hearings that eventually brought Sen. Joseph McCarthy's witch hunts to an end.

During his time in New York Bruno made many close and often lifelong friends, among them the physicists Grace and Larry Spruch and the mathematician Joseph Keller. He also left a deep impression on the NYU physics department. The following letter from Massimo Porrati and his NYU colleagues was read at a memorial service for Bruno held at the University of California, Berkeley, campus on October 24, 2014:

> Bruno Zumino played a crucial role in the development of the physics department at NYU. During several years in the 1960s, he was the chair of the department, which was located on two sites: downtown, near Washington Square; and uptown in the Bronx. During that time, thanks to Bruno's efforts, a number of distinguished senior physicists, as well as young, promising theorists, joined the NYU faculty.
> The first group included Kurt Symanzik and Wolfhart Zimmermann, two of the world's leading field theorists (who received joint appointments in the physics department and the Courant Institute); Henry Stroke, a distinguished atomic and molecular physicist; and Mal Ruderman, a noted astrophysicist. The second group included Alberto Sirlin, Dan Zwanziger, Richard Brandt, and John Lowenstein, who joined the NYU faculty in 1959, 1967, 1969, and 1972, respectively.
> There were also several distinguished visitors. Julius Wess visited NYU a number of times to pursue his seminal work with Bruno that would lead to the foundations of supersymmetry. The group of visitors included also Alfred Mueller, Hirotaka Sugawara, Gianfausto Dell'Antonio, Claudio Orzalesi, Giuliano Preparata, Hector Rubinstein, Guido Altarelli, Giorgio Velo, Antonio Masiero, Jean-Loup Gervais, and Joël Scherk.
> Bruno was a very clear lecturer. In fact, several advanced graduate students at Columbia University used to commute downtown to attend Bruno's lectures on quantum field theory. This is particularly remarkable considering that half of the Columbia physics faculty at the time were already or would become Nobel Prize winners.
> At one stage, Bruno conceived the idea of creating an exchange program between French and NYU physicists. In order to obtain financial support for the program, he approached André Meyer,

> a noted French banker who had moved his operations to New York; he was a trustee of NYU and father of Philippe Meyer, a well-known French physicist. Instead, André Meyer offered to provide substantial funds to construct a new physics building. This turned out to be very important because, at that time, the space and facilities in the downtown department were very inadequate. Bruno's initiative led to the construction of the André and Bella Meyer Hall of Physics that at present houses also the Center for Neural Sciences and the psychology department.
> Bruno left NYU in 1968 to become a permanent staff member of CERN and later a professor at the University of California, Berkeley. However, his memory at NYU is still vivid today and his legacy lives on.

It was during this period in New York that Bruno made his first major contribution to physics: the proof, with Gerhart Lüders, that the combined operations of charge conjugation (turning particles into their antiparticles), parity (mirror reflection), and time reversal left any local quantum field theory invariant. The first in a series of three papers on this proof was written by Lüders alone. Bruno once told me that even though they had actually done the work together, Gerhart wrote it up himself, simply forgetting that Bruno had been involved. Bruno apparently took him at his word and bore him no grudge. They then coauthored the second and third papers. Upon Lüder's death in 1995, Bruno wrote to his widow: "He was a very dear friend who helped me very much, especially when I was young, with his advice and his example. I shall never forget him."

During his NYU years Bruno also wrote influential papers applying symmetry concepts to the study of the strong nuclear interactions; his coauthors included Tsung Dao Lee, Norman Kroll, Julius Wess, Sydney Coleman, and Curtis Callan. In an interview for one of Bruno's obituaries, our friend and colleague Steven Weinberg (the 1979 Nobel laureate) said, "I think Bruno understood earlier than most people how important symmetry was going to be in the development of physics."

In 1968 Bruno married Shirley Noakes. They moved to Geneva in 1969 when Bruno joined the senior staff of CERN, where he served as the Theory Division leader from 1970 to 1973. Bruno and Shirley divorced in 1975, but they remained friends throughout Bruno's life. In a note to me after his death, she described him as a "wonderful, warm, and inspiring man."

It was at CERN that Bruno made major contributions that continue to underlie much of the activity in particle physics today. In 1971, he and Julius Wess wrote an important paper on anomalies — quantum effects that spoil the symmetries of classical field theories. It introduced what came to be known as "Wess-Zumino terms," and had far-reaching ramifications both in physics and mathematics.

Later, in 1974, Bruno and Julius were studying string theory — according to which the point particles of quantum field theory are replaced by tiny two-dimensional objects, or "strings," to describe the strong force. Originally formulated as a theory of bosons — particles with integer spin, or intrinsic angular momentum, string theory was generalized by Pierre Ramond, and by André Neveu and John Schwarz, to include fermions, half-integer spin particles. After these three physicists introduced a symmetry relating bosons and fermions in the two-dimensional system that described the coordinate along the string as a function of time, Julius and Bruno asked themselves if a similar symmetry, now known as supersymmetry — could be realized in our four-dimensional world. They succeeded, then constructed the first interacting theory with such a symmetry, now known as the Wess-Zumino model. And they formulated the supersymmetric version of quantum electrodynamics, the theory of electric and magnetic interactions.

Although Julius and Bruno subsequently learned that their discovery had been anticipated in the Soviet Union, first in Moscow by Yuri Golfand and Evgeny Likhtman, then in Kharkov by Dmitrij Volkov, Vladimir Akulov, and Vyacheslav Soroka — it was the Wess-Zumino papers that put supersymmetry on solid ground and brought it to international attention. The Yuri Golfand Memorial Volume contains a graph that shows the remarkably fast increase in the number of papers on supersymmetry as a function of time after the first three preprints had appeared carrying the names of Wess and Zumino.[37] Bruno went on to write many important papers on the properties of supersymmetry and its implications with Julius, Jean Iliopoulos, and Sergio Ferrara. In 1976, together with Stanley Deser, Bruno formulated supergravity — the supersymmetric version of Einstein's general relativity, which was discovered independently by Ferrara, Daniel Freedman, and Peter van Nieuwenhuizen.

I first met Bruno in 1963, when he was a visiting scientist at CERN. However, we did not collaborate until 1980, after John Ellis, Luciano Maiani, Bruno, and I had become interested in what was known as "$\mathcal{N} = 8$ supergravity". This was a version of Einstein's general relativity with eight supersymmetries, instead of just one, as in the supergravity theory discovered in 1976. This new theory had properties that made it a candidate for the

[37]Shifman, M. 2000. *The many faces of the superworld*: Yuri Golfand memorial volume. Singapore: World Scientific.

realization of Einstein's dream, the unification of all the forces of nature, including gravity. But it also possessed a symmetry under the interchange of electric and magnetic fields, and in order for it to do the task of describing nature as we observe it, there had to be an associated conserved quantum number. Bruno and I found that this was indeed the case, and although $\mathcal{N} = 8$ supergravity was later abandoned as the potential theory uniting all the forces, our paper turned out to have important applications in string theory, now applied to gravity rather than just the strong interaction — and currently the leading candidate theory for realizing Einstein's dream. It was during this collaboration that Bruno and I became a couple; we married after moving to Berkeley.

Bruno and I moved to the University of California, Berkeley, in 1981, with him taking a cut in salary. He did this willingly so that we could stay together; although I myself was a prominent physicist who became the first woman on the Berkeley physics faculty, CERN was unwilling to hire me. (At that time there were no women among the senior scientific staff there.)

During his earlier years at Berkeley, Bruno made important contributions in the areas of differential geometry and quantum groups — an alternative approach to incorporating gravity into the quantum field theory that successfully describes the strong, weak, and electromagnetic interactions — and he was invited to give many talks at conferences during this period. In recent years, together with Ferrara, postdoctoral scholars, and students, he resumed work on the electric-magnetic symmetry that we had studied together, in particular its applications to string theories and the physics of black holes.

Bruno's teaching at Berkeley was every bit as distinguished as it had been at New York University. During his first year he taught a seminar course on supersymmetry and supergravity that was attended not only by advanced graduate students but also by essentially all of the particle theory faculty and Lawrence Berkeley Laboratory staff. He was widely appreciated both for his classroom teaching and for his mentorship of graduate students and postdoctoral scholars. He referred to his research students as his "kids," and they were very devoted to him. Two of his recently graduated students, Anthony Tagliaferro and Shannon McCurdy, made moving tributes to him at his memorial service, in front of the many faculty who where gathered there.

Bruno's concern for students went beyond his own teaching and mentoring. When, together with Julius Wess, he was awarded the Dannie Heineman Prize of the American Physical Society (APS) in 1988, he donated his prize money to the Berkeley physics department for financial support of foreign doctoral students. He considered prize money as "free money" that one shouldn't just spend in an ordinary way. This inspired me to make a similar gesture when I won an APS prize five years later. When we were approached about an initiative to encourage faculty to provide seed funds for graduate-student fellowships, Bruno asked me, "Couldn't we each contribute half?" Thus the Benjamin Lee[38] and Julius Wess Fellowship fund was born.

Bruno and Julius first met when Bruno was visiting the University of Vienna at the invitation of Walter Thirring, who was Julius's doctoral thesis advisor. As Bruno told it, Walter said to him that "there was this very good student" — and could Bruno take care of him? Julius and Bruno became good friends, and they remained very close, both professionally and personally, until Julius's death in 2007. Their collaboration that started in Vienna continued in many other places, most often when Bruno visited Julius in Kalsruhe, Germany, where Julius was a professor for many years (until he moved to the Max Planck Institute in Munich), and also during Julius's frequent visits to CERN. Their joint work earned them several shared awards. I remember many happy moments with Bruno and Julius, including hiking on the French Saléve Mountain near Geneva and, to our surprise, ending up on a newly constructed golf course where we were scolded for being inappropriately attired.

Bruno and I both spoke at Julius's 70th birthday symposium in 2005. Then just three years later, on Election Day in November 2008, we flew to Munich again for his memorial conference. This time I was not a speaker. Jan Louis, who picked us up at the airport, first apologized that we had missed the chance to vote in an historic American election. I assured him that we had already voted by mail. Then he explained that all the speakers would be different from those at the earlier symposium, with the exception of Bruno, who was the highlighted speaker at both events. The loss of his younger and very close friend Julius was a very heavy blow for Bruno, as he wrote to Walter Thirring.

Although Bruno was no longer able to actively engage in research during the last two years of his life, his students visited him on several occasions, and just a month before his death — which occurred in his sleep on June 21, 2014, at the age of 91 — he attended the graduation ceremony for his last student, Darren Shih. Darren wrote, "I will always respect Bruno as one of

[38]Benjamin Lee wrote one paper with Bruno, but he and I collaborated on several influential papers, and were very close friends until his untimely death in 1977.

the greatest persons I have had the honor to work with and to learn from in my life."

Similarly, upon learning of Bruno's passing many of his colleagues wrote that he was a "giant of physics" as well as a warm and generous person whom they remembered for his sense of humor and the "twinkle in his eye".

Bruno's impact on physics was far-reaching. His work influenced the searches for supersymmetry at the Large Hadron Collider, the searches for dark matter in the universe, and many theoretical developments. To quote Steve Weinberg: "Supersymmetry is so beautiful and suggestive that most of us think it has to show up sometime in nature, although so far it hasn't." Lawrence Hall, our colleague at Berkeley, said, "Bruno had an uncanny ability to work on theories important both for their mathematical structure and for their connection to experiments. When he introduced supersymmetry, he could have had no idea that 40 years later there would be a laboratory at CERN with more than 4,000 people looking for evidence of the theory."

At the memorial service at Berkeley, Petr Hořava said that Bruno "changed peoples lives and influenced people so deeply," referring to his scientific contributions. Others spoke about his friendly and unassuming manner. In a similar vein, while most of Bruno's obituaries focused on his scientific achievements, the one in the Berkeley student publication *The Daily Californian* addressed his more personal attributes:

> Though Zumino was internationally lauded for his groundbreaking research, he was well known on campus for his friendly demeanor. "He was a very kind and generous man, and I really just absolutely loved having him around as a colleague," said Raphael Bousso, Zumino's former colleague and a current professor at the Berkeley Center for Theoretical Physics. "He made me feel very welcome when I arrived at Berkeley about 10 years ago, and I'll always be grateful for that."
>
> On Bousso's first day on the job, Bruno walked into his office, leaving him "in awe." The senior professor welcomed his new coworker and then began to pick crumbs up off of the office floor, apparently unhappy that it had been left in that condition. "This unbelievable man, whom I so admired, was standing there cleaning my office," Bousso said, laughing. "But that's the kind of guy he was. He was totally unassuming, and at the same time you could see he was a tremendous intellectual force."
>
> Yasunori Nomura, another professor at the Center for Theoretical Physics, said he didn't detect the slightest hint of arrogance upon first meeting Zumino. He introduced himself by saying "ciao" and

> proceeded to ask Nomura about the Japanese language and culture — which Nomura, then a recent immigrant with poor English skills, said came as something of a relief. "I guessed that he was an old professor, but didn't know he was Bruno Zumino, one of the 'gods' in our field."

These themes were reflected in many of the letters I received after Bruno's death, a sampling of which follows:

> He will always be remembered for his work, but those of us who have had the privilege to know him have lost a dear friend.
> –John Iliopoulos

> He was a brilliant physicist and also a very kind and wonderful person.
> –Won-Yong Lee

> He expressed himself in writing and speaking with great elegance — an elegance that was not a superficial window dressing but a reflection of his deep understanding of physics. And I greatly enjoyed his sense of humor.
> –Kurt Gottfried

> He was for me a friend, a colleague, and a teacher.
> –Sergio Ferrara

> Bruno had a huge impact on many lives. His professional achievements changed the direction of the field, and many people owe him an enormous gratitude for that. In addition, I was always impressed by his kindness and generosity.
> –Nathan Sieberg

> Bruno was one of my heroes in physics not only through his monumental contributions but also for the elegance of his thinking, lecturing, and writing. And thanks to my years in Berkeley I had a chance to appreciate what a wonderful human being he was in many other respects.
> –Fabio Zwirner

> It is hard to imagine a man who could be more humble in the face of his overwhelming accomplishments. I have once heard it said that 'Great people want you to be great.' That would describe Bruno Zumino....When confronted with an error (rare though they were!) or a difficulty, I can picture him simply shrugging and smiling.
> –Brent Nelson

Bruno had no children of his own, but he became very attached to my children from a previous marriage. At Christmas we regularly skied with various members of my family in Verbier, Switzerland, or in Alta, Utah. In the summers we hiked, sometimes with one of my children, in Aspen, Colorado, or in Verbier. When we were invited to visit China in 1985 by Tsung Dao Lee and Zhou Guangzhao, my eldest children, Alain and Dominique, accompanied us. Dominique took advantage of this trip to arrange for her last year at the University of California, Santa Cruz, to take place instead in Beijing, and she then spent several years in Taiwan, where Bruno and I and my youngest son (also named Bruno) visited her.

Bruno Zumino told friends how much he enjoyed being part of my family, and how he loved being called 'Grandpa' by my grandchildren. My youngest grandchildren were particularly attached to him. During a brief informal memorial at our theory group retreat at Lake Tahoe in September, 2014, my 10-year-old granddaughter Kaeli spoke about her part in making the book of family pictures for him when he was hospitalized. Unfortunately, his more than two-month hospital stay took place during the conference celebrating his 90th birthday. With some effort, we were able to convince his medical team to allow him to attend the conference reception, which was just two blocks from the hospital. My son Bruno and a caregiver brought him to the reception in a wheelchair, and he was able to meet most of the conference attendees.

Bruno loved the mountains and hiking. We used to take long walks in Tilden Park, just behind our house in the north Berkeley Hills. He was also a great lover of music in general and of opera in particular. Although he didn't play an instrument himself, he often said that if he hadn't become a physicist he would like to have been a music critic.

In addition to myself, Bruno is survived by his stepchildren Alain, Dominique, and Bruno Gaillard, his nephew Maurizio Saglio and his cousin Salvina Cacciapaglia and their families, and my seven grandchildren, whom he considered as his own.

Selected Bibliography

1. G. Lüders and B. Zumino, Some Consequences of TCP-Invariance, *Phys. Rev.* **106** (1957) 385–386.
2. G. Lüders and B. Zumino, Connection between Spin and Statistics, *Phys. Rev.* **110** (1958) 1450–1453.
3. N. M. Kroll, T. D. Lee, and B. Zumino, Neutral Vector Mesons and the Hadronic Electromagnetic Current, *Phys. Rev.* **157** (1967) 1376–1399.
4. S. R. Coleman, J. Wess, and B. Zumino, Structure of Phenomenological Lagrangians. 1., *Phys. Rev.* **177** (1969) 2239–2247.
5. C. G. Callan, Jr., S. R. Coleman, J. Wess, and B. Zumino, Structure of Phenomenological Lagrangians. 2., *Phys. Rev.* **177** (1969) 2247–2250.

6. J. Wess and B. Zumino, Consequences of Anomalous Ward Identities, *Phys. Lett. B* **37** (1971) 95–97.
7. J. Wess and B. Zumino, A Lagrangian Model Invariant Under Supergauge Transformations, *Phys. Lett. B* **49** (1974) 52.
8. J. Wess and B. Zumino, Supergauge Transformations in Four-Dimensions, *Nucl. Phys. B* **70** (1974) 39–50.
9. J. Iliopoulos and B. Zumino, Broken Supergauge Symmetry and Renormalization, *Nucl. Phys. B* **76** (1974) 310.
10. J. Wess and B. Zumino, Supergauge Invariant Extension of Quantum Electrodynamics, *Nucl. Phys. B* **78** (1974) 1.
11. S. Ferrara and B. Zumino, Transformation Properties of the Supercurrent, *Nucl. Phys. B* **87** (1975) 207.
12. B. Zumino, Supersymmetry and the Vacuum, *Nucl. Phys. B* **89** (1975) 535.
13. S. Deser and B. Zumino, Consistent Supergravity, *Phys. Lett. B* **62** (1976) 335.
14. S. Deser and B. Zumino, A Complete Action for the Spinning String, *Phys. Lett. B* **65** (1976) 369–373.
15. J. Wess and B. Zumino, Superspace Formulation of Supergravity, *Phys. Lett. B* **66** (1977) 361–364.
16. S. Deser and B. Zumino, Broken Supersymmetry and Supergravity, *Phys. Rev. Lett.* **38** (1977) 1433–1436.
17. B. Zumino, Supersymmetry and Kahler Manifolds, *Phys. Lett. B* **87** (1979) 203.
18. M. K. Gaillard and B. Zumino, Duality Rotations for Interacting Fields, *Nucl. Phys. B* **193** (1981) 221–244.
19. B. Zumino, Y.-S. Wu, and A. Zee, Chiral Anomalies, Higher Dimensions, and Differential Geometry, *Nucl. Phys. B* **239** (1984) 477–507.
20. W. A. Bardeen and B. Zumino, Consistent and Covariant Anomalies in Gauge and Gravitational Theories, *Nucl. Phys. B* **244** (1984) 421–453.

2

The Predecessors

P. Ramond

A. Neveu

J. Schwarz

J.-L. Gervais

B. Sakita

Figure 2.1: Pierre Ramond, October 21, 2011. 25-th Solvay Congress in Brussels, Belgium.

EARLY SUPERSYMMETRY ON THE PRAIRIE[39]

P. RAMOND

Institute for Fundamental Theory, Department of Physics
University of Florida, Gainesville, FL 32611-8440, USA
Email: pierre.ramond@gmail.com

My first postdoc started at NAL in September 1969, one of a group of five postdocs, with no senior theorists. NAL, which was to become the Fermi National Accelerator Laboratory, was being energetically constructed under the guidance of Bob Wilson.

Earlier that summer at the ICTP in Trieste, armed with a fresh Ph.D., I had been introduced to the Veneziano model by deriving the triple Reggeon vertex with J. Nuyts and H. Sugawara. We published nothing, as our work, which was cast in an awkward notation, was superseded by Sciuto's paper using the oscillator formalism. Still, I felt attracted to the model by its elegance and because I could understand some of it, and remained determined to work on it.

NAL Theorist

I came back to America and NAL, determined to learn more about these new techniques. In the middle of the Atlantic on the liner "France", I was studying the Fubini-Gordon-Veneziano preprint in the ship's reading room. Imagine my surprise when I saw the *same* preprint on another desk! Its owner was nowhere to be seen, and it was annotated in a different hand. I eagerly waited for the reader. This is how I met André Neveu who was on his way to Princeton University. We decided to keep in touch about each other's progress. My wife Lillian and I were amazed when we asked André to join us at the bar, and found him already in his pajamas at 8 p.m.! I wondered if all physicists were like that.

NAL was more like a summer camp than a laboratory. It was a collection of houses away from the construction site. The five theorists were housed

[39]Editor's note: Revised version containing parts from Ramond's contribution to the First Edition (2000) and arXiv:0708.3656, August 2007. Prepared by the author specifically for the Second Edition and submitted in January 2024. –MS

in one house, experimentalists in another.[40] The director's complex was a bunch of houses put together. Over lunch, Bob Wilson informed us that eventually theorists should be housed in open areas without doors so as to increase communications. I objected, and still remember the silence that followed!

One of the five theorists was Lou Clavelli whom I had already met when he had spent a month in Trieste. David Gordon had worked with Fubini and Veneziano in developing the oscillator formalism, and I looked forward to collaborating with him. David Gordon and I wrote the *first* theory paper out of NAL (THY-1), a not very memorable attempt to include fermions into the Veneziano model.

Jim Swank, a student of Nishijima, and Don Weingarten, a student of Bob Serber completed the quintet. David, Lou, and I shared an interest in studying the Dual Resonance Model amplitudes. I did not yet know about the string connection. In fact, when I told Lou that the mass spectrum looked familiar and reminded me of something I had come across, his response was: "Nambu says it is a string". Lou had been Professor Y. Nambu's graduate student at Chicago, and he soon arranged a meeting over lunch with Nambu at the Quadrangle Club. There the great man gently encouraged us in our studies, asking probing questions with no easy answers that left me totally impressed. He even treated us to the lunch! Brilliance and humility, a combination of traits not often seen in physicists.

We all visited the University of Wisconsin where we met, among others, Bunji Sakita and Miguel Virasoro. They both seemed off-scale, and so way ahead of us in their understanding of Dual Models. In particular, Miguel had just constructed a set of operators which had the potential to decouple the negative norm states found in the amplitudes,[41] but only when the lowest particle was a tachyon and the first excited state massless. He had spent a lot of time trying to get around these "unphysical" constraints, and hesitated to publish his result despite all who have seen his work agreeing that it was too beautiful not to.

Nambu was the only senior theorist with whom we had scientific discussions. Many senior theorists came to NAL to gauge the progress and visit the site. The "NAL Fives" were given guide duties: escorting visiting theorists up a Silo near the ring (then under construction), taking them to lunch,

[40]See http://bama.ua.edu/~lclavell/Weston/ for Lou Clavelli's wonderful account of these early NAL days.

[41]Decoupling was later proven by Brower, Goddard and Thorn.

and setting up a Colloquium/Seminar. For our troubles, we got a free lunch at a restaurant off-site. Very few showed any interest in what we were doing.

André and I had agreed to keep in touch and report on our mutual progress. I soon received a preprint of André Neveu and Joël Scherk who had successfully isolated the infrared divergence of the planar one-loop dual amplitude. It was a feat of mathematical physics, that only those with French training could have achieved: they used the Jacobi imaginary transformation. I invited them to NAL. This is how I met Joël and his first wife. I was impressed by the fact that they had met at Club Med, not the usual hangout of physicists! Joël was as quiet and introspective as André was exhuberant and flashy. They came across as intellectually brilliant, and I took to both right away.

Lou and I found it difficult to work with David Gordon, so we continued our studies of the Dual Resonance Models on our own. It took us a long time before we published our first paper (May 1970). In it we put on firm group-theoretical basis the vertex introduced independently by Nambu and Fubini and Veneziano, using the $SU(1,1)$ operators found by Gliozzi, by Chiu-Matsuda-Rebbi, and by Thorn. From there we understood that fixing the (not yet known as) conformal weight of the vertex was the same as a (free) equation of motion for the emitted particle. We quickly set about to find the vertex for the first excited state. It had never been written down before and this was to be our first truly original work.

Aspen

Early in 1970, Bob Wilson, bless his soul, decreed "All theorists must go to Aspen"! As Bob was the boss, we went, although none of us knew much about the Aspen Center for Physics. This was the best advice an experimentalist ever gave me! The Aspen Center for Physics was at the time a collection of two buildings, set near the music tent at the West end of Aspen, Colorado. The oldest, Stranahan Hall, was made of bricks; the second, Hilbert Hall, was a wooden structure which had been donated to the Center by Bob Wilson after it was used for the 1968 NAL Summer Study.

It was a wonderful stay. Our first day in Aspen, Don Weingarten and I went to dinner at Guido's, where Don confiscated a sign which proclaimed something like "no hippie barefoot people allowed in the restaurant"! Little did we know that the same Guido was known for having chased after hippies with his shotgun! Don (I was told) had achieved early notoriety during student riots by having his picture taken, sitting in the President of Columbia

University's armchair, and smoking the President's cigars! The town was in the afterglow of the hippie era, and my days were spent playing volleyball in Wagner Park, listening to music outside the music tent in the late afternoons, and in other non-scientific activities. In my spare time, I started thinking about the particle spectrum that had been extracted from the Dual Amplitudes. People had already found some sort of position operator $Q_\mu(\tau)$ which appeared in the vertex, and its derivative was like a generalized momentum $P_\mu(\tau)$. Indeed, if one pursued the analogy further, the inverse propagator looked like the square of that generalized momentum. This led me to think of a *Correspondence Principle* by which simple notions of point particles were related to Dual Models:

$$p_\mu = \frac{dx_\mu(\tau)}{d\tau} \qquad \longrightarrow \qquad P_\mu(\tau) = \frac{dQ_\mu(\tau)}{d\tau}$$

with the expansion

$$P_\mu(\tau) = p_\mu + \sum_{n=1}^{\infty} \frac{1}{\sqrt{n\alpha'}} a_\mu^{n\dagger} e^{in\tau} + h.c.$$

The equation of motion then reduced to

$$p_\mu p^\mu \quad \longrightarrow \quad P_\mu(\tau)P^\mu(\tau) = L_0 + \sum_{n=1}^{\infty}(L_n e^{in\tau} + L_{-n}e^{-in\tau})$$

yielding the elements of the Virasoro algebra.

At last, a glimpse of simplicity! Was it the altitude, was it the easy-going atmosphere at the Aspen Center for Physics that enabled me to view the problem through a different lens? I will never know, but that summer stay changed my perspective forever. At the end of my stay in Aspen, I had realized that the correspondence principle could be applied to the Dirac equation, but I needed to develop it in the bosonic case first.

Back at NAL, Lou and I spent the rest of summer refining our results on the excited vertex, and abstracting from it general rules for the construction of Dual Amplitudes, which we would call today conformal theory. We submitted that work to the *Physical Review* in late September. Although interested, Lou did not share my enthusiasm for my analogies and in October, I sent the paper on the bosonic Correspondence Principle to *Physics Letters*.

I then turned to the fermion case and realized the Dirac matrices had to be generalized with anticommutation relations

$$\gamma_\mu, \quad \{\gamma_\mu, \gamma_\nu\} = 2g_{\mu\nu} \quad \longrightarrow \quad \Gamma_\mu(\tau), \quad \{\Gamma_\mu(\tau), \Gamma_\nu(\sigma)\} = 2g_{\mu\nu}\delta(\tau - \sigma),$$
$$\Gamma_\mu = \gamma_\mu + \gamma_5 \sum_n (b_{n\mu} e^{in\tau} + b^\dagger_{n\mu} e^{-in\tau}). \tag{2.1}$$

The resulting Dirac operator

$$\gamma_\mu p^\mu \quad \longrightarrow \quad \Gamma_\mu(\tau) P^\mu(\tau)$$

was written in terms of fermionic operators, F_n which anticommute with one another into a new Virasoro algebra, had the right structure to cancel the fermion zero modes generated by the $b_{n\mu}$ and $b^\dagger_{n\mu}$ with space-time four-vector indices.

This algebra, with both commutators and anticommutators, was of a kind I had never seen, in essence the square-root of the Virasoro algebra. I realized that it was a truly novel algebraic structure, since I could now take the square root of *any* Lie algebra.[42] Tremendously excited, I barely ate, drank and slept for weeks, as every derivation brought more conceptual clarity and more questions. I explained my results to Don Weingarten over lunch and he simply said: "You are set for life". Of course, I did not believe him.

Lillian, an electrical engineer, and fellow NCE alumna, had left for a six week stay at a Western Electric's Training Center near Princeton, and I joined her for a few days, some time in late October, as I recall. I took advantage of the trip to visit Y. Nambu who was spending the Fall at the Institute for Advanced Study. There I also met Jamal Manassah, Satoshi Matsuda, and Mike Green, and sketched my ideas on including the fermions via the Correspondence Principle. I was visiting Lillian (as the only male spouse!) and had no time to see André at the University.

Back in Illinois, a host of distractions caused me to lose focus. First, Ned Goldwasser informed us in early November that our appointments at NAL will terminate in 1971 (except for David Gordon's). This came as a big surprise: when we were hired, we were told to expect longer appointments because of the special circumstances, such as the lack of a senior theorist.

[42]At Yale, I approached the great algebraist, Professor N. Jacobson in the Mathematics Department. I showed him my algebra of commutators and anticommutators, asking if he had seen anything like this. I must not have made a convincing case since he replied negatively and kindly ushered me out of his office.

This was devastating news for us all. My own work had not been manifestly relevant to NAL, but Lou and the others had always kept close to experiments, and worked hard in NAL-sponsored workshops. We all became anxious over finding another job. My last experience had been nerve-wracking, and I was soon freaking-out. Secondly, the Correspondence Principle paper was rejected by *Physics Letters*. I argued but to no avail, and in December I withdrew the paper (Maurice Jacob, the Editor, told me years later that the rejection letter was not so strong, and that it could have been accepted with corrections, but back then I did not know the rules of the game).

Job applications took up most of my time, preventing me from writing up the fermion paper, and it was not until the week after Christmas that I sent it to the *Physical Review*, as well as to André and others. At the same time, I resubmitted the Correspondence Principle paper, this time to *Il Nuovo Cimento*, and went back to the business of job-hunting.

In those days, postdoctoral positions were on the whole awarded at the January Meeting of the American Physical Society in New York. This slave market was a depressing affair for Lou and I: our theoretical teeth and muscles were examined, but with no apparent outcome. The US establishment had no time for Dual Resonance Models/String theory, and showed minimal interest in what we were doing.

Then I got lucky, through a set of fortuitous circumstances orchestrated by Lou Clavelli. Before joining NAL, Lou had post-doctored at Yale and met his future wife, Estelle. As he was to visit his in-laws after the New York meeting, he suggested I join him and use the opportunity to visit the Yale High Energy Theory group, to which I had already applied. That Friday the Yale physicists, Sam McDowell, Charlie Sommerfield, John Harte, and Dick Slansky (Feza Gürsey was out of town) asked me to give an impromptu seminar, my first seminar on the Dual Dirac equation. The same night we flew back to Chicago, and three days later, Charlie Sommerfield offered me a one-year instructor position at Yale with the possibility of a second year! Finally, an early job offer! After talking with Lillian to discuss our next move, I called Charlie just a few hours later, accepting his offer! He told me later he was surprised by the speed of my acceptance, but once again I did not know the rules.

Unfortunately, by that time I was physically and emotionally exhausted. I fell sick with high fever for several days (most unusual for me), and was diagnosed with kidney stones, which were surgically removed. That experience was not all negative. Lillian and I befriended the NAL staff physician, Dr. Cornell, quite a character who had jumped on D-Day with the Airborne in Normandy. Today his favorite chicken recipe, “Chicken à la Cornell” occasionally graces our table. In the DuPage County Hospital, I also became a friend of Jim Ward, a police sergeant from Wheaton, the town we lived in. I organized a visit to NAL (I had much practice by that time) for his younger son Scott, who later credited me for awakening his interest in Physics; he is today a physicist at NREL in Golden, Colorado.

I gave the second (and last) seminar on the dual fermions at the University of Chicago where I found a sympathetic reception. However the outside world did not seem to show a similar interest. My fortunes changed when Professor Stanley Mandelstam (in whose triangle I had toiled as a graduate student) visited NAL. I excitedly told him about the fermions, expecting a reaction; there was none. Later that day, I pressed him for his opinion, and still remember his answer for its honesty: "You claim to have done something that many have tried to do, including my colleagues at Berkeley. I need time to assess it". True to his word, Stanley, together with Nambu, proved to be the most generous advocate of my work.

Three months after my fermion paper, I received a preprint by André Neveu and John Schwarz who proposed the Dual Pion Model, a generalization of the Veneziano model. It had no external fermions, but included fermionic ladder operators with space-time vector indices, just like my model of fermions, with the same name but different labeling. The underlying algebraic structure they proposed was the same as mine. André told me later when I visited him in Princeton (when I first met John Schwarz), that they had indeed been motivated by my paper: they had introduced a Yukawa interaction into my model, computed the amplitude with two fermions and an arbitrary number of bosons. They had great difficulties with the zero modes in the fermion sector (who didn't?), but they factorized the amplitude in the cross channel to extract the Dual Pion Model amplitudes. This result was soon published with proper referencing to my work. To this day I remain baffled by the faint acknowledgment of the seminal role of my work in the dual pion paper.

In 1969, Nambu had suggested that the Veneziano amplitudes should be derivable from a string. Later, Nambu and Goto had independently proposed an action for a relativistic string. It was not linear, and seemed impossible to quantize. When Virasoro found his decoupling coefficients, Nambu had interpreted them as evidence for his string picture, as the generators of the conformal algebra generated by the Fourier coefficients of the energy-momentum tensor of a two-dimensional theory spanned by the world sheet of a relativistic string. His point of view, while intriguing and interesting, was not widely appreciated, as it did not offer any computational advantages over the well-developed and fruitful amplitude approach.

The Spring of 1971 was spent adding electromagnetic interactions to the Dual Dirac equation, and trying to find a meaning for the anticommutators. I understood the anticommuting operators as generators of transformations between bosons and fermions. Such close kinship between fermions and bosons did not surprise me; it was natural in the original formulation of duality applied to the pion-nucleon amplitude, with its implied relations between its fermionic s-channel and bosonic t-channel. Unfortunately, I became confused by the parameters of these transformations (I did not know about Grassmann numbers), and did not get very far.

This symmetry between fermions and bosons was the first manifestation of a new type of symmetry - supersymmetry. The beginning of superstring theory as an impressive incubator of new theoretical ideas.

That summer, I was invited to lecture at the Boulder Summer School on the construction of dual amplitudes for vertices of arbitrary spin that Lou and I had developed. Many visiting senior theorists came through NAL, although the local theorists were largely ignored. There were of course exciting developments taking place elsewhere: Ben Lee, fresh from the Amsterdam Conference, brought the news of a young Dutch theorist who had proved that some massive Yang-Mills theories are renormalizable. This rang a bell because at Syracuse, Joe Schechter had suggested I read Weinberg's great 1967 paper, "A Theory of Leptons", although it was believed at the time that one could not calculate anything beyond tree level. I wondered if the tachyon of the Veneziano model was a signal of spontaneous breaking, but then of what? In the Fall of the same year, at a conference in Rutgers, I tried to interest my fellow theorists in this problem, but to no avail. This was where I learned from Claude Lovelace that cuts disappear from the twisted one-loop diagram when the number of space-time dimensions is twenty six. None of us understood the significance of this result.

I joined Yale in Fall 1971 as an Instructor. I was quite taken by the friendly atmosphere of the High Energy Theory group headed by the Turkish physicist Feza Gürsey. He and his wife Suha provided a great intellectual and humane cocoon into which Lillian and I were readily accepted. It was a wonderful place, and my Yale colleagues have remained life-long friends.

I continued to work in string theory. My student Michael Kalb and I understood the fundamental role of antisymmetric tensors in theories of extended objects, using ideas from Action-at-a-Distance theories where these fields are linked to muti-dimensional world-sheets. Today the two-form B-field is a mainstay of string theories. Little did I know that their study would presage branes and so many other wonderful developments! By the time of the 1974 London Conference, I was working on exceptional groups and their applications to Grand-Unified Theories. My last string publication of the First String Era was with another Yale student, Charles Marshall, on the covariant formulation of String Field Theory.

The decision to leave string theory was driven by the community's lack of interest and subsequent dearth of jobs in this wonderful subject. At the same time Lillian and I had to provide for our two very new daughters. I was also lured away by the intellectual promises of Feza's application of unusual algebraic structures to Physics. I was, needless to say, very pleased when many years later, exceptional groups appeared in the Heterotic superstring theory.

Acknowledgments

I am grateful to the organizers of the "The Birth of String Theory" Filippo Colomo and Paolo Di Vecchia, for giving me the opportunity to present my recollections. I also wish to thank Sudarshan Ananth, Lars Brink and Lou Clavelli for reading of the manuscript and their many helpful suggestions. This work, supported by the DOE Grant No. DE-FG02-97ER41029, was begun at the Aspen Center for Physics which I thank for its hospitality.

SUSY: THE EARLY YEARS (1966–1976)[43]

PIERRE RAMOND
Institute for Fundamental Theory, Department of Physics
University of Florida, Gainesville, FL 32611-8440, USA
pierre.ramond@gmail.com

Abstract: We describe the early evolution of theories with fermion-boson symmetry.

1 Introduction

By the 1940s, physicists had identified two classes of 'elementary' particles with widely different group behavior — bosons and fermions. The prototypic boson is the photon which generates electromagnetic forces; electrons, the essential constituents of matter, are fermions which satisfy Pauli's exclusion principle. This distinction was quickly extended to Yukawa's particle (boson), the generator of strong interactions, and to nucleons (fermions). A compelling characterization followed: matter is built out of fermions, while forces are generated by bosons.

Einstein's premature dream of unifying all constituents of the physical world should have provided a clue for that of fermions and bosons; yet it took physicists a long time to relate them by symmetry. This fermion-boson symmetry is called 'supersymmetry'.

Supersymmetry, a necessary ingredient of string theory, turns out to have further remarkable formal properties when applied to local quantum field theory, by restricting its ultraviolet behavior, and providing unexpected insights into its nonperturbative behavior. It may also play a pragmatic role as the glue that explains the weakness of the elementary forces within the Standard Model of Particle Physics at short distances.

2 Early hint

In 1937, Wigner [1], with some help from his brother-in-law, publishes one of his many famous papers "On Unitary Representations of the Inhomogeneous Lorentz Group." He was then at the University of Wisconsin at Madison, a refugee from Princeton, which had denied him tenure. It was not an easy paper to read, but its results were very simple: there were five types of representations labeled by the values of $P^2 = p_\mu p^\mu = m^2$, one of the Poincaré group's Casimir operator.

All but two representations describe familiar particles found in Nature. Massive particles come with momentum p, spin j, and $2j+1$ states of polarization, e.g. electrons and nucleons with spin 1/2. There are also four types of massless representations with spin replaced by helicity (spin projection along the momentum). The first two describe massless particles with a single helicity (photons with helicity ± 1), or half-odd integer helicity, such as "massless" neutrinos with helicity $+1/2$. The last two representations $O(\Xi)$ and $O'(\Xi)$ describe states which look like massless "objects", particle-like in the sense that they have four-momentum, but with bizarre helicities: each representation contains an infinite tower of helicities, one with integer helicities, the other with half-odd integer helicities. These have no analogues in Nature.[44]

Physicists were slow in recognizing the importance of group representations, even though Pauli provided the first solution of the quantum-mechanical hydrogen atom using group theory. Wigner's paper does not seem to have moved any mountains, and infinite spin representations were simply ignored, except of course by Wigner. Yet, $O(\Xi)$ and $O'(\Xi)$ contained important information: they are "supersymmetric partners" of one another!

3 Hadrons and Mesons

Symmetries were gaining credence among physicists, not as a simplifying device but as a guide to the organization of Nature. Wigner and Stückelberg's "supermultiplet model" unified SU(2) isospin and spin. Once Gell-Mann and Ne'eman generalized isospin to SU(3), it did not take long for Gürsey and Radicati [2], as well as Sakita [3], to propose its unification with spin into SU(6). Pseudoscalar and vector mesons (bosons) were found in the 35 representation of SU(6), while the hadrons (fermions) surprisingly lived in 56, not in 20 [3], as expected by the statistics of the time. This non-relativistic unification proved very successful, both experimentally and conceptually, since it led to the hitherto unsuspected color quantum number.

In 1966, Miyazawa [4] proposed further unification. His aim was to assemble the fermionic 56 and the bosonic 35 into one mathematical structure, such as SU(9) but at the cost of disregarding spin-statistics.

[44] "Infinite spin" representations do not appear in the Poincaré decomposition of the conformal group.

To explain the bounty of strange particle discovered in the 1950s, Sakata had proposed to explain mesons as $\bar{T}T$ bound states of the spin one-half triplet

$$T = (p, n, \Lambda).$$

Miyazawa adds a *pseudoscalar* triplet

$$t = (K^+, K^0, \eta),$$

to the Sakata spinor triplet. The hadron octet would then be described by another bound state, $T\bar{t}$, but he could not describe the spin three-half baryons decimet in the **56**.

He introduces a toy model with two fundamental constituents, a spin one-half and a spin zero particle, $\mathbf{p} = (\alpha_\uparrow, \alpha_\downarrow, \gamma)$. The nine currents

$$\mathbf{p}^\dagger \lambda_i \mathbf{p} = \begin{cases} F_i & i = 0, 1, 2, 3, 8; \\ G_i & i = 4, 5, 6, 7 \end{cases}$$

satisfy a current algebra with both commutators and anticommutators,

$$[F_i,\, F_j] = i f_{ijk} F_k, \quad [F_i,\, G_j] = i f_{ijk} G_k, \quad \{G_i,\, G_j\} = i d_{ijk} F_k,$$

a generalized Jordan algebra which he calls $V(3)$. This is the first example, albeit non-relativistic, of a superalgebra, today called SU(2/1) with even part SU(2) × U(1).

In 1967, he expanded his construction [5], to general superalgebras he calls $V(n, m)$ with the idea of including the decimet. Alas, the phenomenology was not as compelling as that of SU(6); two of the quarks inside a nucleon do not seem to live together in an antitriplet color state. In 1969, Berezin and G. I. Katz [6, 7] show the mathematical consistency of graded Lie algebra which contains both commutators and anticommutators; they give its simplest example generated by the three Pauli matrices $\sigma^+, \sigma^-, \sigma^3$. Physical applications are not discussed, although Berezin's advocacy of Grassmann variables in path integrals was no doubt a motivation.

4 Dual resonance models

In the 1960s, physicists had all but given up on a Lagrangian description of the strong interactions, to be replaced by the S-matrix program: amplitudes were determined from general principles and symmetries, locality, causality, and Lorentz invariance. Further requirements on the amplitudes such as Regge behavior and its consequent bootstrap program were still not sufficient to determine the amplitudes.

In 1967, Dolen *et al.* [8] discovered a peculiar relation in πN scattering. At tree-level, its fermionic s-channel ($\pi N \to \pi N$) is dominated by resonances ($\Delta^{++}, \ldots$), as shown by countless experiments. On the other hand, its bosonic t-channel ($\pi\bar{\pi} \to N\bar{N}$) is dominated by the ρ meson. Using the tools of S-matrix theory in the form of finite energy sum rules, they found that the Regge shadow of the bosonic t-channel's ρ-meson *averaged* the fermionic resonances in the s-channel! This was totally unexpected, since these two contributions, described by different Feynman diagrams, should have been independent. Was this the additional piece of information needed to fully determine the amplitudes of strong interactions? This early example of fermion-boson kinship led, through an unlikely tortuous path, to modern supersymmetry.

An intense theoretical search for amplitudes where the s- and t-channel contributions are automatically related to one another followed. Under the spherical cow principle, spin was set aside and the search for Dolen-Horn-Schmid (DHS)-type amplitudes focused on the purely bosonic process $\omega \to \pi\pi\pi$ [9]. Soon thereafter, Veneziano [10] proposed a four-point amplitude with the desired crossing symmetry,

$$A(s,t) = \frac{\Gamma(-\alpha(s))\Gamma(-\alpha(t))}{\Gamma(-\alpha(s)-\alpha(t))},$$

where $\alpha(x) = \alpha_0 + \alpha' x$ is the linear Regge trajectory. It displays an infinite number of poles in both s-channel $s > 0, t < 0$ and t-channel $s < 0, t > 0$. Veneziano's construction was quickly generalized to n-point "dual" amplitudes. The infinite series of poles were recognized as the vibrations of a string [11–13].

The amplitudes were linear combinations of tree chains which factorize into three-point vertices and propagators. A generalized coordinate emerged [14] from this analysis

$$Q_\mu(\tau) = x_\mu + \tau p_\mu + \sum_{n=1}^{\infty} \frac{1}{\sqrt{2\pi\alpha'}} \left(a_{n\mu} e^{in\tau} - a^\dagger_{n\mu} e^{-in\tau} \right),$$

with an infinite set of oscillators,

$$[a_{n\mu}, a^\dagger_{m\nu}] = \delta_{nm} g_{\mu\nu}.$$

The vertex for emitting a particle of momentum k_μ from the linear chain was simple

$$V(k,\tau) =: e^{ik\cdot Q(\tau)} : .$$

Out of its corresponding generalized momentum

$$P_\mu(\tau) = \frac{dQ_\mu}{d\tau}$$

one derived the operators

$$L_n = \frac{1}{2\pi}\int_{-\pi}^{\pi} d\tau e^{in\tau} : P^\mu P_\mu :\equiv \langle : P^\mu P_\mu : \rangle_n,$$

which satisfy the Virasoro algebra[45]

$$[L_m, L_n] = (m-n)L_{m=n} + \frac{D}{12}m(m^2-1)\delta_{m,-n}. \tag{2.2}$$

Its finite subalgebra, $L_0, L_\pm$, the Gliozzi algebra, generates conformal transformations in two dimensions. The propagator was given by

$$\frac{1}{\alpha' L_0 + 1}.$$

5 Superstrings

The Klein-Gordon equation for a point particle

$$0 = p^2 + m^2 = \langle P^\mu\rangle_0 \langle P_\mu\rangle_0 + m^2,$$

could then be interpreted as a special case of

$$0 = \langle P^\mu P_\mu\rangle_0 + m^2,$$

suggesting a correspondence [15] between point particles and dual amplitudes,

$$\langle A\rangle\langle B\rangle \to \langle AB\rangle.$$

Fermions should satisfy the Dirac equation

$$0 = \gamma_\mu p^\mu + m = \langle \Gamma_\mu\rangle_0 \langle P^\mu\rangle_0 + m.$$

This requires a generalization of the Dirac matrices as dynamical operators

$$\gamma_\mu \to \Gamma_\mu = \gamma_\mu + i\gamma_5 \sum_{n=0}^{\infty}\left(b_{n\mu}e^{in\tau} + b^\dagger_{n\mu}e^{-in\tau}\right)$$

[45] A c-number is added anachronistically.

where the oscillators are *Lorentz vectors*,[46] which satisfy anticommuting relations

$$\{b_{n\mu}, b^{\dagger}_{m\nu}\} = \delta_{nm} g_{\mu\nu},$$

and the sum runs over the positive integers. This led me to propose the string Dirac equation in the winter of 1970 [16], which readily followed from that correspondence:

$$0 = \langle \Gamma_\mu P^\mu \rangle_0 + m.$$

The basic Dirac algebra, $\{\gamma \cdot p, \gamma \cdot p\} = p^2$ is seen to be generalized to an algebra with both commutator and anticommutators

$$\{F_n, F_m\} = 2L_{n+m}, \quad [L_n, F_m] = (2m - n)F_{m+n},$$

where $F_n = \langle \Gamma_\mu P^\mu \rangle_n$, and these new L_n's also satisfy the Virasoro algebra, but with a different c-number.

Neveu and Schwarz then compute the amplitude for a dual fermion emitting three pseudoscalars with the Yukawa vertex

$$\Gamma_5 : e^{ik\cdot Q(\tau)} :, \quad \Gamma_5 = \gamma_5 (-1)^{\Sigma b^{\dagger}_n b_n},$$

and find that the resulting amplitude contains an infinite number of poles in its fermion-antifermion channel, and even identify the residue of the first pole [17]!

A new model with bosonic poles and vertices emerges, written in terms of an infinite tower of anticommuting vector oscillators,

$$\{b_{r\mu}, b^{\dagger}_{s\nu}\} = \delta_{rs} g_{\mu\nu}, \quad r, s = \frac{1}{2}, \frac{3}{2}, \ldots .$$

The triple boson vertex is given by

$$V_{\mathrm{NS}}(k, \tau) k^\mu = H_\mu(\tau) : e^{ik\cdot Q(\tau)} :,$$

where

$$H_\mu(\tau) = \sum_{r=\frac{1}{2}, \frac{3}{2}, \ldots} \left[b_{r\mu} e^{-ir\tau} + b^{\dagger}_{r\mu} e^{ir\tau} \right].$$

---

[46]Later it was realized that this made sense only in ten space-time dimensions where the little group is the spinor-vector schizophrenic SO(8).

These are the building blocks of the "Dual Pion model" [18,19], published in April 1971. The algebraic structure found in the generalized Dirac equation remains the same, producing a super-Virasoro algebra which decouples unwanted modes [20], with Γ_μ replaced by H_μ, through the operators

$$G_r = \langle H \cdot P \rangle_r, \quad r = \frac{1}{2}, \frac{3}{2}, \ldots .$$

The close relation of the two sectors would soon later be formalized by Gervais and Sakita [21], who wrote them in terms of a world-sheet σ-model, with different boundary conditions, symmetric for the fermions, antisymmetric for the bosons. They called the transformations generated by the anticommuting Virasoro operators, *supergauge transformations*, the first time the name 'super' appears in this context.

The following years saw the formulation of the RNS[47] (NSR to some) "Dual Fermion Model", generating dual amplitudes with boson and fermion legs. It lived in ten space-time dimensions, with states determined in terms of transverse fermionic and bosonic harmonic oscillator operators. In the fermionic 'R-sector', the spectrum of states is spanned by the fermionic ground state, $u|0\rangle$ where u is a fixed 32-dimensional spinor, annihilated by both transverse bosonic and fermionic oscillators, a_{ni} and b_{ni}, $i = 1, 2 \ldots, 8$, and integer n. The fermion masses are determined by

$$\alpha' m_R^2 = \sum_{n=1}^{\infty} n \left[a_n^\dagger \cdot a_n + b_n^\dagger \cdot b_n \right].$$

The bosonic NS-sector spectrum starts with a tachyon, $|0\rangle$ annihilated by the same a_{ni}, but also by the NS fermionic oscillators b_{ri}, where r runs over half-integers. The boson masses satisfy

$$\alpha' m_{NS}^2 = \sum_{n=1}^{\infty} n \, a_n^\dagger \cdot a_n + \sum_{r=\frac{1}{2}} r \, b_r^\dagger \cdot b_r - \frac{1}{2}.$$

But there were idiosyncrasies. The correspondence between Neveu-Schwarz and the dual fermion states differed for states with an even number $(G \equiv (-1)^{\sum b_r^\dagger \cdot b_r} = -1)$ $b_r^\dagger$ and states with an odd number, and there is a tachyon in the even number spectrum, at $\alpha' m_{NS}^2 = -\frac{1}{2}$.

In 1976, Gliozzi *et al.* [22] noticed that the NS tachyon can be eliminated by requiring an odd number of anticommuting operators in the bosonic spectrum, $(G = -1)$. The NS ground state $\alpha' m_{NS}^2 = 0 : \; b_{1i}^\dagger|0\rangle$ now consists of

---

[47] Ramond-Neveu-Schwarz.

eight bosons, transforming as the vector (=spinor) SO(8) representation. The first excited states are

$$\alpha' m_{NS}^2 = 1 : \quad b^\dagger_{\frac{1}{2}i} b^\dagger_{\frac{1}{2}j} b^\dagger_{\frac{1}{2}k} |0\rangle, \quad b^\dagger_{\frac{1}{2}i} b^\dagger_{\frac{1}{2}j} |0\rangle, \quad b^\dagger_{\frac{3}{2}i} |0\rangle,$$

that is, 128 = 56(8.7.6/1.2.3)+64(8.8)+8 bosonic states, and so on. In their next step, they show that the R ground state solution could also be reduced to eight fermionic degrees of freedom. In ten dimensions, while a spinor has naturally 32 degrees of freedom, they showed that one can impose both chiral and Majorana (reality) restrictions on it, and reduce the spinor to eight dimensions: the spinor (=vector) SO(8) representation

$$\alpha' m_R^2 = 0 : \quad \psi_\alpha |0\rangle, \quad \alpha = 1, 2, \ldots, 8.$$

The first excited state of the R-sector consists of

$$\alpha' m_R^2 = 1 : \quad b^\dagger_{1i} \psi_\alpha |0\rangle, \quad a^\dagger_{1j} \psi_\alpha |0\rangle,$$

with 128 = 8.8 + 8.8 fermionic states! This was no accident, and using one of Jacobi's most obtuse relations, they showed that this equality obtained at all levels. Indeed this was supersymmetry, with the same number of bosons and fermions, albeit in ten space-time dimensions.

Fermion-boson symmetry, born in its world-sheet realization, reappears as supersymmetry in ten-dimensional spacetime. Meanwhile, behind the iron curtain , . . .

6 Russians

In March 1971, there appears a remarkable and terse paper by Golfand and Likhtman [23,24] who extend the Poincaré algebra generated by P_μ and $M_{\mu\nu}$ to 'bispinor generators', W_α and $\bar{W}_\beta$, which generate spinor translations. Cognizant that spin-statistics requires anticommutating spinors, they arrive at the parity-violating algebra

$$\{W, W\} = [P_\mu, P_\nu] = 0, \quad \{W, \bar{W}\} = \frac{1+\gamma_5}{2} \gamma_\mu P_\mu \tag{2.3}$$

assuming no other subalgebra of the Poincaré group.[48] With little motivation stated,[49] they have written down the $\mathcal{N} = 1$ super-Poincaré algebra

[48]See Addendum, p. 478.

[49]Editor's note: I remember Golfand's narrative that that their original motivation was disproving the validity of the Coleman-Mandula theorem. –MS

in four dimensions! They identify its simplest representation: two 'scalar Hermitean' fields $\phi(x)$ and $\omega(x)$, and one left-handed spinor field ψ_1(x), of equal mass, the earliest mention of the Wess-Zumino supermultiplet. They do not consider auxiliary fields nor display the transformation properties of these fields. However, they show the spinor generators as bilinears in those fields

$$W^0 = \frac{1+\gamma_5}{2} \int d^3x \left[\phi^*(x)\, \overleftrightarrow{\partial_0}\, \psi_1(x) + \omega(x)\, \overleftrightarrow{\partial_0}\, \psi_1^c(x) \right]. \tag{2.4}$$

They also describe that the massive vector multiplet follows with the vector field $A_\mu(x)$, a scalar field $\chi(x)$ and a Dirac spinor field $\psi_2(x)$.

This ground-breaking paper ends with the difficult task of writing interactions, see page 482. Superpotential of the WZ multiplets is presented only by mass terms since the charge assignment for the matter fields in SQED does not allow for cubic terms. The last formula in [23] contains now-familiar features such as the squared D-term. Golfand and Likhtman had firmly planted the flag of supersymmetry in four dimensions.

Interestingly, physicists on both sides of the iron curtain seemed oblivious to this epochal paper. Likhtman [25] seems to be the only one who followed up on this paper. He notices that the vacuum energy cancels out because of the equal number of mass bosons and fermions with the same mass. He finds scalar masses only logarithmically divergent, which he mentions in a later publication [26].

In December 1972, in an equally impressive paper, Volkov and Akulov [27, 28] (see also p. 485) wanted to explain the masslessness of neutrinos in terms of an invariance principle. They note that the neutrino free Dirac equation is invariant under the transformations

$$\psi \to \psi + \zeta, \quad x_\mu \to x_\mu - \frac{a}{2i}\left(\zeta^+\sigma_\mu\psi - \psi^+\sigma_\mu\zeta\right),$$

where ζ is a global spinor. When added to the Poincaré generators, they form a group, of the type Berezin and G. I. Katz had advocated [6, 7] for algebras with commuting and anticommuting parameters. The translation of ψ makes the neutrino akin to a Nambu-Goldstone particle with only derivative couplings.

There follows a Lagrangian that describes its invariant interactions, which we can identify as a non-linear representation of supersymmetry.

The end of their paper contains this remarkable sentence "We note that if one introduces gauge fields corresponding to the(se) transformations, then, as a consequence of the Higgs effect, a massive gauge field with spin 3/2 arises, and the Goldstone particles with spin 1/2 vanish." This remark is followed in October 1973, when Volkov and Soroka [29, 30] generalize their transformations to local parameters and show explicitly that the fermionic Nambu-Goldstone particle indeed becomes a gauge artifact. Thus was born what became known as the "Super Higgs Effect."

7 Wess-Zumino

In October 1973, Wess and Zumino [31] generalize the world-sheet supergauge transformations of the RNS model to four dimensions.

Their's is the paper that launched the massive and systematic study of supersymmetric field theories in four dimensions. The scalar (now called chiral or Wess-Zumino) multiplet is introduced. It consists of two real scalar bosons, A and B, a Weyl (Majorana) fermion ψ and two auxiliary fields F and G. Supergauge transformations generate the algebra

$$\begin{aligned}
&\delta A = i\bar{\alpha}\psi, \quad \delta B = i\alpha\bar{\alpha}\gamma_5\psi,\\
&\delta\psi = \partial_\mu(A - \gamma_5 B)\gamma_\mu\alpha + n(A - \gamma_5 B)\gamma_\mu\partial_\mu\alpha + F\alpha + G\gamma_5\alpha,\\
&\delta F = i\bar{\alpha}\gamma_\mu\partial_\mu\psi + i\left(n - \frac{1}{2}\right)\partial_\mu\bar{\alpha}\gamma^\mu\psi,\\
&\delta G = i\bar{\alpha}\gamma_5\gamma_\mu\partial_\mu\psi + i\left(n - \frac{1}{2}\right)\partial_\mu\bar{\alpha}\gamma_5\gamma^\mu\psi,
\end{aligned}$$

where α is an infinitesimal anticommuting spinor, and n is an integer assigned to the multiplet. With impressive algebraic strength, they are shown to close on both conformal and chiral transformations. In particular, two transformations with parameters α_1 and α_2 result in a shift of x_μ by $i\alpha_1\gamma_\mu\alpha_2$. The free Lagrangian for the scalar multiplet follows:

$$\mathcal{L}_{\mathrm{WZ}} = -\frac{1}{2}\partial_\mu A\partial^\mu A - \frac{1}{2}\partial_\mu B\partial^\mu B - \frac{i}{2}\bar{\psi}\gamma_\mu\partial^\mu\psi + \frac{1}{2}\left(F^2 + G^2\right).$$

It is not invariant under supergauge transformations but since it transforms as a derivative, the action is invariant. To introduce invariant interactions, they derive the calculus necessary to produce covariant interactions, by assembling two scalar multiplets into a third, etc.

They also introduce the vector supermultiplet, consisting of four scalar fields, D, C, M, N, a vector field v_μ, and two spinor fields χ and λ, on which they derive the supergauge transformations. By identifying the vector field with the chiral current generated by a scalar multiplet,

$$v_\mu = B\partial_\mu A - A_\mu B - \frac{i}{2}\bar{\psi}\gamma_5\gamma_\mu\psi,$$

and following it through the algebra, they express all the vector multiplet fields as quadratic combinations of the scalar supermultiplet. In particular $D = 2\mathcal{L}_{\rm WZ}$.

Finally, they notice that one can drop some of these fields, C, N, M, and χ, without affecting the algebra (soon to be called the Wess-Zumino gauge), and write the vector multiplet Lagrangian in a very simple form:

$$\mathcal{L}_V = -\frac{1}{4}v_{\mu\nu}v^{\mu\nu} - \frac{i}{2}\bar{\lambda}\gamma_\mu\partial^\mu\lambda + \frac{1}{2}D^2.$$

This paper contains many of the techniques that were soon to be used in deriving many of the magical properties of supersymmetric theories in four dimensions.

In December 1973, Wess and Zumino present the one-loop analysis [32] of an interacting Wess-Zumino multiplet, and find remarkable regularities: the SUSY tree-level relations are not altered by quantum effects, the vertex correction is finite (in fact later they find that one has only wave function renormalization), and finally the quadratic divergences of the scalar and pseudoscalar fields cancel. As it was realized later, this addresses the "gauge hierarchy problem", and strongly suggests SUSY's application to the Standard Model.

8 Representations

The representations of the supersymmetry algebra were first systematically studied by Gell-Mann and Ne'eman (unpublished). They mapped the algebra in light-cone coordinates to a Fermi oscillator, and they found that in supersymmetry, the massless representations of the Poincaré group assemble into two states with helicities separated by one-half

$$\left(\lambda \pm \frac{1}{2}, \lambda\right),$$

and with the same light-like momentum, yielding an equal number of bosons and fermions.[50] The simplest is $\lambda = 0$, with a real scalar and half a

[50]Editors note: As far as I know, the first published analysis of irreducible representations of $\mathcal{N} = 1$ superalgebra was carried out by Likhtman in 1971 and appeared as Lebedev Physics Institute preprint No. 41 (1971). Its English translation can be found in Nucl. Phys. B (Proc. Suppl.) **101** (2001), page 9 [e-Print: hep-ph/0101209]. –MS

left-handed Weyl fermion. However, CPT symmetric local field theories require the other half of the Weyl fermion, $\left(\frac{1}{2},0\right) + \left(0,-\frac{1}{2}\right)$ which describes one Weyl fermion and a complex scalar boson, the ingredients of the Golfand-Likhtman-Wess-Zumino multiplet.

The massless gauge supermultiplet, $(1,\frac{1}{2}) + (-\frac{1}{2},-1)$ describes a gauge boson, and its companion Weyl (Majorana) fermion describes the gaugino. The supergravity supermultiplet $(2,\frac{3}{2})+(-\frac{3}{2},-2)$ contains the graviton and the gravitino — remarkably the ingredients of interacting supergravity [33,34].

They extend their analysis to the case of $\mathcal{N}$ supersymmetries. Disregarding particles of spin higher than two, they find two cases with manifestly self-conjugate supermultiplets. First, we have an $\mathcal{N}=4$ supermultiplet, with helicities

$$(1) + 4(\tfrac{1}{2}) + 6(0) + 4(-\tfrac{1}{2}) + (-1)$$

which led in 1976 to the $\mathcal{N}=4$ super-Yang-Mills theory [35], which was found much later to have magical properties, such as an enhanced conformal symmetry, and ultraviolet finiteness!

Second, we have $\mathcal{N}=8$ supergravity with helicities

$$(2) + 8(\tfrac{3}{2}) + 28(1) + 56(\tfrac{1}{2}) + 70(0) + 56(-\tfrac{1}{2}) + 28(-1) + 8(-\tfrac{3}{2}) + (-2),$$

which also led to a fully interacting theory, $\mathcal{N}=8$ Supergravity [36].

Massive representations of supersymmetry can be assembled using a group-theoretical Higgs mechanism. The massive vector representation contains a Dirac spinor, a massive vector, and a scalar particle

$$(1,\tfrac{1}{2}) + (-1,-\tfrac{1}{2}) + (0,-\tfrac{1}{2}) + (0,\tfrac{1}{2}),$$

all of equal mass, as considered by Golfand and Likhtman.

9 Towards the supersymmetric standard model

With the Wess-Zumino paper, the flood gates had been opened [37]. In short order, a supersymmetric version [38] of QED is written down, with Abelian gauge invariance, in which the Dirac electron spinor is accompanied by two complex spin zero fields. In January 1974, Salam and Strathdee [39,40] assemble the fields within a supermultiplet into one superfield with the help of anticommuting Grassmann variables. The same authors [41] coin the word "super-symmetry" in a May 1974 paper which generalizes supersymmetry to Non-Abelian gauge interactions.

Before applying supersymmetry to the real world, several conceptual problems must be resolved. Firstly, the absence of fermion-boson symmetry at low energies requires it to be broken. Secondly, its application to the electroweak theory demands the extension of the Higgs mechanism. Finally, the known particles must be assigned to supermultiplets.

In 1974, Fayet and Iliopoulos [42] produce the first paper on spontaneous breaking of supersymmetry in theories with a gauged Abelian symmetry by giving its D auxiliary field a constant value. Their proposal is remarkably simple, just add to the Lagrangian for a U(1) vector multiplet a D-term,

$$\mathcal{L}_V^{\text{FI}} = \mathcal{L}_V + \xi D.$$

This extra term violates neither Abelian gauge invariance nor supergauge invariance, since its supergauge variation is a total derivative. The resulting field equation $\langle D\rangle_0 = \xi$ yields a theory where both gauge and supergauge invariances are broken.

A year later, O'Raifeartaigh [43] invents a different way to spontaneous breaking of supersymmetry, in theories with several interacting scalar supermultiplets. Its simplest model involves three scalar supermultiplets with equations of motion

$$F_1 = -m\phi_2^* - 2\lambda\phi_1^*\phi_3^*, \quad F_2 = -m\phi_1^*, \quad F_3 = \lambda\left(M^2 - \phi_1^*\right)^2$$

where m, M, and λ are parameters. There are no solutions for which all three F_i vanish, and hence supersymmetry is broken. From these two early examples, the auxiliary fields are the order parameters of SUSY breaking.

Both schemes yielded an embarrassing massless Goldstone spinor, which may have impeded the application of supersymmetry.[51] None of these authors were aware of Volkov's papers.

The second hurdle is the generalization of the Higgs mechanism to supersymmetry. This is done in the context of an unusual model by Fayet [44] in December 1974. Like Volkov and Akulov before, Fayet builds models where the electron neutrino is the Goldstone spinor from the breakdown of supersymmetry[52] using the FI mechanism.

Although the model building in this paper did not survive the test of time, two important and more permanent concepts emerged. One is that

[51]In 1976, Weinberg and Gildener note that supersymmetry could explain a low mass scalar boson, but bemoan that it would produce a massless fermion!

[52]In 1974, the Standard Model was not yet "standard", and many authors were still presenting alternatives.

the Higgs mechanism applies, but two scalar supermultiplets are needed to achieve SU(2) × U(1) → U(1) electroweak breaking, in accord with the number of surviving scalars in the massive vector supermultiplets — also the existence of R-symmetry, a new kind of continuous symmetry acting on both the fields and the Grassmann parameters of the superfields.

It was not until July 1976 that Fayet [45] generalizes the Weinberg-Salam (soon to be called the Weinberg-Salam-Glashow, and then Standard) model to SUSY. Its distinctive features are

(i) There are two scalar superfields, S, T, (today, $H_{u,d}$) for EW breaking;
(ii) Leptons and quarks are the fermions inside scalar supermultiplet;
(iii) We have a continuous R-symmetry.

The particle content is the "minimal supersymmetric model" (MSSM). Some kinks still need to be ironed out, having to do with SUSY breaking (à la Fayet-liopoulos in this paper), which produces a massless Goldstone spinor. The continuous R-symmetry in this paper behaves like a "leptonic" number, but it prevents the spinor gluons from acquiring a mass.

Today, we know that SUSY breaking is an active area of theoretical research, even without the presence of a Goldstone fermion, eaten by the Super-Higgs mechanism.

10 SUSY today

By stopping this history of fermion-boson symmetry in 1976, we rob the reader of the many wonderful concepts since discovered, but they are more than adequately recounted elsewhere. The seeds of today's SUSY research were planted in these early papers.

Almost 40 years later, superstring theories have blossomed into a dazzling array of connected theories; the study of $\mathcal{N} = 4$ super-Yang-Mills theories is an active field of research, as is the possible finiteness of $\mathcal{N} = 8$ supergravity.

The Hamiltonian is no longer fundamental, but derived from translations along SUSY's fermionic dimensions.

Few doubt the existence of a deeper connection between bosons and fermions, but opinions differ at which scale it will be revealed: the breaking of supersymmetry remains as mysterious as ever.

Yet, the recent discovery of a low mass Higgs particle suggests that the universe displays more symmetry at shorter distances.

Today, SUSY is unfulfilled, beloved by theorists, but so far undiscovered by experimentalists.

In the words of the late Sergio Fubini, “We do not know if supersymmetry is just a beautiful painting to put on the wall, or something more.”

References

1. E. Wigner, *On unitary representations of the inhomogeneous Lorentz group*, Annals of Mathematics 40(1), 149–204 (1939).
2. F. Gürsey, L.A. Radicati, *Spin and unitary spin independence of strong interactions.* Phys. Rev. Lett. 13, 173–175 (1964).
3. B. Sakita, *Supermultiplets of elementary particles.* Phys. Rev. 136, B1756-B1760 (1964).
4. H. Miyazawa, *Baryon number changing currents*, Progr. Theoret. Phys. (Kyoto) 36, 1266 (1966).
5. H. Miyazawa, *Spinor symmetries and symmetries of Baryons and Mesons*, Phys. Rev. 170, 1586–1590 (1968).
6. F.A. Berezin, G.I. Katz, Mat. Sbornik, Vol. 82 (124) (1970).
7. F.A. Berezin, G.I. Katz, Math. USSR Sbornik 11, 311–325 (1970).
8. R. Dolen, D. Horn, C. Schmid, *Predictions of Regge parameters of ρ poles from low-energy πN data.* Phys. Rev. Lett. 19, 402–407 (1967).
9. M. Ademollo, H.R. Rubinstein, G. Veneziano, M.A. Virasoro, *Bootstrap-like conditions from super-convergence*, Phys. Rev. Lett. 19, 1402–1405 (1967).
10. G. Veneziano, *Construction of a crossing-symmetric, Regge behaved amplitude for linearly rising trajectories.* Nuovo Cim. A 57, 190–197 (1968).
11. Y. Nambu, *Quark model and the factorization of the Veneziano amplitude*, EFI 69-64, in *Symmetries and Quark Model*, ed. by R. Chand (Gordon and Breach, 1970).
12. H. B. Nielsen, *An almost physical interpretation of the integrand of the n-point Veneziano model.* Nordita Preprint 1969, submitted to the *15th International Conference on High energy Physics*, (Kiev, 1970).
13. L. Susskind, *Harmonic-oscillator analogy for the Veneziano model*, Phys. Rev. Lett. 23, 545–547 (1969).
14. S. Fubini, D. Gordon, G. Veneziano, *A general treatment of factorization in dual resonance models*, Phys. Lett. B 29, 679–682 (1969).
15. P. Ramond, *An interpretation of dual theories*, Nuovo Cim. A 4, 544–548 (1971).
16. P. Ramond, *Dual theory for free fermions*, Phys. Rev. D 3, 2415–2418 (1971).
17. A. Neveu, in *The Birth of String Theory*, eds. E. Cappelli, E. Castellani, F. Colombo, P. DiVecchia (Cambridge University Press, Cambridge, 2012).
18. A. Neveu, J.H. Schwarz, *Tachyon-free dual model with a positive intercept trajectory*, Phys. Lett. 34B, 517–518 (1971).
19. A. Neveu, J.H. Schwarz, *Factorizable dual model of pions*, Nucl. Phys. B 31, 86–112 (1971).
20. A. Neveu, J.H. Schwarz, C. Thorn, *Reformulation of the dual pion model*, Phys. Lett. 35B, 529–533 (1971).
21. J.-L. Gervais, B. Sakita, *Field theory interpretation of supergauges in dual models*, Nucl. Phys. B 34, 632–639 (1971).

22. F. Gliozzi, J. Scherk, D.I. Olive, *Supersymmetry, supergravity theories and the dual spinor model*, Nucl. Phys. B 122, 253–290 (1977).
23. Yu. A. Golfand, E. P. Likhtman, *Extension of the algebra of poincare group generators and violation of P invariance*, JETP Lett. 13, 323–326 (1971).
24. Yu.A. Golfand, E.P. Likhtman, Pisma Zh.Eksp. Teor. Fiz. 13, 452–455 (1971).
25. E. P. Likhtman, *Irreducible Representations of the Extension of the Algebra of the Poincaré Generators by the Bispinor Generators*, Report of the Lebedev Institute no. 41 (1971) (retroprinted http://arxiv.org/pdf/hep-ph/0101209.pdf).
26. E.P. Likhtman, JETP Lett. 21, 109 (1975).
27. D.V. Volkov, V.P. Akulov, *Possible universal neutrino interaction*, JETP Lett. 16, 438–440 (1972).
28. D. V. Volkov, V. P. Akulov, Pisma Zh. Eksp. Teor. Fiz. 16, 621–624 (1972).
29. D.V. Volkov, V.A. Soroka, *Higgs effects for goldstone particles with spin 1/2*, JETP Lett. 18, 312–314 (1973).
30. D.V. Volkov, V.A. Soroka, Pisma Zh. Eksp. Teor. Fiz. 18, 529–532 (1973).
31. J. Wess, B. Zumino, *Supergauge transformations in four dimensions*, Nucl. Phys. B 70, 39–50 (1974).
32. J. Wess, B. Zumino *A Lagrangian model invariant under supergauge transformations*, Phys. Lett. B 49, 52 (1974).
33. D.Z. Freedman, P. van Nieuwenhuizen, S. Ferrara, *Progress toward a theory of supergravity*, Phys. Rev. D 13, 3214–3218 (1976).
34. S. Deser, B. Zumino, *Consistent supergravity*, Phys. Lett. 62B, 335 (1976).
35. L. Brink, John H. Schwarz, Joël Scherk, *Supersymmetric Yang- Mills theories*, Nucl. Phys. 121B, 77 (1977).
36. E. Cremmer, B. Julia, *The SO(8) supergravity*, Nucl. Phys. 159B, 141 (1979).
37. S. Ferrara, Review by Pierre Fayet, Phys. Rep. 32, 249–334 (1977).
38. J. Wess, B. Zumino, *Supergauge Invariant Extension of Quantum Electrodynamics*, Nucl. Phys. B **78**, 1 (1974).
39. A. Salam, J.A. Strathdee, *Supergauge transformations*, Nucl. Phys. B 76, 477 (1974).
40. A. Salam, J. A. Strathdee, *On superfields and fermi-bose symmetry*, Phys. Rev. D 11, 1521–1535 (1975).
41. A. Salam, J. A. Strathdee, *Super-symmetry and Nonabelian Gauges*, Phys. Lett. B 51, 353–355 (1974).
42. P. Fayet, J. Iliopoulos, Phys. Lett. B 51, 461–464 (1974).
43. L. O'Raifeartaigh, *Spontaneous Symmetry Breaking for Chiral Scalar Superfields*, Nucl. Phys. B 96, 331 (1975).
44. P. Fayet, *Supergauge Invariant Extension of the Higgs Mechanism and a Model for the Electron and its Neutrino*, Nucl. Phys. B 90, 104–124 (1975).
45. P. Fayet, *Supersymmetry and Weak, Electromagnetic and Strong Interactions*, Phys. Lett. B 64, 159–162 (1976).

Figure 2.2: André Neveu, November 2013.

THIRTY YEARS AGO[53]

ANDRÉ NEVEU

Laboratoire de Physique Mathématique, CNRS–Université Montpellier II, 34095 Montpellier, France

The origins of my involvement in the superstring saga can be traced back to the rather fortuitous and amusing circumstances of my first physical encounter with Pierre Ramond.

Actually, in 1968–1969, while I was preparing my "thèse de troisième cycle" (an analogue of the Ph.D.) in Orsay, I had already heard (in very positive terms) about Pierre from Jean Nuyts (then in Orsay), with whom he had seemingly signed the papers (without having met, if I remember) on crossing-symmetric partial wave amplitudes. Pierre's Ph.D. in Syracuse was devoted to crossing symmetry on which he worked with A. P. Balachandran, his advisor.

In September 1969, I had just finished my four year curriculum at the École Normale Supérieure in Paris. I had been awarded the "Jane Eliza Procter Fellowship" (endowed by Procter and Gamble) which Princeton University reserves each year for an alumnus of the E.N.S.; with it came a Fulbright travel grant which enabled me to choose the ship "France" for my first trans-Atlantic crossing. The ship had a small and pleasant library with a few desks. I was spending many hours there, studying in detail the latest preprints on dual resonance models, as they were called. One afternoon, leaving all my materials spread all over the desk, I walked out of the library, called by an urgent need... During these two minutes when I was absent, Pierre Ramond walked in and looked around for a vacant desk. There appeared to be only one — mine. He walked up to it, realized that it was not really vacant, but was shocked to see on it the Fubini-Veneziano paper on the factorization of dual resonance models, the very same paper he was studying at that moment! He quickly went back to his cabin to make sure that what he had just seen was not his copy! Reassured about his sanity, he came back to the library, wondering who could be the fellow interested in such an esoteric topic. By this time I, too, was back. You can imagine easily the next few hours. This is how we became friends. After spending

[53]Submitted September, 2000.

his summer vacation in France, Pierre was on his way to the National Accelerator Laboratory (now called Fermilab) for his first postdoc.

At the Physics Department of Princeton University, there was no preprint library. No need was felt for it since all the important people there received important preprints themselves, and then disseminated news that might be of any use. Would a short and partly speculative paper by a still relatively obscure postdoc at Fermilab have been considered important? And reach me? Perhaps, but probably not soon enough. As I realized during a visit to Berkeley a few weeks later, the next breakthrough would very likely have been made there, and probably within a couple of months, by one or several of the following: Korkut Bardakçi, Martin Halpern, Michio Kaku, Stanley Mandelstam, Charles Thorn, and ... Miguel Virasoro, who was a postdoc there at that time. But thanks to the crossing on the "France," Pierre had sent John Schwarz and me his preprint on the free Ramond model. We immediately realized that there was something very deep there, and proceeded to build an interacting theory of one fermion line with mesons. By factorizing in the fermion-antifermion channel, we discovered that the mesonic trajectories had many interesting properties, and decided to build a purely bosonic sector embedding what is now called the superconformal algebra. I must confess I was not very comfortable with the complexities of the γ matrices of the fermion sector (for extenuation, I plead youth), and, moreover, what was to be expected of the duality properties of the fermionic amplitudes was far from clear at that time.

So, as for me, there was some amount of luck in our discovery. There was also a certain degree of unconsciousness, as still transpires through a careful analysis of our first paper on the Neveu-Schwarz sector. As far as I remember, I was not even really conscious of the fact that introducing modes with a space-time vector index meant introducing such a huge number of ghosts that one had better first sit back and think of a new symmetry which would give a chance of eliminating them, instead of forging ahead as we did, computing amplitudes and residues almost at random, and realizing that there was, indeed, a miracle. The Ramond superconformal algebra had been our guide in the construction of a few bosonic amplitudes, by factorizing off a free fermion line, but in the Neveu-Schwarz sector, the subtleties of the F_1 and F_2 formalisms could not be guessed in advance, so sitting back and thinking would probably have led us nowhere. I only remember that I was caught by the mathematical beauty of the superconformal algebra, and just hoped, more or less consciously, that, one way or another, in connection with it, a ghost cancelling miracle would occur...

This is an instance where I may quote the French proverb "*Aux innocents les mains pleines*" roughly translated "Hands full for the innocent," remarking that *innocent* in French generally means not just the opposite to guilty, but, in this case, rather refers to the innocence of youth if one is kind, or to feeblemindedness if one is not (the most usual occurrence).

Figure 2.3: John Schwarz, late 2010s.

STRINGS AND THE ADVENT OF SUPERSYMMETRY: THE VIEW FROM PASADENA[54]

JOHN SCHWARZ

Physics Department, California Institute of Technology, Pasadena, CA 91125

Q. To warm up I would like to ask how your early work was related to supersymmetry.

A. The earliest version of string theory (developed in the late 1960's to describe hadron interactions) suffered from various unphysical features. In particular, the spectrum contained a tachyon but no fermions. This motivated the search for a more realistic string theory. The first significant success was made in January 1971 by Pierre Ramond [1] who constructed a string analog of the Dirac equation. At about the same time, Andre Neveu and I were in Princeton constructing a new bosonic string theory. Neveu and I quickly realized [2] that the two constructions were different facets of a single theory and (along with Charles Thorn) we constructed [3] an interacting string theory containing our bosons and Ramond's fermions. In 1972 I showed that consistency of the theory requires that the space-time dimension is 10 and that the ground state fermions are massless.

The Ramond-Neveu-Schwarz string theory actually contains local space-time supersymmetry, but we were slow to realize that. The realization came after supersymmetry in four dimensions was thoroughly studied by others.[55]

Q. What can you say about the string theory today?

A. My impression of the current state of the string theory is that we've now got overwhelming evidence that there is a unique fundamental theory that contains gravity and other forces. But that we only understand this theory in certain special limits, and we don't have a complete understanding of it. It has not been fully formulated. And that until it is fully formulated, that there are many important questions that we are unable to answer. So, while we've made enormous progress in seeing the relations between the limiting cases which we used to think of as different theories, we now realize

[54]These excerpts are based on the interview conducted by Melitta Fitzer and T. Shifman, Santa Barbara, November 19, 1999 (the transcript was prepared by A. Roitman), and on a private letter from J. Schwarz to G.L. Kane, 1999.

[55]See footnote on page xvii.

they are actually different limits of a single theory. That it looks like it is all coming together. But there are really major questions about how we are going to relate this theory to the real world. These questions are partly of a more practical character, and partly of a more theoretical character. There are big qualitative questions that we do not know how they are going to work out. But I think everyone who works on this field is so taken by the beauty of the theory that they have little doubt that the right answer is there to be found, though we do not know exactly what it is. For example, one of the big questions that we do not understand, is what determines how much energy there is associated with completely empty space. This is something referred to as a cosmological constant. From observations we know that this number is extremely small, maybe zero. Yet at a theoretical level we don't see any reason why it should be small. Until we understand that, it's clear that there are big qualitative questions that remain to be explained. Supersymmetry, if it were exact, could explain the vanishing of the cosmological constant. There's been a lot of progress in explaining other qualitative questions about the relationship between strings theory and different quantum field theories, how to learn about the behavior in these theories when the interactions become very strong. This is usually very difficult, but in situations where there is supersymmetry, which is a very special kind of symmetry, one can make some fairly precise statements.

Q. I understand that supersymmetry is a very unusual symmetry. Could you, please, explain why is it so special?

A. Yes, supersymmetry is a very interesting kind of symmetry that has developed within the last 30 years or so. There are different ways of explaining it. It can be thought of as an extension of the symmetry that would be associated with space and time, to some more abstract directions. There are two classes of particles in quantum mechanics, which go by the names of bosons and fermions. Ordinary, more familiar kinds of symmetries — like rotational symmetry of a sphere or something like that — only relate different bosons to one another and different fermions to one another. But supersymmetry can relate these two different classes of particles, the bosons to the fermions. And that is of rather deep significance, because, roughly speaking, you can think of bosons as the particles responsible for forces in nature, and fermions as the particles that are responsible for stable matter. So if you can find some principle to relate these two types of particles then that is quite profound. Maybe, the first experimental evidence for the string theory might arise from the discovery of supersymmetry. The discovery of supersymmetry would not prove that the string theory was right, and its non-discovery would not prove that it was wrong. But, certainly, the discovery would be enlightening and informative, and I think it would be very encouraging for those people who are pursuing string theory.

Q. Thirty years ago, when you started, you were fairly alone in this field, right?

A. Well, I would not put my lonely period back this far. The subject of string theory arose about 30 years ago. It was extremely popular for about five years, there were hundreds of people working on it. And they were working on it for the specific problem of describing strong nuclear forces. And then, around 1973–74, a better theory for the strong nuclear forces arose. It now is referred to as QCD. Everyone became convinced that that was right. String theory was failing in many details to provide a satisfactory theory. So, at that point almost everyone stopped working on the string theory. And I was one of the few who kept working on it despite this failure, just because we were so taken by the mathematical beauty of the subject. We felt that it should be useful for something. One of the problems that we were having with the theory was that it contained a particular kind of particle that did not belong to the strong interactions. Eventually, in 1974, Joël Scherk and I realized that these particles had exactly the right properties to describe gravity.[4] We said: "Let's forget about strong interactions, let's use this theory for completely different purpose than for which it was originally invented, and use it as a theory of gravity. And then other forces would follow."

So my focus completely shifted, almost overnight, at some point in 1974, when we realized that this was the basis for providing a theory of gravity that would be compatible with quantum theory. More traditional approaches to adding gravity into quantum field theory ran into various difficulties. It was clear to us that in the string theory framework those difficulties would be avoided. So the period of loneliness that you refer to was really the subsequent 10 years, from 1974 to 1984, when I was pursuing that program with a number of collaborators, most notably Michael Green. Unfortunately, Joël Scherk died in 1980. I started working with Michael Green just about that time, and Lars Brink collaborated with us on a number of occasions as well. So we were just a handful of people pursuing this program. I was very enthusiastic about it. We were not completely ignored, we spoke at many conferences and so on, but most people were doing other things. Some rather related, but not exactly string theory. Others were working on supersymmetry, supergravity, Kaluza-Klein theory, all of which are parts of what is now considered string theory. But the string element was missing in the work that the others were doing. In any case, there were a number of developments in 1984, one of which Michael Green and I were responsible for, which led to the subject suddenly becoming very popular. And since then it's been a very active subject. It's had its ups and downs since then. There was another flurry of discovery in 1994, ten years later, which further boosted the community into the general acceptance of the subject. By now it is considered pretty mainstream. All the major physics departments are trying to hire new faculty to work in this area. So that's quite a change from the way things used to be, when famous physicists made some rather derogatory statements about it.

Q. When you pursue an idea, when you know within yourself that something feels interesting or right, and there are people around who do not agree with you, how do you deal with this?

A. I've never felt that there's much of a benefit of arguing, at length, with people about what was right and what was wrong. I present my ideas, and they can accept it or reject it, as they wish, I would leave it at that.

In a field like ours, eventually, it seems to me, the power of the idea will carry the day, although it may take some time. I prefer to let the ideas speak for themselves.

Q. I understand that to get grants and promotions the physicists here, in America, have to publish papers which are cited by other researchers. If you work in an area that nobody else is in, how do you solve this problem?

A. Well, I've never had problems of this type, even in the period when there was not much interest in what I was doing. I was employed at Caltech. Murray Gell-Mann, who was an outstanding scientist, was in our group in Caltech. He was responsible for bringing me there. He felt that what I was doing might have some possibility of being right and deserved to be supported. So I had a pretty good situation all along. I didn't have to struggle to get grants.

Q. But emotionally, did you feel that you were separated from others?

A. I tend to forget this with the passage of time. But if I think back, yes, there's no doubt that there were periods in which I felt very frustrated that there wasn't more acceptance of what I was doing. There's a funny thing. When the subject became popular, around 1984, I felt at that time that since I've been working on it for so many years, I had a big advantage over all these people who jumped in. And I would be able to develop many important ideas, more rapidly than they were able to, because I had all this experience. But that turned out to be completely false, and within months I was completely overtaken by all these newcomers.

Q. Sometimes people refer to the string theory as to a theory of everything. Any comments?

A. A theory of everything! I've been quoted on other occasions as being very negative about it, and I am. The phrase was introduced by others, and I think it was very unfortunate for us. In a very limited sense, what we are trying to do is fundamental to everything else, and that is what I guess this phrase was meant to represent. But it overlooks something that is terribly important. Even though, in principle, when you have the most microscopic rules, that determines how the things work on very short scales, as you go to larger scales, there are qualitatively new phenomena that arise, that are almost impossible to understand in terms of the underlying microscopic theory. By knowing string theory, we are not going to make any progress at all in understanding these other subjects. For instance, if you ask about human activities, life, I am certain that a complete understanding of string theory is not going to add any light on that. It is really not what we are addressing.

There is an analogy in the past to that. Around 1930 or so, when quantum mechanics was well understood, there was a famous statement due to Paul Dirac. He said that now all of chemistry is solved, in principle. In the same sense he was correct, the basic equations at that point were known. But getting from there to what the chemists actually observe involves computations that are so complex that they are completely out of the question, even for future computers that might exist in a hundred years from now. What better computers will allow you to do, is to talk about molecules that involve three atoms rather than two, or four atoms rather than three. But to understand macroscopic properties of matter based on understanding these microscopic laws is just unrealistic. And even though the microscopic laws are, in a certain strict sense, controlling what happens at the larger scale, they are not the right way to understand that. And that is why this phrase, "theory of everything," sounds sleazy.

Q. What if it is true that there are extra dimensions of a large size, of the type people discuss since 1996?

A. There are people that are speculating that extra dimensions are much larger than has traditionally been assumed to be the case by people, including myself, who work on this subject. My guess is that it is unlikely that this is the case. I would be quite surprised if the extra dimensions turned out to be as large as these other people suggest. Well, if they are right — strictly speaking, it is not excluded, I mean, experimentally — that would be very exciting, because that would mean that things could actually be studied experimentally. And, maybe, they would be useful for something, I don't know.

Q. You mentioned the beauty of the string theory. Could you elaborate a little bit?

A. Different people see it in different ways. A standard way of explaining this is that you have a rather simple-looking equation that explains a wide range of phenomena. That is very satisfying. Famous examples of that are due to Newton, Einstein, and so forth. We do not actually have a concise formula in string theory that explains a lot of things, so the beauty is of a somewhat different character here. What, I think, captures people in the subject is when they discover that they are dealing with a very tight mathematical system that incorporates things that nobody has understood yet. And then they find a part of that. So there is a deeper structure than we have understood. When you do some complicated calculation, and then you discover that the answer is surprisingly simple, much simpler than you would have expected... And it sort of appears as if there were some sort of a mathematical miracle taking place. Of course, there are no miracles, and when you find something that looks like a miracle, that just means that there is some important concept that you haven't understood yet. When you experience that a few times, you really get seduced by the subject. That's been my experience, and I think of the other people who are now just as enthusiastic, and, maybe, even more so, than I am, have had similar experiences.

Q. Is this a flash of the insight that keeps you in physics? What made you choose physics, to begin with, and what keeps you there?

A. I feel extremely fortunate that I am doing throughout my career exactly what I wanted to do. Nobody has ever tried to make me do anything else. So that's a fantastic situation to be in. And to be lucky enough to just be in the right place at a right time... It is very exciting.

Q. Do you think you ever in your life had a moment when you thought "I'm just throwing it to a dump, and I am going into a different business?"

A. No, I never had that. One thing that occurred to me, at the time when there wasn't interest in what we were doing, but I felt it was important, was that if I tried to look at my own situation very objectively, as an outsider, I don't know how anyone could decide whether or not I was a crackpot. Because, I think, you could find examples of people whose ideas did not hold up as well, that were as convinced and as dedicated to their work as I was. But I've never felt that I was a crackpot...

Selected Bibliography

1. P. Ramond, *Dual theory for free fermions*, *Phys. Rev.* **D3**, 2415–2418 (1971).
2. A. Neveu and J.H. Schwarz, *Factorizable dual model of pions, Nucl. Phys.* **B31**, 86–112 (1971).
3. A. Neveu, J.H. Schwarz, and C.B. Thorn, *Reformulation of the dual pion model, Phys. Lett.* **35B**, 529–533 (1971).
4. J. Scherk and J.H. Schwarz, *Dual models for nonhadrons, Nucl. Phys.* **B81**, 118–144 (1974).

Figure 2.4: Jean-Loup Gervais at SUSY-30, October 2000.

REMEMBERING THE EARLY TIMES OF SUPERSYMMETRY[56]

JEAN-LOUP GERVAIS

Laboratoire de Physique Théorique de l'École Normale Supérieure
24 rue Lhomond, 75231 Paris CÉDEX 05, France[57]

As is well known, supersymmetry started long before the Iron Curtain was dismantled, and thus came into existence separately in the West as well as in the former Soviet Union. In order to speak about what I know personally, I will recall that in the West, superalgebras were first considered by A. Neveu, J. Schwarz and P. Ramond as a basic tool to eliminate negative norm states from the spinning string theories. When Neveu, Schwarz and Ramond first introduced them they were initially referred to as supergauges. These authors used the covariant harmonic oscillator approach, the only known technique at the time, without field-theoretic interpretation. Preparing the present text brought back wonderful memories of the time when my long lasting collaboration and friendship with Bunji Sakita[58] began. In 1970–71 we started to develop the world-sheet interpretation of the spinning string (unknown at that early time), extending earlier discussions of the purely bosonic case begun by H. Hsue, B. Sakita and M. Virasoro [1]. We recognized that the Neveu-Schwarz-Ramond (NSR) models included world-sheet two-dimensional Dirac spinor fields, in addition to the world-sheet scalar fields, common with the Virasoro model. We showed that the supergauges of the NSR models corresponded to the fact that the two-dimensional world-sheet Lagrangian was invariant under transformations with anticommuting parameters which mixed the scalar and spinor fields. This gave the first example of a supersymmetric local Lagrangian (albeit, two-dimensional).

This supersymmetry was closing only on shell for lack of auxiliary fields, and was two-dimensional. In the West, the problems of proceeding off shell

[56]This is a slightly revised excerpt from a paper published in Yuri Golfand Memorial Volume *The Many Faces of the Superworld* (World Scientific, Singapore, 2000).

[57]UMR 8549: Unité Mixte du Centre National de Ia Recherche Scientifique, et de l'École Normale Supérieure.

[58]Then a Visiting Professor both at the Institut des Hautes Etudes, in Bures-sur-Yvette, and at the Laboratoire de Physique Théorique et Hautes Energies, University of Orsay, France.

and to four dimensions were initially solved by Wess and Zumino, as is well known.

These are the historical facts simply stated. It is worth trying to give a more personal picture of the birth of supersymmetry. Sakita himself has already given his own view [2]. As for me, the story really begins during the year 1968–69 which was a sort of a turning point. I had just returned to France after spending two years as a postdoc at New York University (NYU), where I had met J. Wess (a visitor for one year), B. Zumino (then the head of the Theory Group located at the Courant Institute), K. Symanzik, W. Zimmermann (at the time permanent faculty members) and D. Zwanziger (who was about to become a permanent member). Before that, my interest dealt mostly with dispersion relations, Regge poles, and S-matrix theory, but at NYU, I was fully converted to local field theory, and was much impressed by the power of symmetries in that context, be they local or global. Of course, that year saw the beginning of string theory which was, however, initially developed using the covariant operator method within the context of S-matrix theory, giving what looked like a realization of G. Chew's program. Regardless, local field theory also made wonderful progress on its own. The main problems at that time were the Adler-Bell Jackiw anomaly, the spontaneous breaking of symmetries and the quantization of Yang-Mills theory. For the latter, the work of L. Faddeev and V. Popov was gradually becoming more and more popular. It is hardly necessary to say that these topics now belong to textbooks. At that time, the French Government was very generous with temporary positions, and a handful of key visitors came for long visits[59] during that wonderful year, including D. Amati, the late Benjamin Lee, T. Veltman,[60] and B. Zumino. I drew much inspiration from the very stimulating atmosphere they created, together with other permanent members. In particular, together with D. Amati and C. Bouchiat, I devised [3] the now standard method to compute loops in string theories, using coherent states, and with B. Lee I showed [4] how to correctly quantize the linear σ-model in the phase where the spontaneous symmetry breaking takes place.[61]

[59]At the Laboratoire de Physique Théorique et Hautes Energies of Orsay (France) where I was working permanently at the time.

[60]Who had already undertaken to renormalize Yang-Mills theories.

[61]It seems that G. 't Hooft — then a student at Cargese during the following summer — drew much inspiration for quantizing massive Yang-Mills theory, from B. Lee's lecture on spontaneous symmetry breaking.

This is all to say that, when I first met Bunji Sakita in the fall of 1970, I was fully motivated to apply the field theory technique to string theories. Moreover, the year before I had shown [5] that the integrand of the Veneziano model was equal to the vacuum expectations of a product of scalar fields which were local functions of the Koba-Nielsen variables. This result, similar to an independent and better known work [6] by S. Fubini and G. Veneziano, was indeed a strong hint of the world-sheet field theory aspect of string theory. This viewpoint is now commonplace, but at the time it was not at all popular among string theorists. A large majority preferred the operator method, which had achieved striking technical success.

Before we met, Sakita and his collaborators had already made important progress in developing world-sheet field theory technique using path integrals. On the one hand, H. Hsue, B. Sakita, and M. Virasoro [1] had shown how the analog model of H.B. Nielsen could be derived from the path integral over a free scalar two-dimensional field. On the other hand, Sakita had come with a draft of an article where he started to discuss Feynman-like rules for the Veneziano model using the factorization of path integrals over sliced Riemann surfaces. There were many basic problems left, and at the beginning, we spent a lot of time establishing a general scheme. This complicated work was not so well-received, although it contained many precursive results. The most important point was that we essentially relied on the conformal invariance of the path integral representation over scalar free fields in two dimensions [7]. Although we did not really consider the gauge-fixing problem at that time, we were pretty much convinced that conformal invariance of the free world-sheet action is at the origin of the negative norm state elimination. In the spring when we came across the first article of A. Neveu and J. Schwarz, we were motivated [8] to systematically discuss conformal field theories in two dimensions, as a way to classify string theories, by defining what we called "irreducible fields" (now known as primary fields, following A. Belavin, A. Polyakov, and A.B. Zamolodchikov [9]). Considering only quadratic actions we recognized that only spin-zero and spin-one-half fields were possible, covering all existing critical string models of today. This was all well and good, except for one fact: the Neveu-Schwarz model had more ghosts and needed an additional negative-norm-state killing mechanism as compared with the Veneziano model. This motivated these authors, as well as Ramond in his seminal work, to introduce in the operator formalism a set of operators whose anticommutators gave the Virasoro generators. Sakita's visit in France was about to end, bringing with it the well-known burden of moving with a family. The coming deadline stimulated us a great deal, nevertheless. I quickly started to look for the possible symmetries of the action that would be the origin of the additional ghost killing. From the form of the NSR generators it was immediately clear that the transformation of the boson had to be a fermion and vice versa. It was not difficult to envisage that the action could be invariant, except for the mixing between commuting and anticommuting fields, which made everything very confusing. After many hesitations, Sakita and I solved the problem by introducing symmetry transformations with anticommuting parameters, out of which the supersymmetry of our NSR world-sheet action followed very simply. The paper was completed [10], and thus, for us, (conformal) supersymmetry was born, on July 22, 1971, just the day before Sakita departed from France to fill his new prestigious position at City College of New York.

In those days, anticommuting *c*-numbers were far from being understood. They were essentially a formal tool to derive perturbation expansions over fermionic fields from path integrals. This was how we used them in our second article [8], where we derived the tree scattering amplitudes of the Neveu-Schwarz model from world-sheet perturbation. Considering symmetry transformations with anticommuting parameters was, for us, quite a step which we took rather reluctantly. Quite some time passed before our idea was taken seriously — even by us.

In the early 1970s, the world-sheet dynamical approach to string theory was not well appreciated in our community, and our work did not get much attention in general. In December of 1971, Sakita delivered a talk about it at the Conference on Functional Methods in Field Theory and Statistics at the Lebedev Institute in Moscow, organized by E. Fradkin. On the way back, he stopped over in Paris. We wrote a summary of our work for the proceedings, which was sent to the organizers and circulated as a preprint — the complete proceedings themselves were never published.[62] In this way there was some early communication of our work to the Soviet scientific community where apparently it was well received [2].

Scientifically our separation was at a very unfortunate moment for our research program. At that time, of course, there was no email. Telephones were expensive, and airmail, slow. Moreover, Sakita had to deal with his new life and responsibilities, which he did rather successfully. We nevertheless continued our collaboration and met in person as often as possible. I was happy to visit him and his group in New York. However, to my regret,

[62] We later published our text in Ref. [11].

I did not continue pushing very hard further in the direction of supersymmetry. Other problems seemed more pressing. In the meantime the Nambu-Goto action and the Goddard-Goldstone-Rebbi, and Thorn light-cone quantization had come out. Sakita and I[63] showed [12] how the latter may be recovered using path integral with the former action using the Faddeev-Popov method in order to handle reparametrization invariance. This work raised much interest and criticism.[64] The main objection was that our gauge depended upon the external sources, and thus was not easily factorizable, in contrast with our previous path integral formulation [7]. We tried hard to understand what was going on, but failed. The answer was given by S. Mandelstam: the light-cone gauge is not conformally invariant, so that there is only one (preferred) parametrization where factorization holds with our gauge fixing. With this parametrization one sees strings (with lengths proportional to their respective p_+'s) which split and join, and our work played a key role for Mandelstam's subsequent discussion of scattering amplitudes, leading to the light-cone string field theory.

In the spring of 1973, Sakita visited the Niels Bohr Institute, and I made a trip there to meet with him. On the way back home, he went to CERN and gave a talk at which Zumino was present [2]. Sakita reviewed a current work [13] of Y. Iwazaki and K. Kikkawa[65] who were trying to establish the Nambu-Goto type formulation of our world-sheet NS-R dynamics. Later on it appeared that this seminar and a subsequent conversation with Zumino played a key role in leading Wess and Zumino to begin their seminal work on supersymmetry. I also remember that the latter author asked me questions about our works on various occasions. After that the whole subject suddenly exploded, and our contribution was temporarily forgotten for lack of reference,[66] given the fact that we had turned to other research directions.

Selected Bibliography

1. C. Hsue, B. Sakita and M. Virasoro, "Formulation of dual theory in terms of functional integrations", *Phys. Rev.* **D12**, 2857 (1970), reprinted in B. Sakita, *A Quest for Symmetry*, Eds. K. Kikkawa, M. Virasoro, and S. Wadia (World Scientific, Singapore, 1999).

[63]This work was initiated during my one-month visit at City College in October 1972.

[64]In particular, from the referee and at a seminar which Sakita gave at the Institute for Advanced Studies in Princeton in December 1972.

[65]At that time they were both at City College and had much interaction with Sakita.

[66]Our original paper was nevertheless reprinted in the first volume of Ref. [14].

2. See "Reminiscences", in B. Sakita, *A Quest for Symmetry*, Eds. K. Kikkawa, M. Virasoro and S. Wadia (World Scientific, Singapore, 1999) and in the present Volume; also in hep-th/0006083.
3. D. Amati, C. Bouchiat and J.-L. Gervais, "On the building of dual diagrams from unitarity", *Lett. Nuovo Cim.* **2**, 399 (1969).
4. J.-L. Gervais and B.W. Lee, "Renormalization of the σ-model. II. Fermion fields and regulators", *Nucl. Phys.* **B12**, 627 (1969).
5. J.-L. Gervais, "Operator expression for the Koba-Nielsen multi-Veneziano formula and gauge identities", *Nucl. Phys.* **B21**, 192 (1970).
6. S. Fubini and G. Veneziano, "Duality in operator formalism", *Nuovo Cim.* **97**, 29 (1970).
7. J.-L. Gervais and B. Sakita, "Functional integral approach to dual theory", *Phys. Rev.* **D4**, 2291 (1971), reprinted in B. Sakita, *A Quest for Symmetry*, Ref. 1.
8. J.-L. Gervais and B. Sakita, "Generalization of dual models", *Nucl. Phys.* **B34**, 477 (1971), reprinted in B. Sakita, *A Quest for Symmetry*, Ref. 1.
9. A.A. Belavin, A.M. Polyakov, and A.B. Zamolodchikov, "Infinite conformal symmetry in two-dimensional quantum field theory", *Nucl. Phys.* **B241**, 333 (1984).
10. J.-L. Gervais and B. Sakita, "Field theory interpretation of super gauges in dual models", *Nucl. Phys.* **B34**, 832 (1971), reprinted in B. Sakita, *A Quest for Symmetry*, Ref. 1.
11. *Quantum Field Theory and Quantum Statistics*, Essays in honor of the sixtieth birthday of E.S. Fradkin (Adam Hilger, 1987), Vol. 2, p. 435.
12. J.-L. Gervais and B. Sakita, "Ghost-free string picture of Veneziano model", *Phys. Rev. Lett.* **30**, 716 (1973), reprinted in B. Sakita, *A Quest for Symmetry*, Ref. 1.
13. Y. Iwasaki and K. Kikkawa, "Quantization of a string of spinning material Hamiltonian and Lagrangian formulation", *Phys. Rev.* **D8**, 440 (1973).
14. *Superstrings, The First Fifteen Years of Superstring Theory*, Ed. J. Schwarz (World Scientific, Singapore, 1985).

Figure 2.5: Bunji Sakita, October 2000.

REMINISCENCES[67]

BUNJI SAKITA
Physics Department, City College of New York,
New York, NY 10031

In the spring of 1967 I stayed at the ICTP in Trieste for five months. Towards the end of the stay K.C. Wali and I traveled to Israel, specifically the Weizmann Institute for ten days at the invitation of H.J. Lipkin. When we arrived in Israel we found that the atmosphere was extremely tense and people busy preparing for a war with the neighboring Arab countries. Although the touristy places were deserted, we managed to rent a car to visit many places including Jerusalem, Haifa and Acre. Since most of the young Israeli physicists had already been drafted, the physicists working at the Institute were mainly foreigners, among whom were H. Rubinstein, G. Veneziano and M. Virasoro. They were working together on superconvergence relations, which was a subject that I was also interested in at that time. In the discussions we had during this visit, the dual resonance program must have come up, since I remember that afterward in Trieste I started discussing with others about the possibility of constructing scattering amplitudes by summing only the s-channel resonance poles. We left Israel as scheduled on June 4 and the very next day in Ankara, Turkey we heard of the outbreak of the Six Day War.

In 1968, Keiji Kikkawa and Miguel Virasoro, with whom I had become acquainted in the previous trips to Japan and to Israel, joined our group at Wisconsin as research associates. By then I had returned to the University of Wisconsin to resume teaching, which I missed at Argonne, and I was preparing a course, "Advanced Quantum Mechanics", which was essentially a one-year graduate course on quantum field theory. In that summer Virasoro showed up in Madison with a hand-written paper by Veneziano and he explained to us in detail the activities of the Weizmann Institute. At once Goebel and I got interested in the work and all of us started thinking about generalizations. In that fall after Virasoro succeeded in obtaining the five point Veneziano formula, our activity became intensified and within

[67]This is an abbreviated version of the chapter under the same title from Bunji Sakita, *A Quest for Symmetry* (World Scientific, Singapore, 1999).

a few weeks Goebel and I had obtained the N-point Veneziano formula. Then Kikkawa, Virasoro and I started to generalize the formula further to include loops.

At this point we faced a dilemma. Namely, if one considered the Veneziano formula as a narrow resonance approximation of the true amplitude as was commonly assumed at that time, the construction of loop amplitudes based on this approximate amplitude did not make sense. After reviewing the logistics of quantum field theory, we arrived at the conclusion that the construction of the loop amplitudes did make sense if we considered the Veneziano amplitude as a Born term of an unknown amplitude for which we had an expansion similar to the standard Feynman-Dyson expansion in perturbative field theory. With this philosophy in mind we decided to construct a new dynamical theory of strong interactions. First we defined the local duality transformation as the crossing transformation at any four point sub-diagram of a Feynman diagram, and invented a Feynman-like diagram compatible with duality as a diagram which contained all the Feynman diagrams related to each other by the local duality transformations. Then we simply wrote down a prescription for the scattering formula corresponding to each of these Feynman-like diagrams. In practice, we used diagrams which were dual to the Feynman diagrams. A three point vertex of a Feynman diagram corresponds to a triangle in the dual diagram. An N-point Feynman tree diagram corresponds to a specific triangulation of an N polygon in the dual diagram. A local duality transformation in the dual diagram is the transformation of one triangulation of a quadrangle to another triangulation. In terms of the dual diagram, therefore, an N-point Feynman-like tree diagram corresponds to an N polygon. By studying these Feynman-like diagrams, it became clear to us that a dual amplitude corresponded in a one-to-one fashion to a two-dimensional surface with boundaries, and equivalently to a Harari-Rosner quark line diagram, which, by the way, we had also invented independently. In the second paper, we discussed the general Feynman-like diagrams by using the classification of two-dimensional surfaces, and extended the prescription to non-planar diagrams. This classification is the same as that of open string amplitudes.

Kikkawa left Madison in the summer of 1969 for Tokyo, and Virasoro and I left the following summer bound for Berkeley and France respectively. By then the operator formalism of the dual resonance amplitude had been established by S. Fubini, D. Gordon, and G. Veneziano, and independently by Y. Nambu, who further proposed the string interpretation based on this work. When I heard of the string interpretation from Nambu I felt it as natural as if I had known about it beforehand. I remember that I had experienced the same feeling when I had first heard about the Sakata model from Sakata.

At Wisconsin Virasoro had used the operator formalism to analyze the possibility of the negative metric ghost states consistently decoupling from the physical states. He obtained a set of operators, which could be used consistently as the operators of subsidiary conditions on the physical states. These operators were later found to be the generators of conformal transformations on a complex plane. This is the origin of the Virasoro algebra. In discussing this problem with him, I realized that these operators were compactly expressed in terms of a scalar field in a fictitious $1 + 1$ space(finite)-time, and the Veneziano formula itself could be expressed in terms of this scalar field operator. At about this time we received a hand-written paper by H.B. Nielsen: "A physical model for the n-point Veneziano model." Inspecting a few mathematical formulae in the paper, I came up with a functional integral representation of the Veneziano formula. There remained several important points to be clarified, such as the Möbius invariant property of the functional integrand, the connection with the operator formalism, and the calculation of non-planar amplitudes. At the end I, together with Virasoro and my student C.S. Hsue, established the functional path-integral formulation of dual resonance amplitudes, and with Virasoro, a physical model of the dual resonance model based on the "fishnet" diagram.

I stayed in France for one year before I moved to the City College of New York in 1971. To lessen the financial burden on the Institute, Louis Michel had arranged a joint invitation by his Institute at Bures-sur-Yvette and the Bouchiat-Meyer group at Orsay, a group of physicists that later moved to the École Normale Supérieure in Paris. This arrangement turned out to be a very fortunate one for me, as in Orsay I found several young physicists, who were interested in our work. Moreover it was there that I succeeded in starting a long and fruitful collaboration with Jean-Loup Gervais. During this visit, I wrote three papers with Gervais: on the functional integral, conformal field theory, and the super-conformal-symmetry, all in connection with the dual resonance model.

In Wisconsin I had already started working on the factorization of dual resonance amplitudes using the slicing and sewing technique of functional integrals. I had drafted the preliminary results into a paper and had sent it to the Physical Review before I arrived in France. At Orsay, however, I withdrew the paper, as a result of discussions with Gervais, when I convinced myself that a part of the paper was wrong. There were plenty of technical difficulties, on which Gervais and I had to spend another half a year of hard work. In this work we used formally and fully the conformal transformation properties of functional integrals without seriously questioning their validity. Sometime later we suspected the existence of an anomaly, that would explain the critical dimension of the model. I regret that we did not pursue it further.

When I received a paper on the new dual pion model of Neveu and Schwarz in the spring of 1971, I noticed at once that the most important ingredient in the model was the conformal invariance property. One could discuss about the generalization of dual resonance amplitudes in the very general terms of conformally invariant field theories. So, Gervais and I got busy constructing conformally invariant field theories. In this work, we discussed first a general theory of conformal fields by defining the irreducible fields (now known as primary fields) and the conformally invariant Lagrangian, and then we established the functional-integral representation of Neveu-Schwarz model by introducing a fermionic field in the model in addition to the old bosonic field. After the work was completed I wrote a letter to Virasoro (in Berkeley then) informing him of our work, since I heard that he had presented a similar work at a conference in Israel. In the exchange of letters, I learned the Ramond model from Virasoro and that it also could be described by the same Lagrangian simply by changing the boundary condition on the fermionic field.

Gervais and I thought that in the functional-integral representation the elimination of ghost states could be done by factoring out the negative metric components of the fields by using conformal transformations as was done in the standard gauge field theories. The necessary condition for this is, of course, that the Lagrangian is invariant under the conformal transformations. Once we introduced a new field in the new model which generated new ghost states, we had to find out a new set of gauge transformations under which the Lagrangian was invariant. Neveu-Schwarz-Thorn had just published a paper in which they proposed a set of operators to be used as the subsidiary gauge conditions on the physical states of the dual pion model. We tried to interpret these operators as the Fourier modes of the Noether current associated with the new gauge transformations which involved the new fermionic field, and arrived at the superconformal gauge transformations, under which the Lagrangian we had obtained previously was invariant. I believe that these field transformations are the first instances of supersymmetry transformations in a local field theory. The day after we had drafted this paper, I left France for New York. In this work, we had to use anticommuting c-numbers (Grassmann numbers) and functional integration of fermionic variables. These, to us were new concepts and we were initially reluctant to use them. Apparently, others shared this reluctance and this work and the functional-integral work in general, was not appreciated in our circle. However, I received an impression that when I presented the work later in December at the conference on functional integration at the Lebedev Institute in Moscow, it, as well as the use of anti-commuting c-numbers was well appreciated.

When R.E. Marshak became the president of the City College in 1970, I, together with Keiji Kikkawa, accepted a position there. I continued my research on dual resonance theory for a few more years, after I had settled down in the City College. There was a big difference, however, between before and after coming to the City College. Although several faculty members were already there before I came, I was expected to play the role of the leader of the high energy theory group. I felt that it was a great challenge to elevate the group into a quality research group. In a few years, thanks to Marshak's personal connections, we could gather a few talented graduate students into our group. And also we could hire a new faculty member, Michio Kaku and postdocs, such as Yoichi Iwasaki. Moreover, I could invite J.-L. Gervais for short visits on a few occasions. I intentionally spent more time with students, and shared my insights with them.

In the early spring of 1973, I was invited by Ziro Koba to visit the Niels Bohr Institute in Copenhagen for two weeks to deliver a colloquium, and more importantly to discuss the dual resonance string theory with his group, in particular with Holger B. Nielsen and Paul Olesen. By this time at the City College, Gervais and I had already formulated the ghost free Veneziano amplitudes by using the functional-integral representation of the Nambu-Goto string in the light-cone gauge. This work later led to Mandelstam's factorizable functional formulation of light-cone string theory, and eventually to Kaku-Kikkawa's light-cone string field theory. Furthermore in our group at that time, the work of Iwasaki-Kikkawa was near completion. This was an attempt, which I persuaded them to carry out, at a formulation of a light-cone string theory for the Neveu-Schwarz model. I reviewed these activities in Copenhagen. While I was in Copenhagen, David Olive called me up asking me to visit CERN on the way back home. At the CERN seminar, I reviewed the Iwasaki-Kikkawa theory. Later, I was told that this seminar and a conversation after the seminar had led Wess and Zumino to start their seminal work on supersymmetric field theory. I vividly remember the conversation with Zumino at the CERN coffee lounge. When I said, "If you allow me to use anti-commuting c-numbers, Gervais and I have written down a transformation of a fermi-field to a bose-field in the Nuclear Physics paper", he replied, "It's OK to use anti-commuting c-numbers. Schwinger has frequently used them."

3

Discovery of Supersymmetry in Four Dimensions

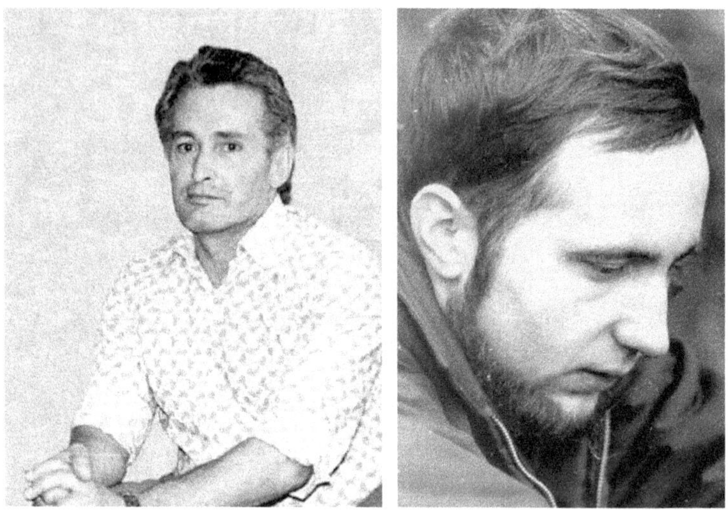

Figure 3.1: Yuri Golfand and Evgeny Likhtman, circa 1980.

NOTES OF AN OLD GRADUATE STUDENT[68]

EVGENY LIKHTMAN

Moscow, Russia

1962: Nine years before the birth of the Superworld in four dimensions. I received a passing grade on my entrance exams to the Department of Physics at Moscow State University but ran into trouble with the medical commission.[69] [1] I gathered my many certificates from winning many university competitions and arranged an audience with the dean of the School of Physics. I explained to him that I wanted to become a theoretical physicist. His answer was: "No one is above the law".

For me, this was nothing less than a tragedy. Science, and especially physics with all its impressive achievements, was an important and respected field in the Soviet Union. My father "applied" physics with enthusiasm in the defense industry, and we often discussed physics problems together. Requirements for admission to the Institute of Physics and Technology and to the Institute for Engineering and Physics were even tougher than those for the physics department at Moscow University.

My teachers arranged for me a meeting with the President of the University. During our interview, I told him about an inaccuracy I had noted in one of the problems in the entrance exam. As a result, once I had passed the entrance exams to the so-called extension division of the Department of Mechanics and Mathematics, I was by order of the President, transferred after one month to the Department of Physics.

In the course of five and a half years of study there, I learned how to take the traces of the products of gamma matrices, learned something about Clebsch-Gordon coefficients, and was fascinated by the beauty of the Noether theorem. I completed my senior project under Yu. M. Shirokov, who worked at the Mathematics Institute of the Academy of Sciences. However, the Graduate School of this Institute turned me down, and the advisor

[68]This essay was first published in the Yuri Golfand Memorial Volume *The Many Faces of the Superworld* (World Scientific, Singapore, 2000). See also Likhtman's talk at SUSY-30, https://arxiv.org/pdf/hep-ph/0101209, which can be viewed as an accompanying article.

[69]Here and below see Editor's comments consolidated at the end.

of my senior project sold me to the Theory Department of the Lebedev Physics Institute (FIAN). This was kind of an "arranged marriage". Thus, it was that at the right time, in the spring of 1968, I found myself in the right place: working with Yuri Abramovich Golfand.

Yuri Abramovich (for that is how I always addressed him, the difference in our ages being more than 20 years [2]) was somewhat short, with long arms that he waved around as he walked, rather like a skier or a speed skater. He was always in a good mood and never raised his voice. And whenever, as it happened, I made some blunder, his face would become radiant — this was a gradual process — with an enchanting smile. I often observed how other people were completely disarmed by this smile of his.

Twice a week, one hour before the beginning of the seminar, I came to the small room of the Theory department with its four desks, and we discussed my results. I wrote out my calculations on the back of the blueprints of old drawings, which were rolled up like wallpaper. Bystanders observed with interest the process of rewinding these rolls. Golfand did not always stay for the seminar after our conversations, he would sometimes slip away with a young female graduate student. I, on the other hand, was very concerned about the development of superscience and strongly disapproved of his lack of dedication.

Superalgebra had been written by Golfand before my appearance in the department. I had to figure out whether less complex superalgebras existed and then to determine whether they had any relation to field theory or high energy physics. The first part didn't take much time — I wrote out fairly quickly all extensions of the algebra of generators of the Poincaré group by bispinor generators. It took significantly longer to put together free field representations: one had to get used to the fact that in one multiplet were unified fields with both integer and half-integer spins. I built a series of representations which included high spins as well. This was a result of some kind, and I wanted to get it published as soon as possible. But instead, Golfand set me the task of building theories of interacting superfields.

1970: The term "supersymmetry" did not yet exist. We didn't know anything about supergroups or superfields with Grassmann parameters. Superinteraction was constructed by the method of indefinite coefficients: the most general form of interaction was written (up to the quartic terms in the fields) and spinor generators (to the third power of fields), after which these polynomials were substituted in the superalgebra, and then the equations determining numerous unknown constants of polynomials were written down with the goal of finding solutions. Among the equations, there were many dependent ones, and this gave hope.

By this time Golfand had finished his doctoral dissertation [3] and ... went off on a ski trip. By doing this, as it seemed to me then, he showed a lack of responsibility and also a certain disregard for appearances.

The first path to the superworld was not the shortest. Now I can no longer remember what specific directions Golfand gave me. More important was the stable, symmetric foundation of the theory and his apparent unshakable faith in its correctness. Superinvariant interaction was eventually constructed [4], and the time came to publish the results. I remember how we got hold of A.D. Sakharov in the corridor of the Theory Department and asked him to recommend our work to the *Reports of the Academy of Sciences*. He talked us out of this idea because of the time-consuming review process in this journal. The *JETP Letters* accepted only short articles, and I wanted badly to set everything out in detail. Golfand abridged my manuscript with a stern hand. Free field theory came out [5] as *Preprint FIAN* #41.

Two reviewers of my candidate dissertation — V.G. Kadyshevskii and V. Manko — made efforts to understand my computations in detail. The results of our work were presented at FIAN, the Institute of Theoretical and Experimental Physics, Moscow University and the Joint Institute for Nuclear Research. A detailed article was also written for the I.E. Tamm Memorial Volume [6]. I defended my dissertation without much ado. At the banquet, Golfand arrived with a big bouquet of flowers, which, to my surprise, was meant not for me, the trailblazer and hero, but for my wife, who stayed at home with our newborn son.

After graduate school, I was rejected by the Theory Department of FIAN, as well as others. The only place where I could get a job was the All-Union Institute of Scientific and Technical Information (VINITI). A certain theoretician, who heard about my place of work, made an analogy to Einstein. At my Institute, conditions were "ideal" for scholarly work: official rules did not prevent me from doing research once a week.

For the Western readers it should be noted that in the Soviet Union one took a job more or less for life. To fire someone legally, three warnings and the agreement of the union were necessary, or ... elimination of the position. In 1972 the Academy of Sciences was required to make personnel cuts. It was a routine campaign. As a rule, administrators of scientific institutes eliminated only unfilled positions, retired, those who had reached pension age, and found various means of retaining actively working colleagues. In the Theory Department of my Institute, Golfand was designated expendable.

As Sakharov wrote later [7] in the popular magazine *Novii mir*, the dismissal of Golfand was unique at FIAN. I argued with administrators for Golfand's retention but was rebuffed.

1973: Now our less frequent meetings occurred first at one, then at another of Golfand's apartments on Lenin Prospect, not far from the Academy of Sciences. Golfand remarried; his young wife looked at him adoringly. After trauma from a skiing accident, he developed a slight limp but didn't like to talk about this. He pretended to be in good mood, and I used to tell him all kinds of things I learned from the grapevine. Only his occasional comments betrayed his attitude toward his former colleagues at the Theory Department of FIAN.

I continued, with the help of obtained techniques, to construct new versions of interactions of the supermultiplets. But all these were far from the real world of elementary particles. I was the first to observe that the number of the fermion and boson degrees of freedom coincided in every supermultiplet, while the infinite energy of vacuum oscillations is cancelled. This discovery did not make much of an impression on those around me. Later I proved that one-loop boson mass diverged not quadratically, but, like the fermion mass, only logarithmically. Even more importantly, I succeeded in constructing a renormalizable theory of massive vector fields. This was not "good news" in the scientific community, since in the Weinberg–Salam model this problem had already been solved. Of importance were nonrenormalizable theories which supersymmetry would make renormalizable. But for the solution of this problem one needed a different technique for constructing superinteraction.

1974: Supersymmetry has been rediscovered by Western scholars. Superfield representation easily allowed them to construct superinvariant interactions. At my institute, VINITI, I was collecting copies of works on supersymmetry, but my file quickly overflowed. I came to the conclusion that I had to find my own topic.

1985: Peter West proposed we write an article for the collection *Supersymmetry: A Decade of Development*. Golfand had something to say about this subject, and he become absorbed by the idea and dictated to me his conditions: he would write the text of the article in English, and I would translate it into Russian for GLAVLIT [8]. During this work he found it very difficult to fulfill all the standard requirements for publication. I remember with what apprehensive cautiousness, fearing refusal, Golfand asked the publisher to alter requirements for the preparation of the manuscript.

1989: L.V. Keldysh — who was head not only of the Theory Department but of the entire Lebedev Physics Institute — nominated our work on supersymmetry for the Academy of Science's Tamm Prize in theoretical physics. The presentation of the award occurred in the impressive setting of the assembly hall of Moscow University. Golfand, accompanied by his wife, and I arrived separately and sat in different parts of the hall. The voice of the Academic Secretary snatched up people to the left and right of me, and they went up to the stage to receive their awards. Then came our turn. I was naturally the second to step forward, and Golfand did not see me. After he received the award, he asked the President of the Academy if he had seen Likhtman. I was struck, yet again, by his childish ingenuousness.

1990: Golfand obtains documents for immigration to Israel with his wife and two lovely daughters. Our farewell dinner in the kitchen. And, as a memento, an album of formulas I had studied back in 1968.

And then I turned away from the mainstream of high energy physics, where supersymmetry runs a great orchestra of fields, and turned onto a different path with a small ensemble of prefields that form extended soliton-like solutions. About the same time I staked my hopes on personal computers, that began appearing at VINITI in the late eighties, and on my developing skills in numerical solutions of variational problems. Within the framework of the Born-Infeld model of nonlinear electrodynamics, I hope to determine the dimensionless characteristic of a soliton that would be independent of the parameters of the model, which is trivially connected with the fine structure constant. But that is another story...

1991: Golfand returned to Moscow for the Sakharov conference. At our meeting, he could not conceal his disappointment with life in Israel, which began with a long search for lost baggage at the airport. He was forced to deal with indifferent clerks, their "European gloss" notwithstanding, as he put it. Despite the fact that the Minister of Science of Israel certainly knew about his contribution to high energy physics, Golfand was unemployed for a long time and lived on welfare. At the time of his trip to Moscow, his life in Israel was just beginning to take shape. And once again our conversations were dominated by his beloved theme: the prospects for receiving a Nobel prize for the discovery of supersymmetry. I was again shocked by his superoptimistic expectations. Now, in hindsight, I understand that despite his seeming autonomy, and scorn for appearances my Teacher desperately needed the support and recognition of other physicists. On the other hand, here again was apparent his unrestricted range of thought in posing super-problems, and this was, perhaps, the last and most important lesson I learned during our work together.

The original written in Russian by Evgeny Pinkhasovich Likhtman in March 1999, was translated in English by A. Liberman and M. Shifman.

Commentary (composed by M. Shifman)

1. This was a standard trick allowing the Soviet authorities to keep quota on Jewish students in universities without ever admitting that the quota existed. The first barrier was the entrance exam itself. "Undesirable" candidates were given problems which were much harder than those given to everybody else. If, against expectations, a given undesirable person was successful in solving them, the medical commission would always find a pretext to declare him/her unfit for studies in the given university for medical reasons.
2. Formal address in Russian is by name and patronymic, rather than by surname.
3. The academic hierarchy in Russia follows the German rather than the Anglo-American pattern. An approximate equivalent of Ph.D. in the US is the so called *candidate* degree. The highest academic degree, doctoral, is analogous to the German *Habilitation.* The doctoral dissertation is usually presented at a mature stage of the academic career; only a fraction of the *candidate* degree holders make it to the doctoral level.
4. Yu.A. Golfand and E.P. Likhtman, *JETP Lett.* **13**, 323 (1971) [Reprinted in *Supersymmetry*, Ed. S. Ferrara (North-Holland/World Scientific, Amsterdam — Singapore, 1987), Vol. 1, page 7; see also p. 478 below, in Appendices].
5. E. Likhtman, Report of the Lebedev Physics Institute # 41, 1971, in Russian. On page 8 of this preprint one can read, in particular: "As is known, in relativistic quantum field theory, in transforming the free energy operator to the normal-ordered form there emerges an infinite term which is interpreted as the vacuum energy. It is also known that the sign of this term is different for the particles subject to the Bose and Fermi statistics. The number of the boson states is always equal to the number of the fermion states. From this it follows that the infinite positive energy of the boson states in any of the representations of the [super-Poincaré] algebra is annihilated by the infinite negative energy of the fermion states." The English translation of the above preprint was published in K. Olive, S. Rudaz, M. Shifman (Editors), SUSY-30, Proc. Int. Symposium celebrating 30 years of supersymmetry, October 13-27, 2000, Nucl. Phys. B, Proc. Suppl., 101 (2001). See Appendix on page 9 following Likhtman's talk. The above quotation is after Eq. (7) of this Appendix.
6. The English translation of this paper, Yu. Golfand and E. Likhtman, *On the Extension of the Algebra of Generators of the Poincaré Group by Bispinor Generators*, I.E. Tamm Memorial Volume *Problems of Theoretical Physics* (Nauka, Moscow 1972), page 37, is published in *The Many Faces of the Superworld*, Ed. M. Shifman (World Scientific, Singapore, 2000), p. 45. The Tamm Memorial Volume article was based in part on a short Likhtman's artilce published in Lebedev Institute journal "Short Communications in Physics", vol. 5, p. 33 (1971) which exists only in Russian, see https://ksf.lebedev.ru/contents.php?post=1&year=1971&number=05&z=0.

7. Here E. Likhtman refers to the memoirs of Academician Andrei Dmitrievich Sakharov. English Edition: A. Sakharov *Memoirs* (Knopf, New York, 1990), translated from Russian by Richard Lourie.
8. GLAVLIT was an agency implementing total censorship in the USSR. Every material intended for publication, including such insignificant stuff as, say, business cards and the like, had first to be cleared through GLAVLIT. The GLAVLIT bureaucrats would require from the authors to supply them with many copies of the paper, in Russian, even in those cases when the paper intended for publication in one of the Western journals had been originally written in English. Since there was no access to photocopying machines, this was a hard task by itself. In general, the process was time and labor-consuming, and I heard from Golfand on several occasions that he hated it.

MY HUSBAND, MY LOVE[70]

NATASHA KORETZ-GOLFAND

Maale Adumim, Israel

Writing this was not easy, for despite the fact that Yuri passed away over five years ago, I still love him. Emotions overwhelm me, making it hard to write something logical. That's why this turned out so chaotic and fragmentary. But I am sending it anyway...

I first met Yuri at the end of 1961, when I was almost 16, and he was almost 40. He was married, and his oldest son was only one and a half years younger than I. Despite all that, I fell in love with him from the first sight, and tried, as much as I could, to excite his interest, gain his attention, and to be close to him even just as a friend. Intuitively I realized right away that we were an ideal couple. For me, it seemed impossible not to see that we were meant for each other, that we had to be together in grief and in happiness and for the rest of our life...

Later on, that is just how it came to be. But for seven years (as in the Bible) we remained friends and nothing more. We belonged to the same "hiking posse." We went to the mountains, canoeing, skiing. But in everyday life we didn't see each other very much.

Yuri worked a lot; he was an excellent theoretician. There were legends about him that he could "crack" any problem in three days, well... perhaps in four, if it was particularly difficult.

A first-class physicist, athletic, very handsome, not only did he like to go skiing, canoeing, and hiking, but he was also one of the first people in Moscow to start practicing yoga. He procured translations and *samizdat*[71] [1] editions of guidebooks, very rare at the time, and taught himself hatha and raja yoga. He knew a lot about Soviet and foreign literature, poetry and art. He was a big fan of *bards* [2] — Bulat Okudzhava, whom he called "the Teacher," and Alexander Galich. I remember how he copied their songs onto his tape-player and later, to the accompaniment of a guitar, sang them with us by the bonfire.

[70]This recollection was first published in the Yuri Golfand Memorial Volume *The Many Faces of the Superworld* (World Scientific, Singapore, 2000).

[71]See Editor's comments consolidated on p. 111.

He was contagiously young, younger than all of my peers, and at the same time it was always interesting to be with him — he always reflected on deep and unusual matters.

Where he found time for all this is difficult to comprehend. To many of his friends and acquaintances he seemed to be a "Mozart in science." His life appeared to be pure enjoyment; but then, "suddenly" and with no apparent efforts, he would publish a new scientific paper. This paper would undoubtedly contain an unexpected and original physics idea which would be later picked up and worked on by many scholars in his and neighboring fields. That was the case with his theory of supersymmetry, — so new that no one in the theoretical department of FIAN [3] was able to understand it for quite some time. Both in science and in life, Yuri was only seriously interested in matters of substance — new and not yet understood by the others.

Yuri had so little tolerance for Soviet propaganda that it could make him physically ill. The "garbage" that came from the media, the "noise," as he called it, wore him out immediately. Because of that, in life, in art, and in his relationships with people, he was very demanding.

But what surprised and attracted me the most was not Yuri's brilliance and erudition, but his absolute freedom in everything. He was born and raised in a totalitarian country; he was educated in regular Soviet schools and universities. He came from a wonderful family and had a relatively happy childhood, at least in contrast with my own. (I was born in one of Stalin's *Gulag* camps, and till the age six lived near "the zone." [4] My parents had been arrested in the infamous 1938, and were able to come back to Moscow only 20 years later, after the 20th Party Congress [5].) Despite his happy childhood, however, Yuri found political hypocrisy and any other lies just as intolerable as I did. He never lived "according to the circumstances," as the majority of the Soviet intelligentsia was forced to live, finding an escape in the "theory of small deeds." According to this theory, one could be a decent person and at the same time remain a loyal Soviet employee (which was the only way to do the work they liked). Many tried to escape into "pure science," "pure art," and narrow professionalism.

Yuri was not afraid *not* to join the Communist Party. He refused to participate in the yearly campaigns of "assistance" to the kolkhoz farmers, when all of Soviet intelligentsia — from university students to renowned professors — were sent to the fields to pick carrots and potatoes [6]. He was not afraid to read and keep forbidden literature (books by Solzhenitsin, Daniel, Brodsky), which was a criminal act according to the Soviet law. He did only that which he wanted to do. At the time many people considered this a defiant and unaffordable luxury, but Yuri simply couldn't live otherwise.

When his friends became dissidents or even were arrested, Yuri didn't break off contact with them. On the contrary, he offered help to them and to their families. After the arrest of the physicist and human rights activist Yuri Orlov (Golfand's old friend and colleague), Yuri participated in the protest pickets against the trial, where no one but the closest relatives and the specially selected KGB "guys" were allowed to attend. He later visited Orlov in exile, accompanying his wife Irina Orlov from Moscow to Kobyai in Siberia and back. Yuri wrote an article about Orlov. He did all this not only to expose the injustice committed by the authorities towards a gifted scholar, but also to promote Orlov's elegant idea of wave logic as it so deserved.

In Yuri's system of values, the notions of decency, honor, and integrity were an indispensable part of existence, despite the fact that these values were being systematically eradicated from the public consciousness by the Soviet regime. Yuri used to say that decency was like fish: it could not be "half-fresh." But in Soviet stores fish was sold not only half, but a "third-fresh," and even for that, one had to stand in a long line. Yuri would never buy such fish, and if there was nothing else, he could go on bread, refusing to adjust to inhuman conditions.

At the time when we became *refuseniks* [7], there was, among the other Moscow dissidents and *refuseniks*, a popular handbook written by Volodya Albrekht [8] on how one should conduct oneself during a KGB interrogation: how to answer or not to answer their questions in order to avoid bringing harm to oneself or to anybody else. This handbook was very popular, for the KGB used these "conversations" to intimidate dissidents. Although the interrogations did not follow any legal procedures and had no judicial grounds whatsoever, to decline the invitation was extremely dangerous and could become the reason for one's arrest. Yuri was invited to these "talks" several times. Not hesitating for a single moment and not afraid of the possible consequences, he would refuse to come to the "meeting" and hung up right away.

We were first denied the right to emigrate to Israel in October 1974, after having submitted the application in December of '73. For almost a whole year we lived in uncertainty, which was unpleasant, to say the least.

Yuri began to learn Hebrew. He brought home a textbook and a tape and studied every day for an hour. At that time I was finishing my studies at the university and was extremely busy. But I also tried not to fall behind. In two months, Yuri could read and write, knew the rules of grammar and tried to listen to the news on the radio, while I was still forming simple sentences!

He was just as fast at mastering computers when he later started working at Haifa's Technion. In two weeks, he learned to work on a computer and used it to lead his correspondence and type all his papers.

There is a stereotype in the public eye of a great scientist fit only for scientific matters and utterly helpless in all practical affairs. Yuri could do almost everything by himself, with his own hands — from cleaning the house to cooking, to serious repairs of electrical systems and appliances. He bragged that in him was "dying" a famous plumber. When we went canoeing or hiking in the mountains, I felt myself as if behind a stone wall. There was never a problem to start a fire in the rain or to patch up a torn coating of the canoe — Yuri always had superglue and dry fuel handy.

When our two daughters were still babies and required constant attention, Yuri helped me in everything — he washed them, changed diapers, did laundry. He took pride in being able to replace me even in nursing them (there was no baby food in Moscow at the time, so that if I was going out, I had to leave him a bottle with my own milk). But when he sat down to work, he would not be distracted by anything. He organized his space not only on his desk, but in the whole house. He said that the work of a scientist is no easier than that of a ditch-digger — before reaching an elegant result, one had to turn and haul over a huge pile of rocks. Indeed, after finishing an article of 10–12 pages, he would throw away 200–300 (sometimes more) sheets of rough drafts.

Yuri knew his worth as a physicist, adequately appraising his contribution to the quantum field theory, realizing that his theory of supersymmetry would forever remain a classic in the history of physics. But he never showed off his achievements and disliked all self-advertisements.

When we finally immigrated to Israel, Yuri was over 65. Many of his friends warned him that he wouldn't be able to find a job: in 1990 the rate of *aliyah* to Israel was at its peak. We heard Mark Azbel explain on the radio that the newly arrived and still unemployed scientists should thank Israel for letting them sweep the streets.

Yuri joined *Ulpan* at the most advanced level and actively began looking for a job. He sent out a list of his publications, gave talks at the universities of Tel-Aviv and Jerusalem, at the Weizmann Institute, in Technion. He got help from his *refusenik* friends, from scholars who had immigrated earlier — Viktor and Irina Brailovsky, Alexander Ioffe, Alexander Voronel. During these difficult times, we got financial support from the Israel Public Council for Soviet Jewry. Six months after our arrival, with the help of Yuval Ne'eman and Misha Marinov (both of whom understood Yuri's significance as a physicist), Yuri received a position as a research fellow in the Physics Department of Technion. There he composed drafts of many new physics papers, but had time to complete and publish only a few of them. He just began the work, that he considered most important for himself; he even found a mathematician — Dov Ramm, with whom he was going to develop a mathematical apparatus.

I am a Humanities person, and the physics of elementary particles is beyond my reach. Nevertheless, Yuri was somehow able to explain to me the subjects of his work. Despite the difference in our occupations, we always found it very interesting to be together. We discussed all kinds of things. During our first years of marriage we would even get into arguments about books or movies, until we finally agreed that everyone had a right to their own opinion.

Yuri thought a lot about the structure of the world — he wondered who created it and how. In a way, all his work was an effort to break through to that kind of knowledge.

Through his practice of yoga, he mastered incredible things — transcendental journeys, controlling his pulse and his blood pressure (he would use this ability during medical exams, when he had to get medically cleared for a skiing trip in Chiget).

Once, still in Moscow, he tried to tell me that the visible reality was only a reflection of our multidimensional world, as a shadow of a lamp post on the ground. We perceive only the shadow and have no idea about the post or the light. (This happened during one of our evening winter walks. On the whiteness of the fresh snow, the shadow from the lamp post appeared very defined and glowed under the bright light). At the time I couldn't comprehend or accept his idea. It frightened me and for some reason seemed inhumane.

After Yuri's death I had the opportunity to read several serious books on mysticism. I was shocked to find great similarity between the model of the universe and of humanity described in the books and Yuri's own ideas. There is no possible way that he could've found such literature in Moscow, so he had to have discovered them completely on his own. Back then, I got scared and closed myself off from this kind of information, but now I deeply regret that I have failed to understand him right away.

Now, when I try as much as I can to comprehend these matters, I become more and more aware of the uniqueness and the power of my husband's personality, and I thank God for the happy and the difficult 32 years that I could spend close to him.

Our life was indeed very happy and very difficult. We had enough trials and tribulations that fell on our path. The *otkaz* [7] years were hard not only because we had to stand against the regime, not only because there was no work and no money. For Yuri, the hardest thing was the isolation from his colleagues — the physicists of FIAN. The people with whom he had worked together for 20 years cut off all communication with him, afraid of approaching a dissident ostracized from the official "Temple of Science."

I don't like to recall all this. Nearly 25 years have elapsed since then. But at the time, the indifference and the cowardice of his former colleagues deeply wounded Yuri. On top of this, at approximately the same time, Yuri had to break off contact with his children from his first marriage, who could not forgive their father for leaving the family. (Thank goodness, this break turned out to be only temporary.)

All this had shaken Yuri up immensely, so that he couldn't work for almost three months. What eventually got him through was his yoga practice and the people who have stuck by us despite everything. We were surrounded by beautiful people, thanks to whom we didn't become completely isolated. Yuri's mother and sister, their friends, my parents, their friends and mine — those people were few, but they were not frightened by our new "status" and tried to support us in every way they could.

In some time, Yuri got introduced to other *refusenik* scientists, who united together to hold seminars. That's how our 16-year-long life in *otkaz* began: talks and conferences with scientists from all over the world; rallies of protest demanding the authorities to let Jews out of Russia; communal celebration of Jewish holidays; house search and arrest; participation in the publication of the magazine "Jews in the USSR;" [9] arguments about the immigration of Jews to Europe and the United States...

Although our Jewish life was very eventful, Yuri's work in physics remained his top priority. He also found happiness in our two growing daughters. He continued to go hiking, although now less frequently and only to the outskirts of Moscow — there was no money for longer trips. He still worked a lot. Thank God a theoretician only needs his head, a desk and some paper to do his job. Of course, he lacked contact with other physicists, but even without it he was able to write and publish several scientific works. During these years Yuri became a member of the New York Academy of Sciences, of the

European and American Physical Societies. The most surprising recognition, however, came from the Soviet Academy of Sciences in 1989, when Yuri and his former student E. Likhtman were awarded the I.E. Tamm Prize for their works on supersymmetry.

Our life in *otkaz* not only made us stronger, but it also cured us from nostalgia for a long time afterwards. Orwell's *1984* gives only a bleak idea of what we went through in dealing with OVIR [10]. I had to subject myself to the torture of interacting with Soviet bureaucrats; Yuri could not stand their arrogance, and refused to reconcile himself to it.

For many years I demanded to know why we were denied the exit visas. The answers that I got were all the same:

- This is not in the interests of the Soviet Union, — was the most frequent answer.
- Divorce your husband and get the hell out of here, if you'd like. But as long as your husband is alive, he is not going anywhere, — was the rudest and the most sincere answer.

I would like to forget. Forget forever the humiliation of endless lines in the halls of OVIR. Forget the time when they took away our passports, making us pay a huge sum for *renouncing* the Soviet citizenship. Forget how we had to spend several sleepless nights at the airport, trying to check in our luggage; and how nonetheless, with the exception of skis and linens, none of our luggage ever made it to Israel... All that is over now. It's been almost ten years — it's time to forget it once and for all... Only Yuri's mother and my parents didn't make it to the day we were finally allowed to leave. They passed away when we were still *refuseniks.* That, we cannot forget. That remains our pain forever...

The difficult years in Russia were followed by the problems of adjustment to our new life in Israel. Of course, these were problems of a different kind, but it was still hard, especially for Yuri. He could not get accustomed to the heat and the dry winds; could not get used to the different landscape, to the small distances between the cities, to the absence of rivers which he so dearly loved back in Russia.

He did not want to adjust to Israeli bureaucracy, and to the "small-town" mentality of Israeli people. But contrary to all expectations, the ultra-orthodox Jews in their theatrical costumes did not bother him. Perhaps it was in those people, and not in the secular part of the population, that lay the reason for the existence of a country like Israel, Yuri thought. Naturally, the fanaticism of the orthodox was alien to him, just like any other form of fanaticism. But he read the Torah and dreamt to read the teachings of the Kabbalah. One day, he hoped to find a very intelligent and not fanatical religious person, who would agree to discuss certain untrivial questions about the religious notions of the world's existence. — That, he didn't find...

He worked, as always, very hard. During the last six months he began to work on a new idea which he imagined to be even more interesting and powerful than the idea of supersymmetry. In the midst of realizing his plans, he had a stroke. Three days later — on February 17, 1994 — he died.

It was impossible to believe that he was no longer with us. Everyone was in shock: his friends, relatives, colleagues. Many people came to say good-bye; those who couldn't make it sent their condolences.

For a long time I believed that it hadn't happened. At night in my dreams, I saw the continuation of our life together. But recently I started feeling that Yuri's beautiful soul is in another world, whatever it could be, and is involved in something very important. He seems fine there, and I am at peace. But I still miss him so...

The original was written in Russian by Natalya Moiseyevna Koretz-Golfand in June 1999, and translated into English by Anya Shifman.

Commentary (composed by M. Marinov, A. Shifman and M. Shifman)

1. *Samizdat* — A huge variety of books were forbidden for publication (or restricted in publication) by Censorship. The class of disfavored books included not only those with political connotations, but, in general, all books which were considered not helpful for the Soviet ideology. Copies of forbidden books were being smuggled in the country illegally. People retyped them, using mechanical type-writers and carbon paper, or photograph them, page by page, using amateur cameras, and then print at home on photopaper, producing huge piles. The process was called samizdat, which can be loosely translated from Russian as self-publishing. *Samizdat* was forbidden by the Soviet law.
2. *Bards* — This word exists in English too, meaning a poet and a singer. In the Soviet Union it was an important cultural phenomenon, beginning in the 1960's. The bards composed their songs, both music and lyrics, presenting a sharp contrast to the every-day lies of the official propaganda. They sang their songs to small groups of people at private apartments, usually accompanied by a guitar. The songs were taped and then retaped many times, from one tape recorder to the next. The best of the bards — Bulat Okudzhava, Alexander Galich, Vladimir Vysotsky and others, were the pride and honor of the nation. People found refuge in their songs — about life and love, honesty and integrity — from the absurdity of the existence in the communist regime. Distributing such tapes was formally outlawed as anti-Soviet propaganda. Alexander Galich was forced into exile, where he shortly died under suspicious circumstances.
3. FIAN — Lebedev Physics Institute of the USSR Academy of Sciences, in Moscow.

4. *The Zone* — In popular jargon, the inner part of the GULAG camps where the inmates were kept in barracks.
5. XXth Congress (of the Communist Party of the Soviet Union) took place in 1956. At this Congress, at a closed meeting, Khrushchev gave a speech revealing and denouncing the mass terror of Stalin's rule. The phenomenon was ascribed not to the general nature of communism, however, but to the so-called Stalin's personality cult. After the XXth Congress many of the GULAG camps were shut down, and millions of innocent people released.
6. *Potato picking* — Each fall all those who were considered unproductive — college students, engineers and scientists, musicians and so on — were sent for a month or so to "help" disintegrating Soviet kolkhozes (collective farms) to collect crops and save at least some of them for the winter. In popular jargon, this routine was commonly referred to as "potato picking," although in practice this could mean any type of agricultural work. As with any slave labor, it was extremely inefficient. The villages usually could not accommodate crowds of people sent there from the cities — the conditions of existence of these *ad hoc* farmers were substandard. This metayage was mandatory; refusal to go "potato picking" could cost one his/her job or place at the university.
7. *Otkaz* (from the Russian "refusal") — the name for social and political conditions under which a large group of people (the so-called *refuseniks*) were forced to exist in the USSR in the 1970's and 80's. The only "crime" committed by these people was that they had applied for and got denied the exit visas to Israel. And yet, they were treated essentially as criminals: fired from jobs and blacklisted, with no access to work (with the exception of low-paid manual labor), constantly intimidated by the KGB, at the verge of arrest. In fact, the most active of them, those who tried to organize and fight back for their rights, were imprisoned.
8. V. Albrekht, a dissident and human rights activist, wrote an article explaining how dissidents could keep their human dignity and resist pressure of the KGB, without entering into conflict with the Soviet laws. The article was distributed as *samizdat.*
9. *Jews in the USSR* — collections of articles on Jewish history and culture, and on important current events. In a sense, this was the chronicle of the Jewish community: information on the struggle of Jews for their rights, on arrests and other persecutions, on *refusenik* seminars and so on, was provided on a regular basis and distributed as *samizdat.* This was one of the successful endeavors of *refusenik* community. A reprint of the full set is available now (the Library of the Jerusalem University).
10. OVIR — Department of Visas and Permits, a section of the Ministry of Interior Affairs which was in charge of exit visas.

LAST INTERVIEW WITH D.V. VOLKOV[72]

D.V. VOLKOV
Theoretical Laboratory, Kharkov Institute of Physics and Technology, UA 310108 Kharkov, Ukraine

Q. Dmitry Vasilyevich, could you tell me, please, what was your path into theoretical physics, how did you become a theoretician, and was this accidental or was there any specific cause.

A. I was born in Leningrad. When I was 16 years old and when I was studying in the 8th grade, the Great Patriotic War began. I was evacuated from Leningrad. These were very difficult years for young people. In this period I came to work on a collective farm and in a military factory. After that I was drafted into the army and took part as a soldier in the war on the Karelian front, above the polar circle. When the war began with Japan, I participated in military action on the Far Eastern front. I want to say that the war had a considerable influence on my attitude to life. In my generation the war created a feeling of responsibility for the country. After the war we carried over the same ideology into civilian life. When the question of a choice of profession arose, many of us thought about how we might be useful to our country. During all the war years I dreamed about going into science, because already in school I was attracted especially to the exact sciences: mathematics and physics. After the demobilization I entered Leningrad State University, in the faculty of physics. At that time prominent scientists such as V.A. Fock and T.P. Kravets were teaching there. The lectures of T.P. Kravets were distinguished by his ability to link the study material with personal moments. He taught us that physics is created by living people and he spoke much about his teacher Lebedev. From the first days Kravets infected us with a deep love for science, and for physics in particular. Aside from that, I listened to the lectures of V.I. Smirnov,[73]

[72]Questions by Yu.N. Ranyuk, reprinted from *Supersymmetry and Quantum Field Theory*, Eds. J. Wess and V. Akulov (Springer Verlag, Berlin, 1998), pp. 1–8, with abbreviations. Translated from Russian by Prof. K. Stelle.

[73]Editor's note: Vladimir Ivanovich Smirnov (1887–1974) was a Russian and Soviet mathematician, the author of the "Course of Higher Mathematics" (vol. 1–5, 1924–1947). –MS

whose widely known multi-volume works were specially intended for theoretical physicists, and, in fact, formed the basis of our whole education. We learned a lot from other mathematicians of his school: O.A. Ladyzhenskaya and M.I. Petrashen. In the final year I received a profound training in the specialization of theoretical physics thanks to the excellent teacher L.E. Gurevich.[74] There were also other teachers, whom I remember to this time with thankfulness.

Unfortunately, a great tragedy happened in Leningrad when I was studying there, and this had an impact on my further destiny. At the end of the 1940's the so-called Leningrad trial took place, as a result of which prominent [Communist] party leaders and representatives of the scientific and cultural community of the city were repressed, and among them were the leading scientists of the Leningrad University. Therefore, the group in which I was studying was dissolved and the scientific line on which I was working ceased to exist there. Here one should give credit to the Kharkov scientists. From the very beginning, the Kharkov scientists had close contacts with their colleagues in Leningrad. In particular, it is known that our Institute, KhFTI, was organized thanks to the arrival in Kharkov of the group of scientists headed by the well-known physicists Kirill Dmitrievich Sinelnikov,[75] Anton Karlovich Walter and others. One of the major concerns at the foundation of the Institute was how to attract young people so that in the future the Institute would be staffed by highly qualified scientists. Therefore a special representative came to Leningrad, to recruit students who were offered the possibility to move to Kharkov and to continue their studies there. I agreed. Together with me came the already well-known scientists E.V. Inopin,[76] K.N. Stepanov,[77] V.F. Aleksin,[78] and some other students.

---

[74]Editor's note: Lev Emmanuilovich Gurevich (1904–1990) was a Soviet theoretical physicist and astrophysicist, the founder of the Soviet school of physical kinetics, professor of physics at the Leningrad University. –MS

[75]Editor's note: Kirill Dmitrievich Sinelnikov (1901–1966) was a Soviet experimental physicist, one of the prominent participants in the creation of the Soviet A bomb. He was Head of Nuclear Physics Department at Kharkov Physical-Technical Institute (KhFTI), later its Director. In 1932, Sinelnikov with A.K. Walter, A.I. Leypunsky and G.D. Latyshev managed to split the lithium nucleus with protons accelerated to 300–400 keV. –MS

[76]Evgeny Inopin (1923–2003) was a theoretical physicist, Head of KhFTI Nuclear Physics Department. –MS

[77]Editor's note: Konstantin Nikolaevich Stepanov (1930–2012) was a Soviet and Ukrainian theoretical physicist specializing in plasma physics. –MS

[78]Editor's note: Vitaly Fedorovich Aleksin was one of the founders of magnetic plasma capture (stellarator research at KhFTI). –MS

Q. Dmitry Vasilyevich, when did you move to Kharkov, and how did your further scientific interests develop?

A. I moved to Kharkov in 1951. Now, about my scientific interests and their development. During my student years, the mathematical aspects of science attracted me even more than the physical ones. At that time I had already developed a deep feeling for the general theory of relativity. Quantum field theory was just beginning. Just at this same time the development of physics received some sharp impacts. On the one hand, quantum electrodynamics was effectively reborn thanks to the theory of renormalization, and, on the other hand, the physics of new particles arose. At this time the charged and neutral π mesons were discovered and discussions on the definition of spin and parity were going on. This was a real revolution in science. Later on, after graduation from university, I began my graduate training. My advisor was A.I. Akhiezer.[79] He organized a small group of graduate students, including R. Polovin, P. Fomin,[80] V. Aleksin and me. All of us actively studied quantum electrodynamics. This was especially important for me, because I continued to work in this area, and quantum electrodynamics became for me a sort of initial example of the theories that I work on now.

I would like to discuss in some more detail the directions of science that interested me most and on which I worked later. I don't know why — somehow intuitively — but already in my years of study my favorite subject was connected with the theory of symmetry groups. At that time, the theory of symmetry groups was already being adopted in physics, but not widely enough. Later on, the application of this theory came into a full flowering, especially when many new elementary particles were discovered and the question of systematizing them on the basis of symmetry groups became very important. When I started to work, there were no powerful group-theoretical methods. The first pieces of work that I consider to be somehow connected to what I am doing now were concerned with the properties of particles with high spin. I would like to emphasize that questions related

[79]Editor's note: Alexander Ilyich Akhiezer (1911–2000) was a Soviet and Ukrainian theoretical physicist known for his contributions to quantum electrodynamics, nuclear physics, plasma physics, and condensed matter physics. He was a recipient of a large number of prizes and awards, Pomeranchuk Prize, Bogolyubov Prize, Mandelshtam Prize, etc. –MS

[80]Editor's note: Pyotr Ivanovich Fomin (1930–2011) was a Soviet and Ukrainian theoretical physicist in the field of astrophysics, quantum field theory and elementary particle physics. –MS

to supersymmetry interested me already at the earliest stage of my activity. How do particles with different spin differ from one another? Why are there bosons, why are there fermions? I remember, in part, that just when I learned about the group SU(3), it amazed me that the multiplets of this group contained simultaneously particles with integral and with half-integral isospin. I was already trying to do something in this direction, replacing isotopic spin with ordinary spin, that is, actually, the idea that now lies at the basis of supersymmetry. An important role in the conception of the idea of supersymmetry was played by work on phenomenological Lagrangians, in which the interactions of Goldstone particles in the presence of spontaneous symmetry breaking are practically uniquely described by geometrical group-theoretical methods. I actively participated in the development of this line; it fascinated me very much. At the same time, the ideology of gauge field started to penetrate actively into physics and I returned to the old questions. It amazed me that all these particles, the Goldstone particles and the gauge fields, were bosons, but the fermions somehow were not involved at all. Here, a kind of inequality appeared: why were some particles — bosons — selected, but others — fermions — not included in this group? This was a key moment, because the very thought that fermions can also be Goldstone particles or gauge fields somehow contained the answer in itself: if it were clear how to build a general scheme of Goldstone particles with integral spin, then upon making the transition to fermions one should replace the corresponding operators with operators firstly carrying spin one-half and secondly being anticommuting, in accordance with their fermionic nature; and in fact I did this. After that, when this was done, I, together with my co-authors V. Akulov and V. Soroka, considered first the global properties of supersymmetric theories and then local properties [1, 2].

I would like to say that I worked not only on the theory of symmetry groups, but I also had pieces of work on nuclear physics and even on the theory of accelerators, but, nevertheless, the theory of symmetry groups has always been my favorite subject. And certainly, my main work was connected with the application of this theory to the physics of elementary particles.

By main, I mean my work on the establishment of parastatistics, work on the discovery of the conspiracy of Regge poles, and on the application of symmetry groups to the classification of hadronic resonances, that is, to baryonic and mesonic resonances. But my most important result, which is the most widely-known one now in the world, is certainly the discovery [1] of a new symmetry group — supersymmetry — and the extension [2] of this discovery to aspects of the general theory of relativity, the so-called supergravity. I can speak about this in some more detail. First, a couple of words about how I came to the discovery of supersymmetry and supergravity. The starting point was the idea of W. Heisenberg. Let me tell you what this idea was. At the end of the 1960s, Heisenberg proposed the idea that all elementary particles can be described by a unified theory on the basis of the nonlinear equation formulated by him [3]. He started this work together with W. Pauli, but sometime later Pauli dissociated himself from this theory and even sharply criticized Heisenberg for his assumptions. But, nevertheless what did Heisenberg contribute to the discovery of supersymmetry? Heisenberg tried to explain the spectrum of all elementary particles on the basis of his theory. And thanks to his great intuition, he guessed the places of the particles and predicted how the photon should appear in this theory. Moreover, he found a place for the neutrino within his theory and assumed that the neutrino emerges as a result of spontaneous symmetry breaking. This was a very unusual assumption, because all known Goldstone particles emerging as a result of spontaneous symmetry breaking had spin zero in all theories known at that time. This idea of Heisenberg was revolutionary, because he was the first to formulate the thought that there might exist Goldstone particles in nature with spin one-half. To tell the truth, he found this particle by an incorrect method, but nonetheless it was an idea that had a strong impact on me and from that time I often, although not constantly, would think about whether this idea could be realized. And when I started to consider how such particle might appear in the theory, I understood that this requires an extension of the usual physical groups, which is the basis for all relativistic processes. That means an extension of the Lorenz group and an extension of the Poincaré group so that new operators would be present, which would correspond to a quantum number of the neutrino. Thus, the main result that we obtained was the construction of such an extension of the Poincaré group, which is now called the Poincaré supergroup.

Later on, we encountered certain difficulties and, if one speaks about the direct application to the neutrino, this idea nonetheless did not work. Why? Because the group that we were considering contained the Poincaré group and it is known that the Poincaré group leads eventually to the general theory of relativity, if one considers the local transformations of the group. That means that the same local transformations of the supergroup would lead to the emergence of certain superpartners of the ordinary gravity field, and these superpartners would totally absorb the Goldstone particles. And thus, when I understood such ideas, I, together with my collaborators, proposed an extension even of Einstein's general theory of relativity, which would include just what we call supersymmetry. So, in fact, spin 3/2 superpartners would emerge. Moreover, in the extension of this theory not only particles with spin 3/2 emerge, but also particles with spin 1, with spin 1/2 and with spin zero. That is, there arose the idea that all elementary particles might be included into the system of supergravity. Now many scientists all over the world are working on it. Different variants of supergravity are being considered and also different variants of superstring theory, which are based on a certain extension of supergravity that takes into account the fact that elementary particles can be not simply point objects but also extended objects. This direction is now the main direction in theoretical physics which suggests that there is a unified theory of all elementary particles, based on one common principle — namely, on local supersymmetry.

Heisenberg's idea was, if one can say so, a physical idea, but in order to give this idea shape in a new precise mathematical theory, a certain mathematical formalism was required. And here I am very thankful to J. Schwinger, whom I personally knew, and who in a number of cases could have discovered supersymmetry. Moreover, I have recently seen an article by Schwinger, in which he writes why he didn't discover supersymmetry in spite of the fact that he was completcly ready for it.

Why do I think that Schwinger could have discovered supersymmetry? Firstly, he introduced the concept of certain anticommuting variables, which played the roles of physical variables in field theory. He was the first to do this. Secondly, when I was saying that in supergravity the superpartner of the gravity field appears, well, these superpartners are particles of spin 3/2, and the theory of such spin 3/2 particles was first developed by Schwinger. It is even called the Rarita-Schwinger theory.

Recently, I received this article by Schwinger in which he writes about the reasons why he did not discover supersymmetry. Schwinger also says that he was very close to the discovery of supersymmetry and explains why. At the same time, in response to a general question as to why he didn't do it, which he received from the audience in the auditorium, he answered that this is a philosophical question and that this philosophy of discoveries is discussed in a number of monographs and that he could give only a general answer to the question.

The next point that played an immense role in the mathematical formulation of supersymmetry was the work of the greatest French mathematician Elie Cartan. He created a special formalism of differential geometry, which seem to be specially appropriate for the system that I was considering.

This particular circumstance, that I knew this work very well, helped me to create the mathematical formalism of supersymmetry. So, I think that there were actually three sources that helped me to develop the theory of supersymmetry and the theory of supergravity: the idea of Heisenberg, the idea of Schwinger and the mathematical idea of the great mathematician Elie Cartan.

Q. Dmitry Vasilyevich, I am not a theoretician, and it is very difficult for me to assess your accomplishments. Could you tell me in popular terms how your work is recognized and about its place in theoretical science.

A. It's very easy to answer this question, mostly because there are many references to our work. The number of references is more than a thousand. This is a very large indicator for scientific works. Besides that, testimony to the fact that this work is recognized in the world is the fact that in 1994 I gave a talk at a very important conference where the results of scientific development for the past 50 years were summarized and where the speakers were the authors who had made fundamental contributions to the development both of theoretical ideas and in the acquisition of new experimental data.

Q. Where was this conference?

A. It was in Erice, in Italy. It was called the "International Conference on the History of Original Ideas and Basic Discoveries in Particle Physics." I was invited to speak on supergravity, and it was a great honor for me. And so I gave my talk at this conference [4].

Q. Dmitry Vasilyevich, who else was studying the same problems that you were working on at that time?

A. At that time, in fact there was one more work that was done at the same time as ours. This was the work by Yu. Golfand and E. Likhtman, who introduced formulations of the theory of supersymmetry, but for completely different reasons. They tried to explain on the basis of supersymmetry the breaking of the parity that exists in nature. Unfortunately, Golfand passed away in 1994. There was no further work. If one talks about supergravity, our first work was done in 1973. The next work in this field appeared in the West only in 1976.

Q. Dmitry Vasilyevich, could you please tell us about your scientific contacts with your colleagues?

A. At KhFTI, when I was starting my work, there was a group studying quantum electrodynamics. Later on, I worked practically alone. But I had very good contacts with physicists from Moscow and Leningrad. I would like in particular to stress the role of Isaac Yakovlevich Pomeranchuk, who was working at that time in Moscow.

Q. Did you know him personally?

A. I knew him personally. Every time I came to Moscow, I always visited Pomeranchuk. We had lengthy discussions together and it was interesting that we had the same way of thinking, because we usually discussed the conceptual part of work without going into formulas. We didn't even write formulas, but clarified what connections are possible between phenomena. I would like to tell about the following episode as an interesting example. It was in 1962 and I had been sent as a member of a delegation from the Soviet Union to CERN. I was lucky to share a hotel room with Isaac Yakovlevich Pomeranchuk. This was a stressful time for me, because just then I had finished some work together with V.N. Gribov about the conspiracy of Regge poles, which I spoke about earlier. This was the most important subject for me at that time. So, I remember lying on a hotel bed I couldn't fall asleep. It was already two o'clock in the morning and a certain idea came to me. I shook Isaac Yakovlevich awake, he awoke and said "What!" and I told him about this idea. After that, we discussed this idea for a couple of hours without turning on the light and we fell asleep only at dawn. This certainly characterizes a person, if he can discuss a topic that interests him at any time of day or night. Such a person was Isaac Yakovlevich Pomeranchuk. [...]

Now, about Kharkov. At first I worked in relative solitude. Then I managed to form a small group of young capable scientists. The personnel of my laboratory consist of about eight people, and I continue working with them.

References

1. D.V. Volkov and V.P. Akulov, *Phys. Lett.* **B46**, 109 (1973).
2. D.V. Volkov and V. Soroka, *JETP Lett.* **18**, 312 (1973).
3. W. Heisenberg, *Introduction to the Unified Field Theory of Elementary Particles* (Interscience Publishers, London 1966).
4. D.V. Volkov, *Supergravity Before 1976*, in *History of Original Ideas and Basic Discoveries in Particle Physics*, Eds. Harvey B. Newman and Thomas Ypsilantis (Plenum Press, New York, 1996) [hep-th/9410024].

NONLINEAR WAY TO SUPERSYMMETRY

VLADIMIR AKULOV
Baruch College, CUNY, 17 Lexington Ave, New York, NY 10010
akulov@gursey.baruch.cuny.edu

This year the physics community will celebrate the 30th anniversary of supersymmetry. I want to tell about the first steps in this direction made by our group headed by Professor Dmitri Volkov, in Kharkov.

Supersymmetry is an extremely interesting mathematical structure which arose in high energy physics in the early seventies independently in Moscow (Golfand and Likhtman), in Kharkov (Volkov and Akulov), and CERN (Wess and Zumino). Supersymmetry differs from previous symmetries in that it unifies (i) bosonic and fermionic particles as members of one supermultiplet, and (ii) the space-time Poincaré symmetries with the internal (isospin) symmetries. The first published mention of the super-Poincaré algebra in four dimensions was in 1970. In this year the International Rochester conference on High Energy Physics was held in Kiev, Ukraine. Prof. Yu. Golfand announced two reports for this conference, but the organizing committee gave him time for only one report, so he preferred to report on the vacuum in QED. The problem of the super-Poincaré algebra which had been already obtained by him at that time seemed too complicated to be presented at the conference. A collection of abstracts of all talks submitted to the Conference was published by the Institute of Theoretical Physics in Kiev, as a rotaprint edition (edited by L. Yenkovsky *et al.*). The first volume carried a picture of Taras Shevchenko on the cover, the second volume the picture of Bogdan Khmelnitsky. Golfand's abstract of the report on the super-Poincaré algebra was there. Yu. Golfand, together with his Ph.D. student E.P. Likhtman, found this algebra trying to unify the parity transformation with the Poincaré group. Unfortunately, now I cannot find this rotaprint edition of the collection of abstracts, while the final volume of the Proceedings did not contain any mention of those talks which had not been delivered.

The mathematical background for supersymmetry had been constructed even earlier: in 1970 Felix Berezin (Moscow) and Georgii Katz[81] (Kiev)

[81]Editor's note: Two famous Russian mathematicians have one and the same surname written in Russian as *Kats*. This surname is transliterated in English in two distinct ways: Georgii *Katz* (1924–1978; worked with Felix Berezin); Victor *Kac* (b. 1943) who is a co-author of the Kac-Moody algebra. –MS

published[82] a paper on groups with commuting and anticommuting parameters. In this way the supergroups appeared in mathematics.

In the 1950's and '60's Dmitri Volkov investigated the connection between spin and statistics and rediscovered (after M. Green, 1953) the parastatistics in 1959. Later he considered fermionic Regge trajectories. The success of the application of the Goldstone theorem to π meson physics stimulated the desire to generalize this theorem to include fermions. Another hint was linked with Heisenberg's (incorrect) idea that the neutrino is possibly a Goldstone particle connected with a (spontaneously broken) discrete symmetry, P parity.

I was a graduate student at this time, and Professor Volkov suggested this topic for my Ph.D. work in 1971.

Using the Berezin-Katz formalism we constructed an extension of the Poincaré group which included fermionic supercharges with anticommuting parameters and $\mathrm{U}(n)$ — the internal group, which is now called the extended Poincaré supergroup. The parameters of this supergroup form a superspace (introduced later by A. Salam).

At first we constructed the exponential representation of the super-Poincaré group, but this approach was very complicated. The last part of the paper by Berezin and Katz contained an example of the matrix realization of supergroup, based on the two-by-two Pauli matrices. We constructed a matrix representation for a supergroup using three-by-three and four-by-four matrices. Then, using the well-known representation for the Poincaré group as the upper triangular matrix, we constructed an extended Poincaré supergroup, which included translations of the spinor Grassmann coordinates and a unitary group of the internal symmetry. The coset space of this supergroup included the Grassmannian coordinates with the unitary index along with the conventional (four-dimensional) Minkowski space.

Based on the extended Poincaré supergroup, we constructed a nontrivial unification of space-time symmetries, like the Lorenz group or the Poincaré group, with the internal symmetries. In this way we bypassed the no-go theorem of Coleman and Mandula.

The above coset was later called *Superspace* by Salam and Strathdee. Our next step was the construction of the action, integral invariant under such group, that would describe a spontaneously broken supersymmetry.

---

[82] F. Berezin and G. Katz, *Mathematicheskiy Sbornik* **83**, 343 (1970) [*Math. USSR — Sbornik* **11**, 311 (1970)].

(Of course, we did not use the word *supersymmetry*, it was put in circulation later.) Thus, a general theory of the spontaneously broken supersymmetry emerged. Our action had a nonlinearly realized supersymmetry. It had the form of the Born-Infeld action in the absence of the gauge F field. As was noted by Renata Kallosh in 1997, from the modern perspective one can say that it was one of the first D3 branes.

During the summer of 1972 we finished this work, and sent the paper to *JETP Letters* and to *Physics Letters.* The latter was a great problem for our Institute, because all paper intended for publication abroad had to be cleared in Moscow. This would normally take 3 months or more. Only after a positive decision from Moscow could we send our papers abroad.

In the autumn of 1972 we attended the International Seminar in FIAN, Moscow, on the μ-e problem. Professor Volkov reported on the possible universal neutrino interaction. Efim Fradkin invited him to give a two-hour talk at the Theoretical Division of FIAN. At that time Professor Victor Isaakovich Ogievetsky told us about the paper by Yu. Golfand and E. Likhtman in *JETP Letters*, that had a similar algebra. Yu. Golfand (who worked at FIAN) was absent during Volkov's talk. Likhtman also did not attend it.

We returned to Kharkov and read the paper by Golfand and Likhtman. This paper contained the action of an Abelian gauge model with the linearly realized supersymmetry. We added the reference to their paper, and sent our detailed report to Moscow, to *Theoreticheskaya i Matematicheskaya Fizika.*[83] This detailed paper contained supergroups as matrix constructions, and the interaction of the Goldstone fermions with other fields.

After global supersymmetry D.V. Volkov considered a local version, and in 1973 (together with V.A. Soroka, because I finished my Ph.D. work then) constructed the first version of supergravity, which consisted of graviton with spin 2, gravitino with spin 3/2 and a Goldstone fermion with spin 1/2. The avalanche began only after the paper by J. Wess and B. Zumino in 1973, but this is another story.

[83] V.P. Akulov, D.V. Volkov, Eng. Transl.: Theor. Math. Phys. **18** (1974) 28.

FROM SYMMETRY TO SUPERSYMMETRY[84]

JULIUS WESS

Sektion Physik der Ludwig-Maximilians-Universität
Theresienstr. 31, D-80333 München
and
Max-Planck-Institut für Physik
(Werner-Heisenberg-Institut)
Föhringer Ring 6, D-80805 München

"... die Wirklichkeit zeigt sich in unseren Erlebnissen und Forschungen nie anders wie durch ein Glas, das teils den Blick durchläßt, teils den Hineinblickenden widerspiegelt."[85]

R. Musil "Der Mann ohne Eigenschaften",
Zweiter Teil, Kapitel "Das Sternbild der Geschwister oder Die Ungetrennten und Nichtvereinten"[86]

Thank you for the invitation and your kind introduction. It is a pleasure to be here, and this for many reasons. Let me mention one explicitly. This is the place where Golfand worked during his last years as a member of this faculty and I would like to dedicate this lecture to his memory. As you know, he was one of the founders of supersymmetry.

It is quite easy to speak about symmetries. Everybody has a notion of symmetry, it is a very deeply rooted and widespread concept, ranging from art to science. In some way or another symmetry is perceived by everybody. I think it is worth mentioning that about thirty years ago there was strong interest in experimenting with apes to see how much they were able to learn [1]. One objective was to see how apes would learn to paint. In one of

---

[84] This is a write-up of the first lecture in the 1999 Technion Distinguished Lecture Series, delivered on March 10, 1999. The draft of the write-up was prepared by Dmitry Gangardt from the tape recording. The lecture was first published in the Yuri Golfand Memorial Volume *The Many Faces of the Superworld* (World Scientific, Singapore, 2000), page 85.

[85] Reality is never shown in our experiences and research other than through a glass that partly lets the view through and partly reflects the person looking into it.

[86] R. Musil "The Man Without Qualities", Second part, chapter "The constellation of the siblings or The Unseparated and Ununited."

these experiments one dot was made at one side of a piece of paper and the ape would then try to make a dot on the other side to balance it symmetrically. That's exactly what we are doing in physics.

It is quite difficult to speak about symmetries. Everybody has his own concept of it, also physicists, and you never know if we communicate about the same thing.

Fortunately, mathematics with its strong capability to abstract has abstracted the concept of symmetries to the concept of groups. Group theory and the theory of representations of groups incorporate many of the different aspects of symmetries as we meet them in nature, in art, in science, etc. When referring to symmetry I mean it in the framework of group theory, representation theory, algebra and differential geometry.

In physics symmetries have been used all along. Group theoretical methods make it much simpler to get information about a system. A problem like the Kepler problem, for example, would be much harder to solve without using rotational symmetry.

Through symmetries you might get information about a system without really understanding the physical laws that govern it or without being able to solve the dynamical laws if you know them. You may think about the trivial example of a scale. Without knowing the theory of gravitation you are quite convinced that, based on the symmetry of a scale, the weight on the left hand side and on the right hand side of the scale is the same. This demonstrates how symmetries can help us find out our way through a system without really knowing its laws.

In addition symmetries have a very strong interplay with experimental physics via conservation laws. Conservation of energy, momentum and angular momentum can be measured experimentally. They are linked to an invariance under time translation, space translation and space rotation. Nœther's theorem states it very precisely: if we know that a system is invariant under some symmetry transformations, then we can show that there are corresponding conservation laws and we know how to find their explicit form. We know that there is a conservation law of the electric charge — we ask for a symmetry and find it in the phase transformation of the Schrödinger wave function. This shows that there is a very strong connection between abstract mathematical concepts and experimental facts. We are very lucky to have such an interplay in physics.

Symmetries in modern physics have taken an even stronger role to such an extent that the laws of modern physics cannot even be formulated without the concept of symmetries. To make the framework of local quantum field

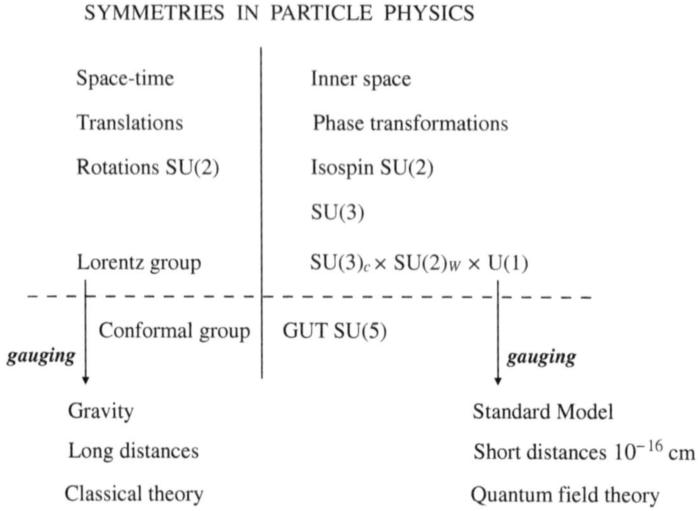

Figure 3.2: Symmetries in particle physics.

theory meaningful, symmetries have to be invoked from the very beginning. It is not that we know the laws and try to find their symmetries, but rather we have to implement the symmetries from the very beginning to be able to formulate these laws in a meaningful way.

I have prepared a table listing some of the fundamental symmetries (Fig. 3.2). It shows two separate columns: one, on the left, for space-time symmetries and one, on the right, for symmetries in an inner space.

At the top of the left column we have the rotation group. It is the group from which a physicist can learn what group theory is about. We know that the laws of electrostatics and magnetostatics are invariant under the rotation group and that Newton's laws allow rotational invariance as well. The rotation group is strongly related to $SU(2)$, the special unitary group in two dimensions, which is connected to the concept of spin. You know all this from quantum mechanics. When electrostatics and magnetostatics are combined to Maxwell's theory of electromagnetism we encounter the Lorentz group. Though electrostatics and magnetostatics describe forces that are different in strength by the order of magnitude of the velocity of light, they nevertheless are part of the same theory. We could say that the unification comes about by generalizing the rotation group to the Lorentz group. This is a good model and yields a good understanding of how theories can be unified by enlarging a group.

The right column shows symmetries in an inner space. I have mentioned the phase transformation of the wave function. Wave functions can span a higher-dimensional space and this is the space which we call inner space. The first step was done by Heisenberg. He had learned what $SU(2)$ is in the framework of the rotation group and spin and put this concept to work in the inner space as isospin. This was later generalized to $SU(3)$ successfully, which we now understand in terms of the quark model. A very successful model, the Standard Model, is based on a group $SU(3)_C$ for color, $SU(2)_W$ for weak interactions and $U(1)$. The Lorentz group together with this $SU(3)_C \times SU(2)_W \times U(1)$ is at the basis of the Standard Model. This model describes physics very well as we know it in our laboratories, down to a scale of 10^{-16} cm.

It is natural to ask if the $SU(3)_C \times SU(2)_W \times U(1)$ symmetry is not part of a larger group, $SU(5)$, for instance. This then gives rise to a theory where strong, weak and electromagnetic interactions would be truly unified. Such a theory we call GUT — for Grand Unified Theory — but what Nature knows about this, we do not yet know. It is the Standard Model based on the Lorentz group and the group $SU(3)_C \times SU(2)_W \times U(1)$ that is supported and by now well tested by experiment.

Looking at the two columns it seems that Nature (or we) has used the same concept of symmetry twice. But has Nature chosen it separately? Has it done the same thing twice? It is natural to ask for a bridge between the two columns.

We have learned from the Maxwell equations that one can make a transformation parameter space-time dependent. This is the property of the gauge transformation in Maxwell theory. We can identify this parameter in Maxwell theory with the parameter of the phase transformation of the Schrödinger wave function and build a gauge theory, as we call it today. Thus the idea of a gauge theory is born. You ask that the theory should be invariant under a group that acts in inner space and has space-time dependent parameters. This is the concept on which the Standard Model is built. The Standard Model is the gauge theory for the group $SU(3)_C \times SU(2)_W \times U(1)$. If the same idea of gauging is used for the space-time symmetry, the Lorentz group, one arrives at Einstein's theory of gravitation.

We know that the Standard Model can be interpreted — via the concept of renormalization — as a Quantum Field Theory and, as such, it is experimentally extremely successful at short distances, the 10^{-16} cm I have mentioned before. We would have liked to find deviations from the

Standard Model experimentally to get a hint where to go next: $SU(3)_C \times SU(2)_W \times U(1)$ seems to be the simplest choice one can make, and it works, and it works and all the experiments verify it again and again. Is Nature not more sophisticated? Doesn't it know about GUT?

As for Einstein's gravitational theory we know that it is a very good theory for long distances. We have no reason to doubt its validity there. Our understanding of space-time at large distances is based on this theory.

We have the situation that the Standard Model and Einstein's theory of gravitation describe the data observed in the laboratory as well as in astronomy and astrophysics very well. Even a cosmology based on these two theories is very reasonable.

Looking at our columns we now have an even more puzzling situation. The same concept of symmetries and gauging them gives, in the left column, a very good classical theory of gravity, defending itself against quantization by an abundant number of singularities. There seems to be a deep conflict between the classical theory of gravitation and quantum field theory. On the other hand the same idea about symmetry and gauging in the right column leads to a very good model of a renormalizable quantum field theory which is mathematically and experimentally successful.

Is gauging all there is or is here a deeper connection between space-time and inner space symmetries? In nuclear physics, a long time ago, Wigner and Hund [2] proposed a group $SU(4)$ that has $SU(2)$ of spin and $SU(2)$ of isospin as a subgroup. This way they unified space-time and inner space and got a good classification of nuclear levels. At the time of $SU(3)$ in particle physics, this idea was generalized to $SU(6)$, incorporating $SU(2)$ of spin and $SU(3)$ of inner space. Reasonable experimental predictions about masses of particles with different spin — spin 0 and spin 1 as well as spin 1/2 and spin 3/2 — can be based on $SU(6)$. This is now better understood on the basis of the quark model. As the quark model was not yet known at the time, many attempts were made to extend the $SU(6)$ model to incorporate the Lorentz group. But this proved to be impossible. It was impossible to build a Lorentz invariant model that at low energies would have $SU(4)$ or $SU(6)$ as a symmetry. If a lot of physicists try something and it does not work, then it might be clever to try to prove that it cannot work. And this is what was done. It started with work by O'Raifeartaigh [3] and came to a very elegant formulation which now is known as the Coleman-Mandula "no-go" theorem [4]. This theorem tells us that for a theory in four dimensions, with the Lorentz group as a symmetry group, and satisfying a certain number of axioms, which I am going to tell you about in a minute, the only possibility for a symmetry group is the direct product of the Lorentz group with some compact inner group. Surprisingly enough, we have a theorem that separates the two columns on the basis of very fundamental axioms.

Now some remarks on the axioms. Apart from the Lorentz invariance, the axioms state that it should be a theory based on quantum mechanics and that it should be local, i.e. it should be a local quantum field theory. Local here means local in the microscopic sense. Two measurements that are separated by a space-like interval cannot influence each other, no matter how small the distance would be. The locality of the theory is based on the locality of the fields. In addition we assume that there is a unique lowest energy state, a ground state which we call vacuum, and that all the other states have larger energy. Probability is supposed to be conserved in the quantum mechanical sense. Finally we assume that there is only a finite number of different particles. These seem altogether very reasonable assumptions, but they imply that you cannot combine the two columns, as we tried by postulating a $SU(4)$ or a $SU(6)$ symmetry.

There is one problem with this set of axioms. With the exception of the free field theory we do not know a local quantum field theory the existence of which we can rigorously prove and which satisfies all the axioms. We have invented very powerful methods of perturbation theory, of separating infinities, going through a renormalization scheme. We are able to extract in this way information that we can test experimentally. In some way it is a kind of art, but it works beautifully. On this basis we understand the models I have been talking about and we make them successful. This is, however, a formulation of a model a mathematician would not like to accept as a theory. But the success in comparing it with experimental data is so strong that we cannot dismiss this type of theories.

Now I would like to make a point. If we try to build a theory on the setting just discussed and we want to have spin-zero, spin-one and spin-one-half fields, then we start from a free field theory and try to arrange the multiplets and couplings such that there should be only a finite number of divergences in perturbation theory. We start with the tree level, study the energy-momentum dependence of the Feynman diagrams and arrange the model in such a way that this dependence is as smooth as possible in order to facilitate integration over energy and momentum variables at the one loop level. The result is a gauge theory with spontaneous symmetry breaking. Thus, without knowing about group theory and symmetries the whole concept of gauge theories could be invented by a good physicist on dynamical grounds. This was shown about twenty years ago by John Cornwall *et al.*,

Llewellyn Smith, and Steve Weinberg [5]. They pioneered this approach. After you know this result it is much easier to take a textbook on group theory, formulate your model based on your knowledge of group theory, and to gauge the symmetry group. In this way you obtain a model and you will find it to be a renormalizable quantum field theory. This can be proved from gauge invariance. Nœther's theorem and the conserved currents are at the heart of this proof. The current has to be a well-defined object and it helps to relate infinite and, therefore, undefined constants such that there is only a finite number left at the end. This defines the renormalization scheme. This is what I meant in the beginning when I said that even in formulating a physical theory we have to use the concept of symmetry. But this also raises the question whether symmetries are very basic or whether they can be derived from the dynamics of the system — based on some reasonable axioms.

Our way of thinking is very much influenced by the deep and widespread notion of symmetry to an extent that if we find a beautiful symmetry in a dynamical system we say that we understand this system. If there is a deviation from the symmetry we say: there must be something going on which we do not understand.

There is another good reason to start from symmetries instead of dynamical requirements. The second way would be difficult and, let me say, ugly. Mathematicians have not developed concepts along this ill-defined line. Thus, we have no mathematical machinery that we can use, contrary to the first approach, where we have all this beautiful mathematics.

There is a surprising way of combining space-time symmetries with internal symmetries, and this is by *supersymmetry.* This is achieved by generalizing the concept of symmetry. As you know, symmetries can be formulated in terms of commutator relations, as the commutator relations of angular momentum in quantum mechanics. A large class of groups — the Lie groups — are related to Lie algebras that are defined by commutator relations. All the symmetries we have mentioned are of this type.

From Dirac we know that not only commutators but also anticommutators are a very useful concept, especially when we deal with particles with half-integer spin.

The idea is to generalize the concept of symmetries to a structure which is formulated in terms of commutators and anticommutators as well. This is not a new idea in mathematics, such structures — graded Lie algebras — have been thoroughly investigated e.g. by Berezin [6]. But can such a concept be realized in a quantum field theory? The answer is yes. There is a

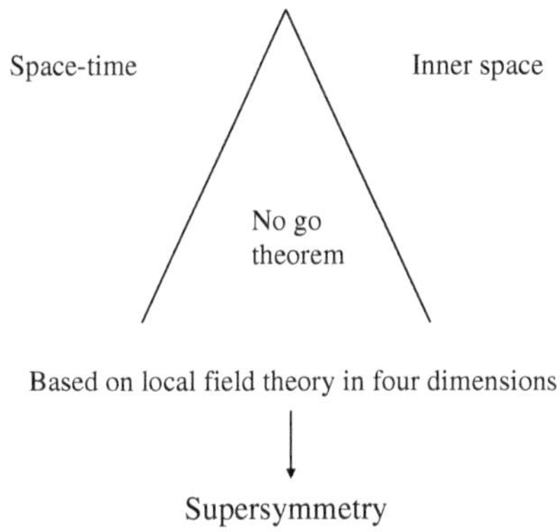

Figure 3.3: Superextension of the Poincaré algebra.

quite unique symmetry and its uniqueness is based on the no-go theorem of Coleman and Mandula (Fig. 3.3). This was shown by Haag, Łopuszański and Sohnius [7]. Here the theorem does not tell you that it does not go, it tells you that it goes in a very unique way expressed by the formula:

$$\{\bar{Q}^N{}_\alpha, Q^M{}_\beta\}_+ = 2\gamma^\nu{}_{\alpha\beta} P_\nu \delta^{NM}. \tag{3.1}$$

The charges ${Q_\alpha}^N$ are spinorial charges, α is a spinor index and ${Q_\alpha}^N$ is a Majorana spinor, N is a free index. If N goes from one to two we speak of $\mathcal{N} = 2$ supersymmetry, if it goes from one to four we speak of $\mathcal{N} = 4$ supersymmetry. The four vector P^ν is the energy-momentum four-vector that generates the translations in space-time. The structure constants of this algebra are the Dirac γ matrices and the Kronecker δ-symbol. This algebra can be combined with the algebra of the Lorentz group, then it is the algebra of supersymmetry.

These spinorial charges can be realized in terms of local fields, and the algebraic relations of supersymmetry hold on the basis of the canonical commutation relations of the fields. The charges are related to currents, which are spin-3/2 currents, in the same way as the energy-momentum vector is related to the energy-momentum density tensor which is a spin-2 object. If the theory is gauged, the currents are the sources of fields; for the electric current this is the photon, with spin one, for the energy momentum tensor it is the graviton, with spin two, and for the supercurrent it is the gravitino with spin 3/2.

Supersymmetric theories have a very encouraging property. As quantum field theories they are less divergent as they would be without supersymmetry.

Supersymmetric theories have a very discouraging property. It follows directly from the algebra that a supersymmetric theory has to have an equal number of bosonic and fermionic degrees of freedom degenerated in mass. This is not how Nature is. We have fermions and bosons but not in the same multiplet structure and in no way degenerated in mass.

The two properties lead to frustration, even more so as they are not independent. Let me explain this with a simple example. Take a bosonic and a fermionic harmonic oscillator, both with the same frequency. The zero point energy will have opposite sign and adding them leads to the cancelation of the zero point energy. A field can be viewed as an infinite sum of harmonic oscillators. The zero point energy will add up to an infinite vacuum energy, except in a theory where the bosonic and fermionic contributions cancel. This was already known to Pauli but he also knew that the world is not like this. Supersymmetry relates that cancelation to an algebraic structure of the theory, and you might be led to believe that it is now based on a deeper property of the theory and nature. Since the days of Pauli we have learned to deal with symmetries that are spontaneously broken. The field-theoretic properties of such theories with spontaneously broken symmetries are maintained, but at a phenomenological level at low energies the symmetry appears to be broken by sizable effects.

The same mechanism that leads to the cancelation of the vacuum energy leads to many other cancelations of divergences. These improved renormalization properties of the theory can be traced back to the cancelation properties of diagrams with bosonic or fermionic internal lines. It can be shown that there are parameters in the theory which do not get any radiative correction, not even a finite one. Even if not a miracle, this is a sensation in quantum field theory. You can introduce parameters like certain masses or couplings that are not changed by radiative corrections at all. Naturally, this has consequences for particle phenomenology based on supersymmetric theories.

Let me first discuss the fact that supersymmetric theories have an equal number of bosons and fermions.

In Nature we know quarks and leptons to be fermions. These are the particles that constitute matter. Each of these fermions has to have two bosonic partners as each spin-one-half particle has two degrees of freedom. We do not know such partners in Nature but we can give them names. The SUSY partners of the quarks we call squarks and the SUSY partners of the leptons we call sleptons. The “s” stands for scalar of supersymmetric partner.

In Nature we know the photon, the vector particles of weak interaction, the gluons, the graviton and the Higgs to be bosons. These are the particles that constitute the forces. Each of these bosons has to have a fermionic partner. We give them names: photino, wino, zino, gluino, gravitino and Higgsino.

With these particles we can build models and consider it a success that we already know half of the particles in such models. But we know also their couplings, which are entirely determined by the couplings of the particles we know. Take a Feynman diagram with the known particles (Fig. 3.4). A line in such a Feynman diagram that either goes from an incoming particle to an outgoing particle or that forms a closed loop can and has to appear in an equivalent diagram where it is replaced by a line that is associated with SUSY partners. Doing this for all the lines you obtain all the diagrams of a certain supersymmetric theory. The coupling constants on the respective vertices are the same as in the theory you started from.

In a truely supersymmetric theory the masses would also be the same. Not so if the symmetry is spontaneously broken. We know that for spontaneously broken gauge symmetries the mass difference within a multiplet might be quite big — like the mass difference between the photon and the W or Z.

It is possible to break supersymmetry as well in such a way that its renormalization properties remain valid, the mass difference between SUSY partners can then be quite big — we call it the SUSY gap.

The radiative corrections, however, will not completely cancel — we can expect finite contributions which then will also be of the order of the SUSY gap. Among the masses and coupling constants of a gauge theory that will have this property are the Higgs masses and the Higgs couplings.

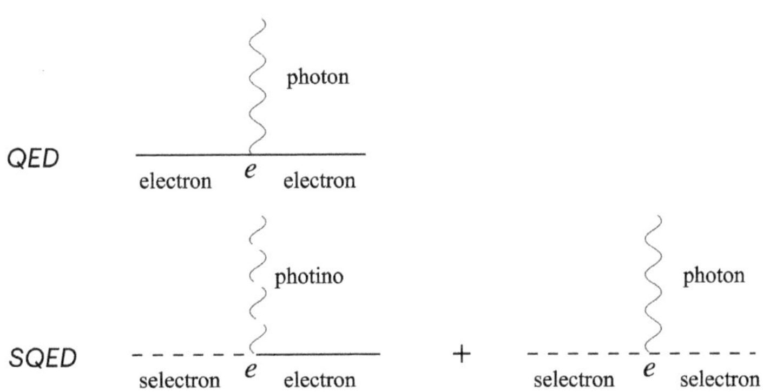

Figure 3.4: Supersymmetric vertices in QED.

Knowing this, one can relate this property to another phenomenon in particle theory. Assume that there is a GUT theory. The unifying gauge group has to be spontaneously broken to render the Standard Model as we know it. The symmetry group $SU(3)_C \times SU(2)_W \times U(1)$ of the standard model is then spontaneously broken to the theory of electromagnetism and weak interactions as we know it at low energies.

This breaking of the symmetries is triggered in a quantum field theoretical model by parameters, the Higgs masses and Higgs couplings, which have to be renormalized in a non-supersymmetric theory by an infinite amount. It is very difficult to understand that the breaking scheme in the two sectors of the breaking is stable and respects the scale. This is the hierarchy problem in particle physics. If, however, the theory is a spontaneously broken supersymmetric theory, then the relevant parameters obtain only radiative corrections of the order of the SUSY gap. If the SUSY gap is of the order of the electroweak breaking scale characterized by the W mass then we would understand the stability of the breaking of the symmetries in a GUT theory. In such a scenario the SUSY gap has to be of the order of one TeV — an order which will be accessible to experiments soon.

My personal belief is that it would be a waste having such a beautiful symmetry as supersymmetry just to solve the hierarchy problem. Supersymmetry might play a much more fundamental role at higher energies and solving the hierarchy problem would just be one of the lower energy remains of supersymmetry.

But let me stress that there are two inputs in this prediction — SUSY and GUT. In any given combination they could be right or wrong.

Now back to the theme mentioned already in the context of gauge theories. We see that supersymmetry has a very strong influence on the dynamical behavior of quantum field theory when it is treated in the framework of renormalization theory. We can start from this framework and ask for a model, e.g. with spin 0 and spin 1/2 particles, that behaves as smoothly as possible not only at the tree level but on the one-loop level as well. We could ask for the cancellation of the infinities in the vacuum energy, in this way we would establish an equal number of fermionic and bosonic degrees of freedom. If we then asked for the non-renormalization properties about which I have been talking before we would construct a supersymmetric theory — with what we call possible soft breaking, a symmetry breaking that does not affect the renormalization behavior. If we asked for the absence of any radiative correction we would obtain a strictly supersymmetric theory.

So again it would be a dynamical concept, now put forward to the level of one-loop diagrams which would have led to the invention of supersymmetry, without knowing about any algebra, about groups or graded groups, just being a good physicist knowing how to handle the Feynman diagrams. Somehow supersymmetry is the next logical step after gauge theory in the framework of renormalizable quantum field theories. You go from the tree level corresponding to a classical theory to the quantized level represented by one loop, apply the same idea once more, and you arrive at supersymmetry. What we do not know is if Nature knows about this way of thinking or if Nature has a different logic, does things differently.

Now to the history of supersymmetry. It started with the work of Golfand and Likhtman [8]. They thought about adding spinorial generators to the Poincaré algebra, in that way enlarging the algebra. This was about 1970 and they were really on the track of supersymmetry. I will come back to this idea at the end of this talk as I think that this is the right question: can we enlarge the algebra, the concept of symmetry, by new algebraic concepts in order to get new types of symmetries?

Then in 1972 there was a paper by Volkov and Akulov [9] which argued along the following line. We know that with spontaneously broken symmetries there are Goldstone particles supposed to be massless. In Nature we know spin one half particles that have, if any, a very small mass, these are the neutrinos. Could these fermions be Goldstone particles of a broken symmetry? Volkov and Akulov constructed a Lagrangian, a non-linear one, that turned out to be supersymmetric. Of course today we know from Haag, Łopuszański and Sohnius that it had to be supersymmetric. But being nonlinear, just as the nonlinear sigma model, the Lagrangian is highly non-renormalizable and does not show any sign of the renormalization properties which we now find so useful and intriguing in supersymmetry.

Another path to supersymmetry came from two-dimensional dual models. Neveu and Schwarz *et al.* [10] had constructed models which had spinorial currents related to supergauge transformations that transform scalar fields into spinor fields. The algebra of the transformation, however, only closed on mass shell. The spinodal currents were called supercurrents and that is where the name "supersymmetry" comes from.

In 1973 Bruno Zumino and I published a paper [11] where we established supersymmetry in four dimensions, constructed renormalizable Lagrangians and exhibited non-renormalization properties at the one-loop level. Our starting points were the supercurrent and the strong belief in Nœther's theorem.

Another paper by another author that could have led to supersymmetry based on the non-renormalization properties of perturbative quantum field theories was never written.

With supersymmetry it is very natural to extend the concept of space-time to the concept of superspace. Energy-momentum generates translations in four-dimensional space-time, so it is natural to have the anticommuting charges generate some translations in an anticommuting space. This new space together with the four-dimensional Minkowski space is called superspace.

Fields will now be functions of the superspace variables and they will incorporate SUSY multiples in a very natural way. This idea was pioneered by Salam and Strathdee [12]. Lagrangians can be formulated very elegantly in terms of superfields and the non-renormalization theorems find very elegant formulations as well, as first shown by Fujikawa and Lang [13].

The spin 0, spin 1/2 multiplets find themselves in the so-called chiral superfields (ϕ) and spin 1, spin 1/2 in the so-called vector superfields (V) (see Table 2).

Any diagram that has only external chiral fields (but no conjugate chiral field ϕ^* or vector field) will not be renormalized (Fig. 3.5). As no external field is in the class of chiral fields only we find the non-renormalization theorem for the vacuum energy. A short look at the superfield Lagrangian

$$\mathcal{L} \sim (\phi^* e^{gV} \phi)_D + (m\phi^2 + \lambda\phi^3)_F \tag{3.2}$$

tells us that the mass terms of the chiral fields are of the form ϕ^2 and the coupling terms of the Higgs type are of the form ϕ^3. These couplings do not get renormalized. The kinetic term of the chiral field is of the form $\phi^*\phi$, there is a wave function renormalization.

Table 2:

spin:	0	1/2	1	3/2	2	
	2	1				ϕ, scalar multiplet
		1	1			V, vector multiplet
				1	1	gravitational multiplet

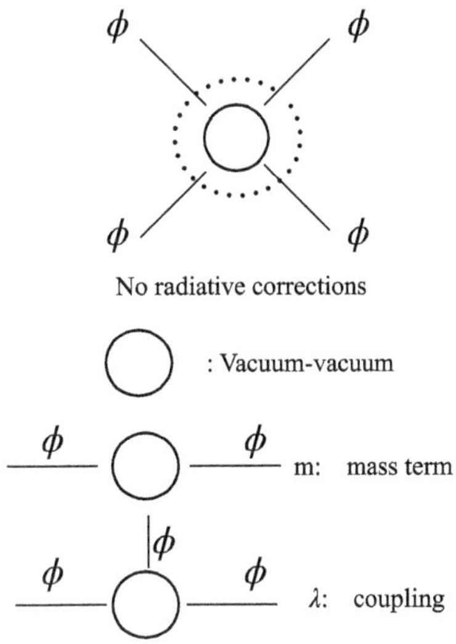

Figure 3.5: Non-renormalization.

The superspace variables also play an important role if we want to gauge an internal symmetry of a supersymmetric theory. Gauging an internal symmetry means formulating a theory that is invariant under transformations with space-time dependent parameters. Space-time by itself is not a supersymmetric concept, we have to replace it by superspace. Gauging an internal symmetry in a supersymmetric theory means to formulate a theory that is invariant under transformations with superspace dependent parameters. This way we know how to supersymmetrize all the known gauge theories.

Gauging supersymmetry finds a natural formulation in superspace. Einstein's theory of gravity can be formulated as a geometrical theory in four-dimensional space-time. Supergravity, the theory that has supersymmetry gauged, finds its formulation as a geometrical theory in superspace. Supergravity incorporates Einstein's theory because supersymmetry incorporates the Lorentz group. It improves the renormalization properties of the usual gravity theory, however, it does not make it a fully renormalizable theory. But it is closer to a dream of gravity as a part of renormalizable quantum field theory as well.

All our interplay with symmetries has centered around the renormalization problem of perturbative quantum field theory. The singularities that have to be renormalized are consequences of the unsatisfying short distance behavior of quantum field theories. Symmetries to some extent improve the situation. But is this really the way how Nature solves the problem of divergences at short distances?

Another possibility is to loosen some of the axioms. But this has to be done in a very controlled way, first so as not to get into conflict with experimental facts, and secondly in order not to have the rules of the theoretical game too wide open and to turn theory into a book-keeping device.

Up till now the only way that meets these requirements is string theory. The concept of a point has changed to the concept of a string — the axiom of locality has been loosened and there are infinitely many particles, the excitations of the string in the theory.

At low energies the string picture might be compatible with our present experimental knowledge summarized in the Standard Model and the Einstein theory of gravity. The theoretical framework of strings is again based on symmetries and differential geometry. Analyticity plays an important role as well. In string theory our concept of space-time is based on differential manifolds — the closest relatives to flat space, however curved as they might be.

In the spirit of Golfand we could ask for a change in the algebraic structure of quantum field theory. We have based it on the canonical structure of quantum mechanics and adapted it to the algebraic possibilities of differentiable manifolds. Non-commutative differential geometry might give a mathematical frame that goes beyond differential manifolds. Attempts in this direction show that this could lead to a lattization of space-time at short distances. This opens possibilities worthwhile exploring. Maybe Golfand was at the track of another theoretically successful development again.

Acknowledgments

This article is based on a talk given at Technion, Haifa, in March 1999. I would like to thank Dmitry Gangardt who wrote the first draft on the basis of a tape recording. Especially I would like to thank Technion and Michael Marinov for their cordial hospitality.

References

1. D. Morris, *The Biology of Art* (Methuen & Co. Ltd, London, 1962).
2. E. Wigner, *Phys. Rev.* **51**, 106–119 (1936); F. Hund, *Z. Phys.* **105**, 202–228 (1937).
3. L. O'Raifeartaigh, *Nucl. Phys.* **B96**, 331 (1975).
4. S. Coleman and J. Mandula, *Phys. Rev.* **159**, 1251 (1967).
5. J.M. Cornwall, D.N. Levin and G. Tiktopoulos, *Phys. Rev.* **D10**, 1145–1167 (1974); C.H. Llewellyn Smith, *Phys. Lett.* **B46**, 233 (1973); S. Weinberg, *Phys. Rev.* **D7**, 2887–2910 (1973).

6. F.A. Berezin, *The Method of Second Quantization* (Academic Press, New York, 1966); F.A. Berezin and G.I. Katz, *Lie groups with commuting and anti-commuting parameters*, in *Mat. Sbornik* (Engl. Transl.: *Math. USSR-Sb.*) **82**, 343 (1970).
7. R. Haag, J. Łopuszański and M. Sohnius, *Nucl. Phys.* **B88**, 257 (1975).
8. Yu.A. Golfand and E.P. Likhtman, *JETP Lett.* **13**, 323 (1971).
9. D.V. Volkov and V.P. Akulov, *JETP Lett.* **16**, 438 (1972).
10. A. Neveu, and J.H. Schwartz, *Nucl. Phys.* **B31**, 86 (1971).
11. J. Wess and B. Zumino, *Nucl. Phys.* **B70**, 39 (1974).
12. A. Salam and J. Strathdee, *Nucl. Phys.* **B76**, 477 (1974).
13. K. Fujikawa and W. Lang, *Nucl. Phys.* **B88**, 61 (1975).

FERMI-BOSE SUPERSYMMETRY[87] (SUPERGAUGE SYMMETRY IN FOUR DIMENSIONS)

BRUNO ZUMINO

CERN

1 Introduction

Fermi-Bose supersymmetry was introduced by Wess and the author [1]. It connects Bosons with Fermions. Its existence was suggested by dual models [2] (when formulated as two-dimensional field theories) and the name supergauge symmetry in four dimensions seemed a natural choice. The supergauge algebra having only a finite number of generators in four dimensions, it seems now reasonable to avoid the word gauge and adopt the expression Fermi-Bose supersymmetry, or simply supersymmetry, suggested recently by Salam and Strathdee.

The supersymmetry algebra is very simple. Let Q_i be a constant Majorana spinor (we use the Majorana representation where the y matrices are real; then Q_i is a hermitian spinor). The algebra is

$$\{Q_i, \bar{Q}_j\} = -2(\gamma^\mu)_{ij} P_\mu, \quad \bar{Q} = \tilde{Q}\gamma^0; \quad [P_\mu, Q_i] = [P_\mu, P_\lambda] = 0$$

where $P^\mu = (H, \vec{P})$ is the energy-momentum. In a supersymmetric theory the spinor charges are

$$Q_i = \int J_i^0 \, \partial^3 x \tag{3.3}$$

where the vector-spinor current is conserved

$$\delta_\mu J_i^\mu = 0. \tag{3.4}$$

Lorentz transformations and parity transform Q_i as a spinor and P^μ as a vector. The algebra (3.3) is not a Lie algebra since it contains both commutators and anti- commutators. If one introduces Majorana spinors α_1, α_2, etc

[87]Talk at 17th International Conference on High-Energy Physics, July 1–10, 1974, London, UK. Published in Proc. ICHEP '74, J. R. Smith and Geoffrey Manning (Eds.), Rutherford Appleton Laboratory. Chilton, 1974, pp. I.254–260.

(they commute with tensors and anticommute with spinors and among themselves), the relation (3.3) can be written as a commutation relation

$$[\bar{\alpha}_1 Q, \bar{\alpha}_1 Q] = -2\tilde{\alpha}_1 \gamma^\mu \alpha_2 P_\mu. \tag{3.5}$$

These objects have been studied in mathematics [3]. From [1] one finds an expression for the total Hamiltonian in terms of the spinor charges

$$H = \frac{1}{4}\sum_{i=1}^{4} Q_i^2 \tag{3.6}$$

valid, in presence of interaction for any supersymmetric theory, and similar expressions for the momentum.

Supersymmetry is not an ordinary Lie algebra and avoids no-go theorems [4] of relativistic SU(6): there exist non trivial (and renormalizable) Lagrangian theories which are exactly invariant under supersymmetry.

Our motivation was to show the feasibility of supermultiplets containing interacting particles with both integral and half integral spin. From a rather different point of view the same algebra (3.3) was considered independently by Volkov and Akulov [5]. They gave a non-linear realization in terms of a single spinor field (see Section 4) and suggested that it may be relevant as a description of the properties of the neutrino. Their Lagrangian is non-renormalizable. In reference [1] an algebra larger than (3.3) was described, which contains also Lorentz transformations, dilatations, conformal [6] and chiral transformations.

That larger algebra was later abandoned [7], to avoid the problems with scale and conformal anomalies.

2 Supermultiplets and Lagrangians

The simplest supermultiplet consists of a scalar field A, a pseudoscalar field B, a Majorana spinor, and two auxiliary fields F and G. Writing

$$\delta A = [\bar{\alpha} Q, A], \text{etc.} \tag{3.7}$$

for an infinitesimal supertransformation, one has

$$\begin{aligned}
&\delta A = i\bar{\alpha}\psi, \quad \delta B = i\bar{\alpha}\gamma\psi; \\
&\delta\psi = \delta_\mu(A - \gamma_5 B)\gamma^\mu \alpha + (F + \gamma_5 G)\alpha, \\
&\delta F = i\bar{\alpha}\gamma^\mu \delta_\mu \psi; \quad \delta G = i\bar{\alpha}\gamma_5\gamma^\mu \delta_\mu \psi.
\end{aligned}$$

The commutator of two supertransformations is a translation. For instance,

$$[\delta_2, \delta_1] = 2i\bar{\alpha}_1\gamma^\mu\alpha_2\delta_\mu A. \tag{3.8}$$

With this supermultiplet Wess and the author [7] constructed the first non trivial supersymmetric model, with Lagrangian

$$L = -\frac{1}{2}\left[(\delta_\mu A)^2 + (\delta_\mu B)^2 + i\bar{\psi}\gamma^\mu\delta_\mu\psi - F^2 - G^2\right] + m\left(FA + GB - \frac{i}{2}\bar{\psi}\psi\right)$$
$$+ g\left(FA^2 - FB^2 + 2GAB - i\bar{\psi}\psi A + i\bar{\psi}\gamma_5\psi B\right). \tag{3.9}$$

The various terms of the Lagrangian each change by a divergence under (3.8): the action integral is invariant. The auxiliary fields F and G can be eliminated by using their own equations of motion and the Lagrangian takes the more familiar form

$$L = -\frac{1}{2}[(\delta_\mu A)^2 + (\delta_\mu B)^2 + i\bar{\psi}\gamma^\mu\delta_\mu\psi + m^2A^2 + m^2B^2 + im\bar{\psi}\psi]$$
$$- gmA(A^2 + B^2) - \frac{g^2}{2}(A^2 + B^2)^2 - ig\bar{\psi}(A - \gamma_5 B)\psi. \tag{3.10}$$

As a consequence of supersymmetry all masses are equal and all couplings are expressed in terms of the single coupling constant g. One verifies that the supercurrent

$$J^\mu = \gamma^\lambda\delta_\lambda(A - \gamma_5 B)\gamma^\mu\psi - (F + \gamma_5 G)\gamma^\mu\psi \tag{3.11}$$

is conserved as a consequence of the equations of motion.

This model has been studied in great detail by Ferrara, Iliopoulos and the author [8]. It can be regularized in a supersymmetric way by introducing higher order derivatives in the kinetic term of the Lagrangian. The Ward identities corresponding to the conservation of the supercurrent (3.11) can be written and used to prove that renormalization does not spoil the relations among masses and coupling constants due to supersymmetry. The model is less divergent than the generic theory with the same kind of couplings. Only one single wave function renormalization constant Z common to all fields is required. The renormalized mass and coupling constant are

$$m_r = Zm_0, \quad g_r = Z^{3/2}g_0. \tag{3.12}$$

The Callan-Symanzik functions β and γ are proportional to each other. As a consequence one can argue that β cannot vanish except at the origin and that the effective coupling increases indefinitely with energy.

Supersymmetry can be broken softly, by adding to the Lagrangian a term proportional to A. Just as in the analogous case of the σ-model, the renormalization program can still be carried out. The masses of the various fields of the multiplet are no longer equal. Instead in the tree approximation

$$m_A^2 + m_B^2 = 2m_\psi^2. \tag{3.13}$$

In higher orders this is corrected by finite terms.

The fact that the supersymmetric model is less divergent than the generic theory of its kind, leads one to ask whether a supersymmetric theory might not be renormalizable even if it does not appear to be so by simple power counting. To answer this question, Laing and Wess [9] have replaced, in the Lagrangian (3.9), the renormalizable interaction proportional to g with the interaction

$$f\left\{FA^3 - GB^3 + 2GA^2B - 3FAB^2 - \frac{3}{2}i(A^2 - B^2)\bar{\psi}\psi + 3iAB\bar{\psi}\gamma_5\psi\right\}.$$

This interaction is supersymmetric but non-renormalizable by power counting. While this question is being investigated, it appears that, in this particular model, supersymmetry, in spite of the compensation of some divergences, is not sufficient to render the theory renormalizable.

3 Gauge Invariance and Supersymmetry

The existence of Lagrangian theories which are both gauge invariant and supersymmetric was first shown by Wess and the author [10]. They use a supermultiplet consisting of a vector field v_μ, a Majorana spinor λ and an auxiliary field D, transforming as

$$\delta v_\mu = i\bar{\alpha}\gamma_\mu\lambda, \quad \delta\lambda = -\frac{1}{4}v_{\mu\nu}\gamma^\mu\gamma^\nu\alpha + D\gamma_5\alpha,$$
$$\delta D = i\bar{\alpha}\gamma_5\gamma^\mu\delta_\mu\lambda, \quad v_{\mu\nu} = \delta_\mu v_\nu - \delta_\nu v_\mu$$

under a supersymmetry transformation. Using real fields the Lagrangian can be written as

$$\begin{aligned} L = &-\frac{1}{4}v_{\mu\nu}^2 - \frac{i}{2}\bar{\lambda}\gamma^\mu\delta_\mu\lambda + \frac{1}{2}D^2 - \frac{1}{2}\sum_{i=1}^{2}\Big[(\delta_\mu A_i)^2 + (\delta_\mu B_i)^2 \\ &+ i\bar{\psi}\gamma^\mu\delta_\mu\psi + m^2(A_i^2 + B_i^2) + im\bar{\psi}_i\psi\Big] + g[D(A_1B_2 - A_2B_1) \\ &- v^\mu(A_1\delta_\mu A_2 - A_2\delta_\mu A_1 + B_1\delta_\mu B_2 - B_2\delta_\mu B_1 - i\bar{\psi}_1\gamma_\mu\psi) \\ &- i\bar{\lambda}\{(A_1 + \gamma_5 B_1)\psi_2 - (A_2 + \gamma_5 B_2)\psi_1\}] - \frac{g^2}{2}v_\mu^2(A_1^2 + A_2^2 + B_1^2 + B_2^2) \end{aligned} \tag{3.14}$$

where the fields F and G ($i = 1, 2$) have already been eliminated. The gauge transformation rotates the fields with the subscripts 1 and 2 into each other and changes v_μ by a four-gradient. The fields λ and D are gauge invariant. The Lagrangian (3.14) is a sort of supersymmetric extension of quantum electrodynamics. All couplings are expressed in terms of the single coupling constant g, as a consequence of super- symmetry of the Lagrangian which has been shown to be renormalizable in the one-loop approximation in a manner consistent with gauge invariance and supersymmetry. A preliminary investigation of higher orders supports this conclusion.

4 Spontaneous Symmetry Breaking

The model described in the previous section can give an example of spontaneous breaking of supersymmetry, with the corresponding emergence of a "Goldstone" spinor. Fayet and Iliopoulos [11] have added to the Lagrangian (3.14) a parity violating, gauge and supersymmetry invariant term $\xi D/g$. Upon elimination of D this gives a term

$$-\xi(A_1 B_2 - A_2 B_1). \tag{3.15}$$

Introducing the new fields

$$\begin{aligned} a_1 &= \frac{1}{\sqrt{2}}(A_1 - B_2), \quad a_2 = \frac{1}{\sqrt{2}}(B_1 + A_2), \\ b_1 &= \frac{1}{\sqrt{2}}(A_1 + B_2), \quad b_2 = \frac{1}{\sqrt{2}}(-B_1 + A_2) \end{aligned} \tag{3.16}$$

this results in the potential (tree approximation)

$$\begin{aligned} &\frac{1}{2}(m^2 - \xi)(a_1^2 + a_2^2) + \frac{1}{2}(m^2 + \xi)(b_1^2 + b_2^2) \\ &+ \frac{g^2}{8}(a_1^2 + a_2^2 - b_1^2 - b_2^2)^2. \end{aligned} \tag{3.17}$$

Supersymmetry is spontaneously broken, since the masses of the fields are no longer equal. For $|\xi| < m^2$ the fields a_i and b_i have vanishing vacuum expectation value, while $\langle D \rangle = -\xi/g$ and, from (3.14)

$$\delta\lambda = \frac{\xi}{g}\gamma_5 \alpha + \cdots \tag{3.18}$$

where the dots denote terms containing other fields. The field λ is a Goldstone Fermion (germion). For $|\xi| > m^2$ one of the quadratic terms in (3.17) has a negative coefficient. The gauge invariance is also spontaneously broken and the vector field acquires a mass $m_v^2 = 2(\xi - m^2)$ (Higgs mechanism).

The Goldstone Fermion which is now a linear combination λ^1 of the fields $\lambda, \gamma_5\psi_1$ and ψ_2, transforms as

$$\delta\lambda^1 = -\frac{m}{g}\sqrt{2\xi - m^2}\gamma_5\alpha + \cdots \tag{3.19}$$

The Goldstone Fermions are massless spinors arising from the spontaneous breaking of supersymmetry, just as Goldstone bosons arise when chiral symmetry is spontaneously broken. In analogy with that case one may ask whether non-linear realizations of super- symmetry exist. The first non-linear realization was given by Volkov and Akulov [5],

$$\delta\chi = \frac{1}{a}\alpha + ia(\bar{\alpha}\gamma^\mu\chi)\delta_\mu\chi. \tag{3.20}$$

A different realization was found by the author:

$$\delta\chi = \frac{1}{a}\alpha + ia(\bar{\alpha}\gamma^\mu\chi)\delta_\mu\chi + ia(\bar{\alpha}\gamma^\mu\chi)\gamma_5\delta_\mu\chi. \tag{3.21}$$

One may inquire about the correction between linear and non-linear realizations. There exist functions of the field χ transforming as in (3.21) which transform linearly as in (3.8).

5 Superspace and Superfields

Supersymmetry representations have been studied with various techniques [12, 13]. Salam and Strathdee [14] have introduced the very interesting concept of superfield and have described supersymmetry transformations as operations on superfields. It has been extended by Ferrara, Wess and the author [15]. A general review is given by Salam and Strathdee [16]. Superspace was considered by Volkov and Akulov [5] in their non-linear realizations.

Consider a space (superspace) whose points are determined by coordinates x_μ, θ and $\bar{\theta}$ where x_μ are the space-time coordinates, θ is a (totally anti-commuting) two-component spinor and $\bar{\theta}$ its complex conjugate.

Supersymmetry transformations are geometrical trans- formations in superspace

$$\begin{aligned} \delta x_\mu &= i\theta\sigma_\mu\bar{\xi} - i\xi\sigma_\mu\bar{\theta}, \quad \sigma_0 = 1, \\ \delta\theta &= \xi, \quad \delta\bar{\theta} = \bar{\xi}. \end{aligned} \tag{3.22}$$

The commutator of two such transformations is a translation — furthermore (3.22) and translations leave invariant the differential form

$$\omega_\mu = \partial x_\mu + i\bar{\theta}\sigma_\mu \partial\theta - i\theta\sigma_\mu \partial\bar{\theta}. \tag{3.23}$$

Adjoining Lorentz transformations the "line element"

$$\omega_\mu \omega^\mu \tag{3.24}$$

is invariant. A superfield is a field in superspace, $V(x, \theta, \bar{\theta})$. Its power series in θ and $\bar{\theta}$, terminates after a finite number of terms,

$$V(x, \theta, \bar{\theta}) = C(x) + i\theta\chi(x) - i\bar{\theta}\bar{\chi}(x) + \cdots + \theta\theta\bar{\theta}\bar{\theta}\frac{1}{2}D(x). \tag{3.25}$$

Therefore a superfield corresponds to a finite supermultiplet of ordinary fields. A superfield is taken to transform as a scalar in superspace under (3.22),

$$V'(x, \theta, \bar{\theta}) = V(x', \theta', \bar{\theta}'), \tag{3.26}$$

and can have spinor or vector indices which determine how it transforms under Lorentz transformations. From (3.26) and (3.25) one derives the transformation of the fields C, χ etc of the supermultiplet. Observe that the new coordinate

$$Z_\mu = x_\mu + i\theta\sigma_\mu\bar{\theta} \tag{3.27}$$

gives

$$\delta z_\mu = 2i\theta\sigma_\mu\bar{\xi}, \tag{3.28}$$

which does not contain $\bar{\theta}$. Therefore, it is consistent to require that a superfield be a function only of z_μ and θ, $S(z, \theta)$, or that it satisfies

$$\bar{D}S = 0 \tag{3.29}$$

where

$$\bar{D} = -\frac{\partial}{\partial\bar{\theta}_z} = -\frac{\partial}{\partial\bar{\theta}} - i\theta\sigma^\mu\delta_\mu, \tag{3.30}$$

$\bar{D}$ is a covariant derivative (under (3.22)) and so is

$$D = \frac{\partial}{\partial\theta_x} + i\sigma^\mu\bar{\theta}\delta_\mu. \tag{3.31}$$

A superfield satisfying (3.29) is called left-handed, one satisfying the covariant constraint

$$DS = 0 \tag{3.32}$$

is called right-handed. A left-handed superfield (together with its right handed complex conjugate) corresponds to the multiplet described in Sect. 2. Volkov and Soroka [18] have developed a description of curved superspace which combines gravitational theory with the interaction of particles of spin $3/2, 1$ and $1/2$. Can a theory of this kind, because of the compensation of divergences due to supersymmetry, provide a renormalizable description of gravitational interactions?

6 Yang-Mills Gauges and Supersymmetry

Using superfields Salam and Strathdee [19] and Ferrara and the author [17] have constructed theories which are both supersymmetric and invariant under non-abelian gauge transformations. We give only the results. The simplest supersymmetric and gauge invariant theory is the ordinary Yang-Mills theory of vectors in interaction with a multiplet of Majorana spinors belonging to the regular (adjoint) representation of the internal symmetry group. For instance, for SU(N), using $N \times N$ matrix notation

$$\begin{aligned} L &= \operatorname{Tr}\left(-\frac{1}{4}v_{\mu\nu}^2 - \frac{1}{2}\bar{\lambda}\gamma^\mu D_\mu\lambda + \frac{1}{2}D^2\right), \\ v_{\mu\nu} &= \delta_\mu v_\nu - \delta_\nu v_\mu + ig[v_\mu, v_\nu], \\ D_\mu\lambda &= \delta_\mu\lambda + ig[v_\mu, \lambda]. \end{aligned} \tag{3.33}$$

One checks that the supercurrent

$$J^\mu = -\frac{1}{4}\operatorname{Tr}\left(v_{\nu\rho}[\gamma^\nu, \gamma^\rho]\gamma^\mu\lambda\right) \tag{3.34}$$

is conserved as a consequence of the equations of motion.

The multiplet v_μ, λ, D can be coupled to "matter multiplets" A, B, ψ, F, G belonging to any representation of the internal group. For the case of a matter multiplet in the regular representation the Lagrangian is given by (3.33)

$$\begin{aligned} &\operatorname{Tr}\left\{-\frac{1}{2}\left[(D_\mu A)^2 + (D_\mu B)^2 + i\bar{\psi}\gamma^\mu D_\mu\psi - F^2 - G^2\right]\right. \\ &\left. + m(FA + GB - \frac{i}{2}\bar{\psi}\psi) + igD[A, B] + g\bar{\lambda}[A + \gamma_5 B, \psi]\right\}. \end{aligned} \tag{3.35}$$

Theories of this kind contain scalar and pseudoscalar fields - since all coupling constants are expressed in terms of the single Yang-Mills coupling g, they will be asymptotically free provided the Callan-Symanrik function g is negative (and provided the theories are renormalizable in accordance with supersymmetry!). For (3.33) plus n matter multiplets like (3.35) it is

$$\beta = -\frac{g^3}{16\pi^2}(3-n)N \tag{3.36}$$

where N referes to SU(N).

If one adds (3.33) and (3.35) for $m = 0$, eliminates the field D and combines the two Majorana fields λ and ψ into a complex spinor

$$\phi = \frac{1}{\sqrt{2}}(\lambda + i\phi) \tag{3.37}$$

one obtains the Lagrangian

$$\begin{aligned}\mathrm{Tr}\bigg\{ & -\frac{1}{4}v_{\mu\nu}^2 - \frac{1}{2}(D_\mu A)^2 - \frac{1}{2}(D_\mu B)^2 - i\bar{\phi}\gamma^\mu D_\mu \phi \\ & - ig\bar{\phi}[A+\gamma_5 B, \phi] - \frac{g}{2}(i[A,B])^2\bigg\}\end{aligned} \tag{3.38}$$

which is invariant under the transformation

$$\phi \to e^{i\omega}\phi \quad \phi^* \to e^{-i\omega}\phi^*. \tag{3.39}$$

The conserved supercurrent for (3.38) is [21]

$$J = \mathrm{Tr}\left\{-\frac{1}{4}v_{\mu\nu}[\gamma^\nu,\gamma^\rho]\gamma^\mu\phi + ig[A,B]\gamma_5\gamma^\mu\phi - i\gamma^\lambda D_\lambda(A-\gamma_5 B)\gamma^\mu\phi\right\}. \tag{3.40}$$

One can also obtain a $V \pm A$ scheme with Fermion number [22]. One need only combine two multiplets $v_\mu^{(1)}$, $\lambda^{(1)}$ and $v_\mu^{(2)}$, $\lambda^{(2)}$ described by Lagrangians like (3.33). The total Lagrangian can be rewritten in terms of the complex spinor

$$\phi = \frac{1}{2}(1-i\gamma_5)\lambda^{(1)} + \frac{1}{2}(1+i\gamma_5)\lambda^{(2)} \tag{3.41}$$

and of the vector and axial vector fields

$$v_\mu = \frac{1}{\sqrt{2}}(v_\mu^{(1)} + v_\mu^{(2)}), \quad a_\mu = \frac{1}{\sqrt{2}}(v_\mu^{(1)} - v_\mu^{(2)}) \tag{3.42}$$

and is again invariant under (3.39).

The examples of this section show that one can now attempt to construct realistic supersymmetric models. The main difficulty at present is that one does not have a device for generating masses without spoiling the renormalizability of the theory. The Fayet-Iliopoulos trick described in Section 4, can be applied only to an invariant Abelian subgroup. Perhaps consideration of the effective potential in higher orders will provide a solution to this problem.

7 Non-trivial Mixing of Internal Symmetry and Supersymmetry

The combination of internal symmetry and supersymmetry described in the previous section attributes all fields of a supermultiplet to the same representation of the internal symmetry group. A more interesting possibility was suggested by Salam and Strathdee [13] and by Wess and the author [23]. It consists in an actual combination of the two symmetries to a new algebra. Imagine the algebra (3.3) written in two component notation, by using a representation of the γ matrices in which γ_5 is diagonal. The supercharges will consist of a two component spinor and its conjugate. One can give these two component spinors an internal symmetry index i or j and write the algebra as

$$\begin{aligned} &\{Q_{\alpha i}, Q_{\beta j}\} = \{\bar{Q}^i_{\dot\alpha}, \bar{Q}^j_{\dot\beta}\} = 0, \\ &\{Q_{\alpha i}, \bar{Q}^j_{\dot\beta}\} = 2\delta^j_i (\sigma_\mu)_{\alpha\dot\beta} P^\mu, \\ &[Q_{\alpha i}, P_\mu] = [P_\lambda, P_\mu] = 0. \end{aligned} \tag{3.43}$$

For instance, the indices i and j could refer to the N-dimensional representation of SU(N) (upper indices to the complex conjugate). The representations of this algebra can be studied by the method of superfields [24], in a superspace labelled by $x_\mu, \theta_{\alpha i}, \bar\theta^j_{\dot\alpha}$. A simpler technique has been suggested by Salam and Strathdee [13]. One can go to the rest frame $P^\mu \to (m, 0, 0, 0)$.

The algebra (3.43) then becomes an algebra of creation and destruction operators. As an example for SU(2) the quantum numbers $(I, J) =$ (isospin, spin) for the fields of the multiplet corresponding to a left-handed superfield, are

$$(0,0),\ (\tfrac{1}{2},\ \tfrac{1}{2}),\ (1,0),\ (0,1),\ (\tfrac{1}{2},\ \tfrac{1}{2}),\ (0,0).$$

The symmetry described by (3.43), or other similar generalizations, seems well worth exploring, by any available techniques. Work along these lines is presently being done by several groups.

8 Conclusion

Three lines of future development come to mind. The first would use supersymmetry as a way of classifying hadronic states and their interactions, a kind of relativistic SU(6), connecting Bosons and Fermions and free of contradictions. The Lagrandian model described in Section 2 gives an extremely simplified version of this kind of theory. The inclusion of internal symmetries could be effected either by introducing them as an additional group commuting with the supersymmetry or by enlarging the algebra as indicated in Section 7. For instance, the indices i, j occurring in (3.43) can be extended to include, besides the physical SU(N), a colour SU(N).

The second line of development is suggested by the gauge invariant model of Section 3. It is very tempting to interpret the supermultiplet (3.14) as containing the photon and the (electron) neutrino. The non-abelian generalizations of the model of Section 3, given in Section 6, are the first step towards a description of weak, electromagnetic and possibly strong interactions.

The third line of development could be a generalized (possibly renormalizable) theory of gravitation, as indicated at the end of Section 5.

References

1. J. Wess and B. Zumino, *Supergauge Transformations in Four-Dimensions,* Nucl. Phys. B **70**, 39–50 (1974).
2. A. Neveu and J. H. Schwarz, *Factorizable dual model of pions,* Nucl. Phys. B **31**, 86–112 (1971); P. Ramond, *Dual Theory for Free Fermions,* Phys. Rev. D **3**, 2415–2418 (1971); Y. Aharonov, A. Casher and L. Susskind, *Dual-parton model for mesons and baryons,* Phys. Lett. B **35**, 512–514 (1971); J. L. Gervais and B. Sakita, *Generalizations of dual models,* Nucl. Phys. B **344**, 477–492 (1971); J. L. Gervais and B. Sakita, *Field Theory Interpretation of Supergauges in Dual Models,* Nucl. Phys. B **34**, 632–639 (1971).
3. F. A. Berezin and G. I. Kats, *Lie groups with commuting and anticommuting parameters,* Mathemat. Sbornik (USSR) **82** (1970) 343–359.
4. See e.g.: L. O'Raifeartaigh, *Mass differences and Lie Algebras of Finite Order,* Phys. Rev. Lett. **14**, no.14, 575 (1965); *Lorentz Invariance and Internal Symmetry,* Phys. Rev. **139**, B1052-B1062 (1965); S. R. Coleman and J. Mandula, *All Possible Symmetries of the S Matrix,* Phys. Rev. **159**, 1251–1256 (1967).
5. D. V. Volkov and V. P. Akulov, *Is the Neutrino a Goldstone Particle?,* Phys. Lett. B **46**, 109–110 (1973).

6. A conformai spinor is the same as the entity called a twistor by Penrose, and there are similarities between the larger algebra of Ref. [1] and twistors. For twistor theory see e.g. R. Penrose and M. A. H. MacCallum, *Twistor theory: An Approach to the quantization of fields and space-time,* Phys. Rept. **6**, 241–316 (1972).
7. J. Wess and B. Zumino, *A Lagrangian Model Invariant Under Supergauge Transformations,* Phys. Lett. B **49**, 52 (1974).
8. J. Iliopoulos and B. Zumino, *Broken Supergauge Symmetry and Renormalization,* Nucl. Phys. B **76**, 310 (1974); S. Ferrara, J. Iliopoulos and B. Zumino, *Supergauge Invariance and the Gell-Mann-Low Eigenvalue,* Nucl. Phys. B **77**, 413 (1974). See also H.S. Tsao, Brandeis University preprint (1974). [*Supergauge Symmetry, Broken Chiral Symmetry and Renormalization*, Phys. Lett. **B 53**, 381–383 (1974) –MS]
9. W. Lang and J. Wess, University of Karlsruhe preprint (1974), submitted to Nucl. Phys. B. [*Investigation of a Nonrenormalizable Lagrangian Model Invariant Under Supertransformations*, Nucl. Phys. **B** 81, 249–252 (1974) –MS].
10. J. Wess and B. Zumino, CERN preprint TH.1857 (1974), to be published in Nucl. Phys. B. [*Supergauge Invariant Extension of Quantum Electrodynamics*, Nucl. Phys. **B** 78, 1, (1974) –MS].
11. P. Fayet and J. Iliopoulos, Orsay preprint (1974) submitted to Phys. Lett. B. [*Spontaneously Broken Supergauge Symmetries and Goldstone Spinors*, Phys. Lett. **B** 51, 461–464, (1974).] Goldstone spinors were discussed earlier by A. Salam and J. A. Strathdee, *On Goldstone Fermions,* Phys. Lett. B **49**, 465–467 (1974), and by J. Iliopoulos and the author in the first paper Ref. [8].
12. S. Ferrara, *Supergauge Transformations on the Six-Dimensional Hypercone,* Nucl. Phys. B **77**, 73–90 (1974).
13. A. Salam and J. Strathdee, Trieste preprint IC/47/16 (1974), to be published in Nucl. Phys. B.
14. A. Salam and J. A. Strathdee, *Supergauge Transformations,* Nucl. Phys. B **76**, 477–482 (1974).
15. S. Ferrara, J. Wess and B. Zumino, CERN preprint TH.1863 (1974), to be published in Nucl. Phys. B. [*Supergauge Multiplets and Superfields*, Phys. Lett. **B 51**, 239, (1974). –MS]
16. A. Salam and J. Strathdee, Trieste preprint IC/74/42 (1974), submitted to Phys. Rev. [*On Superfields and Fermi-Bose Symmetry*, Phys. Rev. **D** 11, 1521–1535, (1975). –MS]
17. S. Ferrara and B. Zumino, in preparation.
18. D. V. Volkov and V. A. Soroka, *Higgs Effect for Goldstone Particles with Spin 1/2,* JETP Lett. **18**, 312–314 (1973).
19. A. Salam and J. Strathdee. Trieste preprint IC/74/36 (1974) [*Supersymmetry and Non-Abelian Gauges*, Phys. Lett. **B** 51, 353–355, (1974). –MS]
20. S. Ferrara and B. Zumino, CERN preprint TH.1866(1974), to be published in Nucl. Phys. B [*Supergauge Invariant Yang-Mills Theories*, Nucl. Phys. **B** 79, 413, (1974). –MS]
21. The supercurrents (3.34) and (3.40) were worked out by the author in collaboration with W. Bardeen.
22. R. Delbourgo, A. Salam and J. Strathdee, Trieste preprint IC/74/45 (1974) [*Supersymmetric v-a Gauges and Fermion Number*, Phys. Lett. **B** 51, 475–478, (1974).]
23. J. Wess and B. Zumino, in preparation.
24. P.H. Dondi and M. Sohnius, University of Karlsruhe preprint (1974) [*Supergauge Transformations with Isospin Symmetry*, Nucl. Phys. **B** 81, 317–329, (1974). –MS]

4

Local Supersymmetry (Supergravity)

Figure 4.1: Dmitri Volkov, Kharkov, mid-1950s.

Figure 4.2: Dmitri Volkov, 1980s.

SUPERGRAVITY BEFORE 1976[88]

DMITRI VOLKOV

Kharkov Institute of Physics and Technology, Kharkov, 310108, Ukraine

Here is my story of how an attempt to realize Werner Heisenberg's idea that neutrinos might be Goldstone particles had led in its development to the discovery of supergravity action.

1 Introduction

There are two kinds of fields which are directly related to continuous symmetries. These are the Goldstone and Yang–Mills fields.

The Goldstone fields are a manifestation of spontaneously broken symmetries. The Yang-Mills fields appear when a symmetry is localized.

The both kinds of fields were predicted theoretically, and at the moment of their invention seemed to have no relation to reality.

However, revealed by Higgs and others, their mutual interplay exhibiting itself as the Higgs effect has thoroughly changed the situation. And nowadays the concept of the gauge fields and of the spontaneously broken symmetries is the backbone of theories unifying all known interactions

<table>
<tr><td rowspan="2">Internal symmetry group G</td><td>→</td><td>spontaneous breakdown</td><td>→</td><td>Goldstone particles, $s = 0$</td><td rowspan="2">→</td><td rowspan="2">Goldstone particles
Gauge fields
Higgs effect</td></tr>
<tr><td>→</td><td>localization</td><td>→</td><td>Y-M gauge fields, $s = 1$</td></tr>
</table>

The modern idea of the unification involves not only internal symmetries but also a new type of symmetry discovered at the beginning of the 1970's, namely, supersymmetry, which is intrinsically intertwines with the former.

Main features of the supersymmetry have been discussed in the talks by S. Dimopoulos and P. Fayet. I shall remind only that supersymmetry was independently discovered by three groups of authors:

Yu. Golfand & E. Likhtman (1971);
D. Volkov & V. Akulov (1972);
J. Wess & B. Zumino (1973).

[88]Proc. History of Original Ideas and Basic Discoveries in Particle Physics, Erice, Italy, July 29-August 4, 1994, (NATO Science Series B: volume 352, 1996), Eds. Harvey B. Newman, Thomas Ypsilantis, pages 663–675 [arXiv:hep-th/9410024].

The motivation and the starting points used by these groups were quite different. The motivation of Gol'fand and Likhtman [1] was to igeneralize the Poincaré group. The starting point of the paper by Akulov and myself [2,3] was the question whether Goldstone particles with spin one half might exist. Wess and Zumino [4] introduced the super-gauge symmetry following the Neveu-Schwarz-Ramond dual models [5,6].

The approach of the Volkov-Akulov papers is the most appropriate for gauging the super-Poincaré group, which was done a little bit later in the papers by Soroka and myself [7,8]. These papers are a natural continuation of the Volkov-Akulov papers since the transformation laws for gauge fields are determined by the same group structure as that used for the description of the Goldstone fields.

As it has already been mentioned, our approach was motivated by the question about possible existence of the spin 1/2 Goldstone particles.

The second question closely related to the first one is: what is the group whose breakdown yields fermion Goldstone fields with the required properties? In some respect this is the reverse of the usual consideration of the Goldstone and gauge fields when an internal symmetry group is given and the properties of the fields are derived from the known symmetry group properties.

May the Goldstone fermion with $s = \frac{1}{2}$ exist?	$\Longrightarrow$	Super-Poincaré group, Supergravity	$\leftarrow$	A group of which kind?

Answering these questions led us to the discovery of the super-Poincaré group.

It is convenient to consider the internal symmetry groups and the super–Poincaré group in parallel since it gives the possibility of learning which features of the both cases are common and which are different.

Goldstone fields Gauge fields Higgs effect	$\Longleftrightarrow$ $\Longleftrightarrow$	Internal symmetries Super Poincaré

As we shall see below the analogy between the two cases gives rather rich intuitive insight which helps one to trivialize the procedure of gauging the super-Poincaré group and getting the supergravity action.

Let us first recall that one of the main directions of the research in theoretical particle physics in the 1960's was the Current Algebra initiated by

M. Gell-Mann [9–11], and the PCAC hypothesis [32], which produced many interesting results in the field of weak and strong interactions, including a number of the so called soft pion, or lowenergy, theorems (Y. Nambu and D. Lurie [12], S. Adler [14], S. Weinberg [15],...). The essential progress was achieved when this direction was goldstonized (Goldstone [16], Nambu, Jona-Lozinio [17]), and especially when S. Weinberg [18] and J. Schwinger [19] proposed the method of Phenomenological Lagrangians which easily reproduced all soft pion theorems with any number of pions.

At the time of the XIVth Conference on High Energy Physics (Vienna, 1968) the problem of the Current Algebra and of the Phenomenological Lagrangians was intensively discussed (see Weinberg's rapporteur talk [20]). There were two papers presented in the current algebra section of the conference in which the generalization of the method of Phenomenological Lagrangians to an arbitrary internal symmetry group had been elaborated. One paper was presented by B. Zumino [21] (co-authors C. Callan, S. Coleman and J. Wess), and another one by myself [22]. The main results of the papers were practically identical. The difference was that in Ref. [22][89] the works of E. Cartan on symmetric spaces and his method of the exterior differential forms were intensively used.

The general procedure for constructing Phenomenological Lagrangians for Goldstone particles of internal symmetry groups [21, 22] is as follows.

A group G is factorized

$$G = KH, \tag{4.1}$$

where H is a stationary subgroup of G, and K is the coset space $K = G/H$.

The g–algebra valued differential 1-forms

$$G^{-1}dG = H^{-1}(K^{-1}dK)H + H^{-1}dH$$

represent the vielbeins $(G^{-1}dG)_k$ and the connection 1-forms $(G^{-1}dG)_h$, where k and h are subspaces of the g algebra corresponding, respectively, to K and H.

The Phenomenological Lagrangian is constructed out of the vielbeins:

$$L = \frac{1}{2}\mathrm{Tr}\,(G^{-1}dG)_k(G^{-1}dG)_k, \tag{4.2}$$

while the connection forms are used to include interaction with other particles which belong to representations of the group G. G multiplets are reduced

[89] Two papers [22, 23] are fused in [24].

to H multiplets with respect to the stationary subgroup H and considered as independent in the approach in question. The splitting of the G multiplets into the H multiplets together with the appearance of Goldstone fields is the main feature of the Phenomenological Lagrangian method.

A little bit later this procedure was generalized by J. Wess and B. Zumino [25] by introducing the so called Wess-Zumino term, and by me for the case of spontaneously broken symmetry groups containing the Poincaré group as a subgroup [22, 26].[90] The Phenomenological Lagrangian for Goldstone fermions, which we now turn to consider, is an example of the latter generalization.

2 Goldstone Fermions, super-Poincaré group and its spontaneous breaking

Now we can return to the question how one can generalize the Poincaré group to the super-Poincaré group by requiring Goldstone particles to have spin one half. As it has been explained in the previous section, the quantum numbers of the Goldstone particles coincide with the quantum numbers of the K generators. Therefore, to ensure the appearance of the Goldstone fermions, the Poincaré group should be generalized in such a way that the generalized version contains generators with spin one–half obeying commutation relations corresponding to the Fermi statistics. From a technical point of view the problem was what representation of the Poincaré group is the most appropriate for such generalization.

The solution to this technical problem was that the following representation of the Poincaré group had all required properties:

$$\mathbf{G}_{\mathbf{Poincare}} = \begin{pmatrix} 1 & iX \\ 0 & 1 \end{pmatrix} \begin{pmatrix} L & 0 \\ 0 & (L^{+})^{-1} \end{pmatrix}, \tag{4.3}$$

where $X_{\alpha\dot{\beta}}$, $L_{\alpha}^{\ \beta}$ and $(\mathbf{L}^{+})^{-\mathbf{1}\dot{\alpha}}_{\ \ \dot{\beta}}$ are $\mathbf{2} \times \mathbf{2}$ matrices.

In the generalization to the super-Poincaré group $K_{\text{transl.}}$ plays the main role.

Let us write it down in a form consisting of four blocks:

$$K = \begin{pmatrix} 1 & iX \\ 0 & 1 \end{pmatrix}. \tag{4.4}$$

[90]The paper [22], which was prepared in autumn of 1971, contains the first mention of the generalization of the Poincaré group to the super-Poincaré group.

Separating the blocks as follows

$$K' = \begin{pmatrix} 1 & - & iX \\ 0 & 1 & - \\ 0 & 0 & 1 \end{pmatrix} \tag{4.5}$$

one can insert into the newly formed hatched blocks Grassmann spinors θ_α and $\bar{\theta}_{\dot{\alpha}}$ so that K' becomes

$$K' = \begin{pmatrix} 1 & \theta & iX' \\ 0 & 1 & \bar{\theta} \\ 0 & 0 & 1 \end{pmatrix}. \tag{4.6}$$

The matrices K' form a group, but only under the condition that X' is complex. To match the reality condition for X' with that of X in Eq. (4) and to conserve the group properties of (6) one should represent X' in the form

$$iX' = iX + \frac{1}{2}\theta\bar{\theta}.$$

The resulting expression for the super-Poincaré group is as follows,

$$G_{\text{SP}} = \begin{pmatrix} 1 & \theta & iX + \frac{1}{2}\theta\bar{\theta} \\ 0 & 1 & \bar{\theta} \\ 0 & 0 & 1 \end{pmatrix} \times \begin{pmatrix} L & 0 & 0 \\ 0 & 1 & 0 \\ 0 & 0 & (L^{+})^{-1} \end{pmatrix}. \tag{4.7}$$

In [2, 3] more general case of $\mathcal{N}$-extended super-Poincaré group was considered. In this case the θ spinors acquire additional internal symmetry index and, in the second factor in (7) the unit matrix is replaced by an internal symmetry group matrix.

From (7) one gets the transformation law for the superspace coordinates:

$$X' = X + i\bar{\theta}\gamma\epsilon, \quad \theta' = \theta + \epsilon, \quad \bar{\theta}' = \bar{\theta} + \bar{\epsilon} \tag{4.8}$$

as well as the following expressions for the left-invariant superspace vielbein 1-forms,

$$e^a = dX^a - i\bar{\theta}\gamma^a d\theta, \qquad \mathbf{e}^\alpha = \mathbf{d}\theta^\alpha. \tag{4.9}$$

The latter are obtained as components of $K^{-1}dK$ corresponding to the generators of K now being the supertranslation subgroup of (7).

The action for the Goldstone fermions is the pullback of a target superspace differential four-form onto the four-dimensional Minkovski subspace (the world space)

$$S = \frac{1}{24}\varepsilon_{absd}e^a e^b e^c e^d. \tag{4.10}$$

In Eq. (10) as well as in further formulas exterior product is implied.

The action (4.10) gives the first example of a method widely applied nowadays for constructing superparticle, superstring and supermembrane actions out of the target superspace vielbein forms with their pulling back onto the worldsheet.

Now we can see what are the differences between (4.10) and (4.2). Firstly, the Lagrangian (4.10) is constructed not out of the 1-forms $\mathbf{e}^\alpha$ in (4.9), as might be expected, but out of the 1-forms e^a in (4.9). Secondly, the Lagrangian (4.10) is not the metric of a Riemann space, but the 4-volume form.

Interactions of the Goldstone fermions with particles of spin 0, $\frac{1}{2}$ and 1 were considered in [3, 23] with the use of the method developed in [23, 24]. Notwithstanding the differences with the internal symmetry case mentioned above, all soft Goldstone particle theorems were reproduced, and we became sure that our generalization of the Poincaré group may have something to do with reality. Our first paper was entitled "Is the neutrino a Goldstone particle?", but since gauging the Poincaré group is directly connected with the Einstein-Cartan theory of gravity, the same should take place for any generalization of the Poincaré group. Therefore, our paper was concluded with the following sentence

> *...[T]he gravitational interaction may be included by means of introducing the gauge fields for the Poincaré group. Note that if the gauge field for the fermionic transformation is also introduced, then as a result of the Higgs effect the massive gauge field with spin three-halves appears and the considered Goldstone particle with spin one-half disappears...*

which was the first (maybe somewhat implicit) mention of the theory of supergravity as a theory containing the Einstein-Cartan action and the action for the Rarita-Schwinger field, the latter being massless in the absence of the spontaneous supersymmetry breaking. We also considered the possibility that the super-Poincaré group might be an approximate symmetry so that (in analogy with π, ρ and A' mesons in hadron physics [14]) the Goldstone particles with nonzero mass and the Higgs effect coexisted.

3 Gauged super-Poincaré group

Now let us go over to the procedure of gauging super-Poincaré group.

Supersymmetry is spontaneously broken symmetry. Therefore, it is convenient to consider its gauged version from the very beginning in the form containing Goldstone fermion fields.

Note that the 1-forms (4.9) are not covariant with respect to the local transformations of the group under consideration. But if local Goldstone and gauge transformation laws are

$$K' = G_L K H, \quad A'(d) = G_L A(d) G_L^{-1} + G_L d G_L^{-1},$$

the sum

$$\tilde{A}(d) = K^{-1} dK + K^{-1} A(d) K$$

transforms as

$$\tilde{A}'(d) = H \tilde{A}(d) H^{-1} + H dH^{-1}.$$

Only $\langle \tilde{\mathbf{A}}(\mathbf{d}) \rangle_{\mathbf{h}}$ are the gauge fields, and $\langle \tilde{\mathbf{A}}(\mathbf{d}) \rangle_{\mathbf{k}}$ are non-gauge fields which have absorbed the Goldstone degrees of freedom and transform covariantly under the subgroup H.

The differential forms presented above can be used as blocks for constructing invariants of the local group, and then summing up all independent invariants one gets the general form of the gauge field action. It is self-evident that the action constructed this way contains not only the terms with spontaneously broken symmetry but a gauge invariant action as well. The latter can be easily extracted by counting the physical degrees of freedom. In other words, one should recall that the basic essence of the Higgs effect is that the Goldstone degrees of freedom though absorbed by the gauge fields are physical degrees of freedom. If some combination of the local invariants considered as an action results in two physical gauge field degrees of freedom, it means that corresponding Goldstone fields do not contribute and they either drop out of the action or become auxiliary fields. The latter, as will be shown in the next section, is the case for $\mathcal{N} = 1$ supergravity.

Here we remind what methods for considering the Higgs effect in the case of internal symmetry groups were used at the end of 1960's and the beginning of 1970's. Such a reminder is useful for the reader to better understand the state of the development of theoretical physics at that time, and, as it has been already said, the consideration of internal symmetries and supersymmetry in parallel helps one to formulate and to solve the problems in the latter case.

In the case of internal symmetry groups the approach for considering the Higgs effect may be roughly divided on to the following stages:

(i) The method described above for constructing the action out of the differential forms with the correlated transformation law of the gauge and Goldstone fields. The advantage of this method is its generality, the drawback is its phenomenological nature which exhibits itself in the presence of a number of arbitrary constants;
(ii) studying simplified models constructed for understanding possible mechanisms of spontaneous symmetry breakdown;
(iii) studying realistic gauge field theories coupled to matter fields which can ensure spontaneous symmetry breaking.

All these stages were applied and are now being applied to supersymmetric field theories.

The gauge fields for the $\mathcal{N}=1$ super-Poincaré group are described by a g_{SUSY}-valued matrix

$$A_{\mathrm{SUSY}}(d) = \begin{pmatrix} \omega(d) & \psi(d) & e(d) \\ 0 & 0 & \bar{\psi}(d) \\ 0 & 0 & \bar{\omega}(d) \end{pmatrix}$$

where $\omega(d)$ is the Lorentz connection, $\psi(d)$ is the Rarita-Schwinger gauge field and $e(d)$ is the vierbein 1-form.

The corresponding $\tilde{A}_{\mathrm{SUSY}}(d)$ forms are

$$\begin{aligned} \tilde{e}(d) &= e(d) + DX + i\Big[(2\psi(d) + D\theta)\bar{\theta} - \theta(2\bar{\psi}(d) + D\bar{\theta})\Big], \\ \tilde{\psi}(d) &= \psi(d) + D\theta, \quad \tilde{\omega}(d) = \omega(d), \\ \tilde{R}(d,d') &= R(d,d') = d\omega(d') - d'\omega(d) + \big[\omega(d), \omega(d')\big], \end{aligned} \tag{4.11}$$

where D is the covariant differential with the connection form $\omega(d)$.

4 Supergravity action and the super-Higgs effect

Contracting the indices of the differential forms (4.11) one can construct the following locally invariant differential four-forms [7, 8]:

$$\begin{aligned} W_1 &= \tilde{R}(d_1, d_2)\tilde{e}(d_3)\tilde{e}(d_4) && \text{Einstein–Cartan action,} \\ W_2 &= D\bar{\tilde{\psi}}(d_1, d_2)\tilde{e}(d_3)\tilde{\psi}(d_4) && \text{Rarita–Schwinger kinetic term,} \\ W_3 &= \tilde{e}(d_1)\tilde{e}(d_2)\tilde{e}(d_3)\tilde{e}(d_4) && \text{Cosmological term,} \\ W_4 &= \bar{\tilde{\psi}}(d_1)\tilde{e}(d_2)\tilde{e}(d_3)\tilde{\psi}(d_4) && \text{Rarita–Schwinger mass term.} \end{aligned}$$

The resulting action for $\mathcal{N} = 1$ supergravity is the sum

$$S = S_1 + S_2 \tag{4.12}$$

with

$$S_1 = a_1 W_1 + a_2 W_2 \tag{4.13}$$

being the pure SUGRA action and

$$S_2 = a_3 W_3 + a_4 W_4 \tag{4.14}$$

being the terms which arise due to the spontaneous breaking of the super-Poincaré group. Let us stress once again that each of the terms W_i and, hence, action S_1 and S_2 obey local supersymmetry.

The action S_1 in (4.13) is the same as one which is now accepted as the action of $\mathcal{N} = 1$ supergravity, and since the action for any newly proposed dynamical system completely determines all its physical properties, the time, when the action was firstly written down is the date of the discovery of the theory. So $\mathcal{N} = 1$ supergravity as a physical theory was discovered in 1973. At the XVII International Conference on High Energy Physics (London, 1974) B. Zumino said [27]:

> *Volkov and Soroka [2] have developed a description of curved superspace which combined gravitational theory with interaction of particles of spin 3/2, 1 and 1/2. Can a theory of this kind, because of compensation of divergences due to supersymmetry, provide a renormalizable description of gravitational interaction?*

Now we proceed to explain Eqs. (12)–(14) by counting the physical degrees of freedom of the gauge fields. It is self-evident that due to the fact that all terms are differential 4-forms the gravitational field has two physical components and corresponds to the unbroken symmetry. The Rarita-Schwinger field corresponds to the broken or unbroken local supersymmetry if it has four or two physical degrees of freedom, respectively. Thus, using this simple argument of counting the degrees of freedom one can find out whether supersymmetry is broken or not. This depends on the values of the coefficients a_i in Eqs. (4.12)–(4.14). To determine the coefficients it is sufficient to consider particular cases, the simplest examples being the Minkowski and the anti-de-Sitter space (if the mass term with a definite ratio of the mass to the curvature of the anti-de-Sitter space is added) as backgrounds for the Rarita-Schwinger field, and then go over to more general cases. It can be shown quite easily that if a background Einstein field satisfies the equation of motion for the term W_1 the Rarita–Schwinger field, in this background, has only two physical components, the Goldstone fermions do not contribute to the first two terms in the sum (4.12) and all effects of the spontaneous breaking of the super-Poincaré group are contained only in the third and the fourth term. If in (4.12), (4.13), (4.14) a_i are arbitrary the Rarita-Schwinger field has four degrees of freedom, so in the general case the Goldstone fermions are physical degrees of freedom, which is the realization of the super-Higgs effect.

The auxiliary Goldstone fermion fields (as well as X^a) can be excluded from the action S_1 without changing its physical contents. But, firstly, this procedure is not unique, and, secondly, which is more important, the off-mass-shell local invariance of S_1 is lost and reduced to the local invariance on mass shell with all its unpleasant consequences that the algebra is not closed, structure "constants" depend on fields etc. And since on mass shell the Lorentz connection $\omega(d)$ becomes functions of $e(d)$ and $\psi(d)$ its transformation law changes and becomes rather complicated. The action S_1, with the local supersymmetry held on mass shell, was regained in 1976 [28,29]. In the review [30] (p. 319) the main result of [29] was formulated as follows:

> *Deser and Zumino [29] have shown that supergravity is nothing but the spin 3/2 Rarita-Schwinger Lagrangian minimally coupled to Cartan first order formalism of general relativity.*

As to the terms S_2 (4.14) in the sum (4.12), they are reproduced in all $\mathcal{N} = 1$ supergravity theories with matter fields specifically added to make supersymmetry spontaneously broken. Of course, in such theories the constants a_3 and a_4 are determined by the choice of the matter fields, their masses and interaction constants.

There are a number of papers where the equivalence of our approach to the conventional one is established. See, for example, [31] and references therein.

Our approach can be applied to any $\mathcal{N} \leq 8$ case as well. In papers [7, 8] we wrote that we considered "the simplest possible local invariants". In the $\mathcal{N} = 1$ case the local invariants W_i constitute the complete set. When $\mathcal{N}$ increases the number of local invariants also increases. If $\mathcal{N} > 2$ not only gauge and Goldstone fields, but also all non-gauge fields which are the superpartners of the gauge fields in the supergravity multiplet, should be used for the construction of the complete set of invariants. The situation is the same as in the case of spontaneously broken internal symmetry with the only difference that in the case of supersymmetry gauge and non-gauge fields may belong to the same supermultiplet.

5 Superspace formulation of Supergravity (First steps into Superspace)

The first superspace formulation of supergravity was attempted by R. Arnowitt, P. Nath and B. Zumino [32]. They proposed the generalization of the Einstein 4-dimensional action to superspace with coordinates (x, θ) in the following form

$$S = \int R\sqrt{\{\mathrm{Ber}\, g\}}\, d^4\theta d^4x.$$

The analysis of this action revealed that it contains a number of drawbacks such as

(a) The presence of fermionic ghosts,
(b) the action does not admit "flat" superspace as a solution;
(c) for any solution torsion is zero, which is a consequence of Riemann geometry chosen;
(d) the holonomy group of the tangent space was OSp(3,1/4) which is too large.

Upon realizing these drawbacks two groups of authors [33,34] generalized the E. Cartan's method of differential geometry to superspace and argued that (i) the holonomy group for superspace curvature is the Lorentz group; (ii) flat superspace has torsion and, hence, the torsion has to be included into the theory of supergravity.

These points are now part of all existing versions of supergravity. The intensive development of the superspace approach to supergravity has been carried out since 1976. At the first stages Wess and Zumino, Ogievetsky and Sokatchev, Siegel, Ferrara and Roček, Scherk and many others have made significant contribution to this development.

6 Resumé

Before 1976 the following general ideas had been proposed and the following results obtained

- Super-Poincaré group ($\mathcal{N} = 1$, and $\mathcal{N} > 1$);
- Flat superspace, supervielbeins;

- Notion of Goldstone fermions;
- Phenomenological Lagrangian for Goldstone fermions interacting with particles of s=0, $\frac{1}{2}$ and 1;
- Gauging the super-Poincaré group;
- Supergravity action, $\mathcal{N} = 1$ (the off-mass-shell formulation);
- Super-Higgs effect, $\mathcal{N} \geq 1$.
- First attempts to the superspace formulation of supergravity.

As far as the techniques developed is concerned, it is mostly connected with the application of E. Cartan's methods of the exterior differential forms and differential geometry. E. Cartan's methods were applied to the component formalism for constructing the Goldstone fermion and supergravity action, and to trace a way to the superspace formulation of supergravity.

Nowadays E. Cartan's methods, being the fundamentals of the fiber bundle theory, are among the main tools used in theoretical physics.

7 Background

As the speakers at the Conference were asked to present not only their ideas and results but also to describe the path along which they went to reach them, I shall describe some of the main impulses and motivation which helped my coauthors and me find and overcome the path to supersymmetry and supergravity.

Of greatest importance was, as I call it, my personal inclination to the problem of connection between the spin and statistics, and related topics. My first papers had been done in this direction of research.

My PhD thesis was devoted to the calculation of radiative corrections to some effects in quantum electrodynamics of scalar particles. I had performed it following Julian Schwinger's papers, and from that time I started to consider him as my teacher by correspondence and, as his student by correspondence, I carefully studied all his papers at stock.

On my only personal meeting with Schwinger in Kiev in 1959 we had a fruitful discussion during which I proposed a possible generalization of an admissible class of variations in Schwinger's quantum-dynamical principle by connecting the variations with the symmetry properties of the action. A little bit later, Schwinger published a paper [35] in which the generalization proposed was elaborated. For me this episode became as an exam which I successfully passed with my teacher by correspondence. Listening the talk by S. Glashow at the Erice Conference I envied his lucky fortune to be in everyday contact with Schwinger.

Of course, the use of Grassmann variables as variations of fermion fields in the Schwinger quantum variation principle was an essential step to supersymmetry, while the Rarita-Schwinger field was a step to supergravity. I am sure that Schwinger had been quite prepared to write down the supergravity action in the form (4.12)–(4.14) in the middle of 1960's or even earlier, and only his engagement in the studies of other problems can explain why he had not done it.

In the late 1950s Werner Heisenberg and Wolfgang Pauli proposed a non-linear theory which was claimed to become the basis for understanding the spectra and interactions of elementary particles. A little bit later, Pauli changed his role — from the coauthor to one of the largest Heisenberg's opponents.[91] But Heisenberg continued his work, and, trying to overcome the mathematical difficulties of the theory, used his great intuition to find the place for each of the existing particles in the scheme. He succeeded in this attempt for a number of particles. Considering the neutrino, Heisenberg proposed that it might be a Goldstone particle connected with the spontaneous breakdown of parity [37]. The assumption impressed me immensely, but at that time I had not been prepared to implement this idea.

As I mentioned in Section 2 my approach to the phenomenological Lagrangians was different in some points from other approaches. The difference was that the starting point of [22] was the action (4.2) with an arbitrary metric tensor on which no symmetry requirements were imposed. Calculating the on-mass-shell scattering amplitudes I got that they depend only on manifestly covariant quantities such as the curvature tensor and its covariant derivatives. Upon imposing the simplest of possible restrictions, namely, that the covariant derivative of the curvature tensor vanishes,

$$R_{absd,f} = 0,$$

I became aware that this condition is the definition of the symmetric spaces proposed by E. Cartan. I began studying his works and they opened for me the mathematical beauty of differential geometry and taught many lessons on how one might apply it to physical theories.

In the previous sections I have stressed the role of Goldstone, Nambu, Schwinger, Weinberg, Higgs and others whose papers introduced me into the world of nonlinearly realized (secret) symmetries.

So the way to the discovery of supersymmetry and supergravity was rather long and not a straight line. And now I see that the seeds of

[91] Heisenberg describes this period in his monograph [36].

supersymmetry had been thrown onto the fertile soil and beautiful mathematical structures had grown out of them, and all of us, theorists and experimentalists, are waiting for the time when we can touch and taste the fruit.

I am very grateful to the Organizing Committee of the Conference for giving me this unique opportunity to illuminate the period in the development of supergravity which is usually omitted exposing the history of its discovery.

References

1. Yu. Golfand and E. Likhtman, *JETP Letters* **13** (1971) 323.
2. D. Volkov and V. Akulov, *JETP Letters* **16** (1972) 438. *Phys. Lett.* **46B** (1973) 109.
3. V. Akulov and D. Volkov, *Theor. Math. Phys.* **18** (1974) 28.
4. J. Wess and B. Zumino, *Nucl. Phys.* **B70** (1974) 39.
5. J.-L. Gervais and B. Sakita, *Nucl. Phys.* **B34** (1971) 633.
6. A. Neveu and J. Schwarz, *Nucl. Phys.* **B31** (1971) 86. P. Ramond, *Phys. Rev.* D3 (1971) 2415.
7. D. Volkov and V. Soroka, *JETP Letters* **18** (1973) 312.
8. D. Volkov and V. Soroka, *Theor. Math. Phys.* **20** (1974) 829.
9. M. Gell–Mann, *Physics* **1** (1964) 63.
10. M. Gell–Mann, see a talk in these Proceedings.
11. D. Gross, see a talk in these Proceedings.
12. Y. Nambu, *Phys. Rev. Lett.* **4** (1960) 380. I. Bernstein *et al.*, *Nuov. Cim.* **17** (1960) 757.
13. Y. Nambu and D. Lurie, *Phys. Rev.* **125** (1962) 1429.
14. S. Adler, *Phys. Rev.* **137B** (1965) 1022; *ibid.* **139B** (1965) 1638.
15. S. Weinberg, *Phys. Rev. Lett.* **18** (1967) 507.
16. J. Goldstone, *Nuov. Cim.* **19** (1961) 155.
17. Y. Nambu and G. Yona–Losinio, *Phys. Rev.* **122** (1961) 345.
18. S. Weinberg, *Phys. Rev.* **166** (1968) 1568.
19. J. Schwinger *Phys. Lett* **24B** (1967) 173.
20. S. Weinberg, Proceedings of the XIV Conference on High Energy Physics, Vienna, 1968.
21. S. Coleman, J. Wess and B. Zumino, *Phys. Rev.* **177** (1969) 2239; C. Callan *et al.* *Phys. Rev.* **177** (1969) 2247.
22. D. Volkov, Kiev preprint ITF–69–75, 1969.
23. D. Volkov, International Seminar on Functional Methods in QFT and Statistics, Preprint of Lebedev Institute (PhIAN) # 141, Moscow, 1971, 4.
24. D. Volkov, *Soviet J. Part. Nucl.* **4** (1973) 1.
25. J. Wess and B. Zumino, *Phys. Lett.* **37B** (1971) 95.
26. V. Akulov and D. Volkov, *JETP Lett.* **17** (1973) 261.
27. B. Zumino in Proceedings of the XVII Conference on High Energy Physics, London, 1974.
28. S. Ferrara, D. Freedman and P. van Niewenhuizen, *Phys. Rev.* **D13** (1976) 3214.
29. S. Deser and B. Zumino, *Phys. Lett.* **62B** (1976) 335.
30. P. Fayet and S. Ferrara, *Phys. Rep.* **32** (1977) 249.
31. E. Ivanov and A. Kapustnikov, *Nucl. Phys.* **333** (1990) 439.

32. R. Arnowitt, P. Nath and B. Zumino, *Phys. Lett.* **56B** (1975) 81.
33. B. Zumino, Proc. Conf. on Gauge Theories and Modern field Theory, Northeastern Univ., 1975, p. 255.
34. V. Akulov, D. Volkov and V. Soroka, *JETP Lett.* **22** (1975) 187.
35. J. Schwinger, *Proc. Nat. Acad. Sci.* **47** (1961) 1075.
36. W. Heisenberg, *Der Teil und Das Ganze*, (Munchen, 1969).
37. W. Heisenberg, *Introduction to the Unified Theory of Elementary Particles*, (Interscience Publishers, London, New-York, Sidney, 1966).

Figure 4.3: Vyacheslav A. Soroka

THE SOURCES OF SUPERGRAVITY

VYACHESLAV A. SOROKA

Institute of Theoretical Physics, National Science Center "Kharkov Institute of Physics and Technology", 310108 Kharkov, Ukraine

When I was an undergraduate student at the Kharkov State University in 1965 my advisor Professor Dmitri Vasilyevich Volkov gave me a task to learn and to study the problem of quantization of the spin-$\frac{3}{2}$ field interacting with other fields. There was a problem with constructing selfconsistent theories for the interacting fields with high spins and to find the methods for quantization of such theories. In particular, Johnson and Sudarshan (1961) encountered the inconsistency under quantization of the spin-$\frac{3}{2}$ field minimally interacting with the external electromagnetic field. At that time I was a starting physicist and did not know the theories with high spins. So, I had to learn two formalisms for description of the spin-$\frac{3}{2}$ field: Bargman-Wigner and Rarita-Schwinger. By the way, from the paper of Rarita and Schwinger (1941) I learned and memorized for the future a remarkable fact that a free massless spin-$\frac{3}{2}$ field in their formalism possesses a gauge invariance with a spinor parameter. I also learned Schwinger's quantum dynamical principle within the framework of which Johnson and Sudarshan had tried to quantize the spin-$\frac{3}{2}$ field.

At that time we had no success in quantization of the spin-$\frac{3}{2}$ fields interacting with other fields. As became clear later on, the problem in that formulation had no solutions at all. It was elucidated only after the discovery of supergravity theory in which the problem was naturally solved. So I was forced to postpone that approach and switched my activity to the derivation of sum rules following from the algebra of fields which had been introduced by Kroll, Lee, Weinberg and Zumino (1967). Since the algebra of fields is based on the gauge theory, I learned the beautiful Yang-Mills theory, and the papers by Utiyama (1956) and Kibble (1961) devoted to the gauge theories with different gauge groups, including, in particular, the gauge space-time symmetries connected with gravity.

Meanwhile Volkov (1969) simultaneously with Callan, Coleman, Wess and Zumino (1969) developed a general method for constructing Lagrangians for the Goldstone particles in the case of an arbitrary spontaneously broken internal symmetry group. Later he extended this method to the case of the spontaneously broken symmetries including the Poincaré group as a subgroup (1971, 1973).

Then Golfand and Likhtman (1971) introduced a super-Poincaré group in connection with a parity violation problem in quantum field theory. Shortly, and independently of them, Volkov formulated the existence problem for the fermionic Goldstone particles with spin 1/2. A successful solution of this problem led Volkov together with Akulov to the discovery of the extended super-Poincaŕe group (1972).

After that Volkov stated the idea of gauging the super-Poincaré group, and suggested I try to realize it. I accepted the suggestion with great enthusiasm, and we started. Then my earlier experience concerning the formalisms for high spin fields and gauge theories was very suitable and useful. We decided to consider the spontaneously broken extended super-Poincaré gauge group in order to study the Higgs effect for the Goldstone particles with spin-$\frac{1}{2}$ which had been recently introduced. Under this investigation very interesting and important features were revealed [1, 2].

- First of all, the real gauge fields with spin-$\frac{3}{2}$ were introduced as graviton superpartners. This fact was very unusual for those times, because till then the gauge fields possessed only integer spins. The graviton superpartners with spin-$\frac{3}{2}$ were later called the gravitino fields.
- Secondly, the extended super-Poincaré gauge group gives a principle possibility for the nontrivial unification of gravity with the interactions based on the internal symmetries.
- Thirdly, the Higgs effect for the Goldstone particles with spin-$\frac{1}{2}$, later called the super-Higgs effect, essentially differs from the Higgs effect for the Goldstone particles with spin-0. It results in not only the gravitinos becoming massive, but also the space-time changes its own metric and topological properties: a nonzero cosmological constant appears.

We have written down our action as a sum of five invariant terms with arbitrary constants before them. These terms contain respectively: the Einstein-Cartan action for gravity, a kinetic term for the gravitino fields, a mass term of the gravitinos, a cosmological term and a term for the Yang-Mills fields with spin 1.

Thus, we see that the main notions and ingredients of supergravity were formulated in our papers [1, 2].

Soon after and independently Wess and Zumino (1974) introduced a four-dimensional superconformal group generalizing the group of two-dimensional supersymmetric transformations present in the dual models. Then they proposed a supersymmetric model (1974) which appeared to be renormalizable.

The main results of our works were used afterwards as starting points for the development of supergravity in the works by Ferrara, Freedman, van Nieuwenhuizen [3] and Deser and Zumino [4]. In particular, their transformation rules of the vierbein, used for the graviton description, and gravitinos, coincided with ours. Explicit use of nonzero torsion, found in supersymmetric theories by us [5] and Zumino [6] also was essential to simplify the derivation of supergravity in Ref. [4]. Moreover, as Volkov explained [7], pure $\mathcal{N} = 1$ supergravity [3,4] can be deduced from our theory [1,2] with a special choice of the constants in front of our invariant terms.

The notions of superspace, introduced by Akulov and Volkov (1972), and superfields, proposed by Salam and Strathdee (1974), were used by Arnowitt, Nath and Zumino in 1975 to initiate a superfield approach to supergravity. In order to overcome the drawbacks they found at their first step in this direction, we undertook [5], simultaneously with Zumino [6], the development of the superspace approach to supergravity. By generalizing to superspace Cartan's methods in differential geometry, we revealed a nonzero torsion of superspace in the supersymmetric theories and found that the true homogeneous holonomy group for the superspace in supergravity is the Lorentz group. These two points are part of the basis of any version of supergravity.

Let me also mention our construction in the superfield formulation of a version of $\mathcal{N} = 1$ supergravity in the linear approximation with the so-called new minimal set of auxiliary fields [8], which we had found earlier than the old minimal set that was obtained by Ferrara, van Nieuwenhuizen, Stelle and West (1978).

So, I recalled the steps in supergravity which had been performed in our works in 1973–1977.

As the complete list of papers on the theme of my reminiscence is very huge and essentially exceeds this note, I refer only to our works and to those very closely related to them. A very extensive list of references concerning the development of supergravity can be found in Ref. [9]. This list contains also our papers [1, 2, 5, 8], however, the references to the latter are omitted in the text of this review.

To finish my note, I want to cite a very surprising (for me) appraisal of supersymmetry given by Yuri A. Golfand during the Conference "Supersymmetry-85" at Kharkov State University in 1985. He said that supersymmetry did not justify his hopes to find a generalization of the Poincaré group such that its representations include particles of different masses. So, Golfand and Likhtman had missed their aim, but had instead found supersymmetry, every representation of which contains fields of different spins. Maybe, somebody will succeed in their original aim too, or maybe both aims at once. Time will tell.

In conclusion I would like to thank the organizers of this Symposium for the opportunity to present my recollections concerning the creation of the main ideas of supergravity at its early stages, which are usually not illuminated.

References

1. D.V. Volkov and V.A. Soroka, *JETP Lett.* **18**, 312 (1973).
2. D.V. Volkov and V.A. Soroka, *Theor. Math. Phys.* **20**, 829 (1974).
3. D.Z. Freedman, P. van Nieuwenhuizen and S. Ferrara, *Phys. Rev.* **D13**, 3214 (1976).
4. S. Deser and B. Zumino, *Phys. Lett.* **62B**, 335 (1976).
5. V.P. Akulov, D.V. Volkov and V.A. Soroka, *JETP Lett.* **22**, 187 (1975).
6. B. Zumino, in Proceedings of the Conference on Gauge Theories and Modern Field Theory, Northeastern University, 1975, p. 255; CERN preprint TH. 2120.
7. D.V. Volkov, in Proceedings of the International Conference on The History of Original Ideas and Basic Discoveries in Particle Physics, Erice, Italy, 29 July–4 August 1994.
8. V.P. Akulov, D.V. Volkov and V.A. Soroka, *Theor. Math. Phys.* **31**, 285 (1977).
9. P. van Nieuwenhuizen, *Phys. Rep.* **68**, 189 (1981).

SUPERGRAVITY AND THE QUEST FOR A UNIFIED THEORY[92]

SERGIO FERRARA

Theory Division, CERN, CH-1211 Genéve 23, Switzerland

It is a great honor and pleasure to be invited to give this Dirac Lecture on the occasion of the 1994 Spring School on String, Gauge Theory and Quantum Gravity. In fact, this School is a continuation of a very successful series initiated by Prof. A. Salam in 1981. Together with J.G. Taylor and P. van Nieuwenhuizen I had the privilege of organizing the first two in the spring of 1981 and the fall of 1982 [1].

At that time, supergravity was in the mainstream of research, namely:

(i) Quantum properties of extended supergravities and their geometric structure;
(ii) Kaluza-Klein supergravity;
(iii) Models for particle physics phenomenology.

These topics were widely covered during the first two schools and workshops. Before going on to discuss supergravity and its subsequent development, let me briefly touch upon the steps taken in the preceding years, when supersymmetry in four dimensions was introduced.

Although the latter, with its algebraic structure, was first mentioned in 1971 by Golfand and Likhtman [2] and in early 1973 by Volkov and Akulov [3] (to explain the masslessness of the neutrino as a Goldstone fermion), it was really brought to the attention of theoretical particle physicists in the second half of 1973, by Wess and Zumino [4]; they had been inspired by a similar structure, found by Gervais and Sakita (1971) [5], already present in two dimensions, in the dual-spinor models constructed in 1971 by Neveu and Schwarz [6] and by Ramond [7]. The relevance of supersymmetry for quantum field theory, especially in view of its remarkable ultraviolet properties and its marriage with Yang-Mills gauge invariance, was soon established in early 1974.

[92]The 1994 Dirac Prize Lecture, delivered at the International Centre for Theoretical Physics, Trieste, on 19 April, 1994.

It is nevertheless curious that it was only, at the time, rather isolated groups that delved into the subject, mainly in Europe: at CERN, the ICTP (Trieste), Karlsruhe, the ENS — Paris, Imperial College — London, Turin University, and essentially two in the United States: Caltech and Stony Brook. The same applies to supergravity and its ramifications in the early years, after its foundation in 1976.

Soon after the very first paper of Wess and Zumino [8] remarkable sequence of events occurred during 1974:

(i) The superspace formulation of supersymmetric field theories (Salam and Strathdee [9] Wess, Zumino, and Ferrara [10]);
(ii) The discovery of non-renormalization theorems (Wess and Zumino [11], Iliopoulos and Zumino [12], Ferrara, Iliopoulos, and Zumino [13]).
(iii) The construction of supersymmetric Yang–Mills theories (Wess and Zumino [14] for the Abelian case; Ferrara and Zumino [15], and Salam and Strathdee [16], for the non-Abelian case);
(iv) The first construction of a renormalizable field theory model with spontaneously broken supersymmetry (Fayet and Iliopoulos [17]);
(v) The construction of a multiplet of currents, including the supercurrent and the stress energy tensor (Ferrara and Zumino [18]), which act as a source for the supergravity gauge fields and had an impact also later, in the classification of anomalies and in the covariant construction of superstring Lagrangians.

In the same year, quite independently of supersymmetry, Scherk and John Schwarz [19] proposed string theories as fundamental theories for quantum gravity and other gauge forces rather than for hadrons, turning the Regge slope from $\alpha' \sim \text{GeV}^{-2}$ to $\alpha' \sim 10^{-34}\ \text{GeV}^{-2}$, the evidence being that any such theory contained a massless spin-2 state with interactions for small momenta as predicted by Einstein's theory of general relativity.

In the following year, many models with spontaneously broken supersymmetry and gauge symmetry were constructed, mainly by Fayet [20] and O'Raifeartaigh [21], and $\mathcal{N} = 2$ gauge theories coupled to matter, which were formulated by Fayet [22].

This was a prelude to two important events, which took place just before and soon after the proposal of supergravity: the discovery of extended superconformal algebras (Ademollo *et al.*, November 1975 [23]) and the finding of evidence for space-time supersymmetry in superstring theory (Gliozzi, Olive, and Scherk, GOS for short, September 1976 and January 1977 [24]). In retrospect, these episodes had a great impact on the subsequent development of string theories in the mid-1980's.

The paper of Ademollo *et al.*, just a few months before supergravity was formulated, was inspired by the fact that it was possible, in higher dimensions ($D = 4$), by endowing the superspace coordinate θ_i with a counting index ($i = 1, \ldots, \mathcal{N}$), to construct extended supersymmetries; indeed, a remarkable theory with $\mathcal{N} = 2$ ($D = 4$) supersymmetry then had just been discovered by Fayet (September of 1975 [22]). In $D = 4$, extended superconformal algebras were accompanied by U(N) gauge algebras [SU(4) for $\mathcal{N} = 4$]. In $D = 2$, superconformal algebras are infinite-dimensional and $\mathcal{N} = 2$ and $\mathcal{N} = 4$ turned out to be accompanied by U(1) and SU(2) Kač–Moody gauge algebras. These algebras, at the time thought of as gauge-fixing of superdiffeomorphisms, were introduced to study new string theories with different critical dimensions [25]. In retrospect, this construction had a major impact on the classification of "internal" superconformal field theories, especially $\mathcal{N} = 2$, as the quantum version of Calabi-Yau manifolds, and on its relation [26] with the existence of space-time supersymmetry in $D < D_{\text{crit}}$.

Meanwhile, in the spring of 1976 [27], supergravity was formulated by Freedman, van Nieuwenhuizen and the author, working at École Normale and at Stony Brook. Soon after, a simplified version (first-order formulation) was presented by Deser and Zumino [28].

While in the second formalism, the spin-$\frac{3}{2}$ four-fermion interaction has the meaning of a contact term (similar to seagull terms in scalar electrodynamics or non-Abelian gauge theories) required by fermionic gauge invariance, in the first-order formalism it has the meaning of a torsion contribution to the spin connection from "spin-$\frac{3}{2}$ matter." This discovery also implied that any supersymmetric system coupled to gravity should manifest local supersymmetry.

This observation eventually led some physicists to go deeper in string theory in order to explore whether the "dual spinor model" could accommodate target-space supersymmetry. The GOS paper (September 1976 and January 1977 [24]) gave dramatic evidence for space-time supersymmetry in the superstring theory (called at that time the dual-spinor model) [29] by cutting out the G-odd parity states in the Neveu-Schwarz sector and comparing its bosonic spectrum with the fermion spectrum of the Ramond sector. In the proof, they used an identity that had been proved by Jacobi in 1829 (*Aequatio identica satis abstrusa*[93])! This paper came out following some sequential developments in supergravity, just after its first construction in the spring of 1976, namely the first matter-coupling to Maxwell theory (Ferrara, Scherk, and van Nieuwenhuizen, August 1976 [30]) to Yang-Mills theory and chiral multiplets [31] and the first formulation of extended supergravity [$\mathcal{N} = 2$] (Ferrara and van Nieuwenhuizen, September 1976 [32]). It is interesting to note that two of the GOS authors (G and S) also took part in some of the above-mentioned supergravity papers.

The hypothesis of GOS (later proved in great detail by Green and John Schwarz [33]) also implied the existence of an $\mathcal{N} = 4$ Yang–Mills theory, eventually coupled to an $\mathcal{N} = 4$ extended supergravity. This was implied by a dimensional reduction of the ten-dimensional spectrum. The full $N = 4$ supergravity contained in this reduction was found a year later (Cremmer, Scherk, and Ferrara, December 1977 [34]) and it was shown to contain an $\mathrm{SU}(4) \times \mathrm{SU}(1,1)$ symmetry. Meanwhile, three other important developments were announced at the end of 1976. The construction of $N = 3$ supergravity (Freedman; Ferrara, Scherk, and Zumino, November 1976 [35]) and the discovery of (Abelian and non-Abelian) duality symmetries, generalizing the electromagnetic duality $\mathrm{F} \to \tilde{\mathrm{F}}$ in $N = 2$ [U(1)] and $\mathcal{N} = 3$ [U(3)] supergravity (December 1976 [36]). This duality generalizes to $\mathrm{SO}(6) \times \mathrm{SU}(1,1)$ in pure $N = 4$ supergravity and to $\mathrm{SO}(6,n) \times \mathrm{SU}(1,1)$ in $\mathcal{N} = 4$ supergravity coupled to n matter (Yang-Mills) multiplets.

In retrospect, these symmetries play a crucial role in compactified superstrings, where the manifold

$$\frac{\mathrm{SO}(6,N)}{\mathrm{SO}(6) \times \mathrm{SO}(N)} \times \frac{\mathrm{SU}(1,1)}{\mathrm{U}(1)}$$

(modded out by some discrete symmetries) describes the moduli space of toroidally compactified ten-dimensional strings, according to the analysis of Narain, Sarmadi, and Witten [37]).

In September 1976, also the covariant world-sheet formulation of the spinning string was presented in two papers [38] by Brink, Di Vecchia and Howe and by Deser and Zumino. In retrospect this can be considered as a crucial ingredient for the Polyakov formulation [39] of spinning strings with arbitrary world-sheet topology. In this respect, $(p+1)$ supergravity is necessary for the consistent formulation of any p-dimensional extended object coupled to fermions.

[93] A rather obscure identity. –Lat.

In the subsequent years all higher extended four-dimensional supergravities with $\mathcal{N} = 5$, 6 and 8 were constructed. [

The maximally extended supergravity ($N = 8$) was found by Cremmer and Julia [40], by dimensional reduction of eleven-dimensional supergravity previously obtained by the same authors with Scherk (1978 [41]), and its gauged version, accompanied with an SO(8) Yang-Mills symmetry, by de Wit and Nicolai (1982 [42]). Gell-Mann had earlier observed that SO(8) cannot accommodate the observable gauge symmetry $SU(3) \times SU(2) \times U(1)$ of electroweak and strong interactions. However, it was later observed by Ellis, Gaillard, Maiani and Zumino (1982 [43]) that a hidden local SU(8) symmetry (found by Cremmer and Julia) could be identified as a viable Grand Unified Theory (GUT) for non-gravitational interactions. The basic assumption was that the degrees of freedom of the SU(8) gauge bosons could be generated at the quantum level, as it was known to occur in certain two-dimensional CT-models, following the analysis of Di Vecchia, D'Adda and Lüscher [44]. However, contrary to two-dimensional σ-models, which are renormalizable and therefore consistent quantum field theories, it turned out later that $\mathcal{N} = 8$ supergravity in $D = 4$, which is also a kind of generalized σ-model, is unlikely to enjoy a similar property. This is one of the reasons why this attempt was abandoned. Another reason was closely related to the forthcoming string revolution, when Green and Schwarz (GS) (1984 [45]) proved that $D = 10$, $N = 1$ supergravity, coupled to supersymmetric Yang–Mills matter, could be embedded in a consistent superstring theory for a particular choice of gauge groups (SO(32) and $E_8 \times E_8$).

The GOS and GS papers gave strong evidence that superstrings consistent with space-time supersymmetry containing supergravity + matter (rather than pure higher extended supergravity), in the massless sector, were a possible candidate for a theory of quantum gravity, encompassing the other gauge interactions and free from unphysical ultraviolet divergences. On the contrary, in the context of point-field theories, these systems, even if the symmetries dictated in an almost unique way all the couplings, were found to be non-renormalizable, already at one loop, when standard perturbative techniques were applied to them (Grisaru, van Nieuwenhuizen, and Vermaseren, 1976 [46]). Indeed it was later shown that this was also the case for pure supergravities at and beyond three loops. [These theories had, however, the remarkable property of being one- and two-loop finite (Grisaru, van Nieuwenhuizen, and Vermaseren [46]; Grisaru [47]; Tomboulis [47]).] Pioneering work, in the late 70's, was also the analysis of spontaneous supersymmetry breaking in global and local supersymmetry. In rigid supersymmetry, Fayet opened the way [48] to the construction of the minimal supersymmetric extension of the Standard Model (MSSM), which in particular demanded two Higgs doublets. However, the gauge and supersymmetry breaking introduced by him required more degrees of freedom than the MSSM.

When supersymmetry is gauged, i.e. in supergravity, the supersymmetric version of the Higgs mechanism appears (super-Higgs), i.e. the Goldstino is eaten up by the spin-$\frac{3}{2}$ gravitino (the gauge fermion of supergravity, the partner of the gravitons), which then becomes massive.

The possibility of having spontaneously broken supergravity with vanishing cosmological constant was shown by Deser and Zumino (April 1977 [49]) and proved in detail by Cremmer *et al.* (August 1978 [50]), by studying the most general matter coupling $\mathcal{N} = 1$ supergravity for a chiral multiplet, whose superpotential triggers a nonvanishing gravitino mass. The Higgs effect for Goldstone fermions had also been considered earlier by Volkov and Soroka [51].

Another important result at that time, found by Zumino (August 1979 [52]), was the fact that the most general supergravity couplings of chiral multiplets (with two-derivative action) were described by Kählerian σ-models.

Again, in retrospect, this Kählerian structure and the generalization thereof have played a role in superstring theory from both the world-sheet and target-space points of view.

Although in the 1970's the work done in supersymmetric models for particle physics (using renormalizable Lagrangians with spontaneously broken supersymmetry) and that towards a deeper understanding of the structure of supergravity theories (off-shell formulations, matter couplings, etc.) went in parallel, with small intersections, they came closer and became eventually deeply connected after two major developments were made in the early 1980's.

The first was the call made upon supersymmetry breaking near the electroweak scale, to solve the so-called hierarchy problem of gauge theories with fundamental Higgs scalars (Gildener, Weinberg; Veltman; Witten; Maiani [53, 54]).

This development and general properties of criteria for supersymmetry breaking, contained in two pivotal papers by Witten (April 1981, January 1982 [54]), opened up the field of supersymmetry and supergravity as main stream research in the United States and in the rest of the world.

The hierarchy problem is connected to the unnaturalness of the hierarchy E_{F}/E_X (E_{F} being the Fermi scale) in any renormalizable theory with fundamental scalars, whose vacuum expectation value triggers the electroweak gauge symmetry breaking at a scale $E_{\rm F}$ much lower than any other (cut-off) scale E_X.

This is due to the quadratic dependence on the cut-off Λ of the effective potential, which, at one loop, manifests itself in a term

$$\sum_{J_i}(-)^{2J_i}(2J_i+1)\mathcal{M}^2_{J_i}(\phi)\Lambda^2,$$

where $M^2_{J_i}(\phi)$ are the (scalar) field-dependent masses of particle species with spin J_i. In an arbitrary supersymmetric renormalizable field theory with no traceful Abelian gauge group factor, the expression multiplying Λ^2 identically vanishes (owing to the special relation between boson and fermion couplings), as was shown by Girardello, Palumbo and the author (April 1979 [55]). This is also true for matter-coupled $\mathcal{N}=1$ supergravity with a single chiral multiplet on a flat Kähler manifold (1978 [50]) and in spontaneously broken extended supergravity via the Scherk-Schwarz mechanism (1979 [56]).

However, a closer look at boson-fermion mass matrices revealed that this property made models previously considered by Fayet more problematic, since they tended either to give an unrealistic spectrum with some scalar superpartners of quarks and leptons lighter than their fermionic counterparts, or to need a traceful additional U(1) gauge interaction, plagued with triangular anomalies. Cancelling these anomalies usually needed extra fields, which eventually allowed vacua with broken color or charge symmetry.

However, when the most general coupling of $\mathcal{N}=1$ supergravity to an arbitrary matter system, with arbitrary gauge interactions, became available (Cremmer, Ferrara, Girardello, and Van Proeyen, 1982 [57]), it was realized that, provided $m_{3/2} \ll M_{\rm Pl}$ and possibly $\simeq O(\text{TeV})$, mass terms for any observable scalar $O(m_{3/2})$ were easily generated, thus resolving the partner–spartner splitting problem, which generally occurred in spontaneously broken rigid theories.

There is an alternative way of phrasing this: in the Fayet-type models, the goldstino has coupling to the observable sector $O(1)$ and the gravitino mass is very tiny, $m_{3/2} \sim 10^{-13}$ GeV, while in supergravity models with $m_{3/2} \gtrsim O(m_{\rm Z})$ the Goldstino coupling is highly suppressed $[O(m_{3/2}/M_{\rm Pl})]$, which implies that the gravitino only carries gravitational interactions (Fayet [58]).

In the limit in which $m_{3/2}$ is kept fixed and couplings $O(1/M_{\rm Pl})$ are neglected, spontaneously broken supergravity models behave as globally supersymmetric models with softly broken terms, i.e. terms with dimension ≤ 3, which do not induce quadratic divergences in the low-energy effective theory.

These terms had been classified in 1981 by Girardello and Grisaru [59]. A generalization of non-renormalization theorems for superpotential terms in a generic theory were also found using superspace techniques, by Grisaru, Siegel, and Roček (June 1979 [60]).

Softly broken terms and renormalization theorems were used to construct viable supersymmetric GUTs, including the MSSM as their low-energy effective theory, with no hierarchy problem (the first of these was constructed by Georgi and Dimopoulos in the summer of 1981 [61]). Soon after, realistic electroweak and GUT models, with spontaneously broken supersymmetry triggered by the supergravity couplings at the tree level, were constructed (Barbieri, Ferrara, and Savoy, 1982 [62]; Chamseddine, Nath, and Arnowitt, 1982 [63]; Hall, Lykken, and Weinberg, 1983 [64]). A general feature of these models is that the messengers of supersymmetry breaking to the observable sector (encompassing electroweak and strong interactions) are a set of neutral chiral multiplets (called the hidden sector), which have only gravitational interactions and decouple from the low-energy theory; in the latter, the only trace of them is to produce the soft-breaking terms, then having the effect of modifying the supertrace formula of global supersymmetry with an additional (field-independent) constant (with no physical consequences on the theory decoupled from gravity).

Nowadays, in the MSSM, the electroweak symmetry is broken through radiative corrections, through a Coleman-Weinberg mechanism, while supersymmetry is broken at the tree level through the soft-breaking terms.

Considering the initial condition for the couplings as given at the Planck scale and evolving them through the renormalization group equations (Ibàñez, Ross in 1981 [65], Alvarez-Gaumé, Polchinski, and Wise in 1982 [66]) in a region of the parameter space, the electroweak symmetry was indeed found to be spontaneously broken with a Higgs mass of the same order of magnitude as the gravitino mass. There is a particular subclass of spontaneously broken supergravity models, called no-scale supergravities (Cremmer, Ferrara, Kounnas and Nanopoulos, 1983 [67]; Ellis *et al.* [68]), where the supergravity-breaking scale $m_{3/2}$ is arbitrary at the tree level (owing to flat directions in the superpotential). In these models, radiative corrections may generate the hierarchy $m_{3/2} = M_{\rm Pl}\, {\rm e}^{-c/g^2}$, then explaining how a small scale can be created in a theory whose only original dimensionful scale is $M_{\rm Pl}$.

It was later shown by Witten [69] that many four-dimensional superstring models have, in the point-field limit, a no-scale structure; therefore, after supersymmetry breaking, they may give rise to a dynamical hierarchy.

Nowadays almost every particle physicist knows what $\tan\beta$, A, B represent in the general parametrization of the soft-breaking terms of the MSSM.

The second breakthrough was on physics at the Planck scale (Green, Schwarz, September 1984 [45]), namely the discovery of anomaly-free ten-dimensional supergravity coupled to Yang-Mills matter or consistent superstring theories, for specific gauge group choices (in open and heterotic strings) (Gross, Harvey, Martinec, and Rohm, November 1984 [70]). Heterotic string theories, after suitable compactification of six extra dimensions (on some compact manifolds with special properties), led to $\mathcal{N} = 1$ effective supergravity theories, with a spectrum of charged chiral multiplets (chiral with respect to the surviving gauge group $G' \supset \mathrm{SU}(3) \times \mathrm{SU}(2) \times \mathrm{U}(1)$ (after compactification) and accommodating families of quarks and leptons with the $\mathrm{SU}(3) \times \mathrm{SU}(2) \times \mathrm{U}(1)$ assignment of the Standard Model.

The use of ten-dimensional Yang-Mills fields, prior to compactification, is crucial to overcome previous difficulties encountered in Kaluza-Klein supergravities (Freund and Rubin, 1980 [71]; Witten, 1981 [72]; Duff, Nilsson, and Pope, 1986 [73]), where attempts were made at obtaining the $\mathrm{SU}(3) \times \mathrm{SU}(2) \times \mathrm{U}(1)$ gauge symmetries from the isometries of the internal manifold. In fact, even if in some cases the desired gauge group could be obtained (Witten, 1981 [72]; Castellani, D'Auria, and Fré, 1984 [74]), these failed because the resulting fermion spectrum was not chiral with respect to the electroweak gauge symmetry.

In models where supersymmetry breaking occurs via a non-trivial dilaton superpotential, the neutral fields coming from the internal components of the metric tensor G_{IJ} are natural candidates for flat directions, at least in the limit of manifolds that are "large" with respect to the string scale.

In particular, in four-dimensional heterotic superstring theories, compactified on Calabi-Yau manifolds (Candelas, Horowitz, Strominger, and Witten, January 1985 [75]), or their "orbifold limit," [76] a natural identification of the hidden sector occurs with a set of "moduli fields", which parametrize the deformations of the Kähler class and complex structure of the manifold (generalization of radial deformations of simple tori) [77]. A popular scenario for a non-perturbative dilaton superpotential is the gaugino-condensation mechanism (Ferrara, Girardello, and Nilles [78]) in the hidden sector gauge group (Derendinger, Ibáñez, and Nilles; Dine, Rohm,

Seiberg, and Witten [79]). The fact that some moduli remain large (with respect to the string scale) could result in a sliding gravitino mass, which could then be stabilized through radiative corrections in the observable sector with a dynamical suppression with respect to $M_{\rm Pl}$.

In recent years a suggestion has been made (Duff [80]; Strominger [81]) that strings may, in the strong coupling regime, have a simpler formulation in terms of a dual theory (five-brane) in the weak coupling. These theories, in $D = 10$, have the same field theory limit, namely, $D = 10$ supergravity (which is unique because of supersymmetry), but electrically charged massive string states correspond to "magnetically" charged five-brane states (and vice versa) with a similar relation as it occurs between electric and magnetic charge in Dirac monopole quantization [82].

This is an explicit manifestation of the general fact that a $(p+1)$ form gauge field is, in D dimensions, naturally coupled to a p-dimensional extended object, and that its "dual" potential (which is a $D-p-3$ form) is naturally coupled to a $D-p-4$ extended object. From topological arguments, similar to Dirac's, the product of the two couplings must be quantized.

In toroidal compactifications, it has indeed been shown (Sen and Schwarz [83]) that the spectrum of both electrically and magnetically charged states (the latter obtained from the low-energy effective field theory, i.e. an $\mathcal{N} = 4$ supergravity coupled to Yang-Mills) satisfies an SL(2, Z) duality for the dilaton chiral multiplet $S = (1/g^2) + i\theta$ (here g^2 is the field-dependent gauge coupling and θ is the field-dependent "θ angle"). This may therefore suggest a "modular-invariant" dilaton potential [84]. This symmetry is similar to the "moduli duality," which occurs in weakly coupled strings as a consequence of the world-sheet non-trivial topology [85].

This approach is worth exploring, even if it is difficult to imagine that a microscopic, consistent quantum theory describing five-brane propagation could possibly exist.

Finally, let me conclude by making some remarks about the possible indirect experimental signals, indicating that supersymmetry may be just around the corner. With an optimistic attitude, these are:

1) The non-observation of proton decay within the limit of a lifetime of 10^{32} years in the main channel $p \to \pi^0 e^+$, excluding minimal GUT unification.
2) The LEP precision measurements, which are incompatible with gauge-coupling unification for conventional minimal GUTs, but are in reasonable agreement with minimal supersymmetric GUTs, with supersymmetry broken at the TeV scale.

3) The top Yukawa coupling, unusually large with respect to the one of other quarks and leptons.
4) The possible resolution of the dark-matter problem, with some of the neutral supersymmetric particles as natural dark-matter candidates.

Although none of these facts is *per se* a compelling reason for supersymmetry and may find alternative explanations, it is fair to say that they can all be explained in the context of a supersymmetric extension of ordinary gauge theories.

Whatever the final theory (strings?) for quantum gravity will be, supergravity [86] remains a deep and non-trivial extension of the principle of general covariance and Yang-Mills gauge symmetry, which played such an important role in the description of natural phenomena.

Let us hope that nature has used this fascinating structure!

References

1. S. Ferrara and J.G. Taylor (Eds.), *Supergravity 81* (Cambridge Univ. Press, Cambridge, 1981); S. Ferrara, J.G. Taylor and P. van Nieuwenhuizen (Eds.), *Supersymmetry and Supergravity 82* (World Scientific, Singapore, 1982).
2. Yu.A. Gol'fand and E.P. Likhtman, *JETP Lett.* **13**, 323 (1971).
3. D.V. Volkov and V.P. Akulov, *Phys. Lett.* **46B**, 109 (1973).
4. B. Zumino, in *Renormalization and Invariance in Quantum Field Theory*, ed. E. Caianello (Plenum Press, New York, 1974), p. 367.
5. J.-L. Gervais and B. Sakita, *Nucl. Phys.* **B34**, 633 (1971).
6. A. Neveu and J. Schwarz, *Nucl. Phys.* **B31**, 86 (1971).
7. P. Ramond, *Phys. Rev.* **D3**, 2415 (1971).
8. J. Wess and B. Zumino, *Nucl. Phys.* **B70**, 39 (1974).
9. A. Salam and J. Strathdee, *Nucl. Phys.* **B76**, 477 (1974).
10. S. Ferrara, B. Zumino and J. Wess, *Phys. Lett.* **51B**, 239 (1974).
11. J. Wess and B. Zumino, *Phys. Lett.* **49B**, 52 (1974).
12. J. Iliopoulos and B. Zumino, *Nucl. Phys.* **B76**, 310 (1974).
13. S. Ferrara, J. Iliopoulos and B. Zumino, *Nucl. Phys.* **B77**, 413 (1974).
14. J. Wess and B. Zumino, *Nucl. Phys.* **B78**, 1 (1974).
15. S. Ferrara and B. Zumino, *Nucl. Phys.* B **79**, 413 (1974).
16. A. Salam and J. Strathdee, *Phys. Lett.* **51B**, 353 (1974).
17. P. Fayet and J. Iliopoulos, *Phys. Lett.* **51B**, 461 (1974).
18. S. Ferrara and B. Zumino, *Nucl. Phys.* **B87**, 207 (1975).
19. J. Scherk and J. Schwarz, *Nucl. Phys.* **B81**, 118 (1974).
20. P. Fayet, *Phys. Lett.* **58B**, 67 (1975).
21. L. O'Raifeartaigh, *Nucl. Phys.* **B6**, 331 (1975).
22. P. Fayet, *Nucl. Phys.* **B113**, 135 (1976).
23. M. Ademollo, L. Brink, A. D'Adda, R. D'Auria, E. Napolitano, S. Sciuto, E. del Giudice, P. Di Vecchia, S. Ferrara, F. Gliozzi, R. Musto and R. Pettorino, *Phys. Lett.* **62B**, 105 (1976).
24. G. Gliozzi, J. Scherk and D. Olive, *Phys. Lett.* **65B**, 282 (1976); *Nucl. Phys.* **B122**, 253 (1977).

25. M. Ademollo, L. Brink, A. D'Adda, R. D'Auria, E. Napolitano, S. Sciuto, E. Del Giudice, P. Di Vecchia, S. Ferrara, F. Gliozzi, R. Musto, R. Pettorino and J. Schwarz, *Nucl. Phys.* **B111**, 77 (1976).
26. D. Gepner, *Nucl. Phys.* **B296**, 57 (1988); *Phys. Lett.* **199B**, 477 (1987); W. Lerche, D. Lüst and A.N. Schellekens, *Nucl. Phys.* **B287**, 477 (1987); H. Kawai, D.C. Lwellen and S.-H.H. Tye, *Nucl. Phys.* **B298**, 1 (1987); L. Antoniadis, C.P. Bachas and C. Kounnas, *Nucl. Phys.* **B289**, 87 (1987).
27. D.Z. Freedman, P. van Nieuwenhuizen and S. Ferrara, *Phys. Rev.* **D13**, 3214 (1976).
28. S. Deser and B. Zumino, *Phys. Lett.* **62B**, 335 (1976).
29. This nomenclature was inherited from the "dual models" that had started with the seminal paper by G. Veneziano, *Nuovo Cim.* **57A**, 190 (1968).
30. S. Ferrara, J. Scherk and P. van Nieuwenhuizen, *Phys. Rev. Lett.* **37**, 1035 (1976).
31. S. Ferrara, F. Gliozzi, J. Scherk and P. van Nieuwenhuizen, *Nucl. Phys.* **B117**, 333 (1976); D.Z. Freedman and J.H. Schwarz, *Phys. Rev.* **D15**, 1007 (1977); S. Ferrara, D.Z. Freedman, P. van Nieuwenhuizen, P. Breitenlohner, F. Gliozzi and J. Scherk, *Phys. Rev.* **D15**, 1013 (1973).
32. S. Ferrara and P. van Nieuwenhuizen, *Phys. Rev. Lett.* **37**, 1669 (1976).
33. M.B. Green and J.H. Schwarz, *Nucl. Phys.* **B181**, 502 (1981); *Phys. Lett.* **109B**, 444 (1982) and **136B**, 367 (1984).
34. E. Cremmer, J. Scherk and S. Ferrara, *Phys. Lett.* **68B**, 234 (1977).
35. D.Z. Freedman, *Phys. Rev. Lett.* **38**, 105 (1977); S. Ferrara, J. Scherk and B. Zumino, *Phys. Lett.* **66B**, 35 (1977).
36. S. Ferrara, J. Scherk and B. Zumino, *Nucl. Phys.* **B121**, 393 (1977).
37. K. Narain, *Phys. Lett.* **169B**, 41 (1986); K. Narain, M.H. Sarmadi and E. Witten, *Nucl. Phys.* **B279**, 96 (1986).
38. L. Brink, P. Di Vecchia and P. Howe, *Phys. Lett.* **65B**, 471 (1976); S. Deser and B. Zumino, *Phys. Lett.* **65B**, 369 (1976).
39. A.M. Polyakov, *Phys. Lett.* **103B** (1981) 207 and 211.
40. E. Cremmer and B. Julia, *Phys. Lett.* **80B**, 48(1978); *Nucl. Phys.* **B159**, 141 (1979).
41. E. Cremmer, B. Julia and J. Scherk, *Phys. Lett.* **76B**, 409 (1978).
42. B. de Wit and H. Nicolai, *Nucl. Phys.* **B208**, 323 (1982).
43. J. Ellis, M.K. Gaillard, L. Maiani and B. Zumino, in *Unification of Fundamental Interactions*, eds. S. Ferrara, J. Ellis and P. van Nieuwenhuizen [Plenum Press (Ettore Majorana Series), New York, 1980], p. 69; J. Ellis, M.K. Gaillard and B. Zumino, *Phys. Lett.* **94B**, 343 (1980).
44. A. D'Adda, P. Di Vecchia and M. Luscher, *Nucl. Phys.* **146**, 63 (1978) and **B152**, 125 (1979).
45. M. Green and J. Schwarz, *Phys. Lett.* **149B**, 117 (1984).
46. M.T. Grisaru, P. van Nieuwenhuizen and J.A.M. Vermaseren, *Phys. Rev. Lett.* **37**, 1662 (1976); P. van Nieuwenhuizen and J.A.M. Vermaseren, *Phys. Lett.* **65B**, 263 (1976).
47. M.T. Grisaru, *Phys. Lett.* **B66**, 75 (1977); E.T. Tomboulis, *Phys. Lett.* **67B**, 417 (1977).
48. P. Fayet, *Phys. Lett.* **69B**, 489 (1977).
49. S. Deser and B. Zumino, *Phys. Rev. Lett.* **38**, 1433 (1977).
50. E. Cremmer, B. Julia, J. Scherk, S. Ferrara, L. Girardello and P. van Nieuwenhuizen, *Phys. Lett.* **79B**, 231 (1978); *Nucl. Phys.* **B147**, 105 (1979).
51. D.V. Volkov and V.A. Soroka, *JETP Lett.* **18**, 312 (1973).
52. B. Zumino, *Phys. Lett.* **87B**, 203 (1979).

53. E. Gildener and S. Weinberg, *Phys. Rev.* **D13**, 3333 (1976); S. Weinberg, *Phys. Lett.* **82B**, 387 (1979); M. Veltman, *Acta Phys. Polon.* **B12**, 437 (1981); L. Maiani, *Proc. Summer School of Gif-sur-Yvette* 1979 (IN2P3, Paris, 1979), p. 1.
54. E. Witten, *Nucl. Phys.* **B185**, 513 (1981) and **B202**, 253 (1982).
55. S. Ferrara, L. Girardello and F. Palumbo, *Phys. Rev.* **D20**, 403 (1979).
56. J. Scherk and J. Schwarz, *Nucl. Phys.* **B153**, 61 (1979); E. Cremmer, J. Scherk and J. Schwarz, *Phys. Lett.* **82B**, 60 (1979).
57. E. Cremmer, S. Ferrara, L. Girardello and A. van Proeyen, *Phys. Lett.* **116B** 231 (1982); *Nucl. Phys.* **B212**, 413 (1983).
58. P. Fayet, *Phys. Lett.* **70B**, 461 (1977).
59. L. Girardello and M.T. Grisaru, *Nucl. Phys.* **B194**, 65 (1982).
60. M.T. Grisaru, W. Siegel and M. Roček, *Nucl. Phys.* **B159**, 429 (1979).
61. S. Dimopoulos and H. Georgi, *Nucl. Phys.* **B193**, 150 (1981).
62. R. Barbieri, S. Ferrara and C. Savoy, *Phys. Lett.* **199B**, 343 (1982).
63. A.H. Chamseddine, R. Arnowitt and P. Nath, *Phys. Rev. Lett.* **49**, 970 (1982).
64. L. Hall, J. Lykken and S. Weinberg, *Phys. Rev.* **D27**, 2359 (1983).
65. L. Ibáñez and G.G. Ross, *Phys. Lett.* **105B**, 439 (1981) and **110B**, 215 (1982).
66. L. Alvarez-Gaumé, J. Polchinski and M. Wise, *Nucl. Phys.* **B221**, 495 (1983).
67. E. Cremmer, S. Ferrara, C. Kounnas and D.V. Nanopoulos, *Phys. Lett.* **133B**, 61 (1983).
68. J. Ellis, A.B. Lahanas, D.V. Nanopoulos and K. Tamvakis, *Phys. Lett.* **134B**, 429 (1984); J. Ellis, C. Kounnas and D.V. Nanopoulos, *Nucl. Phys.* **B247**, 373 (1984).
69. E. Witten, *Phys. Lett.* **155B**, 151 (1985).
70. D.J. Gross, J.A. Harvey, E. Martinec and R. Rohm, *Phys. Rev. Lett.* **54**, 502 (1985).
71. P.G.O. Freund and M.A. Rubin, *Phys. Lett.* **97B**, 233 (1980).
72. E. Witten, *Nucl. Phys.* **B186**, 412 (1981).
73. M.J. Duff, B.E.W. Nilsson and C.N. Pope, *Phys. Rep.* **130**, 1 (1986).
74. L. Castellani, R. D'Auria, and P. Fré, *Nucl. Phys.* **B239**, 60 (1984).
75. P. Candelas, G. Horowitz, A. Strominger and E. Witten, *Nucl. Phys.* **B258**, 46 (1985).
76. L. Dixon, J.A. Harvey, C. Vafa and E. Witten, *Nucl. Phys.* **B261**, 678 (1985).
77. N. Seiberg, *Nucl. Phys.* **B303**, 286 (1988); S. Cecotti, S. Ferrara and L. Girardello, *Int. J. Mod. Phys.* **4**, 2475 (1989); P. Candelas, P. Green and T. Hübsch, *Nucl. Phys.* **B330**, 49 (1990); P. Candelas and X.C. de la Ossa, *Nucl. Phys.* **B342**, 246 (1990); S. Ferrara and A. Strominger, in *String '89*, eds. R. Arnowitt, R. Brijan, M.J. Duff and D.V. Nanopoulos (World Scientific, Singapore, 1989), p. 245; A. Strominger, *Comm. Math. Phys.* **133**, 42 (1990); L. Dixon, V.S. Kaplunovsky and J. Louis, *Nucl. Phys.* **B329**, 27 (1990); S. Cecotti and C. Vafa, *Nucl. Phys.* **B367**, 359 (1991); A. Ceresole, R. D'Auria, S. Ferrara, W. Lerche and J. Louis, *Int. Mod. Phys.* **8A**, 79 (1993).
78. S. Ferrara, L. Girardello and H.P. Nilles, *Phys. Lett.* **125B**, 457 (1983).
79. J.-P. Derendinger, L.E. Ibáñez and H.P. Nilles, *Phys. Lett.* **155B**, 65 (1985); M. Dine, R. Rohm, N. Seiberg and E. Witten, *Phys. Lett.* **156B**, 55 (1985).
80. M.J. Duff, *Class. Quant. Grav.* **5**, 189 (1988); M.J. Duff and J.X. Lu, *Phys. Rev. Lett.* **66**, 1402 (1991).
81. A. Strominger, *Nucl. Phys.* **B343**, 167 (1990); C.G. Callan, J.A. Harvey and A. Strominger, *Nucl. Phys.* **B359**, 611 (1991).
82. C. Montonen and D. Olive, *Phys. Lett.* **B72**, 117 (1977).
83. A. Sen, *Mod. Phys. Lett.* **A8**, 2023 (1993); J.H. Schwarz and A. Sen, *Nucl. Phys.* **B411**, 35 (1994); *Phys. Lett.* **B312**, 105 (1993).
84. A. Font, L. Ibáñez, D. Lüst and F. Quevedo, *Phys. Lett.* **B249**, 35 (1990).

85. K. Kikkawa and M. Yamasaki, *Phys. Lett.* **149B**, 357 (1984); N. Sakai and L. Senda, *Progr. Theor. Phys.* **75**, 692 (1986); V.P. Nair, A. Shapere, A. Strominger and F. Wilczek, *Nucl. Phys.* **B287**, 402 (1987); A. Giveon, E. Rabinovici and G. Veneziano, *Nucl. Phys.* **B322**, 167 (1989); A. Shapere and F. Wilczek, *Nucl. Phys.* **B320**, 167 (1989); M. Dine, P. Huet and N. Seiberg, *Nucl. Phys.* **B322**, 301 (1989); J. Molera and B. Ovrut, *Phys. Rev.* **D40**, 1150 (1989); J. Lauer, J. Mas and H.P. Nilles, *Phys. Lett.* **B226**, 251 (1989); *Nucl. Phys.* **B351**, 353 (1991); W. Lerche, D. Lüst and N.P. Warner, *Phys. Lett.* **B231**, 417 (1989); M. Duff, *Nucl. Phys.* **B335**, 610 (1990); A. Giveon and M. Porrati, *Phys. Lett.* **B246**, 54 (1990); *Nucl. Phys.* **B355**, 442 (1991); A. Giveon, N. Malkin and E. Rabinovici, *Phys. Lett.* **B238**, 57 (1990); J. Erler, D. Jungnickel and H.P. Nilles, *Phys. Lett.* **B276**, 303 (1992); S. Ferrara, D. Lüst, A. Shapere and S. Theisen, *Phys. Lett.* **B233**, 147 (1989); L. Dixon, V. Kaplunovsky and J. Louis, *Nucl. Phys.* **B355**, 649 (1991); G.L. Cardoso and B. Ovrut, *Nucl. Phys.* **B369**, 351 (1992) and **B392**, 315 (1993); J.-P. Derendinger, S. Ferrara, C. Kounnas and F. Zwirner, *Nucl. Phys.* **B372**, 145 (1992); L. Antoniadis, K. Narain and T. Taylor, *Phys. Lett.* **B267**, 37 (1991); L. Antoniadis, E. Gavai and K. Narain, *Nucl. Phys.* **B365**, 93 (1992).
86. Reviews on the developments of supergravity in the 80's can be found in: P. van Nieuwenhuizen, *Phys. Rep.* **68**, 189 (1981); H.P. Nilles, *Phys. Rep.* **110**, 1 (1984); S. Ferrara, Vol. 2 of *Supersymmetry* (North Holland, Amsterdam/World Scientific, Singapore, 1987).

Figure 4.4: Young Sergio Ferrara, Daniel Freedman, and Peter van Nieuwenhuizen. 1976.

Figure 4.5: Special 2019 Breakthrough Prize in Fundamental Physics. Acceptance Ceremony. Left to right: Peter van Nieuwenhuizen, Sergio Ferrara, and Daniel Freedman.

SUPERGRAVITY AND ITS LEGACY[94]

SERGIO FERRARA
Department of Theoretical Physics
CERN
Espl. des Particules 1, 1211 Meyrin, Switzerland;
INFN — Laboratori Nazionali di Frascati
Via Enrich Fermi 40, I-00044 Frascati, Italy;
Department of Physics and Astronomy and
Mani L. Bhaumik Institute for Theoretical Physics,
UCLA, Los Angeles CA 90095-1547, USA
Email: Sergio.Ferrara@cern.ch

1 The Prelude

In the early 1970s I was a staff member at the Frascati National Laboratories of CNEN (then the National Nuclear Energy Agency). With my colleagues Aurelio Grillo and Giorgio Parisi we were investigating, under the leadership of Raoul Gatto (later Professor at the University of Geneva) the consequences of the application of Conformal Invariance to Quantum field Theory (QFT), stimulated by the ongoing experiments at SLAC where an unexpected Bjorken scaling was observed in inclusive electron-proton cross sections, which was suggesting a larger space-time symmetry in processes dominated by short distance physics.

In parallel with Alexander Polyakov, at the time in the Soviet Union, we formulated Conformally Invariant Operator Product Expansions (OPE) and proposed the Conformal Bootstrap as a non-perturbative approach to QFT.

Conformal Invariance, OPEs and the Conformal Bootstrap have again become fashionable subjects in recent times, because of the introduction of efficient new methods to solve the Bootstrap Equations (Riccardo Rattazzi, Slava Rychkov, Eric Tonni and Alessandro Vichi), and mostly because of their role in the AdS/CFT correspondence.

The latter, pioneered by Juan Maldacena, Edward Witten, Steve Gubser, Igor Klebanov and Polyakov, can be regarded, to some extent, as one of the great legacies of higher-dimensional Supergravity. It can be used to gain information on strongly coupled gauge theories, and provides a variety

[94]Ettore Majorana Medal address at International School of Subnuclear Physics, 54th Course, Erice, 14 June 14–23, 2016, ArXiv 1701.04317.

of applications to Particle Physics, String Theory and Condensed Matter Physics.

In 1973 I was awarded a Fellowship at the CERN Theory Division, which was considered the most prestigious place a young European physicist could go. Subsequent events revealed that going to Geneva was a truly timely choice, since when I started my Fellowship, in the fall of 1973, Julius Wess and Bruno Zumino had just formulated supergauge invariant (now called supersymmetric) QFT in four dimensions, and then found that milder divergences occur in these theories with respect to standard renormalizable QFT.

The original name was inherited from fermionic strings, where a local (conformal) fermionic symmetry on the world sheet was introduced by André Neveu and John H. Schwarz, Pierre Ramond, and Jean-Loup Gervais and Benji Sakita.

During my two-year period as a CERN Fellow I wrote four papers with Zumino, two of which were also with Wess and Jean Iliopoulos (at the time a visiting scientist at CERN). With Zumino we formulated four dimensional Supersymmetric Yang Mills Theories (the new name for supergauges introduced by Abdus Salam and John Strathdee), thus opening the way to consider supersymmetric extensions of the Standard Model, the Interstellar medium (ISM) and alike.

The merging of supersymmetry with Yang-Mills gauge symmetry was a non-trivial step, and the potential link with the Standard Model was motivated by the so-called *Hierarchy Problem* and *Naturalness of Scales*, which in these new theories were alleviated thanks to so-called *Non-Renormalization Theorems.*

Many people contributed to these developments. Aside from the previously mentioned collaborators of Zumino, I can surely mention Kelly Stelle, Peter West, Warren Siegel, Jim Gates, Marc Grisaru and Martin Roček.

In retrospect, the most influential of my joint papers with Zumino, written during my period as a Fellow, was the one on the "Supercurrent multiplet," a multiplet which extended Noether's theorem to Supersymmetry and whose components include the energy momentum tensor and a vector-spinor fermionic current, as well as a vector, a scalar and a pseudoscalar. The considerations that follow were among the starting points which led Daniel Freedman, Peter van Nieuwenhuizen and myself to formulate Supergravity in four dimensions as the supersymmetric extension of General Relativity (GR).

Since in theories with local (gauge) symmetries Noether currents couple, to first order, to gauge fields, the fact that the stress tensor couples to the metric tensor in GR indicated that the spinor current should have coupled to a spin-3/2 massless gauge field, of the type introduced by William Rarita and Julian Schwinger (RS) in 1941, and promptly used by Shuichi Kusaka as a hypothetical spin-3/2 neutrino.

The structure of the super-current multiplet indicated two facts that were later exploited, namely that a supersymmetric extension of GR should have involved only two particles of helicities (2,3/2), and that the field corresponding to the latter should have been a Rarita-Schwinger field, to be coupled to GR. This seemed in contrast with a superspace formulation of a Super Riemannian Geometry, which was originally pursued by Pran Nath, Richard Arnowitt and Zumino, where many ordinary fields with different spins were needed.

In the fall of 1975, even if I had offers from the US, following a suggestion of Jacques Prentki, at the time CERN-TH Division leader, I accepted a CNRS position at the École Normale Supérieure in Paris.

There I had the chance to meet Daniel Freedman, who was also trying to work out a consistent theory for spin 3/2 coupled to GR.

Peter van Nieuwenhuizen was then at his colleague at Stony Brook. He was also collaborating with us, since he was investigating the quantum properties of fields of different spins coupled to GR, and the possibility of ultraviolet cancellations similar to those encountered in super-Yang-Mills theories.

In the spring of 1976 the breakthrough was achieved. Using the Noether procedure, Fierz rearrangements and Riemann tensor identities (and a computer at the Brookhaven National Laboratory), we proved that there is a unique Lagrangian, with only spin 2 and 3/2, and a unique set of supersymmetry transformation rules, which defines pure Supergravity in four dimensions, the supersymmetric extension of GR.

General coordinate transformations, when combined with Supersymmetry, promoted the latter to a gauge symmetry as well, since spinors in curved space must necessarily depend on the space-time point due to the local Lorentz symmetry.

2 The Play

The construction of Supergravity as the gauge theory of local supersymmetry prompted Stanley Deser and Zumino to formulate this theory as the Einstein-Cartan version of GR minimally coupled to a (spinor valued) one-form field, the RS gravitino (word coined by Sidney Coleman).

In their work, a quartic spin-3/2 coupling with gravitational strength, required in our formulation by local supersymmetry, originated as a torsion contribution to the spin-connection when treated as an independent field. This extends the so-called first-order or Palatini formulation of GR.

A subsequent superspace formulation of Supergravity by Wess and Zumino in 1978, where an appropriate super-torsion constraint was introduced to reproduce the flat super-geometry of the super-Poincaré algebra, revealed that our second-order approach, rather than the Palatini-Einstein-Cartan formulation, is equivalent to the Wess-Zumino curved Superspace. In particular, it reproduces the so called auxiliary fields, needed to have an off-shell formulation and thus a tensor calculus as in standard GR, introduced by Peter van Nieuwhenuizen as well as myself and Stelle and West.

In the months following the Spring of 1976 several important results were obtained, both during the Summer Institute organized by the École Normale Supérieure in Paris and during subsequent visits of some of us to Stony Brook and to CERN.

In particular, also in collaborations with Eugene Cremmer, Joel Scherk, Bernard Julia, Ferdinando Gliozzi and Peter Breitenlohner, the first matter couplings were obtained and, with Peter van Nieuwnhuizen, the first extended $\mathcal{N} = 2$ Supergravity formulated.

In a fundamental parallel work, Gliozzi, David Olive and Scherk discovered that the dual spinor model, after a GSO projection, becomes supersymmetric in space-time and not only on the string world-sheet. Moreover, they found that $\mathcal{N} = 4$ Supergravity Coupled to $\mathcal{N} = 4$ Yang-Mills is the low energy effective theory of the $10D$ dual spinor model. Meanwhile, Scherk, Lars Brink and Schwarz classified super Yang-Mills theories in all dimensions.

In these milestones of Superstring Theory, Supergravity in diverse dimensions started to emerge as a low-energy limit.

The years 1977 and 1978 were the most performing for Supergravity because the discovery of off-shell formulations allowed to write general matter-coupled Lagrangians and to begin a systematic study of the super-Higgs effect, the mass generation for the gravitino as a consequence of the spontaneous breaking of local Supersymmetry. This includes a joint work with Cremmer, Julia, Scherk, Girardello and van Nieuwenhuizen.

Moreover, the ultraviolet properties of Supergravity theories started to be investigated and pure supergravity was shown to be finite at one and two loops. This is to be contrasted with the later, fundamental work of Marc Goroff and Augusto Sagnotti, where GR was shown to diverge at two loops. Three-loop counter terms in different forms of Supergravity were, however, exhibited by different authors, including Renata Kallosh, Grisaru and Deser, Jason Kay and Stelle.

In a pivotal step, in those years examples of all extended supergravities in four dimensions were derived and the appearance of electric magnetic (e.m.) duality symmetries was noticed, with duality groups that act in a non-linear way on the scalar fields of the theory.

The $\mathcal{N} = 8$ supergravity in 4D was constructed by Cremmer and Julia, and shown to possess a very large duality group, the exceptional group E7(7).

Gauged $\mathcal{N} = 8$ Supergravity in 4D was constructed by Bernard de Wit and Hermann Nicolai, and gauged Supergravity in diverse dimensions was studied by many people, including Michael Duff, Murat Günaydin, Chris Hull, Nicholas Warner, Paul Townsend, Leonardo Castellani, Riccardo D'Auria and Pietro Fré. E.m. dualities were later promoted to string U-dualities by Hull and Townsend.

Cremmer, Julia and Scherk constructed Supergravity in the highest possible number of dimensions, eleven, following a previous classification of Super-Poincaré algebras by Werner Nahm.

This theory led Witten to formulate, in the mid-1990s, the remarkable M-theory hypothesis, thus identifying an elusive theory in eleven spacetime dimensions via the strong coupling limit of the Type-IIA superstring. Through webs of dualities, this step allowed to unify the different form of $10D$ Superstrings, thus also connecting different types of extended objects, called generically branes, which emerge as solitonic solutions from higher dimensional supergravity.

Duff had been the first, in the 1980s, to advocate for the central role of generic extended objects in an overall picture.

D-branes in string theories first appeared in the classic works of Massimo Bianchi, Gianfranco Pradisi and Sagnotti, and of Joseph Polchinski.

The advent of M-theory marked the second string revolution. The 1980s witnessed extensive investigations of Kaluza-Klein (KK) theories, largely aimed at classifying possible KK consistent reductions (spontaneous compactifications).

General Lagrangians including soft breaking terms of gravitational and/or gauge origin were derived by Cremmer, Luciano Girardello, Antoine van Proeyen and myself, and were applied to supersymmetric models of particle physics.

No-scale supergravity, a theory with a naturally vanishing cosmological constant, was built by Cremmer, Costas Kounnas, Dimitri Nanopoulos and myself, and was readily applied to physics beyond the Standard Model with

John Ellis. Twenty years later these models were linked to string vacua in flux compactifications by Steve Giddings, Shamit Kachru and Polchinski.

All these developments made it possible to derive parameter spaces of masses and couplings for supersymmetric extensions of the Standard Model, which can be explored in experimental searches for new physics, as was done at LEP experiments in the past and is currently done at the LHC collider.

Supersymmetric extensions of the Standard Model were introduced and widely investigated by Pierre Fayet and by Savas Dimopoulos and Howard Georgi, relying heavily the soft-breaking terms proposed by Girardello and Grisaru and derived in supergravity by Riccardo Barbieri, Carlos Savoy and myself, and independently by Lawrence Hall, Joseph Lykken and Steven Weinberg, and by Arnowitt, Ali Chamseddine and Nath.

"Split supersymmetry," a mechanism advocated in order to get large splitting among MSSM super partners while keeping the nice predictions of gauge coupling unification was later introduced by James Wells, Nima Arkani-Hamed, Dimopoulos, Gian Giudice, Andrea Romanino, and was further developed by the same authors.

A new wave of activity in Supergravity was spurred in 1984, when Michael Green and Schwarz ignited the "first superstring revolution" with the discovery of their cancellation mechanism (for gauge groups E8×E8 or/and SO(32)), making heterotic and Type-I superstrings (which can only afford the latter option) and their point limit, $10D$ (1,0) supergravity, free of potential gauge anomalies and also of the gravitational anomalies previously studied by Luis Alvarez-Gaumé and Witten.

The anomaly cancellation mechanism placed restrictions on compactification manifolds, uncovering enticing connections between the supergravity analysis and algebraic geometry.

The Calabi-Yau compactifications investigated by Philip Candelas, Gary Horowitz, Andrew Strominger and Witten resulted in a paradigmatic example of $\mathcal{N} = 2$ supergravity analysis in terms of Special Geometry and Mirror Symmetry.

All this machinery of compactifications was thus reconsidered, taking into account restrictions coming from the Green-Schwarz mechanism. Important implications exist for compactifications to six dimensions preserving eight supercharges, where some anomaly canceling terms modify the low energy couplings of tensor multiplets to gauge fields.

In the 1990s, supergravity evolved into a major tool for the study of black holes (BH), and in particular of their BPS properties, given their emergence as supersymmetric solitons in four-dimensional $\mathcal{N}$-extended Supergravity.

An interesting interplay between charged black holes and the moduli-space geometry of the underlying supergravity emerged through the so-called "attractor mechanism," a property noticed by Kallosh, Strominger and myself when we met at an Aspen Summer Institute, in Colorado, in 1995.

In $\mathcal{N} = 2$ supergravity supersymmetric black holes enjoy a universal formula for their Bekenstein-Hawking entropy, in terms of the (square modulus) of the central charge computed at its extremum in moduli space (see also work with Gary Gibbons).

The name attractors was coined since these solitons follow trajectories in the moduli space of scalar fields which, independently of the initial conditions, reach the same extremal point, which coincides with the value of field at the BH Horizon. This explains why the moduli dependent Arnowitt-Deser-Misner (ADM) mass is a continuous parameter while the entropy is quantized in terms of the BH charges.

These results were then extended to theories with $\mathcal{N}$-extended supersymmetry and in higher dimensions in work including my collaborators and friends Kallosh, Paolo Aschieri, Bianca Cerchiai, Laura Andrianopoli, Stefano Bellucci, Anna Ceresole, Chamseddine, Gianguido Dall'Agata, D'Auria, Fré, Murat Günaydin, Maria Lledó, Alessio Marrani, Mario Trigiante, Armen Yeranyan and Zumino.

Partial supersymmetry breaking in $\mathcal{N} = 2$ supergravity was studied in collaboration with Girardello and Massimo Porrati, with inspiration from the work of Ignatios Antoniadis, Hervé Partouche and Tomasz Taylor, and a minimal model with flat potential was explored.

In 1997 supergravity found itself again at center stage in the work of Juan Maldacena, amusingly entitled "The Large-N Limit of Superconformal Field Theories and Supergravity", where the famous AdS/CFT duality was formulated, which was soon followed by important additions by Witten, Gubser, Klebanov and Polyakov. This might be called the "third string revolution," and would have been impossible without a particular compactification of type IIB supergravity on AdS5×S5 derived in 1985 by Hyiung-Jin Kim, Larry Romans and Peter van Nieuwenhuizen.

This completes my personal recollection of many past years devoted to supergravity. The recent years (2013–2016) witnessed a revival of spontaneous breaking of local supersymmetry, boosted by applications to cosmology, and in particular to the inflationary Universe with a de Sitter phase, with an eye to observational results of the Planck Mission.

A prominent role in this work was played by Andrei Linde, a proponent of Inflationary Cosmology, and indeed Daniel Freedman and I collaborated independently with Kallosh and Linde. I also collaborated with a number of colleagues, including Antoniadis, Dall'Agata, Emilian Dudas, Alex Kehagias, Porrati, Sagnotti, Jesse Thaler, Nieuwenhuizen Proeyen, Timm Wrase and Fabio Zwirner.

Supergravity has also found applications in higher-spin field theories, via the AdS/CFT correspondence. Supersymmetric versions of OPEs and of the Conformal Bootstrap for $\mathcal{N}$-extended superconformal field theories are currently under investigation. In the quantum domain $\mathcal{N} = 8$ supergravity in four dimensions was proved to be ultraviolet finite at very high loop order, following the pioneering work of Zvi Bern, Lance Dixon and David Kosower. These properties seem to descend from some amazing relations with the rigid $\mathcal{N} = 4$ Yang-Mills theory, which possesses exceptional conformal and integrability properties. It is not yet excluded that $\mathcal{N} = 8$ supergravity be a perturbatively finite theory of Quantum Gravity. In any case this search is going to teach us new lessons on the ultraviolet cancellations which can occur in theories including General Relativity.

Acknowledgments: The author was supported in part by the CERN Department of Theoretical Physics and by INFN (IS CSN4-GSS-PI).

Figure 4.6: Stanely Deser, celebrating Bruno Zumino's 90th Birthday May 2–4, 2013, Berkeley

A BRIEF HISTORY (AND GEOGRAPHY) OF SUPERGRAVITY: THE FIRST THREE WEEKS AFTER...[95]

S. DESER

Walter Burke Institute for Theoretical Physics
California Institute of Technology, Pasadena, CA 91125;
Physics Department, Brandeis University, Waltham, MA 02454

I summarize, at its 41st — and what would have been Bruno Zumino's 94th birthday — the history of the discoveries of Supergravity, and some of its structure and later developments.

1 Background

Briefly, the relevant background — and the ideas in the air — prior to Supergravity's (SUGRA's) inception lay in two new, if quite different, realms: supersymmetry (SUSY), and the, also emerging, difficulties in achieving consistent gravity-higher ($s > 1$) spin gauge field interaction. Indeed, the Western discoverers of SUSY, Julius Wess and Bruno Zumino [1], would frequently visit Boston from NYU[96] to spread the SUSY gospel, which did get even our blasé attention after a while, especially since the simplest SUSY multiplet pattern $(s, s + 1/2)$ linking adjoining Fermi-Bose fields had no obvious reason to stop at the then studied $s = 0$ and $s = 1/2$ models. [The earlier, Eastern, discovery is [2].] The specific idea of a SUGRA was also in the air, both in terms of spacetime-dependent SUSY parameters [3] and of (albeit too general at the time) superspace approaches [2]. Separately, I had been intrigued for some time by the woes of higher spin gauge fields interacting — as they must — with gravity [3,4], and was particularly struck by Buchdahl's [5] early, if classical, study of the first difficult, $s = 3/2$, case.

A rather different, but extremely potent motivation was the striking improvement observed in the UV behavior of SUSY models, where infinities from the Bose/Fermi components miraculously cancel each other. It had been shown a bit earlier that quantum gravity coupled to matter was in dire

---

[95] Eur. Phys. J. H **43**, no. 3, 281–291 (2018) [arXiv:1704.05886 [physics.hist-ph]]. Abbreviated.

[96] New York University.

need of miraculous cancellations, being non-renormalizable already at 1-loop level for spin $(0, \frac{1}{2}, 1)$ — both Maxwell and Yang-Mills matter sources [6], while the source-free system was known to diverge at the next, 2-loop, level [7]. That UV hope was an enticing — if still unresolved — carrot; indeed, SUGRA not only shares the one-loop finiteness of pure GR [8] — the only "matter" field to do so, but stays finite at 2 loops because there are no ($\sim$Riemann^3) invariant counterterms there. The (Riemann^4) stick, found a bit later [9], only strikes from 3 loops on. On a personal level, I had invited Zumino to lecture at the last — 1970 — Brandeis Theory Summer school. We stayed in touch thereafter, and his presence was a strong motivation to spend my Sabbatical term at CERN. After our initial SUGRA work, we wrote a number of related papers in the field, including one on the 2D basis of "Superstring SUGRA" [10].

2 Action

When I arrived on the evening of April 1, 1976, Bruno was awaiting for me in the CERN cafeteria and we instantly began our nonstop 3-week endeavor — especially at night, when we filled the CERN lecture hall's immense blackboards with our seemingly endless Fierz identity slogs! [We were once startled by another insomniac, Claud Lovelace, lurking in the curtains.] I had to go back to the US at dawn on the 22nd for a short stay; fortunately, we were done with all calculations and writeup by then, and Bruno handed our manuscript [11] to Raoul Gatto, the local Editor of Phys. Lett. B, CERN's then "house organ". One reason our work succeeded so rapidly was that we were both familiar with two papers by the great mathematician Hermann Weyl on coupling (Dirac) spinors to gravity [12]. (Amusingly, this too was a West-East discovery; see [13]). The first and oldest of these, showing how to do it, is well-known; the second — directly relevant for us — some two decades later, in 1950, was, and remains, an obscure gloss on the first. That was the year I began my student subscription to the Physical Review and pretended to read the fancy theory papers, in particular that one. Weyl noticed, in that short afterthought, that spinors can couple to affinities two ways, namely "first" and "second" order: either the affinities are regarded as independent variables or expressed as vierbein affinities, and that those two ways differed by (non-derivative) terms quartic in the spinors — the torsion. Weyl's notation was rather cumbersome, and ours were real Majorana vector — rather than Weyl's complex Dirac spinors — so we worked that dictionary out for ourselves, even though our aim was precisely to avoid dealing with the horrors of torsion, by sticking to first order. I was of course a rabid first order fan, as that was the basis of the 1959–62 "ADM" (Arnowitt-Deser-Misner) formulation of gravity, and Bruno was ambidextrous as well. [Indeed, we could have saved a lot of chalk by not worrying about the transformation properties of the independent (in first order) affinity, which is essentially just a Lagrange multiplier.] The beauty of our approach (pace Boltzmann's dictum that elegance is for tailors, not physicists) is that the entire action in [11] is the two-term sum of the (first order) Einstein and (the modern version of) minimally gravitationally coupled Rarita-Schwinger-Davydov-Ginzburg actions (yet another West-East discovery). The latter part just involves a covariant curl of the vector-spinor using the independent, non-metric affinity, as does the Einstein ("Palatini") action. That's it — neither quartic spinor nor auxiliary field debris!

This deceptively naive form of the coupled field equations obeyed the required "Bianchi" identities — that their divergences vanished, both for the gravitational and spin $\frac{3}{2}$ field equations. That of the latter is mandated by SUGRA's invariance under the local, fermionic gauge transformation generalizing that of the flat space massless $\frac{3}{2}$ field (again, simply replacing an ordinary partial by a covariant derivative) while the vierbein transforms like all SUSY bosons, with no derivatives of the gauge parameter. The vanishing of the spin $\frac{3}{2}$ field equation's divergence is in fact related to the other influence I mentioned at the start, Buchdahl's remark [5] that the massless spin $\frac{3}{2}$ equation is only consistent in Ricci-flat spaces, because its divergence is proportional to the 2-index Ricci tensor that arises from the resulting covariant derivative commutator. He thought this meant this matter field could not consistently live in a General Relativity background (except as a test field), since the Ricci tensor was proportional to its stress tensor, so could not vanish. What this classical study omitted of course was the power of Fierz identities in (necessarily) quantized vector-spinor fields to show that this stress-tensor contribution vanishes identically (as it had better do, not being gauge-invariant by itself). It also sealed the fate, as was later shown explicitly [14], of "hypergravities", systems like $(2, 5/2)$ because the spin $5/2$ tensor-spinor field equation's divergence has more indices, hence unavoidably brings in the full Riemann (rather than just Ricci) tensor, whose Weyl tensor part is left undetermined by the Einstein equations. This completes the (admittedly all too compressed, but accurate) survey of the motivation, history, and not least — conceptual plus technical — tools (and sweat, at our already advanced physics ages) involved in [11].

3 Re-action

When we started, Bruno had just received (in those pre-internet days) a preprint of the initial version of the other SUGRA group [15], attempting to stake a claim to ownership of SUGRA, based on lowest order coupling in the traditional second order formalism. However, there is a long, instructive, history of would-be extended gauge invariant systems: Lowest order coupling always works, only to fail at next order — the real test of consistency only occurs after the presumed first step's effect impinges on the next one; indeed, the $(2, \frac{5}{2})$ model is of this type. Further, the $(\frac{3}{2}, 2)$ idea was in any case already very much in play, as mentioned earlier. After we had finished, Bruno phoned me to report that a second version of [15] had just arrived, stating that an extensive computer calculation of their quartic (in the fermions) remainder terms had finally shown all of them to vanish; we instantly, and unilaterally, decided to cite their results in our published version. As we have seen, both results were entirely on a par in timing, while all aspects of the model's constructions were mutually independent technically and conceptually. This was yet another example of simultaneous scientific discoveries, unsurprisingly when the right ideas are in the air. Most familiar in recent times are the three separate inventions of the "Higgs" in the sixties, and most relevant to our theme, all the East-West groups mentioned earlier.

There is usually a reasonably fair consensus in such cases as to who did what when and deserves what share of the credit. To be sure, at the essential, scientific, level, each group is intrinsically rewarded just by seeing the dazzling new light for the first time, and the real truth is graven in the Platonic heavens. Here on earth, however, a Whig twist occasionally prevails, as it seems to have done here. This anniversary provides a chance to correct it, as Bruno often wished. In the Appendices,[97] we provide some of the technical details to help non-experts appreciate the subject.

4 The post-1976 flowering of SUGRA

It is totally impossible to summarize, let alone discuss, the ~15,000 papers on SUGRA generated to date. Instead, I can only very briefly highlight some of the directions that have been explored, at that with many omissions of important work! These include: 1. Different dimensions, 2. Higher internal symmetrics, 3. Quantum effects: fighting the divergences, 4. Relation to

[97] The original publication [16] contains three appendices two of which we omit in this abbreviated version.

superstring theory. The original early papers are fortunately gathered in a 2-volume annotated anthology [17] that provide an encyclopedic (in the authors' words) guide for the years 1976 to 1989; there also exist other compilations, but an updated list would be a welcome addition.

SUGRA models are highly dimension-dependent, because the number of Bose and Fermi degrees of freedom (DoF) must match: in the original $D = 4$ models, this was guaranteed because all massless spin > 0 fields have two DoF there. Perhaps the first $D \neq 4$, and certainly the lowest-dimensional, system was the, $D = 1$, spinning particle [18], whose bosonic part $x^\mu(t)$ and fermionic $\psi(t)$ represents its spin. It is of course also the simplest, and is amusingly the (1st order) particle plus spinor action coupled to $D = 1$ SUGRA, which has no dynamics, or even action, of its own but ensures that the particle's action is both coordinate- and local super-invariant. Its simplicity makes it is an excellent starting-point to experience SUGRA: the action is:

$$A = \int d\tau \left[\pi\dot{x} - \frac{1}{2}e\pi^2 - \frac{i}{2}\psi\dot{\psi} - \frac{i}{2}\chi\psi x\right], \tag{4.15}$$

where e is the "einbein" and χ its SUGRA partner "spin 3/2" fermion, here enforcing the on-shell conditions $\pi^2 = 0 = \dot{\psi}x$; we have omitted the vector world indices on x^μ and ψ_μ, understood to be summed throughout. Amusingly, there is again room for a "cosmological" plus particle mass term, exactly as in $D = 4$.

The next case, $D = 2$, is of special interest because it describes the superstring in its intrinsic motion as a $D = 2$ surface moving in spacetime [11]. The variables are the two-component $(i = t, x)$ position $A_i(t, x)$ and Majorana vector-spinor $\chi_i(t, x)$, whose action is required to obey on-shell conditions just like the $D = 1$ case, and which will be enforced again by coupling to $D = 2$ SUGRA, which again has no dynamics (recall that Gauss's theorem says the Riemann scalar density is a total divergence in $2D$!). We will not repeat the covariant action here, but note that it is a relatively straightforward extension to a $2D$ curved space and that it has the additional Weyl symmetry under conformal transformations beyond its coordinate and local SUGRA invariance. Thus the superstring may be thought of as the matter field coupled to $D = 2$ SUGRA; needless to say this formulation does not directly probe the deep facts that the space-time must have $D = 10$ for quantum reasons.

Moving on to $D = 3$, a dimension where SUGRA is still not dynamical because the Ricci tensor is equivalent (double dual of) Riemann, so it has no freedom — the Einstein equations imply spacetime is everywhere flat in absence of sources and completely determined by them otherwise. But there is a new and unexpected model here, topologically massive gravity (TMG) [19], which does have one massive DoF with quite unusual properties: it consists of the usual Einstein-Hilbert term plus the Chern-Simons (CS) invariant, an integral of third derivative (but harmless) order, the sum now being the sum of Einstein and CS, schematically $A(\text{grav}) \sim \int[R(\omega, e) + \mu^{-1}\omega^3]$, where ω is again an independent field and μ has mass dimensions. The Fermionic companion is the field strength $f^\mu = \epsilon^{\mu\nu\tau}(D_\nu\psi_\tau)$ and the $s = 3/2$ part of its action is [20],

$$A(3/2) \sim \int [\bar{\psi}D(\omega)\psi + \mu^{-1}f\bar{f}] \tag{4.16}$$

suitably adorned with gamma matrices and dreibeine. The total action represents one DoF of each species, is parity-variant but SUGRA-invariant (for ω independent!), When μ goes to infinity, one recovers the dull pure Einstein DoF-less $3D$ SUGRA. We have already done $D = 4$; the successive steps of higher D, while each somewhat different, reveal no qualitative change — until, that is, one hits $D = 11$. Recall that there are two requirements on a SUGRA theory: the Bose/Fermi excitations must balance and the spin cannot exceed 2 because of the "Buchdahl" problem that the "Bianchi" identities on the higher spin equations cannot be maintained. We will return to the balance question in the next section, but it suffices to note that $D = 11$ is the highest D that obeys these constraints, as was shown by a group-theoretic counting in [21]]; furthermore, $D = 11$ SUGRA is unique and can be compactly written as the sum of Einstein, spin 3/2 and a new, 3-form Bosonic field [22]; note that this is one dimension higher than superstrings live in. The new, 3-form part, is relatively simple in terms of the field strength $F_{\mu\nu\iota\sigma} = \partial_{[\mu}A_{\nu\iota\sigma]}$, namely a "Maxwell" term $A \sim -\int F^2$ plus a CS-like $A = \int \mu^{-1}\epsilon^{1\ldots11}F_{1..4}F_{5..8}A_{9..11}$, and their Fermion vector-spinor counterparts. Needless to say, there is a whole industry devoted to dimensional reduction and the relation of various rungs on the D-ladder. It is also perhaps deep that while Bosonic gravity exists in any D, SUGRA puts a relatively low "cap" on it. There is no mystery why: Nature (or at least theory) abhors spin > 2.

Our next topic is that of extended SUGRAs, another chapter in the Bose/Fermi balance: consider lower spin Supersymmetric (SUSY) models, $(s, s + 1/2)$ etc. When these are appropriately coupled to SUGRA, say in $D = 4$, the resulting SUGRA theory now has more components, labelled by the "internal" symmetries. In particular the widest range would include spins $(0, \frac{1}{2}, 1, \frac{3}{2}, 2)$ with various multiplicities — however ending with just one graviton. But there is a way to double the seeming "$N = 4$" structure. It essentially consists of "running down the scale" again to make $\mathcal{N} = 8$ with a single maximal spin 2; remember that in $D = 4$, each $s > 0$ (and Majorana spinor) has 2 modes. The totals are $(70, 56, 28, 8, 1)$ fields of spin $(0, \frac{1}{2}, 1, \frac{3}{2}, 2)$ with 128 modes of each species (the 70 scalars have only 70 excitations). Note that unlike lower $\mathcal{N}$ models, $\mathcal{N} = 8$ is unique [23] in that it cannot couple to lower spin SUSY matter; this is also true of the $D = 11$ model, where there simply is no "matter" but only the three SUGRA fields. This of course also differs violently from bosonic gravity, that accepts all matter fields (of $s < 2$). Higher $\mathcal{N}$ SUGRA has also generated enormous work in various directions, that I cannot cover here, but can easily be traced via arXiv; we will, however, briefly discuss the quantum aspects of various SUGRAs. The sheer algebraic complication of handling high $\mathcal{N}$ systems has led to the use and elaboration of superfields that simultaneously represent many components, much like (hyper) complex numbers unify two variables. A dynamical extension is to study the possible gauging of these systems, for example the internal coupling of the various fields to the vector components, rather than simply leave them free. Both of these are now major subfields.

Finally, we turn to the loop aspects of SUGRA. It is — by construction — a quantum theory because the spinor components must be quantized, hence also the Bosonic ones — since they rotate into each other. This means it has to face the problem of perturbative non-renormalizability of GR and its dimensional coupling — Newtonian/Planck — constant. But SUSY models have been spectacularly successful in the cancelation of infinities between Bose/Fermi modes, culminating in the, not only renormalizability, but finiteness, of $\mathcal{N} = 4$ Super Yang-Mills theory [24], so there is still hope *a priori*! This has been a roller-coaster, still unfinished, saga, as we shall see. First, a reminder of $D = 4$ Bosonic GR's problems. On purely dimensional grounds, its one-loop counterterms are quadratic in curvature; however, as pointed out in [8], the Gauss-Bonnet identity expresses $\int \text{Riem}^2$ as the combination $\int [4\text{Ricci}^2 - R^2]$. Hence these terms can all be removed, to the one-loop order at hand, by a field redefinition $\Delta g_{\mu\nu} = \kappa^2[aR_{\mu\nu} + bRg_{\mu\nu}]$, whatever the specific coefficients, using the tree level action via $\int \Delta g_{\mu\nu} G^{\mu\nu}$. However, this salvation becomes unavailable as soon as matter coupling, in [8] to scalar fields, is included. This in turn led to a search for the effects of the other then known matter systems, spinors, and vectors (but not yet vector-spinors!) — both Maxwell and Yang-Mills. Rather arduous calculations [6] showed that none of them, nor mixtures, produced infinities that were redefinable away. At two loops, GR was shown to become infinite, by heroic calculation, because those infinities explicitly included Riem^3 terms, that were not saved by Gauss-Bonnet [7]. It was here that the second miracle of SUGRA took place; the first was that at one loop it IS field-redefinably finite — basically, SUGRA is a single system like pure GR, so its infinities are also combinations of the field equations, using "Super Gauss-Bonnet". The miraculous aspect of 2 loops is that there is NO cubic super-invariant, unlike the quadratic one [9,25]. After our realization of this second miracle, we were tempted to use Fermi's dictum, that if something is true to first and second order it's always true, but fortunately held off before submitting. Instead, we remembered that there is an old four-index quantity called the Bel-Robinson tensor, $B_{\mu\nu\sigma\tau} \sim$ Riem^2 that is conserved, and symmetric, and whose square $\sim$Riem^4 can be supersymmetrically extended to be a viable counterterm at 3 loops [9]! (Basically, B is a tensorial version of the Einstein, and its SUSY extension is a version of the spin 3/2, "stress tensor"). Indeed, it has become the bane of all subsequent attempts at healing some version of SUGRA. But there is yet another, modern, twist: Thanks to great improvement in computational techniques, the maximal, $N = 8$, system has been studied to many loop order–well beyond the R^4 3 loop hurdle–and remains finite so far! For the latest review as of now, see [26]. Whatever the outcome, is has been an exciting saga–it may stop at some as yet unknown loop order, or the system may yet be finite, in which case it will be even more mysterious: In the real world, SUGRA is "maximally" broken, so what would be the moral of a finite result? There is clearly room for progress here! Separately, there have been many intriguing connections between $D = 10$ and 11 SUGRA and $D = 10$ string theory; I cannot expand on this vast subject here except to note the interest of any deep connection between string and quantum field theories, especially since strings are finite while the divergences of SUGRA become, if anything, worse at higher D because the the loop integrals are higher power and the Planck constant is higher dimensional!

In conclusion, I quote Chou-En-Lai's reply to a query as to the effects of the then 200 year-old French revolution: "it's too early to tell": SUGRA is a beautiful set of theories, very much broken in the real world, yet with many useful and unexpected lessons in theory-building past and future.

5 Important update, September 2019

Quite recently, a private organization, the Fundamental Physics Prize Foundation, announced its "Breakthrough" award for SUGRA, but only to one group, with no mention whatever of Zumino's and my contribution. This blatant Soviet-style airbrushing out of history is particularly dishonest in view of: all the relevant available dated documents, Zumino's widow's (the distinguished theoretician Mary K. Gaillard) explicit eyewitness testimony, and the universal literature citations of both papers on a par (even in a major book by the award Committee's Chair and two of its members, and by the other group's authors). Absolutely no reason was (nor indeed could be) given for this blatant Breakdown.

Acknowledgements

This work was supported by grants NSF PHY-1266107 and DOE#desc0011632. I thank my moderating encouragers, Mike Duff and Mary K. Gaillard, my unique translator Joel Franklin, and Paul Townsend for useful correspondence. A condensed version of this work was posted on CQG+. This journal's founding editor, W. Bieglböck (especially) and a referee both suggested useful improvements.

References

1. J. Wess and B. Zumino, Nucl, Phys. **B** 70 (1974) 39.
2. Y.A. Gol'fand and E.P. Lichtman, JETP Letters, 13, 323 (1971); D.V. Volkov and V.P. Akulov, ibid. **16**, 438 (1972).
3. B. Zumino and S. Deser, in Proceedings of the Conference on Gauge Theories and Modern field Theory, Northeastern University, September 1975, Eds. R. Arnowitt and P. Nath (MIT Press, 1976).
4. C. Aragone and S. Deser, Nuov. Cim. **3A**, 709 (1971).
5. H. A. Buchdahl, Nuov. Cim. **10**, 96 (1958).
6. S. Deser and P. van Nieuwenhuizen, Phys. Rev. **D 10**, 401; 411; 3337 (1974).
7. M.H. Goroff and A. Sagnotti, Phys. Lett. **B 160**, 81 (1985).
8. G. 't Hooft and M.J.G. Veltman, Ann. Inst. H. Poincare Phys.Theor. **A 20**, 69–94 (1974).
9. S. Deser, J.H. Kay, K.S. Stelle, Phys. Rev. Lett. **38**, 527 (1977); arxiv:1506.03757.
10. S. Deser and B. Zumino, Phys.Lett. **B 65**, 369 (1976).
11. S. Deser and B. Zumino, Phys. Lett. **B 62**, 335 (1976).
12. H. Weyl, Z. Phys. **56**, 330 (1929); Phys. Rev. **77**, 699 (1950).
13. V. A. Fock, Z. Phys. **56**, 261–277 (1929).
14. C. Aragone and S. Deser, Phys. Lett. **86B**, 161 (1979); Nucl. Phys. **B170**, 329 (1980).
15. D.Z. Freedman, P. van Nieuwenhuizen and S. Ferrara, Phys.Rev. **D** 13, 3214 (1976).
16. S. Deser, *A brief history (and geography) of Supergravity: the first 3 weeks... and after,* Eur. Phys. J. H **43**, no.3, 281–291 (2018) [arXiv:1704.05886 [physics.hist-ph]].

17. A. Salam and E. Sezgin, "Supergravities in diverse dimensions" Vol. 2, North Holland/ World Scientific (1989).
18. L. Brink, S. Deser, B. Zumino, P. DiVecchia and P. Howe Phys. Lett. **B 64**, 435 (1976).
19. S. Deser, R. Jackiw and S. Templeton, Annals Phys. **140**, 372–411 (1982); Phys. Rev. Lett. **48**, 975 (1982).
20. S. Deser and J.H. Kay, Phys. Lett. **B 120**, 97 (1983).
21. W. Nahm, Nucl. Phys. **B 135**, 149 (1978).
22. E.Cremmer, B. Julia and J. Scherk, Phys. Lett. **B 76**, 409 (1978).
23. E. Cremmer and B. Julia, Phys. Lett. B 80 (1978) 48.
24. L. Brink, O. Lindgren and B.E.W. Nilsson, Phys. Lett. **B 123**, 323 (1983).
25. M.T. Grisaru, Phys. Lett. **B 66** 75 (1977).
26. Z. Bern *et al.*, Phys. Rev. **D 96** 126012 (2017); Phys. Rev. **D 96**, 126012 (2017), and to appear.

A BIG YEAR[98]

S. DESER
Walter Burke Institute for Theoretical Physics
California Institute of Technology, Pasadena, CA 91125;
Physics Department, Brandeis University, Waltham, MA 02454

In 1975–1976, it was time for another leave from Brandeis. My wife and I decided not to uproot our children from their schools so the first semester I stayed local but split the second one between Kings College, London, and CERN. The family would join me as soon as school was over, so the girls could spend the summer in and around Geneva. This turned out to be an *annus mirabilis* for my research, one in which many things came together in unexpected ways. That first semester I was appointed Loeb Lecturer at Harvard on Paul Martin's nomination. This was prestigious and fun. The Loeb is an endowed set of lectures in its Physics department, intended to explain frontier results in one's field; in my case, the state of the art in Gravity. I was also invited by Goldberger to Princeton as a short-term visiting Professor, where I worked (separately) with two then-lowly instructors, Claudio Teitelboim and Frank Wilczek. The collaboration with Claudio continued over the years, including a simple proof of the positive energy of both Supergravity (SUGRA for short) and General Relativity shortly after the former's birth — see below.

Around New Year's Day, I flew to King's College, one of the original bastions of modern General Relativity. I stayed in a "council flat", a proletarian apartment run by one of the boroughs of the City London in one of the only remaining bombed-out areas not yet reconstructed. It had been pretty proletarian all along. One had to walk through a full-scale Hollywood wartime movie set to get to it from the Tube station and climb a steep staircase, often adorned with drug users. Being alone, I worked a great deal and often came home quite late. The only place I encounter on the way was a Fish and Chips shop, where the food was always wrapped in yesterday's newspaper. I didn't eat there too often, but then again food was not (yet) the reason one came to London. At King's were a couple of very bright young people,

[98]Stanley Deser, *Forks in the Road: A Life in Physics*, (World Scientific, 2022), excerpt, pages 68–73.

Mike Duff and Chris Isham, whom I'd met before. We immediately got into the swing with our research. This was the beginning of something called the study of anomalies, in particular conformal anomalies, which are a popular and important subject of Quantum field Theory to this day. The research dealt with the distorting effects of quantizing a system that is classically perfectly acceptable. The Physics building was an example of the worst kind of London post-war construction, located just off the Waterloo Bridge on the banks of the Thames. My office was on the fourth floor. The Strand widens there; in the middle, like an island, was the ancient church of St. Mary's-in-the-Strand. The view from my room was of its steeple. What made it less pleasant was the notorious London winter chill that easily permeates flimsy post-war buildings. At some point it became so cold, even for my relatively young age, that I couldn't work. Finally, one of the custodians came by, wearing his standard grey smock and carrying a whirligig-like instrument. He held a handle, at right angles to which was a little object that rotated. I inquired if he was sent to warm up the molecules by his motion. He replied indignantly that the device was for measuring the average temperature. After a few moments of its agitation, he took a reading, and said, "Sir, the temperature is 40°F (5°C)." I inquired if that was "a little bit below normal." He said, "Yes. We will try to do what we can." In due course, the temperature went up to 50°F (10°C). I never really warmed up until I left in the spring. At some point, I was giving a seminar and there appeared Hermann Bondi about to become Sir Hermann. He was Her Majesty's Chief Scientific Advisor to the Department of Energy. He arrived late with one of those battered government briefcases chained to his wrist. I was told that was a rare sighting. He was a smart, stimulating man who led some of the early advances in the field.

It was with regret that I watched April 1st approaching. On the other hand it was not exactly torture to be moving on to Geneva's springtime. I went via Paris to recover my old Citroën, which we had bought several years before and left in Europe. After a long day's drive, on pre-autoroute road, I arrived at the CERN cafeteria around 9 p.m. The experimentalists worked all night so they had to have a place to get food and stay awake. The theoreticians often worked late, so they showed up as well. There was Bruno Zumino, one of my old friends and colleagues. Indeed, I had invited him to give a set of lectures at the last of the Brandeis summer schools, in 1970. These lectures turned out to be extremely prophetic. I was in touch with Bruno a great deal. He and Julius Wess, his close collaborator, who was also at NYU, discovered something called Supersymmetry (SUSY) around

1973. Actually, they had independently re-discovered it, since Golfand and Likhtman in the Soviet Union first did in 1970, but no one in the West paid attention. No one in the West paid attention to Wess and Zumino either. No one, including myself, understood what point it might have. This is something that is now a main object of searches at the Large Hadron Collider. Indeed, it was one of the reasons the LHC was commissioned: an interesting example of how Physics functions. At the cafeteria, Bruno said he wanted to talk to me about something, though I was pretty exhausted. The next three hours we sat and discussed the possibilities of Supergravity. I had also thought a bit about it, but for entirely different reasons. He had been awaiting my arrival because I would be the one person at CERN, who would be able to collaborate. We spent 18 hours a day at the blackboards. He was eight years older than I, so that was an even greater task for him. Our best work was done in the major CERN conference auditorium, a huge room whose advantage to us was the infinity of blackboards where we could write our extremely complicated formulas and work them out. In a period of just three weeks, to our amazement, we had a consistent theory. It hinged partly on the fact that both he and I had learned things in our past that nobody else knew. We were thereby able to short-circuit enormous calculations. There was competition: three other people were working very hard on this and were just about finishing up; it was neck-and-neck.

I had arranged to go home and visit my family on April 23rd. At the time, some one-plane airline was offering flights from Geneva to Luxembourg, where the new Icelandic Airlines had its European headquarters. There was barely room for us two passengers. The pilot and copilot navigated using Michelin road maps, trying to find landmarks below. Terrible thunderstorms meant we couldn't fly very high. Both of us were in mortal terror until we finally touched down in Luxembourg. The rest of the trip was, by comparison, an Air Force One experience. Bruno and I had just finished our (quite short and entirely analytic) manuscript as I left; it was typed up by the incomparable Tatiana Fabergé, chief secretary of our division. Bruno handed it to Raoul Gatto, Editor of Physics Letters, while the other trio also finished their lengthy, and computer-assisted, calculations. There were some rather unpleasant priority aspects to all this, which to this day have left a nasty aura. The other team claimed that we had merely improved on their results, rather than independently achieved them, and employed various non-Physics means to try to convince people of this manifestly absurd and self-serving claim. I freely admit it rankled us both for quite a long time, especially since it gained credence among people who should have known better, simply by reading the papers! Bruno and I wrote other works on Supergravity; one was the first formulation of Superstrings.

CERN expanded enormously over the summer when people from all over the world come and make it a very pleasant milieu. The population of the Theory division used to swell even more. In the office next to mine was a very bright, hardworking applied mathematician from Harvard, T. T. Wu, who told me that he and Yang had a problem that was driving them crazy. This was the question of proving that there could be no static solution of Yang-Mills Theory, just as there could be no static solution of Maxwell's Theory. It is often said that you learn subjects in Physics by teaching them, certainly true in this case. I remembered having taught students in my mechanics course that a static system cannot have self-stress. I realized, as I sat in some seminar, that this 19th Century theorem provided a one-line proof for the Yang-Mills system. The theorem had sort of been proven by Coleman and by Derrick around the same time, much more ponderously.

Back in Boston, I had two very bright graduate students, Kellogg Stelle, now professor at Imperial College, and John Kay. Kelly was finishing his thesis. He came to Brandeis in an unusual way. He was an undergraduate at Harvard, and wrote us asking for an early decision to be admitted as a graduate student as he was about to go off for a year to the South Pole. He would stay until the ice melted and since there was no internet, he would not be getting mail, if at all, on any reliable basis. We happily took him on. John was also getting close to finishing. We started a variety of Supergravity projects. In particular, one of the reasons for betting on Supergravity was that it might allay, if not entirely remove, the infinities that made the direct quantization of General Relativity inconsistent. Stelle and Kay were very bright, so they picked it up and got excited about the project. There are successive steps to it, and Pure Gravity passes only the first one. The next was a different story. We discovered that the next step was also safe because there were no possible two-loop infinities in Supergravity, unlike, as had been very laboriously shown, in pure General Relativity. So Supergravity was an enormous improvement from Pure Gravity: it was finite in the first two orders. We were ready to send off a paper to Physical Review Letters, the prime rapid publication journal, saying Supergravity is finite. We relied on Fermi's "Theorem": if its first and second approximations work, then a claim must be true. This is of course a rather weak "theorem", so I decided we had better wait till morning and sleep on it, literally. This illustrates the sometimes Byzantine paths/forks involved in seeking new results. During my sleepless night, I recalled that matter (SUSY), not SUGRA, systems have supermultiplets led by the system's stress tensor, hence invariants starting with its square. Gravity has no stress tensor, however; its nearest avatar is the Bel-Robinson tensor quadratic in curvature and only covariantly conserved. I guessed that it must have a SUGRA continuation, in which case the above SUSY picture (which I reproduced with a resulting scalar $\sim$(Bel-Robinson)^2, quartic in curvature), was just the feared three-loop death knell. Those nightmare bad parallels also came to Stelle's mind, as he told me, and resulted in his and Kay's feverish searches for quartic terms in the then-existing General Relativity textbooks. It is now known as "the notorious Debney-Kerr-Schild curvature^4 term": our brilliant delay saved us from ignominy. Those were days of great fervor. In the new American Institute of Physics keyword classifications, Supergravity and Supersymmetry, acquired a niche of their own.

Neither Supergravity nor Supersymmetry show any sign of being used by nature; with every passing year that seem less likely: It does not have to use anything we invent. We all like to say that elegance is so important because in its economy and clarity it leads us towards the light. Dirac always believed that. Sometimes, however, we only recognize a theory's elegance after it has been proven to be correct. On the other hand, another great hero of times gone by, Ludwig Boltzmann, one of the fathers of Statistical Mechanics amongst other things, used to say "Eleganz ist für Schneider;" — elegance is for tailors. He didn't care how much heavy lifting you had to do to get your story right. Ours were very elegant theories and they have kept a lot of people busy. It must be remembered that all of this research in the mid-1970s was in the shadow of the final developments of the Standard Model of Strong and Weak Interactions, including quarks, asymptotic freedom, the Weinberg Electroweak Theory and a number of others. This was also the glory time of CERN as they discovered the weak neutral currents, the most convincing proof of the Standard Model, at a detector aptly named Gargamelle, the mother of Rabelais's Gargantua. Super's future is hard to predict, despite the by now more than 15,000 papers on SUGRA alone.

To use Chou-en-Lai's legendary reply when asked about the effects of the French revolution, "it's too early to tell"

A TRIBUTE TO S. DESER: FROM HIS PENULTIMATE POSTDOC[99]

BAYRAM TEKIN

Department of Physics
Middle East Technical University, 06800 Ankara, Turkey
Email: btekin@metu.edu.tr

I had the good fortune to be the penultimate post-doc of S. Deser for two years (2001–2003) and have maintained close communication until he passed in 2023. Here I describe a brief account of our interaction.

1 A historical note about my interaction with S. Deser

It is very common for tributes written for well-established scientists like Deser, but what is perhaps not common is this being done by unknown physicists like myself, hence my hesitation. In the end, I decided that my vantage point in this interaction would do no harm and perhaps would even be of some benefit to some people. The age gap between Stanley and myself was over 40 years, so I can in no way claim that I have known him. When we first shook hands, he was 70 and I was 28. Nevertheless, we did have a very frequent interaction, I have been to his house and his wonderful wife Elsbeth (Oscar Klein's daughter) cooked dinner for us, and we have maintained good communication in the last two decades. In the first decade, we talked mostly about physics, and in the last one, our conversation centered around the evaporation of the trace amount of democracy we used to have in Turkey. Based on his own experience before and during WWII, he was urging me to leave. But here writing about that subject in more detail would offer me an undesired visit to a prosecutor's office. Instead, I will write a little bit about the two years I spent at Brandeis.

1.1 *There is something in a name. "Write my name as S. Deser!"*

In September of 2001, when I finished writing our first joint work with Deser that summarized our talks on the board in my office, I wrote "Stanley Deser"

[99]e-Print: 2307.12758, abbreviated by the author in July 2023.

as an author and printed a copy for him to suggest corrections. To be perfectly honest, the paper was not a very impressive one, but still, I was quite proud to have produced something publishable with Stanley whose works in the Hamiltonian formulation of gravity, supergravity, topologically massive gravity, and gauge theory had been written in stone. I had turned down a potentially much more fruitful post-doc offer and came to Brandeis just to work with Stanley. So joint work with him was exciting for me. After a glance at the cover page of the manuscript, he immediately requested me to shorten his name to "S. Deser" and noted that he had always signed all his works as such. I did not pay much attention to this incident and wrote his name as S. Deser in our joint works [1–12]. At the time his insistence on writing S. seemed to me like some kind of representation of humility. My understanding was this: as an author, acknowledge that you are the author, but otherwise remove yourself from the room between the reader and the paper. For me on the other hand, as I come from a barren land as far as science is concerned, I explicitly write my name as there has been no other "Bayram" in gravity or high energy theory: B. would be too cryptic and could not have been guessed. Back to S.: only, in 2020 or 2021, I came to understand his (almost) passion in shortening his name. My explanation could be poetic, but still, I think it could be true. In 2020 or 2021 Deser sent me an early draft of his book "*Forks in the Road, A Life in Physics*" [13]. I read the book immediately the night I received the draft and made some suggestions. The book is a must-read. I learned from that book that he, as a Jewish boy, started his life first with the name "Salomon" in a Polish city (now in Ukraine) and left for Palestine with his name intact; but his name was changed to "Lucien" when the family moved to France when he was a refugee. His name was kept as Lucien when the family escaped to Spain and Portugal, but when he ended up in New York; his name was changed one more time to "Stanley". I think that old man that I worked with who insisted that his name be kept as S. Deser in the papers was always attached to the little Salomon, and that S. also referred to Salomon. The Nazis brought out the worst in humanity, forcing a boy to flee his name and home many times. This could happen again, it was not a one-time event in human history. We must be vigilant.

1.2 *September 11 2001: S. says "We must evacuate!"*

On September 11, 2001, I was working in my office at the Brandeis University physics department. Stanley rushed into my effectively windowless office in a panic. He said "I don't think Palestinians can do something this big, but in any case, we have to evacuate the university". (Brandeis is the largest Jewish campus in the States, and at the initial hours of the September 11 attacks, it was not clear who did it and where else was a target.) I had no idea what he was talking about, I had not heard the news. I did not know what happened. As he rushed out, I did not ask him. But I packed my bag and left and took the train to Boston from Waltham and went home. That home was no home really: I had arrived in Boston from Oxford (UK) about 20 days earlier. I rented an empty house in Sommerville. I was sleeping on the wooden floor. I had no phone, no computer, no TV, and no money. I had no idea what was going on in the USA, even though Boston was in some sense at the epicenter of the events. But clearly, something big was taking place. As all the planes that hit the buildings took off from the Boston airport, something interesting happened: my landlord started to check on me. As a Middle Eastern-looking man, living in his house, with zero furniture, no bed, with just a large duffle bag and weird papers scattered over the floor, I sure was a suspect. He did not know that a poor theorist post-doc lived in less fortunate conditions than a well-funded terrorist. My salary was \$2000 per month while my rent was \$1400. Anyway, I decided to go to Harvard Square to understand what was going on. On the subway train, someone shouted that everybody followed him and sang "God Bless America". That was interesting, I did not know the lyrics, I had lived in Oxford (UK) and had a slight inclination to sing "God Save the Queen". I learned about the terrible situation at Harvard Square and immediately realized the consequences: America will fall into the nationalism trap that I am all too familiar with from my own country which is de facto the landlord of that trap.

Back to Stanley. As a graduate student in Minnesota, I applied to post-doc jobs, Stanley offered me a position to start in September 1999. But, I went to the UK. After two years, amazingly, Stanley was kind enough to offer me again. I arrived in Boston at about the end of August 2001. Stanley arranged a guestroom at Brandeis for me to stay just for a week. He was on vacation in Sweden. I started working. When he got back, after about a couple of minutes of introduction, he asked what I was working on. I explained on the board in the office my idea on Newtonian limits of massive gravity with a cosmological constant. Anyway, before the end of September, the paper was ready. And I told that story above.

1.3 *Time Travel: S. says "good luck!"*

Sometime in 2002, I was at work in Stanley's office. We were discussing, perhaps fine-tuning a paper that we were writing. His office phone rang.

A journalist (perhaps, from the Boston Globe) was calling to ask about time travel in physics. What was the reason for such a call? A physics professor from the University of Connecticut had claimed recently that time travel was possible and he had been building a time machine. The journalist wanted to know what Deser thought of this. Deser with Roman Jackiw (sadly, Jackiw, a giant of physics, passed away 2 months after Stanley) had written a paper on time travel and I guess the journalist found out about that paper. Anyway, Deser at one point on the phone mentioned me to the journalist. He said, "My post-doc Dr. Tekin is here, if you want you can talk to him about this". Of course, the journalist did not want that. Anyway, in the end, Deser said "Time travel requires organizing all the mass in the universe in a very particular configuration. So good luck!" What he was referring to was Gödel's solution of Einstein's equations, the so-called Gödel universe. That solution allows time travel to the past. Gödel was afraid of dying and so invented a universe (a solution to Einstein's equations) in which going around a loop in space (not all loops do this), brings you back in time. He wrote the paper for the occasion of Einstein's birthday celebrations. If I had picked up the phone, I would have told that journalist the following: you want to go back in time in a relatively cheap way, just travel to East, say to North East Anatolia. You would be going back about 2–3 centuries.

1.4 *Deser's Wit and Languages*

Stanley was a witty and humorous person; he, most of the time, would not say something directly but instead, encase his message in a terse joke. One had to know him a little bit to understand what he was saying. I have hundreds of emails from him in my mailbox, and some of them made me emotional as I went through them while writing this. He tried to help me out when I was struggling with the authorities here. He offered his help several times and searched for possible solutions.

One day Stanley brought me two preprints one in French and one in German. I do not remember exactly what we were discussing. He asked me to read them. I said that I did not read French and I could barely read German with a good dictionary. He seemed surprised because he spoke many languages and asked why I did not know these languages. I said that I came from a region where people barely spoke one language and learning a little bit of English was considered already a success.

Stanley was amazingly quick with his response to emails, one would have thought that he was always sitting in front of his computer. He was not so patient with some people who omitted some obvious references. But, since he would write his messages in the subject line with shortened words, very often the receivers would be baffled and ask questions as to what he meant. That usually turned into an interesting, rather slowly converging, exchange with Stanley answering in shortened words and hard-to-decipher sentences for those who did not know his style.

I must say this: when he did not like the introduction that I wrote for a paper, he wrote his version on some yellow paper with a stylographic pen. When he gave his version to me I looked at it for a second and before I said anything he said "Of course you cannot read it, but the secretary can." Stanley's handwriting was simply unreadable. I remember him writing the address on an envelope that we were sending to the journal "Classical and Quantum Gravity" and I was thinking to myself "There is no way this mail will arrive there."

As it could be common with his generation, corrections on an accepted paper never ended with Stanley: I remember that "Physical Review Letters" accepted our paper without any correction, but Stanley started a marathon of corrections that seemed to have no end. When I wrote a slightly angry email about this (as he was in Sweden at the time), he took his time to reply. After 3 days of silence, he replied with a very kind email and urged me to ask his way of writing to Roman Jackiw. Of course, I did not do that.

Finally, let me note the following: our major joint work, conserved charges in generic gravity theories, was summarized in [3, 4]. The work started as follows: Stanley brought the paper [14] to me and said that he thought the paper was not correct. This is a very interesting paper about purely quadratic gravity in four dimensions where it was shown that "exact solutions representing isolated systems have precisely zero energy" and so there is a confinement of energy in the purely quadratic theory. Stanley could not accept such a confinement of energy in the sense that certain localized energy-momentum tensors could not affect the spacetime at larger distances. For him, there could be no confinement of energy. For me, it was possible. As per Stanley's request, I started working on how one might consistently define global charges in quadratic gravity theories for asymptotically flat and constant curvature spacetimes. Our first results were published as [3] and a more detailed version was published in [4] where we included a section discussing the zero energy issue which was the original impetus for that work.

What was our common interest that Stanley offered me a post-doc position? I never asked him this question, but I have a guess. I did my PhD in Minnesota between the years 1994 and 1999 working with my advisor Yutaka Hosotani on various perturbative and non-perturbative problems in quantum field theory. I learned a great deal from Yutaka, especially in tackling hard and long computations. Minnesota had remarkable experts on quantum field theory and particle physics. But nobody was working on gravity. While working on the confinement problem in the 2+1 dimensional Georgi-Glashow model with a Chern-Simons term, I got hooked to a paper Witten wrote [15] on 2+1 dimensional quantum gravity and its equivalence of Chern-Simons theory. So I started working on the subject; and submitted two manuscripts to the physics ArXiv on this before I applied for post-doc jobs. I think Stanley offered me a post-doc job twice due to these works, and my main research shifted towards gravity. I am grateful to him for the subtle things I learned in our endeavor. We had several unfinished discussions with him on the degenerate vacua of gravity. He was hard to persuade and so I left those works unfinished. I will remember S. Deser for the rest of my life as a teacher and a colleague.

References

1. S. Deser and B. Tekin, *Newtonian counterparts of spin-2 massless discontinuities,* Class. Quant. Grav. **18** (2001), L171.
2. S. Deser and B. Tekin, *Massive, topologically massive, models,* Class. Quant. Grav. **19** (2002), L97-L100.
3. S. Deser and B. Tekin, *Gravitational energy in quadratic curvature gravities,* Phys. Rev. Lett. **89** (2002), 101101.
4. S. Deser and B. Tekin, *Energy in generic higher curvature gravity theories,* Phys. Rev. D **67** (2003), 084009.
5. S. Deser and B. Tekin, *Shortcuts to high symmetry solutions in gravitational theories,* Class. Quant. Grav. **20** (2003), 4877–4884.
6. S. Deser and B. Tekin, *Energy in topologically massive gravity,* Class. Quant. Grav. **20** (2003), L259.
7. S. Deser, J. Franklin and B. Tekin, *Shortcuts to spherically symmetric solutions: A Cautionary note,* Class. Quant. Grav. **21** (2004), 5295–5296.
8. S. Deser, I. Kanik and B. Tekin, *Conserved charges of higher D Kerr-AdS spacetimes,* Class. Quant. Grav. **22** (2005), 3383–3390.
9. S. Deser and B. Tekin, *Conformal Properties of Charges in Scalar-Tensor Gravities,* Class. Quant. Grav. **23** (2006), 7479–7482.
10. S. Deser and B. Tekin, *New energy definition for higher curvature gravities,* Phys. Rev. D **75** (2007), 084032.
11. S. Deser, O. Sarioglu and B. Tekin, *Spherically symmetric solutions of Einstein + non-polynomial gravities,* Gen. Rel. Grav. **40** (2008), 1–7.

12. S. Deser, H. Liu, H. Lu, C. N. Pope, T. C. Sisman and B. Tekin, *Critical Points of D-Dimensional Extended Gravities,* Phys. Rev. D **83** (2011), 061502.
13. S. Deser, *Forks in the Road,* World Scientific, (2021).
14. D. G. Boulware, G. T. Horowitz and A. Strominger, *Zero Energy Theorem for Scale Invariant Gravity,* Phys. Rev. Lett. **50**, 1726 (1983).
15. E. Witten, *(2+1)-Dimensional Gravity as an Exactly Soluble System,* Nucl. Phys. B **311**, 46 (1988).

2000 ZEROS AND THE BIRTH OF SUPERGRAVITY IN 1976

PETER VAN NIEUWENHUIZEN

Department of Physics and Astronomy
Stony Brook University, Stony Brook, NY 11794, USA
vannieu@gmail.com

Theoretical physicists usually do research in extensions of the two great theories of the beginning of the 20th century: relativity and quantum mechanics. However, it sometimes happens that a radically new theory is created; not often, but it does happen. This is a personal account of the highs and the lows of someone who was involved in the creation of a new theory. First I will provide some background, then my recollections.

The theory of Special Relativity (1905) is a theory for the non-gravitational interactions, and the theory of General Relativity (1916) is a theory for gravitation. Both were created in large part by Albert Einstein, and both have been tested to great precision in several experiments, most recently the tests which became possible when in 2015 gravitational waves were discovered. The theory of Quantum Mechanics was created in 1925 and early 1926 by Heisenberg and Schrödinger, and was soon after applied by Dirac to the electromagnetic fields of Maxwell in a non-relativistic way in 1926. Two years later, Dirac discovered a relativistic equation for the electron, and using it in Maxwell's theory of electromagnetism, he started (very successfully) what is nowadays called relativistic quantum field theory. Thus Dirac unified Special Relativity and Quantum Mechanics in a model for electrons and photons called QED (Quantum ElectroDynamics). Quantum corrections in that model were a huge problem in the 1930's because there were infinities, but in 1947 a consistent method for removing these infinities was created by Kramers, Tomonaga, Bethe, Schwinger, Feynman, Weisskopf, Dyson, and many others. This method is called renormalization and has been spectacularly successful in the past 70 years. It was applied to the "Standard Model" (SM) of the non-gravitational interactions (the theory of quarks, leptons and gauge fields), and not only were all corrections finite after renormalization, but to date no discrepancy between the predictions of the theory and experimental results is known.

The implications of General Relativity (GR) were meanwhile further worked out by relativity physicists (not by particle physicists) and today it forms the framework in which the huge amount of data on cosmology is analyzed. All this work treats the gravitational field as "classical", it does not use a quantized gravitational field. Of course, several physicists tried in the fifty years between 1926 and 1976 to construct a theory of quantum gravity, but the method of renormalization, so successful for the non-gravitational interactions, does not work for the gravitational interactions because the gravitational coupling constant, the Newton constant G, is dimensional, and this leads in general to infinities which cannot be removed by renormalization. (The functional form of the divergences is different from the classical action, they contain squares, cubes, etc. of curvatures). Thus, in quantum gravity one can only hope that infinities cancel "miraculously", and if that does not happen, the theory is doomed. The word "miraculously" has a meaning for insiders, it means that further hidden symmetries are present in addition to those of Einstein's theory of gravity (General Relativity) which cancel the otherwise fatal infinities.

In the early 1970's, a series of discoveries were made which put the problem of quantum gravity in a new perspective. First of all, 't Hooft and Veltman (1971) were able to construct an extension of the old method of renormalization which was consistent (meaning: it got rid of infinities while preserving the symmetries) when applied to Yang-Mills theory (the theory of gauge fields such as those that appear in the Standard Model). Thus, they created the Quantum Standard Model, just as in 1947 the theory of QED was created. Their work pointed the way to further experiments, which in due time discovered the remaining quarks of the SM, and finally the last building block of the SM: the discovery of the Higgs boson in 2012. So, today, Quantum Field Theory and the SM "reign supreme", but the quotation marks indicate that something is lacking. Indeed: **why** the Standard Model is as it is, is a mystery, and furthermore it contains 28 parameters whose values are only known from experiments but not from theoretical insight. Many (but not all) physicists believe that the Standard Model is a shadow of an as yet unknown theory whose presence will become clear if one does sufficiently accurate experiments. One can only speculate what such a theory might be.

The other discovery in the early 1970's was Supersymmetry. Two Soviet groups stumbled on a very interesting new concept, bose-fermi symmetry, a new symmetry between the two kinds of particles in our universe: bosons (such as photons, and the other gauge particles of the SM) and fermions (electrons, neutrinos, quarks). Golfand and Likhtman (1970) wrote down what would be called today an interacting theory of an $\mathcal{N} = 1$ massive multiplet containing a scalar, two spinors, and a vector field, all of the same mass.

They made an incorrect statement about parity, but in the process they discovered an extension of the mathematics used to describe symmetries. (This branch of mathematics is called Group Theory and was developed in the 19th century. They were the first to write down what one now calls a Superalgebra, with commutators (standard) and anticommutators (very non-standard). They were in a certain sense the successors of Jordan, who in 1928 had introduced anticommutators for the quantization of the electron). The other group of Soviet physicists, Volkov and Akulov, proposed in 1973 an explanation of why neutrinos are massless (as they were supposed to be at that time): just as the theory for the Higgs boson needs and contains extra massless fields (would-be Goldstone bosons) which are eaten by gauge fields, which then become massive as a result of this banquet, they proposed that the neutrinos were the fermionic equivalents of these massless fields (would-be Goldstone fermions). Closer study revealed that the neutrinos of our world do not have the properties of would-be Goldstone fermions, but the idea of a bose-fermi symmetry (Supersymmetry) had been launched. Then Volkov and Soroka (1973) constructed a "nonlinear realization of a Supersymmetric theory with a super Higgs effect": spin-$\frac{3}{2}$ fields eat spin-$\frac{1}{2}$ fields and become massive. These Soviet papers were unknown in Europe and the US and were only rediscovered after the very influential work of Wess and Zumino. Julius Wess and Bruno Zumino wrote in 1973 a series of papers (after Gervais and Sakita had constructed in 1971 a two-dimensional relativistic quantum field theory for strings with Supersymmetry) in which they were able to incorporate Supersymmetry into four-dimensional relativistic quantum field theory. Thus by the mid-70's Quantum Field Theory had returned as the most successful description of the non-gravitational forces, and Supersymmetry had made its appearance. The time was ripe for the next step: combining gravity with Supersymmetry in the framework of relativistic quantum field theory.

At this point I will start describing how I came to the problem of introducing Supersymmetry into Einstein gravity. This theory is now called Supergravity, and physicists all over the world are actively working out its implications. More than 6200 articles have been published with "Supergravity" in the title, and more than 19700 articles have Supergravity as a keyword. Since its inception in 1976 it has been recognized that Supergravity is the foundation on which String Theory is constructed (so that one now speaks of Superstrings). Supergravity has also been used in mathematics to solve long-standing problems. However, the theory requires that there exist superparticles: for every known particle, there should exist a superpartner, just as Dirac concluded in the 1930's that for every electron there should be an anti-electron. Superparticles are not the same as antiparticles: for every antiparticle there should exist a superpartner. Very extensive experiments looking for these superpartners have been performed at Fermilab in the US and at CERN in Europe, but so far not a single superpartner has been found. The verdict is still out: new accelerators at higher energies may find them. If they are not found, the legacy of Supergravity will lie in mathematics (where it is alive and well) and in physics as a general framework, but not as a realistic Theory of Nature.

In 1973 I came to the US to study as a postdoctoral fellow with Stanley Deser at Brandeis University (near Boston). 't Hooft and Veltman had just published an article in which they applied their quantization methods (which were developed for the Quantum Standard Model) to Einstein's theory of gravity, and to the coupling of this theory to a scalar field (such as the Higgs boson). They wrote: "A certain exhaustion however prevents us from further investigation, for the time being.", a clear hint for a young physicist to throw his hat in the ring. Deser and I studied the quantization and finiteness of many gravitational theories: GR coupled to Dirac's electron, to the gauge fields of Yang and Mills, and with Marc Grisaru and others I extended these calculations to other cases (QED). But try as we would, the infinities never canceled, and thus all these beautiful models were doomed by quantum mechanics. I still hoped to find the Holy Grail: a model ("the" model?) in which the infinities cancel. I tried an old theory of Hermann Weyl of the 1910's ("local scale invariance", now a key ingredient of string theory) but I did not complete this work (and would probably have failed; for insiders: because one needs ghosts for ghosts). At that time (1974) I went to a summer workshop in Paris at the École Normale (where I had been a Joliot-Curie fellow for a couple of years and to which I have kept my ties). There I met Daniel Z. Freedman (Dan), a theoretical physicist from Stony Brook, and we hit it off straight away. (Dan later told me that he had heard me the year before give a seminar on my work on quantum gravity at Princeton and had found it very interesting). After the workshop, Dan went back to Stony Brook, and I received an offer to go there as an Assistant Professor. In those days there were more jobs than postdocs, and since Stony Brook lies in New York, and I had heard from postdoc colleagues that New York was dangerous, I hesitated. But I went to visit Stony Brook. With some misgivings, I entered the Physics Department, and all my worries were confirmed in dramatic fashion: the first thing I saw was a student running away from an older person, but the older person tackled the student and a scuffle ensued. (Later I found out that the student had been caught committing fraud, impersonating another student at a written exam, and the older person was a professor who tried to stop him.) I went back to Boston and demurred accepting the offer. A second visit was proposed, and this time things went initially much better. I described my progress in Quantum Gravity (progress, not success), and after my talk a lot of faculty stood around me asking further questions. But then suddenly everybody left me and walked quickly to a corner in the room where somebody had just entered with a piece of paper showing stunning results just discovered at Brookhaven National Laboratory (half an hour away from Stony Brook). These results were later interpreted as the discovery of the charmed quark but that was not yet known as I stood there alone. I went back to Boston, but I did accept the offer, probably one of the best things that ever happened to me.

So in the fall of 1975 I came to Stony Brook to work on Quantum Gravity. The past year Dan had studied with Bernard de Wit the articles on Supersymmetry for non-gravitational models, so what was more natural than to collaborate on Supersymmetry in Gravity? Wess and Zumino were working on it at CERN, and Zumino gave a lecture on it at a conference in Boston in the early fall of 1975. Arnowitt and Nath at Northeastern had published a theory which was a very direct extension of GR but which turned out (later) to be inconsistent. And Volkov and Soroka had "gauged" the Superalgebra mentioned above, to obtain the super-Higgs effect. So it was clear that there was competition. Dan went to Paris for a few months that fall on a sabbatical (again at the École Normale) where he met Sergio Ferrara from CERN. Sergio had come to Paris to work on another project, and in a discussion with Dan he made important suggestions about coupling spin 3/2 spin fields directly to gravity. Such was the situation when Dan returned at the end of the fall. We started working very hard, with Dan contributing his insights about supersymmetry (about which I did not know much) and myself contributing my experience in Quantum Gravity, in particular my expertise in how to describe fermions in gravitational fields. We exchanged frequent airmail letters with Sergio about our work (there was of course no internet and transatlantic phone calls were expensive and thus discouraged). The next summer I finally met him in Paris, and he became for many years my collaborator and friend. As mentioned above, for a supersymmetric theory of gravity one needs a new kind of fermion, with spin 3/2 instead of spin 1/2. We looked in the literature for articles on spin 3/2 and found articles by Bargman, and Pauli and Fierz, but those theories were quite complicated, with many fields and constraints on those fields. But then we discovered that massive spin-$\frac{3}{2}$ fields had been studied by Rarita and Schwinger in 1941. Their spin-$\frac{3}{2}$ field was a vector-spinor field ψ_μ^α (later called gravitino by Sydney Coleman), and there were no constraints in their theory.

(In a companion article, Kusaka used their theory to find out whether neutrinos have spin 3/2 or spin 1/2. The results clearly showed that neutrinos do not have spin 3/2. We now know that they have spin 1/2. Incidentally, when I was looking for models of GR and other particles with extra symmetries in which the divergences would cancel, one senior physicist (Salam) mentioned to me supersymmetry, and another senior physicist (Veltman, my PhD advisor) suggested that I might study spin-$\frac{3}{2}$ fields. At that time I did not follow up on their suggestions.)

We worked very hard, every day till late, all weekends, no free time. Sometimes things moved forward and at other times we hit roadblocks. Dan had a sign on his office door with "Supergravity lives" if things went OK, and "Supergravity is dead" when we got stuck. There was much support of the members of our Institute, in particular C.N. Yang (Frank Yang), its director (I later succeeded him as director). But support was not enough, we needed a way to handle the enormously complicated equations we discovered. In the next paragraph, I shall now describe what is meant by "handle the equations". This part of the story is of necessity more technical than the rest, but it plays an essential role.

The issue we faced was to show that all supersymmetry variations of the action vanish. We assumed that the gravitational field (represented by e_μ^m instead of $g_{\mu\nu}$ because a fermionic field ψ_μ^α was present) transformed as $\delta e_\mu^m = \bar{\varepsilon}\gamma^m\psi_\mu$ (the alternative $\delta e_\mu^m = \bar{\varepsilon}\gamma_\mu\psi^m$ was rejected because it was nonlinear in e_μ^m). The e_μ^m were later called vielbein fields by Gell-Mann to generalize the name vierbein fields (vier=four and viel=many in German). For the gravitino we began with $\delta\psi_\mu^\alpha = \partial_\mu\varepsilon^\alpha$ and discovered that a gauge-invariant action for ψ_μ^α was unique and equal to the action Rarita and Schwinger had written down (but for massive gravitino, without the gauge invariance $\delta\psi_\mu^\alpha = \partial_\mu\varepsilon^\alpha$). We then coupled this action to gravity with a spin connection (a composite gauge field for local Lorentz symmetry that is made from vielbein fields), which was given in Weinberg's book on Gravitation. In this form of the theory the linear variations (with one spin 3/2 field) vanished by a "heart-warming" calculation. However, terms of the form $\varepsilon\psi\psi\partial\psi$ remained. They did not vanish but, inspired by a review of Friedrich Hehl on torsion, we modified our starting point by adding terms quadratic in gravitinos to the spin-$\frac{3}{2}$ transformation law, and terms quartic in gravitinos to the action. After much work we were able to show that the cubic variations did cancel, but new terms of fifth order in spin-$\frac{3}{2}$ fields turned up in the variation of the modified action. We showed (essentially by dimensional analysis) that further modifications of the starting point (terms with four gravitinos in the spin-$\frac{3}{2}$ transformation law, and terms with six gravitinos in the action) could not help in canceling the remaining variations, thus, we wrote, "...it is critical to verify that it vanishes". Dan, Sergio and I decided to publish our results, and submitted our article "Progress towards a Theory of Supergravity" to the Physical Review (the premier American physics journal) on 29 March 1976.

The issue was now of course whether the quintic variations did vanish. At that point we decided to use the computer. In Utrecht (where I got my PhD), I had learned to handle symbolic manipulation programs, and had used them in Paris and Brandeis to investigate quantum gravity. Using a simple Fortran program, not even a symbolic manipulation program, I started applying it to the equations. In order that the theory of Supergravity (the name was obvious: Supersymmetric Gravity = SuperGravity) existed, there were about 2000 terms whose coefficients had to cancel. According to the theory of Supersymmetry, these coefficients were all integer numbers, so the computer program did not have to be accurate in decimals. But we did not yet know the theory so we tried various possibilities, each time with negative results. There were also computer bugs, and we spent days on end to simplify the computer program to bring the cost down. (The total bill was in the tens of dollars, a ridiculously small amount of money compared with the millions of dollars now spent on computers in physics). After a lot of work we came to a point where there were no further ambiguities: either the 2000 coefficients all vanished, or "Supergravity was definitely dead". One night (I worked every night) I was sitting in the small computer room in our Institute. The result of the big (big for those days) CDC computer at Brookhaven came in over the phone. I realized it was now up or down. In the room was also a colleague who had recently been told he would not get tenure, and who was for obvious reasons depressed. In those gloomy surroundings I waited for the results. If any of the 2000 coefficients were nonzero, it would mean that half a year of frenetic work was in vain. And then the zeros started coming, in batches of 20. The first dozens of them were indeed all zero, but that we knew already, the crucial point was the last series of batches. More zeros, rising tension. Zeros near the end. I held my breath. And then it was all over: all 2000 numbers were zero: "Supergravity lives". I phoned Dan who was in a hotel for a conference in Chicago (it was by now the middle of the night) and told him the news. I heard a sleepy voice say, "That is great", followed by a yawn. I went home, and I felt depressed. That may sound strange to outsiders, but I have noted that other physicists also sometimes feel depressed when after long and arduous work they finally succeed. It is in my opinion the scientific equivalent of post-partum depression. However, in the days after, the clouds lifted, and we knew that we had constructed a fundamental new theory of physics, which combined Einstein's General Relativity with a symmetry between the two classes of particles in our universe. We sent a "Note added" to our article with the message that all variations now canceled.

Up to this point I have described my day-to-day experiences in Stony Brook, but the world had not stood still. Our paper had been submitted on 29 March. A few weeks later, Dan received a letter from Bruno Zumino dated April 17 which he has kept in which Bruno congratulated us on our success and compared our work with work he had been doing with Wess on the same problem. And, again two weeks later, on 28 April, Deser and Zumino submitted a paper entitled "Consistent Supergravity" to the leading European journal Physics Letters B. In this article they referred to our work, but proposed a reformulation which stressed the role of "torsion". The notion of torsion has a long history in GR. Originally it was proposed by the mathematician Elie Cartan (of the École Normale in Paris) in 1922, as a generalization of Einstein's theory of GR. Since that time many relativists have studied torsion further, but it never played a significant role. In Supergravity, on the other hand, it is an important option because it simplifies the formulation, eliminating the need to consider the quintic variations explicitly. Years later, I found with Paul Townsend of Cambridge University in England a third formulation of Supergravity which combines the virtues of the formulation of FNF (Freedman, van Nieuwenhuizen, and Ferrara) and that of DZ (Deser and Zumino). Their formulation with an independent torsion field is called the first-order formulation, and our formulation where torsion is due to pairs of fermionic spin-$\frac{3}{2}$ fields is called second order formulation. Dan called our third formulation the 1.5 order formalism. It is theoretically rather obvious, but for practical calculations it has turned out to be indispensable.

So what about the divergences in quantum supergravity? 't Hooft and Veltman had computed the divergences in the coupled scalar field — Einstein gravity theory, and had found that they did not cancel. Soon afterwards (1974) Deser and I repeated this calculation for the coupled Maxwell-Einstein theory, and we found that the sum of all one-loop divergences was equal to 137/60 times a standard divergence (137, not 137.035999084!).

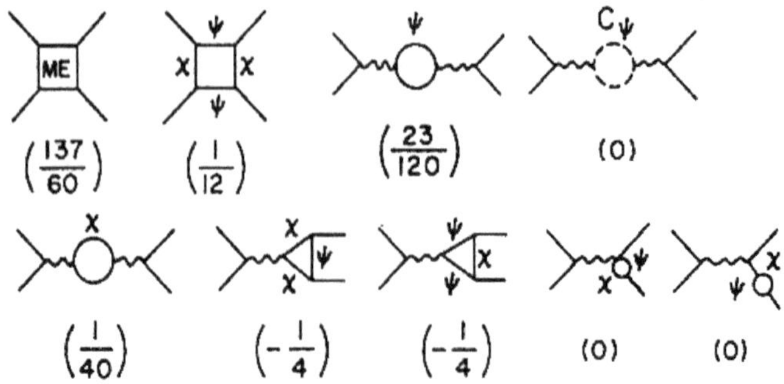

Figure 4.7: The divergences of photon-photon scattering in the Maxwell-Einstein theory with spin (2,3/2) supergravity coupled to the matter supersymmetric spin (1,1/2) Maxwell multiplet. The sum of divergences does not cancel, it equals (25/12) times the standard divergence $(n-4)^{-1}\kappa^4\,[T_{\mu\nu}(\gamma)]^2$ where $T_{\mu\nu}(\gamma)$ is the photon energy tensor. ME stands for all graphs present in ordinary (non-supersymmetric) Maxwell-Einstein theory.

So, as soon as the $\mathcal{N}=1$ supergravity and its coupling to the supersymmetric Maxwell multiplet was constructed (a system with spin 2, 3/2, 1, 1/2), I calculated with Jos Vermaseren in 1976 its divergences. This was the first calculation of quantum corrections to supergravity, and we found that the supersymmetry of this model was not strong enough to cancel all divergences.

However, soon after, I constructed with Sergio Ferrara the first extended supergravity theory, the $\mathcal{N}=2$ model in which the graviton, the photon, and two gravitinos form a single supermultiplet, and with Jos Vermaseren and Marc Grisaru I repeated the one-loop calculation in this model, and found to my everlasting delight that all divergences canceled. This proved two things: first that the number 137/60 in the calculation with Deser was right, and second that supersymmetry could lead to a finite theory of quantum gravity. Clearly, the extended supergravities (pure marble in Einstein's words, not mere wood) were the way to go. Then Marc Grisaru, through a very clever argument, was able to show that 2-loop divergences in extended supergravities cancel, and the loop race was on. Stanley Deser with John Kay and Kelly Stelle showed that there existed a locally supersymmetric 3-loop invariant, so the issue was whether its coefficient would vanish.

Several other physicists joined, and it was concluded that 3-loop was finite. Since then, more and more extended supergravities (in particular the largest model, the $\mathcal{N}=8$ model) have been studied, and at present the $\mathcal{N}=8$ model has been shown to be finite up to but not yet including 7 loops. Also the divergences of extended supergravities in other dimensions have been calculated.

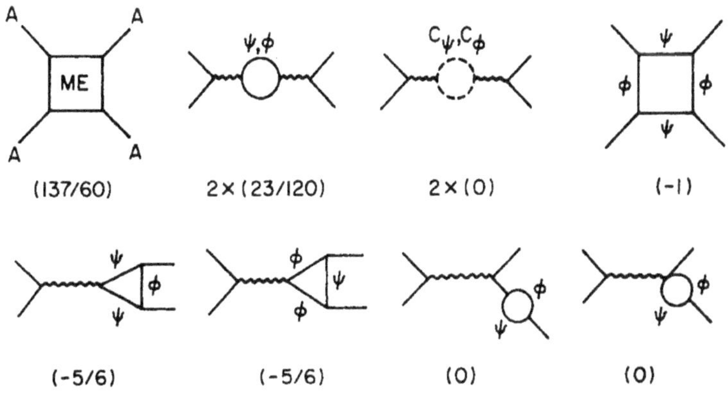

Figure 4.8: The one-loop ultraviolet divergences of photon-photon scattering in the $\mathcal{N} = 2$ extended supergravity theory. The number below the graphs denote the coefficients of the standard divergence $(n-4)^{-1}\kappa^4\,[T_{\mu\nu}(\gamma)]^2$, while ME stands for all graphs present in ordinary (non-supersymmetric) Maxwell-Einstein theory. The sum of all divergences cancels.

In the meantime, string theory came of age, and it was shown to be all-loop finite. Clearly, supergravity and superstring theory have changed the landscape of divergences for the better.

Epilogue. Since these stressful but exciting weeks in 1976, Supergravity has grown tremendously. It has become part of string theory, more precisely string theory becomes Supergravity at energies low enough that experiments can be done to test whether it exists and if so, its properties. On one hand, it is not the Holy Grail because it contains (probably, the issue is not settled) infinities of quantum field theory. On the other hand, it is full of new hidden symmetries (dualities) which have led to major changes in our thinking. In 1998 it was discovered by Maldacena that Supergravity in five dimensions corresponds to a certain non-gravitational but supersymmetric theory in four dimensions (our world), and this "AdS/CFT correspondence" has been extended to many other models.

I am now an 85 year old professor who (still) teaches advanced graduate courses. Sometimes when my thoughts are drifting, I dream of the day that the discovery of supersymmetric particles is announced. At other times I worry that I am in the same position as old professors in the early 1900's who kept believing in the ether. Let us hope that Nature is aware of our efforts.

LECTURE FOR THE FORTIETH ANNIVERSARY OF SUPERGRAVITY

DANIEL Z. FREEDMAN
Stanford Institute for Theoretical Physics and Department of Physics
Stanford University, Stanford, CA 94305, USA
Department of Mathematics and Center for Theoretical Physics
Massachusetts Institute of Technology, Cambridge, MA 02139, USA
dzf@math.mit.edu

The first paper on supergravity in $D = 4$ spacetime dimensions [1] was submitted to the Physical Review in late March, 1976. It was a great honor for me that the fortieth anniversary of this event was one of the features of the 54th Course at the Ettore Majorana Foundation and Centre for Scientific Culture in June, 1976. This note contains some of the material from my lectures there. The first part focuses on the most basic ideas of the subjects of supersymmetry and supergravity, ideas which I hope will be interesting for aspiring physics students.

Fundamentals of SUSY and SUGRA

Supergravity is built upon the principle of supersymmetry. Global supersymmetry was discovered in the early 1970's, see [2–4], and rapid development then followed. The theory postulates a symmetry between fermions and bosons. The symmetry is generated by conserved charges that are spacetime spinors Q_α, $\alpha = 1, 2, 3, 4$. Since they are spinors, it is natural that they enjoy simple properties under anti-commutation, and the basic supersymmetry algebra is given by

$$\{Q_\alpha, \bar{Q}^\beta\} = -(\gamma_\mu)_\alpha{}^\beta P^\mu, \tag{4.1}$$

where P^μ, $\mu = 0, 1, 2, 3$ are the generators of spacetime translations. In a supersymmetric theory the fields and the particles they describe are grouped in multiplets which contain both bosons and fermions. Supersymmetry is the only framework that permits the unification of particles of different spins within a single algebraic structure, called a superalgebra.

The parameters of *global* SUSY transformations are *constant* spinors ϵ_α. The appearance of translation generators P^μ on the right side of 4.1 suggests a relation to the geometry of spacetime. From this one can envisage a supersymmetric theory of gravity in which the graviton acquires a fermionic partner, the gravitino. In a gravitational theory the constant spinors must be replaced by arbitrary functions $\epsilon_\alpha(x)$ of the spacetime coordinates. In this way supergravity becomes the gauge theory of supersymmetry.

I read the path-breaking paper of Wess and Zumino [4] when it first appeared in early 1974. I was deeply impressed, and no other single paper has so transformed my subsequent professional life. The idea that there is a consistent symmetry principle that relates particles of different spin and statistics came as a thunderbolt. SUSY and SUGRA are attractive to me because the component formalism makes close contact with the basic principles of quantum field theory and their close relation with geometry. These principles include:

(i) Relativistic treatment of spin via Dirac γ-matrices. By this I mean the complete set

$$I,\ \gamma^\mu,\ \gamma^{\mu\nu},\ \gamma^{\mu\nu\rho}\ \ldots, \tag{4.2}$$

and their algebraic relations, e.g. $\gamma^\mu\gamma^{\nu\rho} = \gamma^{\mu\nu\rho} + \eta^{\mu\nu}\gamma^\rho - \eta^{\mu\rho}\gamma^\nu$. Note that these matrices of the Dirac-Clifford basis are totally antisymmetric in their tensor indices.

(ii) Basic equations of the geometry of gauge fields and Riemannian geometry, e.g. gauge field Bianchi identity

$$\partial_\mu F_{\nu\rho} + \partial_\nu F_{\rho\mu} + \partial_\rho F_{\mu\nu} = 0. \tag{4.3}$$

(iii) Spin and statistics: $\{\psi_\alpha(x), \psi_\beta(y)\} = 0$ and $\{\epsilon_\alpha, \epsilon_\beta\} = 0$ even at the level of classical manipulations.

As the first example of the interplay of these principles, let's consider the action of the free Maxwell-Dirac theory:

$$S = -\int d^Dx \left[\frac{1}{4}F^{\mu\nu}F_{\mu\nu} + \bar{\psi}\gamma^\mu\partial_\mu\psi\right]. \tag{4.4}$$

We will show that it is invariant under SUSY transformation rules in any spacetime dimension D,

$$\delta A_\mu = -\bar{\epsilon}\gamma_\mu\psi + \bar{\psi}\gamma_\mu\epsilon, \quad \delta\psi = \frac{1}{2}\gamma^{\nu\rho}F_{\nu\rho}\epsilon, \quad \delta\bar{\psi} = -\frac{1}{2}\bar{\epsilon}\gamma^{\nu\rho}F_{\nu\rho}. \tag{4.5}$$

In the proof it suffices to consider only the terms containing ϵ, since the terms with $\bar{\epsilon}$ are related by complex conjugation. The separate variations of the Maxwell and Dirac actions are

$$\delta S_1 = -\int d^D x F^{\mu\nu} \partial_\mu \delta A_\nu = \int d^D x\, \partial_\mu F^{\mu\nu} \delta A_\nu \overset{\text{SUSY}}{\longrightarrow} \int d^D x\, \partial_\mu F^{\mu\nu} \, (\bar{\psi}\gamma_\nu \epsilon),$$

$$\delta S_{1/2} \overset{\text{SUSY}}{\longrightarrow} -\frac{1}{2}\int d^D x\, (\bar{\psi}\gamma^\mu \gamma^{\nu\rho}\epsilon)\, \partial_\mu F_{\nu\rho}. \tag{4.6}$$

We need to show that the sum vanishes. This requires the γ-matrix algebra relation given above which we insert in $\delta S_{1/2}$:

$$\delta S_1 = \int d^D x\, \partial_\mu F^{\mu\nu}\, \bar{\psi}\gamma_\nu \epsilon,$$

$$dS_{1/2} = -\frac{1}{2}\int d^D x \left(\bar{\psi}\left[\gamma^{\mu\nu\rho} + \eta^{\mu\nu}\gamma^\rho - \eta^{\mu\rho}\gamma^\nu\right]\epsilon\right) \partial_\mu F_{\nu\rho}. \tag{4.7}$$

The last two terms cancel with δS_1, so first term must cancel by itself. Indeed, it does cancel using the gauge field Bianchi identity which is repeated from above for full clarity:

$$\gamma^{\mu\nu\rho}\partial_\mu F_{\nu\rho} = \frac{1}{3}\gamma^{\mu\nu\rho}(\partial_\mu F_{\nu\rho} + \partial_\mu F_{\nu\rho} + \partial_\mu F_{\nu\rho}) = 0.$$

The relation between spin and statistics played no role in this proof because we dealt with a *free* theory. In *interacting* SUSY theories it enters in a big way. For example, the proof of invariance of any interacting SUSY requires a manipulation of cubic terms in the fermions known as a Fierz rearrangement. Fierzing can be tricky, so I will just say that the basic Fierz relation is based on the completeness of the full set of Dirac γ-matrices in the particular spacetime dimension under study.

Actually, the basic spinor field in 4 spacetime dimensions is not a complex Dirac spinor $\psi_\alpha(x)$. Rather it is a Majorana spinor[100] $\ell_\alpha(x)$. Suppose we assume that the components of $\ell_\alpha(x)$ are ordinary numbers, i.e. $[\ell_\alpha(x), \ell_\beta(y)] = 0$. One can show using the reality property that the basic free spinor kinetic and mass terms vanish:

$$S_{\text{Majorana}} = -\frac{1}{2}\int d^4 x \bar{\ell}(\gamma^\mu \partial_\mu - m)\ell = -\frac{1}{4}\int d^4 x \partial_\mu(\bar{\ell}\gamma^\mu \ell) = 0, \tag{4.8}$$

so a free Majorana theory of the wrong statistics is vacuous.

[100]One key discovery of Ettore Majorana was that in spacetime dimension $D = 4$ the basic fermion field $\ell_\alpha(x)$ (and the Dirac γ_μ matrices) can be chosen to be real. The chiral projections $L\ell = \frac{1}{2}(I + \gamma_5)\ell$ and $R\ell = \frac{1}{2}(I - \gamma_5)\ell$ are effectively two-component Weyl spinors. Supersymmetric theories can be formulated using either Majorana or Weyl spinors.

Let's move on to something more sophisticated. In the (highly recommended) book [5], it is called "the universal part of supergravity." It comprises the first steps in the proof of invariance of local supersymmetry which are valid in any spacetime dimension and for any type of spinor. It is my dream that these steps one day become a standard part of the curriculum in quantum field theory courses.

In supergravity the fermionic partner of the spin 2 graviton is the spin 3/2 gravitino. This is described by a spinor-vector field $\psi_\mu(x)$, whose suppressed spinor index α takes $2^{[D/2]}$ values in D dimensions. This field was first considered by Rarita and Schwinger in 1941. When spinors are present in a gravity theory, the basic bosonic variable becomes the frame field $e^a_\mu(x)$. It is related to the metric tensor by $g_{\mu\nu} = e^a_\mu \eta_{ab} e^b_\nu$, where η_{ab} is the Minkowski metric of flat spacetime. I must assume that readers have some basic knowledge of the way connections and curvatures work in the frame field formalism. (If not they can consult [5].)

We begin by writing the action of the universal part of supergravity. It is the sum of the standard Hilbert action S_2 plus the massless gravitino action $S_{3/2}$. Note that this involves the third rank element $\gamma^{\mu\nu\rho}$ of the Dirac algebra,

$$S_2 = \frac{1}{2}\int d^D x\, e\, e^{a\mu} e^{b\nu} R_{\mu\nu ab} = \frac{1}{2}\int d^D x \sqrt{-g}\, g^{\mu\nu} R_{\mu\nu} \tag{4.9}$$

$$S_{3/2} = -\int d^D x\, e\, \bar{\psi}_\mu \gamma^{\mu\nu\rho} D_\nu \psi_\rho, \quad D_\nu \psi_\rho \equiv (\partial_\nu + \frac{1}{4}\omega_{\nu ab}\gamma^{ab})\psi_\rho. \tag{4.10}$$

We need the local SUSY variations of these fields. For the gravitino we take

$$\delta\psi_\rho = D_\rho \epsilon(x) = \left(\partial_\rho + \frac{1}{4}\omega_{\rho ab}\gamma^{ab}\right)\epsilon(x), \quad \delta\bar{\psi}_\rho = \bar{\epsilon}(x)\overleftarrow{D}_\rho\,. \tag{4.11}$$

This bears a close resemblance to $\delta A_\rho = \partial_\rho \theta(x)$ in Maxwell gauge theory and its non-Abelian extension. Indeed one of the good things that happens in supergravity is that it contains a new (in 1976) spin 3/2 gauge principle which joins the spin 1 gauge principle of electromagnetism and Yang-Mills theory and the spin 2 gauge principle of gravitation. These are the only gauge theories which allow consistent interactions.[101] The frame field and metric variations are

$$\delta e^a_\mu = -\frac{1}{2}\bar{\psi}_\mu \gamma^a \epsilon + \text{H.c.} \implies \delta g^{\mu\nu} = \frac{1}{2}(\bar{\psi}^\mu \gamma^\nu + \bar{\psi}^\nu \gamma^\mu)\epsilon + \text{H.c.} \tag{4.12}$$

[101] Vasiliev theories which involve an infinite number of higher spin gauge fields are also consistent [6].

We assume that the gravitino field is complex because that case should be more familiar for readers. In our proof that $\delta(S_2 + S_{3/2})$ vanishes to *lowest order* in ψ, it is sufficient to study terms containing ϵ and $\bar{\psi}$. If these terms vanish, so do the conjugate terms.

Let us start with S_2 and write its standard Euler variation,

$$\begin{aligned}\delta S_2 &= \frac{1}{2}\int d^D x\sqrt{-g}\,\delta g^{\mu\nu}\,(R_{\mu\nu} - \frac{1}{2}g_{\mu\nu}R) \\ &= \frac{1}{2}\int d^D x\sqrt{-g}\left[(\bar{\psi}^{\mu}\gamma^{\nu}\epsilon)\,(R_{\mu\nu} - \frac{1}{2}g_{\mu\nu}R) + \text{H.c.}\right]. \qquad (4.13)\end{aligned}$$

For $S_{3/2}$ we permute indices, $\mu\nu\rho \to \rho\mu\nu$, insert $\delta\psi_{\nu}$, and note that the commutator of covariant derivatives gives us the Riemann curvature tensor. This relation is called the Ricci identity of differential geometry,

$$\begin{aligned}\delta S_{3/2} &= -\frac{1}{2}\int d^D x\sqrt{-g}\bar{\psi}_{\rho}\gamma^{\rho\mu\nu}[D_{\mu}, D_{\nu}]\epsilon \\ &= -\frac{1}{8}\int d^D x\sqrt{-g}\bar{\psi}_{\rho}\gamma^{\rho\mu\nu}\gamma^{ab}R_{\mu\nu ab}\epsilon. \qquad (4.14)\end{aligned}$$

Somehow we have to show that the spinor bilinear with the product of five γ-matrices in 4.14 simplifies and then cancels the single γ-bilinear in 4.13. This involves γ algebra that is more difficult than previous but doable with some practice. The product $\gamma^{\rho\mu\nu}\gamma^{ab}$ can be expressed uniquely as the sum of 5th rank, 3rd rank, and 1st rank γ's:

$$\begin{aligned}&\gamma^{\rho\mu\nu}\gamma^{ab}R_{\mu\nu ab} = [\gamma^{(5)} + \gamma^{(3)} + \gamma^{(1)}]R_{....} \\ &\quad = \gamma^{\rho\mu\nu ab}R_{\mu\nu ab} + 2\gamma^{\mu\nu b}R_{\mu\nu\rho b} + 4\gamma^{\nu\rho b}R_{\mu\nu}{}^{\mu}{}_{b} + 4\gamma^{\mu}R_{\mu\nu}{}^{\rho\nu} - 2\gamma^{\rho}R_{\mu\nu}{}^{\mu\nu}. \\ &\qquad (4.15)\end{aligned}$$

The 5th rank γ-matrix vanishes in four dimensions. The matrix is present for $D \geq 5$, but there is a cyclic contraction of indices which produces the first Bianchi identity of the curvature tensor, so the 5th rank term vanishes. The first rank-3 term also vanishes by the Bianchi identity, and second one vanishes because $\gamma^{\rho\nu\sigma}$ is anti-symmetric, while the Ricci tensor $R_{\nu\sigma}$ is symmetric.[102] We are left with the rank 1 terms in the last line. It is easy to substitute them in 4.14 and find that the sum $\delta S_{3/2} + \delta S_2 = 0$, which establishes the result we wanted.

[102] We converted from frame indices to coordinate indices to make this statement, e.g. $\gamma^{\sigma} = \gamma^{b}e^{\sigma}_{b}$.

The deed is done! I hope that the readers noticed how Dirac algebra and Riemannian geometry conspired to make it happen. This is the universal part of supergravity, but it just the first step of a complete proof. For example, in the simplest case of pure $\mathcal{N} = 1,\ D = 4$ supergravity, there are order $\psi^3\epsilon$ terms that arise from the variation

$$\delta\gamma^\mu = \gamma^a \delta e_a^\mu = -\tfrac{1}{2}\gamma^a(\bar{\psi}_a \gamma^\mu \epsilon + \text{H.c.})$$

Order $\psi^5\epsilon$ terms also arose within the framework of the techniques used in [1], but we have learned how to avoid them [7–9]. Cancellation of such higher order terms is a delicate business.

It is even more delicate in extended SUGRA theories in $D = 4$ dimensions in which the graviton has $\mathcal{N} \geq 2$ gravitino partners and lower spin fields are also needed to complete the bose-fermi supermultiplet. Sophisticated calculations are also needed to construct SUGRA theories for $D \geq 5$. It turns out that extended SUGRA theories exist up to $\mathcal{N} = 8$ in $D = 4$, and there is also a limit $D = 11$ for higher dimensional SUGRA. The two maximal theories have been constructed in [10] and [11] and are closely related. The $D = 11$ theory can be dimensionally reduced on the 7-dimensional torus, and this process gives $\mathcal{N} = 8,\ D = 4$ supergravity. The limits $\mathcal{N} = 8$ and $D = 11$ are related to the remarks above concerning consistent gauge principles. The supermultiplets needed to exceed $\mathcal{N} = 8$ and $D = 11$ necessarily contain higher spin gauge fields such as the symmetric tensor spinor $\psi_{\mu\nu}(x)$, and there are no consistent interacting theories for such fields.

To conclude, a few brief comments are in order about what has been achieved within the framework of supersymmetry and supergravity and what has not been achieved. What has not yet been achieved is clear experimental support; the necessary superpartners of the known elementary particles have not been found up to the scale $M \approx 1 - 2$ TeV accessible at the LHC accelerator. Arguments based on the idea of "naturalness" of electroweak symmetry breaking suggested that evidence for SUSY would be found at the LHC, but Nature apparently does not know about this form of "naturalness."

What has been achieved is a vast theoretical framework which has greatly influenced modern theory. Global supersymmetry in $D = 4$ gives significant control of strong coupling effects [12]. The maximal $\mathcal{N} = 4, D = 4$ global theory [13] is ultraviolet finite. It is a superconformal theory. Techniques have been developed [14] to calculate its on-shell amplitudes by methods that are far simpler than the evaluation of many Feynman diagrams.

The study of $\mathcal{N} = 2,\ D = 4$ supergravity has led to the development of "special geometry" of considerable influence in both physics and differential geometry [15]. Theorists have speculated that the maximal $\mathcal{N} = 8$, $D = 4$ SUGRA theory is an ultraviolet finite theory of quantum gravity. This idea has passed calculation tests up to 4-loop order [16]. The on-shell amplitudes program [17] can be applied to this theory and results [18, 19] show that divergences cancel through 6-loop order, but cannot be excluded beyond this level. If the theory is finite to all orders, a new principle that goes beyond the obvious symmetries must be found. Another important line of investigation concerns black hole solutions of supergravity theories [20]. One interesting feature is the attractor mechanism [21] for solutions in theories with scalars. Finally there is a close relation between 10- and 11-dimensional supergravity and superstring theory and M-theory. Roughly speaking supergravity is the low energy limit of superstring theory, and many papers that address ideas in string theory actually work at the supergravity level. There are D-brane solutions in supergravity, and much of the calculational support for the AdS/CFT correspondence comes from calculations that compare quantities in maximal $\mathcal{N} = 8$, $D = 5$ SUGRA [22] with the $\mathcal{N} = 4, D = 4$ superconformal theory.

The comments above are far too brief to cover this vast area of theoretical work. References are meant to be suggestive and are far from complete. I was told a few months ago that there are about 14,000 papers on the SPIRES archive with the word "supergravity" in the title or as a keyword.[103]

Acknowledgments

The author thanks Prof. A. Zichichi for the opportunity to present these ideas and his warm hospitality. He is supported in part by the US National Science Foundation Grant No. PHY 1620045 and as Templeton Visiting Professor at Stanford.

References

1. D. Z. Freedman, P. van Nieuwenhuizen and S. Ferrara, *Progress Toward a Theory of Supergravity,* Phys. Rev. D **13**, 3214 (1976).
2. Y. A. Golfand and E. P. Likhtman, *Extension of the Algebra of Poincare Group Generators and Violation of p Invariance,* JETP Lett. **13**, 323 (1971).
3. D. V. Volkov and V. P. Akulov, *Is the Neutrino a Goldstone Particle?,* Phys. Lett. **46B**, 109 (1973).
4. J. Wess and B. Zumino, *Supergauge Transformations in Four-Dimensions,* Nucl. Phys. B **70**, 39 (1974).

[103] Martin Rocek, private communication.

5. D. Z. Freedman and A. Van Proeyen, *Supergravity,* (Cambridge University Press, 2012).
6. M. A. Vasiliev, *Consistent equation for interacting gauge fields of all spins in (3+1)-dimensions,* Phys. Lett. B **243**, 378 (1990).
7. S. Deser and B. Zumino, *Consistent Supergravity,* Phys. Lett. **62B**, 335 (1976).
8. P. K. Townsend and P. van Nieuwenhuizen, *Geometrical Interpretation of Extended Supergravity,* Phys. Lett. **67B**, 439 (1977).
9. A. H. Chamseddine and P. C. West, *Supergravity as a Gauge Theory of Supersymmetry,* Nucl. Phys. B **129**, 39 (1977).
10. E. Cremmer and B. Julia, *The SO(8) Supergravity,* Nucl. Phys. B **159**, 141 (1979).
11. E. Cremmer, B. Julia and J. Scherk, *Supergravity Theory in Eleven-Dimensions,* Phys. Lett. **76B**, 409 (1978).
12. N. Seiberg and E. Witten, *Electric-magnetic duality, monopole condensation, and confinement in $\mathcal{N} = 2$ supersymmetric Yang-Mills theory,* Nucl. Phys. B **426**, 19 (1994) Erratum: [Nucl. Phys. B **430**, 485 (1994)] [hep-th/9407087].
13. L. Brink, J. H. Schwarz and J. Scherk, *Supersymmetric Yang-Mills Theories,* Nucl. Phys. B **121**, 77 (1977).
14. R. Britto, F. Cachazo, B. Feng and E. Witten, *Direct proof of tree-level recursion relation in Yang-Mills theory,* Phys. Rev. Lett. **94**, 181602 (2005) [hep-th/0501052].
15. B. de Wit and A. Van Proeyen, *Potentials and Symmetries of General Gauged N=2 Supergravity: Yang-Mills Models,* Nucl. Phys. B **245**, 89 (1984).
16. Z. Bern, J. J. Carrasco, L. J. Dixon, H. Johansson and R. Roiban, *The Ultraviolet Behavior of $\mathcal{N} = 8$ Supergravity at Four Loops,* Phys. Rev. Lett. **103**, 081301 (2009) [arXiv:0905.2326 [hep-th]].
17. H. Elvang and Y. t. Huang, *Scattering Amplitudes in Gauge Theory and Gravity,* (Cambridge University Press, 2015).
18. J. M. Drummond, P. J. Heslop and P. S. Howe, *A Note on $\mathcal{N} = 8$ counterterms,* arXiv:1008.4939 [hep-th].
19. N. Beisert, H. Elvang, D. Z. Freedman, M. Kiermaier, A. Morales and S. Stieberger, *E7(7) constraints on counterterms in $\mathcal{N} = 8$ supergravity,* Phys. Lett. B **694**, 265 (2011) [arXiv:1009.1643 [hep-th]].
20. G. T. Horowitz and A. Strominger, *Black strings and P-branes,* Nucl. Phys. B **360**, 197 (1991).
21. S. Ferrara, R. Kallosh and A. Strominger, *$\mathcal{N} = 2$ extremal black holes,* Phys. Rev. D **52**, R5412 (1995) [hep-th/9508072].
22. M. Gunaydin, L. J. Romans and N. P. Warner, *Gauged $\mathcal{N} = 8$ Supergravity in Five Dimensions,* Phys. Lett. **154B**, 268 (1985).

1976 WAS A VERY GOOD YEAR[104]

DANIEL Z. FREEDMAN
Stanford Institute for Theoretical Physics and Department of Physics
Stanford University, Stanford, CA 94305, USA
Department of Mathematics and Center for Theoretical Physics
Massachusetts Institute of Technology, Cambridge, MA 02139, USA
dzfmit@gmail.com
June 24, 2016

1. I am happy to be here, and I want to thank Jian, and Luis, and Wolfgang for organizing this event. It is a great feeling when one's research is recognized by the community in which one works. It is especially appropriate at CERN, where so much work on supergravity has been done through the years.

2. Time is a funny thing. It can appear to go slowly, but I can tell you that somehow 40 years goes by very fast. What I plan today is to convey my recollections of the time when Sergio, Peter, and I were developing Supergravity. I chose the title "1976 WAS A VERY GOOD YEAR," after the song made famous by Frank Sinatra. It is a song about chasing girls. Of course, Peter, Sergio, and I were not chasing girls, we were chasing the truth.

3. I want to start a few years earlier and describe what it was like to work in the ITP at Stony Brook under the direction of Professor C. N. Yang. The leaders of the state of New York were thrilled to convince Frank (as we all called him) to leave the IAS in Princeton and come to a brand new university, and they gave him everything that he wanted. This created great working conditions for us and a stream of distinguished visitors such as Paul Adrien Maurice Dirac. One of my proud possessions is a copy of his "Quantum Mechanics" book with his signature, just plain P. A.M. Dirac.

One day (a couple years after the birth of supergravity) I went to a conference. For the last session, a panel of distinguished theorists sat on the stage. They were supposed to comment on the future of theoretical physics and to take questions from the audience. One question was directed

[104]Unpublished talk at the Commemoration of the 40th Anniversary of Supergravity at CERN, July 2016.

to Frank Yang and asked whether he thought that Supersymmetry was a fundamental advance in physics. Frank thought a moment and remarked that he always thought that a fundamental concept in physics should be linked to a fundamental concept in mathematics. So he had gone to his friend Prof. S.S. Chern in Berkeley and asked him whether "anti-commuting c-numbers" or "Grassmann-valued functions" were a profound or superficial development. Chern answered that he thought that they were rather superficial.

So I went home somewhat discouraged, and I remember going to Peter's office and asking "What are we doing here? Our boss thinks what we work on isn't worthwhile." But I was wrong. Frank was a person who was able to hold his own views on physics and at the same time to be an encouraging and supportive administrator. And I cherish the memories of my 12 years at Stony Brook.

4. In the ITP we had a reduced teaching load, and that made it easy to travel. Thus I was able to accept an invitation to visit the École Normale Supérieure in Paris in November-December 1975.

Just before leaving the US, I had been to a conference in Boston where Bruno Zumino spoke. Superspace methods had become very popular. Bruno spoke about his work with Julius Wess. Their goal was to find a local version of supersymmetry, which is what supergravity is, using superspace geometry. He said that there were dificulties because the geometry was complicated. It was a geometry with both curvature and torsion, and they had not yet found the right way to do things.

I had great respect for Wess and Zumino, and I knew that sooner or later they would succeed in formulating supergravity, and that I should prepare for this. I knew that, however complicated the mathematics, the physics was that the spin-2 graviton would interact with a partner, the spin-$\frac{3}{2}$ gravitino. So in my early weeks in Paris, I studied the basic literature on the vector-spinor Rarita-Schwinger field. That field had 16 components, and is written $\psi_{\mu\alpha}$ where $\mu = 0; 1; 2; 3$ is a spacetime index, and $\alpha = 1; 2; 3; 4$ is a spinor index à la Dirac. Most of their 1941 paper was about fields with mass, but at the end there was a brief remark to the effect that the massless case was gauge invariant, just like the electromagnetic field. That was a pregnant remark, and the modern form of the Rarita-Schwinger Lagrangian is what we use today. Sergio Ferrara visited ENS in mid-December. I was eager to meet him, since I was already familiar with his very good papers. The purpose of his visit was to work with Gliozzi on what eventually turned out to be a paper on the $\mathcal{N} = 2$ superstring, a paper with 13 authors, namely Lars Brink, John Schwarz and 11 Italians. Sergio and Ferdinando worked in the office next to mine. The wall wasn't very good, so lots of very loud Italian penetrated the wall. Fortunately, I didn't understand it, so it sounded like white noise to me and didn't disturb my work. Lots of cigar smoke also penetrated the wall, and that didn't bother me either, since I was a cigar smoker for some years until it was made illegal in all workplaces in Cambridge, MA.

At some point in his visit, Sergio and I talked and I told him what I was thinking about. He suggested that I should try to formulate supergravity directly in terms of the physical fields for the graviton and gravitino. He had seen my then recent paper with Bernard de Wit on SYM theory in terms of physical fields, rather than superfields which were then so popular. (Our approach appears in Steve Weinberg's book on SUSY.)

So I wrote down the Einstein-Hilbert action for gravity and added a covariant version of the massless Rarita-Schwinger action for the gravitino. It was fairly easy to guess the lowest order transformation rules. Within a day or two I had shown that the sum of the spin 2 and spin 3/2 actions was invariant under local SUSY to first order in the field. It turned out to be a significant first step toward SUGRA. I showed the calculation to Sergio just before he went home for the holidays, and I don't remember his reaction. But the theory wasn't complete. There would certainly be order $(\psi_{\mu\alpha})^3$ terms from varying the curved space γ-matrices and the spin connection. So Sergio went home to Geneva, and I went home to Stony Brook convinced that I could find the modifications very soon.

5. I tried to work on it for perhaps 1–2 weeks in January. But nothing good happened; the fields involved had too many indices. So I asked Peter to join me. We worked very hard for several weeks. Again nothing really worked. We communicated as well as we could with Sergio who was at CERN. In 1976, there was no internet, no email, and no fax machines. Telephone calls were too expensive, and they were discouraged. We had to modify the transformation rules to 2nd order in (ψ) and to add new 4th order terms to the Lagrangian. Then we had to add up all their effects and hope to get zero. We failed. We knew a little about torsion, but we didn't know how to use it in the most efficient way. So we decided that I should give an internal group seminar on our work. I described the lowest order success and the higher order failures. My colleagues were not impressed, yet somehow that seminar was a turning point.

After the talk, Peter and I sat down for a strategy session. We decided to be very systematic. We wrote down the most general ansatz consistent with symmetries and dimensional constraints for the transformation rules and same for the Lagrangian. That gave us a manageable number of parameters. Perhaps 5–6. We then worked out the effects of the ansatz on δS to 3rd order in (ψ_μ). It was still very complicated, because there were many Fierz rearrangements to perform and that technique was well known to be messy. It took a few days to get it right, but we found a unique set of parameters that made it vanish.

6. But the theory still wasn't complete, because the modifications that made the cubic (ψ_μ) term vanish, produced 5th order terms from several sources. Those terms looked impossible to control; there were even more indices. But we also showed that vanishing at order $(\psi)^5$ was a life-or-death issue for supergravity, because no further modifications could make it vanish. But it still looked impossible, so we decided to write up our work and tell the story as it stood. We did this and submitted our paper to Physical Review on 29 March 1976. It was a short paper, only 5 pages in the journal.

7. A few weeks later, I received an Air Mail envelope from CERN. It contained a hand-written letter from Zumino with the date 17 April. He congratulated us on our good work and went on to say how it was related to his unpublished treatment of $D = 2$ supergravity, a much easier and much less interesting case. He also made some remarks about its likely relation to his current superspace work with Wess.

8. A short while after that Peter came to me with the idea that we could use a computer to examine the 5th order term. We would ask the computer to do explicitly the anti-symmetrizations needed to simulate the anti-commutation of fermionic fields, and the computer could do the explicit sums over the indices. We used a numerical code in Fortran language, since symbolic manipulation was far too primitive at that point in time. As I remember it took about a week to get the code running. But it did work and it eventually spit out a long string of zeros, one for every choice of trial values of the $\psi_\mu(x)$ inputs.

9. A short while after that we received a preprint from Deser and Zumino. (Submission date — April 28.) They used the Palatini formalism to show that the spin connection in supergravity carried torsion. In that setup there simply is no 5th order variation, thus no need for explicit computation. That was certainly a useful simplification. In the Palatini formalism the spin connection $\omega_{\mu\, ab}$ is an independent field, so one must find its local SUSY transformation. This is an awkward feature, especially when the theory lower

matter multiplets. Most work in SUGRA is now done by combining the best features of the technique of Deser and Zumino with the techniques we developed at Stony Brook. It is called the 3/2 order formalism, found by Peter together with Paul Townsend.

10. I will describe how the future development looked to us in 1976. We had constructed pure $\mathcal{N} = 1$ SUGRA. The fields it contained were just the graviton and ONE gravitino. It was clear that this gravity multiplet could be coupled to lower spin $\mathcal{N} = 1$ matter multiplets. We also knew from the representation theory of SUSY algebras that extended SUGRA theories should exist with up to $\mathcal{N} = 8$ gravitini.

Beyond $\mathcal{N} = 8$, higher spin fields with $s \geq 5/2$ would be needed, and there would be new and fundamental dificulties. Indeed, those dificulties are still unsolved. We knew that the $\mathcal{N} = 8$ limit in four space-time dimensions corresponded to a limit $D \leq 11$ on the spacetime dimension. At that point, the same higher spin dificulty would occur. The theories that we envisaged were all constructed by the end of 1978. It was a remarkably exciting period, and many people contributed.

One thing that I did not anticipate was the close relation between SUGRA and superstring theory. Roughly speaking SUGRA is the low energy limit of superstring theory. The paper that started the first superstring revolution appeared in 1984 by Michael Green and John Schwarz. They showed that anomalies would cancel. If you look at the calculations that were done in their paper, they were all done in the limit in which SUGRA is an accurate approximation to string theory. In fact the same is true of many papers on string theory and many papers based on the AdS/CFT correspondence.

Peter's 75th birthday was celebrated with a conference at Stony Brook in January 2015. It was a happy occasion. One of Peter's colleagues had asked the SPIRES archive to count the number of papers that had the word SUPERGRAVITY in the title or as a keyword. There were more than 14,000. That is not bad for a theory with no experimental evidence, yet. Our best hope up to rectify the situation is the LHC collider here at CERN!

11. I have told you how I spent a bit of 1975 and the first few months of 1976. But I want to end by describing an event that occurred during my sabbatical leave from MIT in 1986 which I spent visiting the IAS in Princeton. It was an event which led to one of the finest moments of my career.

One evening there was a cocktail party and Julian Schwinger was there. In his prime he was one of the giants of theoretical physics. He was a fairly shy man and I saw him standing in a corner sipping some wine. I am also fairly shy man, and I was in another corner sipping some wine. I said to myself "Dan this is your one and only chance, do it!" So I did it. I went over to his corner and introduced myself. The conversation started with a few platitudes about the weather, but I soon asked him what he thought about SUSY. He responded loud and clear and with no hesitation "I SHOULD HAVE INVENTED SUPERGRAVITY!" That is indeed one of the finest moments of my career!

5

The Pioneers

A GLIMPSE

JOHN STRATHDEE

New Zealand

At this point all that comes to my mind is my annoyance at having our note[105] describing the unitary representations rejected by *Physical Review Letters* as being of "insufficient interest." I suppose I really suspected that the editor was quite right, and so was annoyed with myself for getting excited about such a far-fetched speculation.

Of course, I now realize, like everyone else, that the subject of supersymmetry is truly interesting ...

April 29, 2000

[105]Editors' note: The paper A. Salam and J. Strathdee, *Unitary Representations of Supergauge Symmetries*, was published in *Nucl. Phys.* **B80**, 499 (1974). The authors develop the ideas put forward in their previous paper [*Nucl. Phys.* **B76**, 477 (1974)] where the superfield formalism had been introduced. They present an algebraic construction of the unitary representations of the super-Poincaré algebra. In *Nucl. Phys.* **B80**, 499 (1974) they also introduce what is called in modern terminology "extended superalgebras," and find some representations of the extended superalgebras. –MS

Editor's note: Abdus Salam (1926–1996) was a Pakistani-born theoretical physicist who shared the 1979 Nobel Prize in Physics with Glashow and Weinberg for the invention of what is now called *the Standard Model.* Salam was the first Pakistani and the first from an Islamic country to receive a Nobel Prize in science. His rich and busy life was an endless quest for insights in physics.

Not only was he one of the most influential physicists of the 20th century but also a generous humanitarian who set up the world-famous International Center for Theoretical Physics in Trieste (which is now named after him) for educating third-world scientists. From a remote village in Pakistan to the prestigious colleges of Cambridge and then to the vibrant atmosphere of London to Trieste — that was his road to Olympus.

Salam left Pakistan for London in 1974, after the Pakistani parliament passed the Second Amendment to Constitution that declared Ahmadiyya Muslim Community to be non-Muslim. Abdus Salam was an adherent of this relatively small sect; his departure was a gesture of protest –MS.[106]

Figure 5.1: Abdus Salam, 1970s.

[106] In addition to the articles by Martin Sohnius, Peter West and Robert Delbourgo in this Collection the reader is also recommended to read the detailed essay by M. J. Duff, *Abdus Salam at Imperial College*, Int. J. Mod. Phys. **A 32** (2017) 08, 1741001.

MEMORIES OF ABDUS SALAM AND THE EARLY DAYS OF SUPERSYMMETRY[107]

PETER WEST

Department of Mathematics,
King's College London and Mathematical Institute,
University of Oxford, UK
peter.west540@gmail.com

I thank Misha Shifman for the suggestion to write this article. Its purpose is not to give an account of the life of Abdus Salam or be a systematic account of the early years of supersymmetry. Rather, it is to give an impression of what it was like to be a PhD student of Abdus Salam and also to take part during the early stages of the development of supersymmetry.

After an enjoyable undergraduate degree at Imperial college I was accepted for PhD studies at Imperial in 1973 but who would be my supervisor was not specified. I had a choice of two possible supervisors, Tom Kibble or Abdus Salam, or Professor Kibble and Professor Salam as us students referred to them. It was a measure of the strength of Imperial College in theoretical physics at that time that a student could have a choice between two professors who had both already laid the cornerstones of the standard model and who had also made other very important contributions. Although at that time there was not much talk of the standard model, at least among us students or in our lectures. Professor Kibble was very friendly and polite but he was rather reserved and did not say anything he had not thought carefully about. Indeed he did not say very much at all. His deep understanding of physics was obvious during the lecture course that he gave us.

In contrast Professor Salam was very lively, had charisma and an enthusiasm for physics but he also had a friendly disposition which made one feel at home, at least that was the case for me. As he explained some point of physics he usually had a faint playful smile on his face and one could not help wanting to understand what he was saying and join in. I got the sense that studying under Professor Salam would be very exciting and one might find results of interest. Being keen to make my way in theoretical physics

[107]Partly an excerpt from P. West's talk at Abdus Salam's 90th Birthday meeting, 25–28 January 2016, NTU, Singapore, Editors L. Brink, M. Duff, and K. Phua, (Proc: World Scientific, Singapore, 2017). See also *A Brief Review of E Theory*, Int. J. Mod. Phys. **A 31** (2016) 26, 1630043; [ArXiv 1609.06863]. Updated on March 5, 2024.

I choose Professor Salam despite the fact that I was warned that many PhD students of Professor Salam did not prosper. Ali Chamseddine also began with Professor Salam at the same time.

For my first year I did not see much of Professor Salam as I was taking my preparatory courses, but once the summer came I had finished my courses and so I went to see Professor Salam to find out what I would work on. At that time Professor Salam had an office on the fifth floor of the physics building but later on he moved office to the newly built Huxley Building. In the former office his door would be open but a large screen had been placed near to the door which blocked the view of his desk where he worked. While in the latter office the door would be only slightly ajar and it did not permit a view of this desk. A knock at the door resulted in a soft voice beckoning you in.

To my surprise he asked me what I wanted to do. I said that the infinities in quantum field theory were very ugly and so I would like to work on general relativity which was more aesthetically pleasing. Rather than explain the flaws in this naive approach, he suggested that there would not be so much to do in general relativity and that I might like to look at his very recent paper with John Strathdee in which they had discovered superspace and also super-Feynman rules [1]. With this he handed me the ICTP preprint with its characteristic pale blue boarder. As he sometimes did on later occasions he did not say I might like to look at this humble paper. I do not remember that he said much about the paper but only that I should look at it. Professor Salam had previously written papers on constructing spacetimes from cosets, for example the one with Mack [2] on viewing the conformal group from the viewpoint of a coset. As such it was natural that he and Strathdee should construct superspace as the coset of the super Poincaré group modulo the Lorentz group. Within days I was captured by the ideas in this paper and began working on infinities in supersymmetric theories. From the perspective of today I realise that I had been subject to his great charm and diplomacy, a skill which he had used to such great effect all over the world.

Professor Salam was not often at Imperial as he spent most of his time at the centre he had created in Trieste. Little did I, as a student, know of the many endeavours for the good of science in the third world that he was undertaking. Us students never knew when we would next see Professor Salam. He would come about once a month and the first we realised that he was in was that the door of his room was slightly ajar as we came into work. I would knock and he would welcome me in. Remarkably he was usually sitting at his desk studying some document by himself. He was always very cheerful, friendly and seemed to have time to talk as if he had all the time in the world. Since he was away most of the time we had no one to watch over us and as a result we did not work very hard unless we really thought something was going to work out in a very good way.

Supersymmetry had been discovered in 1971 in the very remarkable paper of Golfand and Liktman [3] and a non-linear realisation was later found by Volkov and Akulov [4]. These papers went essentially unnoticed even though the former paper did contain a four dimensional model of supersymmetry. Supersymmetry also appeared in the superstring whose fermionic sector was found by Ramond [5] and the bosonic sector in the later paper of Neveu and Schwarz [6]. Hidden in these papers was the two dimensional supersymmetry algebra and this was extracted by Gervais and Sakita [7]. Wess and Zumino generalised this algebra to four dimensions and constructed the invariant Wess-Zumino model [8].

It was only with this last paper that a few physicists took an interest in supersymmetry. They were located at a only a few institutions including CERN in Geneva, the École Normale Supérieure in Paris, Karlsruhe and the two institutes to which Salam belonged, namely Imperial College and the International Centre for Theoretical Physics in Trieste. Thus in 1973 there were only a few people working on supersymmetry. Being a student of Salam I was one of them as were indeed a number of other young people who were the PhD students of those few who thought that supersymmetry was of interest.

In trying to read the paper of Salam and Strathdee it quickly became clear that supersymmetry required very advanced manipulations in the matrix algebra associated with fermions. There were no courses on supersymmetry in 1973, at least not at Imperial College, nor were there any papers where one could look this, and a number of other difficult matters, up. Fortunately Professor Salam's long term collaborator Bob Delbourgo had an office nearby and he provided me with all the technical help I needed. He even had a problem I might work on involving the calculation of superspace Feynman diagrams in the Wess-Zumino model. Understandably he was unable to resist the temptation do this himself and in a matter of days he had found the result [9] before I had even begun to think about the problem. Using the superspace Feynman rules he found that the Wess-Zumino model only needed a wavefunction renormalisation as had previously been found using the component formalism by Wess and Zumino [8] and Iliopoulos and Zumino [10].

At that time the big problem was to spontaneously break supersymmetry and hence hopefully find a realistic model of nature that was supersymmetric. Particles in a supersymmetric theory had the same mass and although one could spontaneously break supersymmetry at the classical level [11, 12], the pattern of masses it led to was not consistent with nature. Professor Salam thought that radiative corrections would break supersymmetry and lead to more promising results. Such corrections had been used to good effect in a paper of Coleman and Weinberg [13] in very different context, namely scalar electrodynamics.

Professor Salam and his long term collaborator John Strathdee, who was in Trieste, produced a series of models and it was my job to compute their one-loop effective potentials and see if supersymmetry was spontaneously broken and what pattern of masses they lead to. The first models did not work and as time went on the models became more and more complicated involving very many fields. If I had not completely finished computing with a given model by the time I next meet with Professor Salam it was not a problem, there was always a much better model to look at instead.

Eventually Bob Delbourgo and I looked at one of the proposed models that required us to diagonalise a 26 by 26 matrix. After much work we did do this but the entires were not real numbers meaning that the minimum of the potential which we had also given was not actually a minimum. When I told Professor Salam about this result he was not at all perturbed and quickly produced another model which he must have brought with him from Trieste. Fearing the same fate for this model I suggested there was a problem with the model to which Salam gave a quick reply. It was not clear who was giving the worst argument. Shortly afterwards Salam spoke to Bob Delbourgo and asked him to tell me that all was well.

Eventually I realised that if supersymmetry was preserved at the classical level then the effective potential vanished in the most general $\mathcal{N} = 1$ theory invariant under rigid supersymmetry [14]. This meant that one could not spontaneously break supersymmetry using perturbative quantum corrections, although one could still hope that it was spontaneously broken by non-perturbative corrections. The problem of breaking supersymmetry in a natural way is still largely unsolved. The result had another more favourable consequence, as was pointed out by others [15], namely that supersymmetry did solve the hierarchy problem, at least technically. In a supersymmetric version of the standard model the Higgs mass would not be swept up to some large unified scale by quantum corrections as long as supersymmetry was not broken much above the weak scale. This in turn lead to the hope that supersymmetry might be found at the LHC (Large Hadron Collider at CERN). My article in this volume gives a more complete account of these developments. As we now know the LHC did not find supersymmetry, at least not so far. While the above argument was plausible there is nothing wrong with supersymmetry being broken at a very high scale and the problem of naturalness in the standard model is resolved in some other, as yet unknown, way. A more general non-renormalisation theorem was later found by Grisaru, W. Siegel and M. Roček [16] using some reformulated superFeynman rules.

In those days there was no internet and so no arXiv. Once a result was written up the paper was sent to the journal but it was usually a long time before it would be published even if the referee comments were favourable. To enable more speedy access to the paper copies of it, called preprints, were produced and these were sent out by mail to those institutes whose members might be interested in it. Since there were not many people who were working on supersymmetry this did not involve much effort. People who had heard of the result and who did not have access to the preprint would send a post card requesting a copy which was then sent directly to them. Each week the preprints that had arrived at Imperial would be placed on some racks in the corridor. Each paper had a piece of paper attached to it on which one could write ones name which resulted in it being passed to you but only after the others who were higher up the list had read it. The consequence was that people working on supersymmetry were not always aware of what every one else was doing. One advantage was that people could pursue lines of independent research without too much influence from what others were doing.

After three years I wrote up my results in a thesis and the time came for my PhD examination. Professor Salam had enrolled Professor John Taylor from Cambridge University to be the external examiner. He was rather softly spoken, very polite and had found important results in how symmetries were realised in quantum field theory. They assembled in Salam's office and I was called in. Professor Salam indicated that John Taylor should ask me some questions on my thesis which he then did, whereupon Professor Salam began working at his desk on some no doubt important task. This lasted half an hour at most and then there was silence which Professor Salam eventually noticed. He turned to me and said I should explain to Professor Taylor some results on gauging the super Poincaré group I had found with Ali Chamseddine. With this Professor Salam resumed the study of the document he was reading and I began writing on the board. This lasted at most twenty minutes whereupon silence returned which Professor Salam again noticed. At this point the exam was declared over and, after a glance between them, they told me I had passed.

After the Wess-Zumino paper an obvious problem was to find a theory of gravity which was supersymmetric, that is, a generalisation of Einstein's theory to contain spin two as well as a spin three-halves particle. There was an attempt [17] to construct supergravity by generalising the geometry of Einstein's theory but it was not clear how to do this. After about two years the first supergravity theory was found by Ferrara, Freedman and van Nieuwenhuizen [18]. They used the low key, but very powerful Noether technique that had been discussed in the context of gravity by others. The advantage of this approach was that it did not require any preconceived geometrical notions, the disadvantage was that it required a lot of difficult algebra. Indeed they were only able to show that the supergravity theory they had found was invariant under supersymmetry by using a computer. The problem was that the result could not be reproduced by others. Deser and Zumino then showed that the spin three-halves particle did, unlike in other theories, propagate in a causal manner and they also discussed the theory from the viewpoint of first order formalism [19].

There then followed two papers coupling the new supergravity theory to the matter supermultiplet containing spin one and spin one half particles [20]. The next paper on supergravity was the result of Ali Chamseddine and myself [21] which Salam had asked me to explain to Professor Taylor. We had taken it in our heads to gauge the super Poincaré group in the style of Yang and Mills. Of course even Einstein's gravity is not a Yang-Mills theory but we were students who did not know any better. After only a few lines many of the features of the new supergravity theory emerged. Indeed one could write down an action which was just that found in the paper of Ferrara, Freedman and van Nieuwenhuizen. One great advantage was that one could prove using the usual algebraic manipulations, and in a relatively few lines, that the supergravity action was invariant under the supersymmetry transformations. In this way the discovery of supergravity theory was completed and it was in this way that the all subsequent supergravity theories were shown to be invariant. Our approach had a significant drawback, it required one to adopt a condition that broke the gauge symmetry, but miraculously once one took this unconventional step everything worked very well and the desired result emerged. Professor Kibble suggested in his understated way that we should understand this step better. Our proof of the invariance of the action was further elucidated in a paper by Paul Townsend and van Niewenhuizen [22].

Of course we had also derived Einstein gravity as a gauge theory of the Poincaré group. The gauging technique was used to find the conformal supergravity theories [23] and higher spin theories emerge if one took a much bigger gauge group. Ali Chamseddine and myself also gauged the OSp(1,4) group and found the transformations of the vierbein, spin connection and gravitino. The results were reported in thesis of Ali Chamseddine [24]. An action was later found by MacDowel and Mansouri [25] which agreed with the then known supergravity theory with a cosmological constant [26]. One problem was that our paper was rejected by Nuclear Physics B and being students we did not know what to do next. As a result it took quite some time to resubmit it after removing some of the fibre bundle language which really had no proper place in the paper. An account of previous work on gauge symmetry and gravity by Sciama, Kibble, Trautman and others and the connection to our work is explained in detail in reference [27].

The time came to get a job. I applied to the research council of the UK for a post doctoral fellowship. The resulting interview did not go well. The panel did not think much of the new fangled idea of supersymmetry and wanted to know how it would affect current particle physics experiments. Neither I nor anyone else, had any real idea of what was the answer to this question and I did not get a fellowship. The Royal Society did offer me a fellowship to take to the École normale supérieure. No doubt I had Salam and others at Imperial to thank for this. Most of the young people who worked on supersymmetry had to wait a long time to gain a permanent job although I had better luck.

There was another problem with the original supergravity theory found by Ferrara, Freedman and van Nieuwenhuizen. It had a supersymmetry algebra which was tied to the supergravity theory rather than being an independent algebra. Up to this point a symmetry of a theory existed in a way that was independent of that theory. In other words the transformations formed a group which had an independent life. For example, the Poincaré transformations exist independent of Maxwell's theory and so can be used as a guide to construct other Poincaré invariant theories. As such to construct the coupling of super matter to the original supergravity was very difficult since this required a new supersymmetry algebra and this was only achieved for a few very specific models [20, 28].

For about two years it was unclear what to do. The key step was the observation that the fields of a supergravity theory were in a one to one correspondence with the currents in a supersymmetric theory and this had already been worked out by Ferrara and Zumino [29]. This supermultiplet contained the energy–momentum tensor and the supercurrent, but there were other objects which had to correspond to additional fields in the supergravity theory. To find the theory we used the same Noether procedure. It was a very long but enjoyable calculation. The extension of the original supergravity theory which did have a supersymmetry algebra that was independent of the supergravity theory was found by Kelly Stelle and myself [30] and independently by Ferrara and van Nieuwenhuizen [31]. This theory had additional fields that did not lead to physical degrees of freedom.

Equipped with this new formulation of supergravity it was possible to construct a tensor calculus for supergravity [32, 33] and then it was relatively easy to find the most general coupling of matter to supergravity [34–36] which played such a key role in the construction of realistic models of supersymmetry using the ideas of Chamseddine, Arnowitt and Nath [34, 37]. Other ways to find the most general coupling of matter to supergravity have proliferated but they perhaps lack the simplicity and efficiency of the original approach.

Perhaps on the basis of these results Professor Salam invited me to visit him in Trieste for some time. When I arrive at the centre in the morning I would find Professor Salam sitting at his desk studying what I thought were the latest papers. Talking to Professor Salam you could not escape his great enthusiasm for physics; you came to understand that it was a lot of fun to do physics and that it was good to work in a very relaxed and free thinking way. Once we began our discussion he would think of a vast number of ways to proceed in the quest to find new things. He was always most interested in very new ideas and while not all of his ideas worked they included many of the deepest ideas that have come to dominate the subject. As one of his students I was, perhaps, able to absorb some of these qualities. Certainly, it was due to him that I began working on supersymmetry rather than on some uninteresting direction.

At the end of my visit to Trieste the time came for me to leave for the airport. Salam realised that I would be travelling at the same time as the Italian Minister for Science, who was visiting the centre, and so he suggested we could share the same car to the airport. This was met with a frown by the organiser of the visit who, no doubt correctly, thought that a scruffy post-doc with a rucksack might dent the carefully created image that the centre wanted to portray. Salam's suggestion was quietly shelved and I went to the airport by train.

I did not meet Julius Wess or Bruno Zumino during my doctorate studies but their presence was always with me and, I suspect, the other young physicists working on supersymmetry. Their results and the clarity of their papers was a benchmark to which we could but try to attain. When we did meet they were very friendly and encouraging. One aspect of there being so few working on supersymmetry was that there was a feeling of a common purpose and being enguaged in an exciting endeavour. Of course supersymmetry was great to work on. The calculations were usually difficult but often worked out in a spectacular manner and this encouraged us to continue despite the lack of support from the community at large.

The superspace formulation of supergravity using geometrical concepts analogous to those used in Einstein's theory was found by Wess and Zumino in [38]. It turned out that the curvatures and torsion were subject to very strong constraints which were allowed by the fact that the tangent space group was only the Lorentz group under which they formed a very reducible representation. We suspected that they had used the fact that it must lead, when expressed in components, to above formulation of supergravity to guide them to this result. There then appeared a paper by Warren Siegel claiming that their constraints were so strong that there were no physical degrees of freedom left [39]. Wess and Zumino explained that this paper was wrong but they had a worried look. Of course we found it very amusing that a young unknown, at least to us, post-doc had dared to contradict a paper of the gods. Eventually it was clear that the paper of Warren Siegel was indeed wrong but he then wrote a paper [40] solving the constraints on the torsions and curvatures. A pedagogical and further elucidation of this work was given in a paper by Gates and Siegel [41]. It was a very considerable feat involving technical leaps of such difficulty that I suspect few have worked through it.

The last time I meet Bruno Zumino was in the Newton Centre at the University of Cambridge in 1997. David Olive and I were puzzled by a point what to do with a coset construction and we discussed it with Bruno late in the afternoon. It was not a crucial point but it would have been good to understand it. The next day when I arrived quite early I found a hand written note on my desk from Bruno which had worked it all out in great detail and in a very clearly way. David also found a note on his desk.

I end with an account of two meetings with Abdus Salam that display his warmth and humanity.

I met Abdus Salam in his office in London a few days after he had won the Nobel prize. I asked him what was it like to win such a prize, he reassured me that he was just the same. He then suggested that we go for coffee in the common room in the old physics building at Imperial. To get there we had to go through a number of doors and he insisted that I go first through each door despite my protests.

During the time that Salam was very ill there was a conference in his honour at Trieste, but he was not well enough to go to all the talks. I saw him sitting at the very back of the big auditorium. I asked if it would be alright to say hello, but I was told that he might not recognise me. Since this might be the last time I would see him I went anyway. I said hello, he put up his hand and I shook it. He then immediately said how was Sue. Sue is my wife's name who he had met only once many years before.

Acknowledgements: I want to thank Misha Shifman for suggesting I write this article. I also thank Ali Chamseddine for comments.

References for this section

1. A. Salam and J. Strathdee, *On Superfields and Fermi-Bose Symmetry*, Phys. Rev. **D11** (1975) 1521.
2. G. Mack and A. Salam, *Finite-Component Field Representations of the Conformal Group*, Annals of Physics **53**, (1969), 174–202.
3. Y.A. Golfand and E.S. Likhtman, *Extension of the Algebra of Poincaré Group Generators and Violation of P Invariance*, *JETP Lett.* **13**, 323 (1971).
4. D.V. Volkov and V.P. Akulov, *Possible universal neutrino interaction*, *Pis'ma Zh. Eksp. Teor. Fiz.* **16**, 621 (1972); *Phys. Lett.* **46B**, 109 (1973).
5. P. Ramond, *Dual theory for free fermions*, Phys. Rev. **D3** (1971) 2415.
6. A. Neveu and J.H. Schwarz, *Factorizable dual model of pions*, Nucl. Phys. **B31** (1971) 86; A. Neveu and J.H. Schwarz, *Quark Model of Dual Pions*, Phys. Rev. **D4** (1971) 1109.
7. J. L. Gervais and B. Sakita, *Field Theory Interpretation Of Supergauges In Dual Models*, Nucl. Phys. **B34** (1971) 632.
8. J. Wess and B. Zumino, *Supergauge transformations in four dimensions*, Nucl. Phys. **B70** (1974) 139; A Lagrangian Model Invariant Under Supergauge Transformations, Phys. Lett. **49B** (1974) 52.
9. R. Delbourgo, *Superfield Perturbation Theory and Renormalization*, Nuovo Cim. **A25** (1975) 646.
10. J. Iliopoulos and B. Zumino, *Broken supergauge symmetry and renormalization*, Nucl. Phys. B76 (1974) 310.
11. L. O'Raifeartaigh, *Spontaneous Symmetry Breaking for Chiral Scalar Superfields*, Nucl. Phys. **B96** (1975) 331.
12. P. Fayet, *Spontaneous Supersymmetry Breaking Without Gauge Invariance*, Phys. Lett. **58B** (1975) 67.
13. S. Coleman and E. Weinberg. *Radiative Corrections as the Origin of Spontaneous Symmetry Breaking*, Physical Review **D7** (1973) 1888.
14. P. West, *Supersymmetric Effective Potential*, Nucl. Phys. **B106** (1976) 219.
15. E. Witten, *Dynamical Breaking of Supersymmetry*, Nucl. Phys. **B188** (1981) 513.

16. M. Grisaru, W. Siegel and M. Rocek, *Improved Methods for Supergraphs*, Nucl. Phys. B **159** (1979) 429.
17. P. Nath and R. Arnowitt, Phys lett **56B** (1975) 177, R. Arnowitt, P. Nath and B. Zumino, Phys. Lett **56B**, (1975) 81,
18. D. Freedman, P. van Nieuwenhuizen and S. Ferrara, *Progress toward a theory of supergravity*, Phys. Rev. **D13**, 3214 (1976); D. Freedman and P. van Nieuwenhuizen, *Properties of supergravity theory*, Phys. Rev. **D14**, 912 (1976).
19. S. Deser and B. Zumino, *Consistent Supergravity, Phys. Lett.* **62B**, 335 (1976).
20. S. Ferrara, J. Scherk and P. van Nieuwenhuizen, *Locally Supersymmetric Maxwell-Einstein Theory, Phys. Rev. Lett.* **37**, 1035 (1976); S. Ferrara, F. Gliozzi, J. Scherk and P. van Nieuwenhuizen, *Matter Couplings in. Supergravity Theory, Phys. Rev.* **D15**, 1013 (1977).
21. A. Chamseddine and P. West, *Supergravity as a gauge theory of supersymmetry, Nucl. Phys.* **B129**, 39 (1977). This paper was received on the 28 September 1976 but was initially rejected for publication. The revised version, received on 10 June 1977, was essentially the same except for two paragraphs explaining in more detail its relation to references [18] and [19]. It was also published as the Imperial preprint ICTP/75/22, September 1976.
22. P. Townsend and P. van Nieuwenhuizen, *Geometrical interpretation of extended supergravity, Phys. Lett.* **B67**, 439 (1977).
23. P. van Nieuwenhuizen, M. Kaku, and P. Townsend, *Properties of conformal supergravity*, Phys. Rev. **D17**, (1978) 1501.
24. A. H. Chamseddine, *Supersymmetry and higher spin fields*, Ph.D. Thesis, defended September 1976, Imperial College, London University.
Chapter 6 of the thesis is stated as collaborative work with P. West. The thesis can be found at https://drive.google.com/file/d/0B8tITtoqQkxfeHQ3cFJ0WUxIcHM/view?usp=sharing.
25. S. MacDowell and F. Mansouri, *Unified geometric theory of gravity and supergravity*, Phys. Rev. Lett. **38** (1977) 739. This paper was received on the 9 February 1977.
26. Paul K. Townsend, *Cosmological constant in supergravity*, Phys. Rev. **D15** (1977) 2802.
27. A. Chamseddine and P. West, *The role of the 1.5 order formalism and the gauging of spacetime groups in the development of gravity and supergravity theories*, Mod. Phys. Lett. **A37** (2022) 08, 2230005, arXiv:2201.06874.
28. P. Breitenlohner, D.Z. Freedman, *Supergravity with axial-gauge invariance, Phys. Rev.* **D15**, 1173 (1977).
29. S. Ferrara and B. Zumino, *Transformation properties of the supercurrent*, Nuclear Physics **B87** (1975) 207.
30. K. Stelle and P. West, *Minimal auxiliary fields for supergravity, Phys. Lett.* **B74**, 569. 330 (1978).
31. S. Ferrara and P. van Nieuwenhuizen, *The auxiliary fields of supergravity*, Phys. Lett. **B74**, 333 (1978).
32. K.S. Stelle and P. West, *Tensor calculus for the vector multiplet coupled to supergravity, Phys. Lett.* **77B**, 376 (1978); *Relation between vector and scalar multiplets and gauge invariance in supergravity*, Nucl. Phys. **B145** (1978) 175.
33. S. Ferrara and P. van Nieuwenhuizen, *Tensor calculus for supergravity*, Phys. Lett. **76B** (1978) 404;
34. *Structure of supergravity*, Phys. Lett. **78B**, 573 (1978). A. H. Chamseddine, R. Arnowitt and P. Nath, *Locally supersymmetric grand unification*, Phys. Rev. Lett. **49** (1982) 970.

35. E. Cremmer, S. Ferrara, L. Girardello and A. Van Proeyen, *Coupling supersymmetric Yang-Mills gauge theories to supergravity*, Phys. Lett. **116B** (1982) 231.
36. P. Nath, R. Arnowitt and A. H. Chamseddine, *Applied N=1 Supergravity,* ICTP series in Theoretical Physics, Volume 1, (1982), World Scientific, Singapore.
37. P. Nath, R. Arnowitt and A. H. Chamseddine, *Gauge hierarchy in supergravity in supergravity Guts*, Nucl. Phys. **B227** (1983) 121.
38. J. Wess and B. Zumino, *Superspace formulation of supergravity*, Phys Lett **66B** (1971) 361
39. W. Siegel, it A Comment on the Wess-Zumino Formulation of Supergravity 1977, Harvard University preprint.
40. W. Siegel, Solution to Constraints in Wess-Zumino Supergravity Formalism, Nucl. Phys. **B142** (1978) 301–305; *Supergravity Superfields Without a Supermetric*, Harvard preprint HUTP-771 A068, *Nucl. Phys.* **B142**, 301 (1978);
41. S. James Gates, Jr and W. Siegel, *Understanding Constraints in Superspace Formulations of Supergravity*, Nucl. Phys. **B163** (1980) 519.

Figure 5.2: Robert Delbourgo.

LIFE WITH SALAM (1959–1976)[108]

ROBERT DELBOURGO
School of Physical Sciences, University of Tasmania, Hobart, Tasmania 7001, Australia
bob.delbourgo@utas.edu.au

Having significantly interacted over 17 years with Abdus Salam, as an undergraduate, postgraduate, postdoc, and eventually as an academic colleague, I will try to paint a personal picture of Salam which may convey something about the man, his greatness and his humanity. This conference commemorates the life and achievements of Abdus Salam. I know that some of you have worked with him in some capacity at some stage, but I suspect that only very few of you will have had the privilege of interacting 17 years with Abdus Salam as I have done. It is therefore incumbent upon me to begin by drawing a picture of the man with a few private reminiscences that may convey something about his greatness, genius and humanity.

I shall cover the period 1959–1976 when I was closely involved with him.

Salam was a man in a hurry; his reputation preceded him everywhere. As a lowly undergraduate student I first came across him in 1959 when we had to choose our third year specialty by making a selective tour of the various research departments at Imperial College (IC). At that time Salam was housed in the Mathematics section (before moving to Physics in 1960) and I can vividly remember his verve and vivacity as he explained to us his latest pet project, which happened to be chiral symmetry and γ^5 invariance for favouring massless left-handed neutrinos and leading to parity violation. That discourse went right over all our heads at the time but they coaxed me at least to turn to theoretical physics for my final year specialty. I am sure others have succumbed to Salam's persuasive abilities on whatever topic he expounded. That very year he taught us advanced quantum mechanics à la Dirac, whom he idolized; his lectures seemed pretty good to me, so in 1960 I embarked on my PhD, with Salam acting as my supervisor from 1961 onwards. During those years Salam's areas of interest were on vanishing of renormalization constants for composite systems, Lie Groups and on the "Gauge Technique." He took a keen interest in my research topic and would

[108]Talk delivered at Abdus Salam Memorial Meeting (Singapore, January 25–28, 2016). Reprinted from Int. J. Mod. Phys. **A32**, No. 8 (2017) 1741005.

enquire every morning as to what progress I had made — putting great pressure on me, as he did on all his other students. He was always bubbling with new ideas and postgraduates found it very hard to survive his changes of tack or emphasis; but it definitely steeled us.

That period saw the development of the eightfold way and I very well remember Salam's heated arguments with Ne'eman that emanated from his office across the corridor. Salam lost out on the birth of SU(3) because of his insistence on a fundamental (Sakata) triplet so you can say that he tripped up on that. However in hindsight was he that far out? Think quarks and you will agree that his intuition was amazing. Yes, he could make mistakes — and who does not — but on most things his inspiration was spot on. When I sometimes asked him where and how he got his latest idea, he would give a wicked smile and point upwards. He always moved on to something new when an old idea was established and played out; he was never one for pot-boilers and he never suffered fools gladly, in private anyway. When he became somewhat contemptuous of the work of some scientists he referred to them as 'tom-tits' or 'broken reeds' or 'youths'. However in public he was always polite and he encouraged anyone who presented a new concept. If there is one lesson that Salam has taught me it is that one should not be ashamed to move on if a concept is not bearing fruit. That may explain why there was always great anticipation whenever he delivered a lecture on some topic: the expectation was that he would spring something new on the audience.

Salam was a demon for hard work. For instance, in the summer of 1967 he had an appendectomy; I visited him in hospital two days after his operation and it was not long before he launched into discussing multiquark states and their current algebras, despite his obvious physical discomfort. He was well travelled and especially during 1962–65 when in the process of setting up ICTP.[109] Funding problems beset him for a good while and he would rile at politicians who opposed his initiative, including the Australian representatives in UNESCO! Owing to his regrettable experiences in Pakistan after leaving Cambridge[110] and his constant bemoaning of the decline of Islamic science, he felt a driving need to found such a centre to assist third world countries. He had a special affinity for isolated scientific personnel who, like himself at first, struggled to keep abreast of the latest advances. He believed that initially concentrating on theoretical studies would serve the purpose as

[109]International Centre for Theoretical Physics, Trieste, Italy

[110]See Editor's note on page 249.

it would cost relatively little but represented the forefront of physics; later on the Centre could be used as a launchpad for other branches of science. At Imperial College, and later at Trieste, Salam became a major magnet for Pakistani students as well as those from Africa and Latin America. The place was abuzz with them and Salam took great pains to foster their work. His initiatives and his constant movement gave him little time for relaxation and I vividly remember several meetings that John Strathdee and I had with him in the hotel lobby at Trieste, after his energysapping perambulations. I think he was able to maintain his stamina because he was an early riser and went to bed early too. My request that he not ring me before 7 am, unless there was an emergency, must have tested his self-control.

Always the perfect host, he warmly welcomed new visitors to IC and ICTP by inviting them to dinner at his home or elsewhere. He was not in the least pretentious about the venue. I recall one occasion at IC when Bruno Zumino came to give us a lecture. Instead of taking him to the Staff Club for lunch he opted for the College cafeteria so he could mingle with the 'plebs' and sample the canteen fare, which he rather savoured! The thing that most impressed John and I about his eating habits was when we were consuming fish; Salam would crunch his way through the spine and bones, leaving only the head and tail! If waiters were tardy with producing the bill, Salam would get up and leave the 'trattoria' when his patience ran out; the sight of the 'camerieri' scurrying after him with 'il conto' was pure comedy. At coffee he would often ruminate about the heyday of Arabic science and how vital it was in the Middle Ages for passing on the Greek scientific legacy to Europe via Spain.

I have been asked by the organizers to comment upon the the birth of the standard model during 1967 and Salam's prominent role in it. This is an excellent occasion to set the record straight and recount my view of its history; if nothing else to refute innuendos which have occasionally surfaced during the 1970s that Salam was not deserving of the Nobel Prize.

That autumn of 1967 I had been in charge of organizing the seminars at IC. Because Salam was constantly on the move and hardly spent more than one month at a stretch in London, I arranged with him to give a couple of lectures on his recent research (in October, to the best of my recollection) during his spell at IC to kick off the seminar season, as it was early in the academic year. He agreed to do so even though the audience attending those talks was somewhat thin. Paul Matthews was certainly present, but Tom Kibble was away in sabbatical in the USA. My memory of his lectures is a bit indistinct nowadays, but I do remember that he kept on invoking these k-meson tadpoles which disappeared into the vacuum which induced the spontaneous breaking of the gauge symmetry: what we now know as the expectation value of the Higgs boson. The resulting model looked rather ugly — and it still is — and I admit that I paid little attention to it; nor do I think that Salam himself was especially enraptured by the model's beauty. A week or so later, I wandered into the Physics Library and came across Steven Weinberg's Physical Review Letter, which I noticed looked suspiciously like Salam's attempt. I showed the article to Salam, who was rather troubled that it was almost the same as his own research, but which was of course entirely independent. Matthews and I urged him to publish his work at the earliest opportunity and this happened to be the upcoming Nobel Symposium. As they say, "the rest is history." I hope that this account of the events at the time scotches all aspersions that Salam should not have been a prize recipient.

Above all Salam impressed upon us the importance of tackling challenging problems: to prospect for new scientific fields and abandon raking over old coals, if one was to make one's mark. He lived up to that precept throughout his life, in spite of accusations that he had a scattergun approach to physics. It is a lesson that some young scientists today should heed.

In his last years, despite his grave illness, he addressed the puzzle as to why certain life forms have particular handedness; what could be more fundamental or significant than that? I know that I miss his wise words, his friendship, his guidance, his generosity and his humanity. This conference is therefore a personal acknowledgement of how much he helped to shape my own career. More widely, it is a timely reminder of how much the scientific landscape and international developments owe to him. The ICTP is a permanent testament to that.

Figure 5.3: Martin Sohnius at SUSY-30, 2000.

RECOLLECTIONS OF A YOUNG CONTRIBUTOR

MARTIN F. SOHNIUS
Novell European Services Ltd.,
1, Arlington Square, Bracknell, England
Msohnius@novell.com

Having been asked to write some "personal recollections" of the early days of supersymmetry research, I must start with the peculiar apology that they are indeed going to be recollections! Literally. You see, being too young to be writing my memoirs quite yet, I do not have "my papers ordered" to do proper research into my own past. On the other hand, I have been "retired" from academic research for exactly 12 years and 13 days (as of the date of the symposium[111]) — just long enough not to have names, institutions or, indeed, the latest gossip at my fingertips. As well as to have forgotten a good deal of the physics, and not to be quite sure whether what I do remember was ever correct, or is still considered to be correct. I am sitting here writing in an office of Novell Inc. with lovely high-bandwidth access to the Internet, but no access at all to a library containing *Physics Letters* or *Nuclear Physics B*. Sadly, 30-year old high-energy physics papers tend not to be on-line! Which, of course, means that my current "paper" will not have any references, and will mention names only where my middle-aged brain can recall both them and at least a fair approximation of their spelling. It'll all be quite personal.

In fact, I don't even have my own publication list handy. When Misha Shifman wrote in his invitation email that my paper in *Nucl. Phys.* **B88**, 257 (1975) would qualify me as a "pioneer of supersymmetry," I was at a loss: which of those early papers might he be referring to? I could, of course, guess (I didn't write that many famous papers, after all), but I wanted to make sure, and was therefore rather dismayed when the first thing "Yahoo!" came up with was a Japanese web page with the phrase "*Nucl. Phys.* **B88**, 257 (1975)" as the sole Latin content in what was otherwise a sea of Kanji. Well, in the end I could confirm, on the Web, the identity of that reference as the paper I wrote with Rudolf Haag and Jan Łopuszański

[111]International Symposium "30 Years of Supersymmetry," Theoretical Physics Institute, University of Minnesota, October 13–15, 2000.

about graded extensions of the Poincaré and conformal algebras (as, with hindsight, I would title it today). Actually, I always prefer to say "the paper that Haag and Łopuszański wrote with me" — it more accurately reflects the relationship of a young graduate student to a pair of world-renowned senior physicists. The following recollections of how it all came about are necessarily from a very personal point of view — moreover, the point of view of a Ph.D. student with not much of an inkling of what was going on in the big world of physics outside Julius Wess' Karlsruhe Institute. So, in the end I have come full circle: I did not know then what was going on in physics, and here I am back at that same state of affairs today!

Which, of course, gives me the immense advantage of a fool's license: none of you will in the near future referee any paper or grant application of mine, nor do I need to compete with any of you folks anymore. (I'll stick to competing with Bill Gates' minions instead. That's hard work, too, believe me!)

Supersymmetry all started, as far as we were concerned (and we were not aware of Golfand and Likhtman's work), with Julius Wess scurrying about secretively in his office during the autumn of 1973. He kept hiding away for hours on end with this good-looking Italian guy. Both of my subsequent wives have been great fans of Bruno over the years, so I know what I am talking about! I was busy with other things at the time: studying for my *Diplom* exam was one, and getting married for the first time was another. Once both of these were duly out of the way, Julius (a) kindly agreed to keep me on as a Ph.D. student, and (b) announced an institute seminar talk by himself with the title "Super-gauge Symmetry in Four Dimensions (in English)," or some such. Little did I know that 27 years later the paper he was going to talk about would be one of the most widely quoted in the literature.

The next thing that happened was the arrival of a couple of preprints from Imperial College. Abdus Salam and John Strathdee had taken up the ball, started running with it, and had shown that Wess and Zumino's algebra was $SU(2,2|1)$, as well as invented superspace while they were at it. Personally, I was most impressed by their proposal to drop the "-gauge" from Wess and Zumino's baby and call it plain "super-symmetry" — with the hyphen. The "-gauge" bit had been a left-over from the two-dimensional conformal super-gauge symmetry (which has infinitely many generators) of old fermionic string theory. Thus, a new name was born! I don't know the current whereabouts of John, but I sure hope to see him and renew our acquaintance at the symposium. The last I heard, in the '80s, he was off to

New Zealand to raise sheep, only coming back once in a while to write a paper with Salam. (Or was the sheep thing a joke?)

Abdus Salam, as we all know, is sadly not with us anymore. A lot has been written and said about him since his illness and much too early death in 1996, but that death is now long enough ago that I may be permitted to resurrect the image of the living man as well as gossip a little. Talk about two wives! In 1982, my first wife came to England to visit Judith and me. As it happened, Abdus was in town, and I had a short conversation alone with him in his office at Imperial. I mentioned with the self-confidence of relative youth that Eva was visiting us and cheerily — I thought — added: "you see, I too can get along with two wives." Whereupon Abdus's face broadened into the kind of all-over wrinkly smile only seen on the faces of bearded Punjabis of a certain age, and with a sparkle in his eyes he leaned forward, touched my arm, and said: "But, my friend, you don't sleep with both of them." Advantage Professor Salam!

Back to serious things. By the beginning of 1974, I don't think Professor Wess was quite sure what to do with me. As I had a fairly prestigious doctoral scholarship, and was married to a wife who earned a good salary, it was probably the right thing for him to give it a long shot, not actually bother about a thesis problem at all, but rather let me try my hand at primary research right away. Salam and Strathdee had hinted that it should be possible to extend $SU(2, 2|1)$ into $SU(2, 2|2)$. Julius told Peter Dondi, a postdoc who recently arrived from DAMTP in Cambridge, to have a go at that and to let me in on the fun. Which he did. As a consequence, I ended up with a case of Repetitive Strain Injury to my wrists from multiplying 6×6 matrices in all possible ways by means of pencil and paper, and with my first publication. It was the sixth in the new field of Supersymmetry (not counting Golfand and Likhtman, of whom we were blissfully ignorant, or Volkov and Akulov whose stuff we did not understand).

In early summer, Jan Łopuszański came to visit from Wroclaw for a few weeks before he was due to go on to visit CERN for the rest of the summer. Wess told him what he had been working on, only to draw the disbelieving comment from Jan that O'Raifeartaigh and others had shown that such a thing could not be done. Jan was a stubborn man, and I don't think that in those initial conversations he quite grasped at first the major "new thing," the anti-commutators. In the end, Julius became impatient, and — quite politely, of course, as behooves an Austrian of the old school when talking to a distinguished member of the Polish gentry — he told Jan to go away and "talk to Martin who has done a lot of this algebra stuff." My good luck!

Jan quickly changed tack. Obviously, the S-matrix guys (Sidney Coleman, Jeff Mandula, Lochlainn O'Raifeartaigh, and possibly others) hadn't been wrong; it was just that they had gone past the particular crossroad leading to fermions without paying much attention. They had squarely aimed at Pati, Salam, and Strathdee's earlier attempt to grandly unify particle symmetries with space-time, and had blown it out of the sky quite spectacularly — hence, of course, that immediate interest of the Imperial College group in "super-symmetry"! Jan set to work, with a bit of help from me, to examine what was left over of the older no-go theorems when taking the possibility of supersymmetry into account. Together, we produced a rather confused though almost certainly correct paper which, I believe, neither of us fully understood. (Well, let me be fair: all I can definitely say is that I, for one, did not understand the axiomatic field-theory bits of it.) The paper was submitted, and Jan left for CERN. Within a couple of weeks two surprising things happened: the paper was accepted unchanged by *Nuclear Physics B* almost by return of post, and Jan withdrew it. I was flabbergasted. What had happened?

Well, Jan had run into Rudolf Haag, on a sabbatical from Hamburg to CERN that year. Together, they had taken the thing, torn it into its constituent parts, thrown those up into the air, and had them land on the ground with a solid, satisfying thud: a self-consistent, beautifully reasoned, clear piece of work, hewn into granite and set in concrete. As Haag later said to me when I was trying to slip an iffy calculation into the paper: "I have a reputation to lose." Bless him! Of course, the original Łopuszański-Sohnius paper had been shrunk to about half its size.

Nowadays, of course, one would make sure to add two or even three papers to one's list of publications in a situation like this. Not so in 1974. I was working with a pair of senior full professors, secure in their positions, to whom their professional pride counted far more than an extra notch on their already well-worn belts. So the original paper was withdrawn. Moreover, Haag was very interested at the time in the conformal-invariance aspects of Wess & Zumino. He wanted to extend the paper to include that, too. Being a master of analytic S-matrix theory, he also knew exactly what could and what could not survive of the no-go theorems in the zero-mass limit.

Let me slip in a vignette about Haag, the physicist. Ask anyone, and they'll probably say that physicists don't come much more mathematical than him. And yet: for a couple of years after 1974, he and I worked on and off on a paper to explore the consequences of conformal supersymmetry on the S-matrix. The idea was that there should be a lot of form-factors that result from the underlying symmetry — Clebsch-Gordan coefficients for $SU(2, 2|n)$, so to speak. A lovely mathematical problem, and since this was a space-time symmetry, those "Clebsches" would make predictions about scattering angles and all sorts of wonderful things. Then, quite suddenly, Rudolf lost interest, and nothing ever happened to that work. (I published a tiny bit of it, with his permission, much later in Jan's 60th-birthday volume.) In the '80s, I asked Rudolf why he had so suddenly cooled off. He answered "because you told me that we had been wrong." Huh? I had told him we were wrong? When? How? "Well, it was you who pointed out to me that the zero-mass approximation meant that not only the electron mass had to be small against the scattering energy, but also the W-boson mass. And that feat even LEP will not be able to achieve, so why bother calculating scattering dynamics for a non-existing physical situation?" So, the great man had made a silly mistake, the Ph.D. student had always been aware of the mistake, but discarded it because he thought that life was long and that the maths was fun independent of the physics. But when he told the great man, the physicist in Haag won out over the mathematician and he lost interest totally and completely.

By the end of the summer 1974, Jan's time had run out — back home to the Workers' Paradise, where the first stirrings were faintly heard from the shipyards in Gdansk around that time! So, next thing, recent graduate Martin S. was put on a train to Geneva (my first business trip ever), picked up at the station by an elderly (about my age now) and never-before seen gentleman who chain-smoked, chewed on a blade of grass and rather slowly drove a completely clapped out estate-car ("station wagon" to some of you). He took me to his flat, and we spent the next week or so calculating and writing at the kitchen table for about 17 hours a day and sharing the bedroom in exhausted sleep for the remaining seven. (I didn't snore then and luckily neither did he — I wish he could let me into the secret of how to manage that as a man of over 50! Should I start chewing on blades of grass?)

Thus, in three easy stages, was conceived *Nucl. Phys.* **B88**, 275 (1975).

The summer was over, a new academic year about to begin, and there was a new kid on the block. In fact, the new kid had been on my block for some time — he was the kid from next door, literally, from number 23 to the Sohnius' number 21. For many years we had seen him walk past the house on his way from school, a gangly, somewhat shy but very obviously bright teenager. Our parents were colleagues and friends, occasional dinners, neighborly barbecues and glasses of wine — that sort of thing. He started reading physics at the local uni, just as I had done a couple of years earlier, and in due course had shown up in the exercise classes that went with Julius' lecture course on Theoretical Physics. Talk about a bright kid! Most of my readers will have taught undergraduates and be familiar with the kind of messy pieces of paper you often get back from them: sloppy argumentation, half-guesses, wild detours leading to the correct result, or not. Not so this guy. Straight to the bone, three tight lines of maths zeroing in on the correct result, each time. I could hardly believe it! Then he asked Julius to be his thesis adviser at just about the same time as our parents fell out over some domestic nonsense. Let's just say it was neither his nor his parents' fault. Consequently, stupidly, and utterly unprofessionally, I asked Wess to be allowed to stay at arm's length from the new kid. My bad luck! His name was, and is, Hermann Nicolai. Is that why that second volume of my Physics Report, the one on Supergravity, never got written? I don't know.

The rest, they say, is history ...

Figure 5.4: Pierre Fayet, 2014.

ON SUPERPARTNERS AND THE ORIGINS OF THE SUPERSYMMETRIC STANDARD MODEL

PIERRE FAYET

Laboratoire de Physique de l'École normale supérieure, ENS-PSL, CNRS, Sorbonne Univ., Univ. Paris Cité, Paris, France

pierre.fayet@phys.ens.fr

We recall the obstacles which seemed, long ago, to prevent supersymmetry from possibly being a fundamental symmetry of Nature. Which bosons and fermions could be related? Is spontaneous supersymmetry breaking possible? Where is the spin-$\frac{1}{2}$ Goldstone fermion of supersymmetry? Can one define conserved baryon and lepton numbers in such theories, although they systematically involve self-conjugate Majorana fermions?, etc. We then recall how an early attempt to relate the photon with a "neutrino" led to the introduction of R-symmetry, but that this "neutrino" had to be reinterpreted as a new particle, the *photino.* This led us to the Supersymmetric Standard Model, involving the $\mathrm{SU}(3) \times \mathrm{SU}(2) \times \mathrm{U}(1)$ gauge interactions of chiral quark and lepton superfields, and of two doublet Higgs superfields responsible for the electroweak breaking and the generation of quark and lepton masses. The original continuous $U(1)_R$ symmetry was then abandoned in favor of its discrete version, R-parity — reexpressed as $(-1)^{2S}\,(-1)^{(3B+L)}$ — so that the gravitino and gluinos can acquire masses. We also comment about supersymmetry breaking.

1 Introduction

Where is the idea of a Superworld coming from? Could half of the particles, at least, have escaped our direct observations? According to common wisdom supersymmetry is an algebraic structure which allows in principle to relate particles of half-integer spins, namely fermions, with particles of integer spins, which are bosons. Could one relate the fermions, constituents of matter, with the bosons, which appear as the messengers of interactions, and then arrive at some sort of unification between Matter and Forces? This would certainly be very attractive, but unfortunately things are not so simple.

Let's go back in time, now about fifty years ago. The algebraic structure of supersymmetry in four dimensions was introduced in the beginning of the seventies by Golfand and Likhtman, [1] Volkov and Akulov, [2] and

Wess and Zumino, [3] as recalled in various contributions to this book. It involves a spin-$\frac{1}{2}$ fermionic symmetry generator, called the supersymmetry generator, satisfying anticommutation relations. This supersymmetry generator Q is defined so as to relate fermionic with bosonic fields, in supersymmetric relativistic quantum field theories.

At that time it was not at all clear if — and even less how — supersymmetry could actually be used to relate fermions and bosons, in a physical theory of particles. While very interesting from the point of view of relativistic field theory, supersymmetry seemed clearly inappropriate for a description of our physical world. In particular one could not identify physical bosons and fermions that might be related under such a symmetry. It even seemed initially that supersymmetry could not be spontaneously broken at all — in contrast with ordinary symmetries — which would imply that bosons and fermions be systematically degenerated in mass! Supersymmetric theories also involve, systematically, self-conjugate Majorana spinors — unknown in Nature — while the fermions that we know all appear as Dirac fermions carrying conserved (B and L) quantum numbers. In addition, how could we account for the conservation of the fermionic numbers B and L (only carried by fermions) in a supersymmetric theory in which fermions are related to bosons? Most physicists were then considering supersymmetry as irrelevant for "real physics."

Still, this algebraic structure could actually be taken seriously as a possible symmetry of the physics of fundamental particles and interactions, once we understood that the above obstacles preventing the application of supersymmetry to the real world could be overcome. After an initial attempt illustrating how far one could go in trying to relate known particles together, in particular the photon with a "neutrino," and the $W^{\pm}$ bosons with charged "leptons" — and the limitations of this approach — in a spontaneously broken SU(2) × U(1) electroweak theory involving two chiral doublet (Brout-Englert) Higgs superfields [4], we were quickly led to reinterpret the fermions of this model, which all possess a conserved R quantum number carried by the supersymmetry generator, as belonging to a new class of particles. The would-be "neutrino" ought to be considered as a new particle distinct from ν_e, ν_μ, and later ν_τ, a "photonic neutrino." A name which I transformed in 1977 into *photino*, also calling at the same time *gluinos* the fermionic partners of the colored gluons (distinct from the quarks), and so on. More generally, this led us to postulate the existence of new R-odd "superpartners" for all ordinary particles and consider them seriously, despite their rather non-conventional properties: e.g. new bosons carrying "fermion" number, now known as *sleptons* and *squarks*, or Majorana fermions transforming as an SU(3) color octet, which are precisely the *gluinos*, etc. [5,6] In addition the electroweak breaking must be induced by *a pair* of electroweak Higgs doublets, not just a single one as in the Standard Model, which requires the existence of *charged Higgs bosons*, and of several neutral ones.

The still-hypothetical superpartners may be distinguished by a new quantum number called R-parity, [7] associated with a Z_2 remnant of the continuous $U(1)_R$ symmetry, which may be multiplicatively conserved in a natural way, and is especially useful to guarantee the absence of unwanted interactions mediated by squark or slepton exchanges. The conservation (or non-conservation) of R-parity is closely related with the conservation (or non-conservation) of baryon and lepton numbers, B and L, as illustrated by the well-known formula [8] reexpressing R-parity as $(-1)^{2S}\,(-1)^{3B+L}$. The finding of the basic building blocks of what we now call the Supersymmetric Standard Model (whether "minimal" or non-minimal) allowed for the experimental searches for "supersymmetric particles," which started with the first searches for gluinos and photinos, selectrons and smuons, in the years 1978–1980, and have been going on continuously since. These searches often rely on the "missing energy" signature, corresponding to energy-momentum carried away by unobserved neutralinos [5, 8–10]. A conserved R-parity also ensures the stability of the "lightest supersymmetric particle," a good candidate to constitute the non-baryonic Dark Matter that seems to be present in our Universe. The general opinion of the scientific community towards supersymmetry and supersymmetric extensions of the Standard Model has considerably changed since the early days, and it is now widely admitted that supersymmetry may well be the next fundamental symmetry to be discovered in the physics of fundamental particles and interactions, although this remains to be experimentally proven.

2 Nature does not seem to be supersymmetric!

Let us now travel back in time, and think about the supersymmetry algebra, and the way it might be realized in Nature. This supersymmetry algebra

$$\begin{cases} \{Q,\ \bar{Q}\} = -2\,\gamma_\mu P^\mu, \\ [Q,\ P^\mu] = 0, \end{cases} \tag{5.1}$$

was introduced, in the years 1971–1973, by three different groups, with quite different motivations. Golfand and Likhtman [1], in their remarkable work published in 1971, first introduced it with the apparent hope of understanding parity violation: when the Majorana supersymmetry generator Q_α is written as a two-component chiral Dirac spinor (say Q_L), one may have the impression that the supersymmetry algebra, which then involves a chiral projector in the right-hand side of the anticommutation relation (5.1), is intrinsically parity violating (which, however, is not the case); they suggested that such (supersymmetric) models must therefore necessarily violate parity, probably thinking that this could lead to an explanation for parity violation in weak interactions. Volkov and Akulov [2] hoped to explain the masslessness of the neutrino from a possible interpretation as a spin-$\frac{1}{2}$ Goldstone particle, while Wess and Zumino [3] wrote the algebra by extending to four dimensions the "supergauge" (i.e. supersymmetry) transformations, [11] and algebra, [12] acting on the two-dimensional string world sheet. However, the mathematical existence of an algebraic structure does not imply that it has to play a role as an invariance of the fundamental laws of Nature.[112]

Indeed many obstacles seemed, long ago, to prevent supersymmetry from possibly being a fundamental symmetry of Nature. Which bosons and fermions could be related by supersymmetry? Maybe supersymmetry could act at the level of composite objects, e.g. as relating baryons with mesons? Or should it act at a fundamental level, i.e. at the level of quarks and gluons? But quarks are color triplets, and electrically charged, while gluons transform as an SU(3) color octet, and are electrically neutral. Is spontaneous supersymmetry breaking possible at all? If yes, where is the spin-$\frac{1}{2}$ Goldstone fermion of supersymmetry if it cannot be identified as one of the known neutrinos? Can we use supersymmetry to relate directly known bosons and fermions? And, if not, why?

If known bosons and fermions cannot be directly related by supersymmetry, do we have to accept this as the sign that supersymmetry is *not* a symmetry of the fundamental laws of Nature? If we still insist to work within the framework of supersymmetry, how could it be possible to define conserved baryon and lepton numbers in such theories, which systematically involve *self-conjugate* Majorana fermions, unknown in Nature, while B and

[112]Incidentally while supersymmetry is commonly referred to as "relating fermions with bosons," its algebra (5.1) does not even require the existence of fundamental bosons! (With non-linear realizations of supersymmetry a fermionic field can be transformed into a *composite* bosonic field made of fermionic ones [2]; but we shall work within the framework of the linear realizations of the supersymmetry algebra, which allows for renormalizable supersymmetric field theories.) The supersymmetry algebra (5.1) certainly does not imply by itself the existence of the superpartners. Just as the mathematical existence of the SU(2) group does not imply the physical existence of the isospin or electroweak symmetries, the existence of SU(3) does not imply that of the strange quark, and of the flavor or color symmetries; the existence of SU(4) does not require technicolor, nor that of SU(5), grand unification.

L are carried only by fundamental (Dirac) fermions — not by bosons? And, once we are finally led to postulate the existence of new bosons carrying B and L — the new spin-0 squarks and sleptons — can we prevent them from mediating new unwanted interactions?

While bosons and fermions should have equal masses in a supersymmetric theory, this is certainly not the case in Nature. Supersymmetry should then clearly be broken. But spontaneous supersymmetry breaking is notoriously difficult to achieve, to the point that it was even initially thought to be impossible! Why is it so? Supersymmetry is a special symmetry, since the Hamiltonian, which appears in the right-hand side of the anticommutation relations (5.1), can be expressed proportionally to the sum of the squares of the components of the supersymmetry generator, as $H = \frac{1}{4} \sum_\alpha {Q_\alpha}^2$. This implies that a supersymmetry preserving vacuum state must have vanishing energy [13], while any candidate for a "vacuum state" which would not be invariant under supersymmetry may naïvely be expected to have a larger, positive, energy.[113] As a result, potential candidates for supersymmetry breaking vacuum states seemed to be necessarily unstable, leading to the question:

$$\text{Q1:} \quad \textit{Is spontaneous supersymmetry breaking possible at all?} \qquad (5.2)$$

As it turned out, and despite the above argument, several ways of breaking spontaneously global or local supersymmetry have been found. [14–16] But spontaneous supersymmetry breaking remains, in general, rather difficult to obtain, since theories tend to prefer, for energy reasons, supersymmetric vacuum states. Only in very exceptional situations can the existence of such states be completely avoided!

As explained above in global supersymmetry a non-supersymmetric state has, in principle, always more energy than a supersymmetric one; it then seems that it should always be unstable, the stable vacuum state being, necessarily, a supersymmetric one! Still it is possible to escape this general result — and this is the key to spontaneous supersymmetry breaking — if one can arrange to be in one of those rare situations for which *no supersymmetric state exists at all* — the set of equations for the auxiliary field VEV's $\langle D\rangle$'s $= \langle F\rangle$'s $= \langle G\rangle$'s $= 0$ having *no solution at all.* But these situations are in general quite exceptional. (This is in sharp contrast with

[113] Such a would-be supersymmetry breaking state corresponds, in global supersymmetry, to a *strictly positive* energy density — the scalar potential being expressed proportionally to the sum of the squares of the auxiliary D, F and G components, as $V = \frac{1}{2} \sum (D^2 + F^2 + G^2)$.

ordinary symmetries, in particular gauge symmetries, for which one only has to arrange for non-symmetric states to have less energy than symmetric ones in order to get spontaneous symmetry breaking.)

These rare situations usually involve an Abelian U(1) gauge group [14], allowing for a gauge-invariant linear "ξD" term to be included in the Lagrangian density[114]; and/or an appropriate set of chiral superfields with special superpotential interactions which must be very carefully chosen (so as to get "F-breaking") [15], preferentially with the help of additional symmetries such as R-symmetries. In local supersymmetry [17], which includes gravity, one also has to arrange, at the price of a very severe fine-tuning, for the energy density of the vacuum to vanish exactly [16], or almost exactly, to an extremely good accuracy, so as not to generate an unacceptably large value of the cosmological constant Λ.

Whatever the mechanism of supersymmetry breaking, we have to get — if this is indeed possible — a physical world which looks like ours (which will precisely lead to postulate the existence of superpartners for all ordinary particles). Of course just accepting the possibility of explicit supersymmetry breaking without worrying too much about the origin of supersymmetry breaking terms, as is frequently done now, makes things much easier — but also at the price of introducing a large number of arbitrary parameters, coefficients of these supersymmetry breaking terms. In any case such terms must have their origin in a spontaneous supersymmetry breaking mechanism, if we want supersymmetry to play a fundamental role, especially if it is to be realized as a local fermionic gauge symmetry, as in the framework of supergravity theories. We shall come back to this question of supersymmetry breaking later. In between, we note that the spontaneous breaking of the global supersymmetry must in any case generate a massless spin-$\frac{1}{2}$ Goldstone particle, leading to the next question,

$$\text{Q2:}\quad \textit{Where is the spin-}\tfrac{1}{2}\textit{ Goldstone fermion of supersymmetry?} \tag{5.3}$$

Could it be [2] one of the known neutrinos? A first attempt at implementing this idea within a $\mathrm{SU}(2) \times \mathrm{U}(1)$ electroweak model of "leptons" [4] quickly illustrated that it could not be pursued very far. Actually, the "leptons" of this first electroweak model were soon reinterpreted to become the "charginos" and "neutralinos" of the Supersymmetric Standard Model.

---

[114]Even in the presence of such a term, one frequently does not get a spontaneous breaking of the supersymmetry: one has to be very careful so as to avoid the presence of supersymmetry restoring vacuum states, which generally tend to exist.

If the Goldstone fermion of supersymmetry is not one of the known neutrinos, why hasn't it been observed? Today we tend not to think at all about the question, since: 1) the generalized use of soft terms breaking *explicitly* the supersymmetry seems to make this question irrelevant; 2) since supersymmetry has to be realized locally anyway, within the framework of supergravity, [17] the massless spin-$\frac{1}{2}$ Goldstone fermion ("goldstino") should in any case be eliminated in favor of extra degrees of freedom for a massive spin-$\frac{3}{2}$ gravitino [6, 16].

But where is the gravitino, and why has no one ever seen a fundamental spin-$\frac{3}{2}$ particle? Should this already be taken as an argument against supersymmetry and supergravity theories? And should one consider that the crucial test of such theories should be the discovery of a spin-$\frac{3}{2}$ particle? In that case, how could it manifest its presence? In fact to discuss this question properly we need to know how this spin-$\frac{3}{2}$ particle should couple to the other particles, which requires us to know which bosons and fermions could be associated under supersymmetry [5]. In any case, even without knowing that, we might already anticipate that the interactions of the gravitino, being proportional to the square root of the Newton constant $\sqrt{G_N} \simeq 10^{-19}$ GeV^{-1}, should be absolutely negligible in particle physics experiments.

Quite surprisingly, however, this may not necessarily be true! We might be in a situation for which the gravitino is light, maybe even extremely light, so that this spin-$\frac{3}{2}$ particle would still interact very much like the massless spin-$\frac{1}{2}$ Goldstone fermion of global supersymmetry, according to the "equivalence theorem" of supersymmetry [6]. In that case we are led back to our initial question, where is the spin-$\frac{1}{2}$ Goldstone fermion of supersymmetry? But at this point we are in a position to answer, the direct detectability of the gravitino depending crucially on the value of its mass $m_{3/2}$, which is fixed [6, 18] by the supersymmetry breaking scale parameter $\sqrt{d}$, or Λ_{SS}.

In any case, much before getting to the Supersymmetric Standard Model, and irrespective of the question of supersymmetry breaking, the crucial question, if supersymmetry is to be relevant in particle physics, is:

Q3: *Which bosons and fermions could be related by supersymmetry?* (5.4)

But there seems to be no answer since known bosons and fermions do not appear to have much in common — excepted, maybe, for the photon and the neutrino. This track deserved to be explored [4], but one cannot really go very far in this direction. In a more general way the number of (known) degrees of freedom is significantly larger for the fermions (now 90, for three families of quarks and leptons) than for the bosons (27 for the gluons, the photon and the $W^\pm$ and Z gauge bosons, ignoring for the moment the spin-2 graviton, and the spin-0 Brout-Englert-Higgs boson, that got discovered much later, in 2012). And these fermions and bosons have very different gauge symmetry properties.

As we said supersymmetric theories involve, systematically, self-conjugate Majorana spinors — unknown in Nature — while the fermions that we know all appear as Dirac fermions carrying conserved B and L quantum numbers. This leads to the question

Q4: *How could one define (conserved) baryon and lepton numbers, in a supersymmetric theory?* (5.5)

These quantum numbers, presently known to be carried by fundamental fermions only, not by bosons, seem to appear in Nature as *intrinsically-fermionic* numbers. Such a feature cannot be maintained in a supersymmetric theory, and one has to accept the (then rather heretic) idea of attributing baryon and lepton numbers to fundamental bosons, as well as to fermions. These new bosons carrying B or L are the superpartners of the spin-$\frac{1}{2}$ quarks and leptons, namely the now-familiar (although still unobserved) spin-0 *squarks* and *sleptons*. Altogether, all known particles should be associated with new *superpartners* [5].

Of course nowadays we are so used to dealing with spin-0 squarks and sleptons, carrying baryon and lepton numbers almost by definition, that we can hardly imagine this could once have appeared as a problem. Its solution went through the acceptance of the idea of attributing baryon or lepton numbers to a large number of new fundamental bosons. But if such new spin-0 squarks and sleptons are introduced, their direct (Yukawa) exchanges between ordinary quarks and leptons, if allowed, could lead to an immediate disaster, preventing us from getting a theory of electroweak and strong interactions mediated by spin-1 gauge bosons, and not spin-0 particles, with conserved B and L quantum numbers! This may be expressed by the following question

Q5: *How can we avoid unwanted interactions mediated by spin-0 squark and slepton exchanges?* (5.6)

Fortunately, we can naturally avoid such unwanted interactions, thanks to R-parity (a Z_2 remnant of the continuous $U(1)_R$ symmetry) which, if present, guarantees that squarks and sleptons *cannot* be directly exchanged between ordinary quarks and leptons, allowing for conserved baryon and lepton numbers in supersymmetric theories.

3 $U(1)_R$ symmetry and electroweak breaking (from an early attempt to relate the photon and the neutrino)

Let us now return to an early attempt at relating *existing* bosons and fermions together [4],[115] also at the origin of the definition of the continuous R-symmetry $U(1)_R$, the discrete version of which leading to R-parity. It also showed how one can break spontaneously the SU(2) × U(1) electroweak gauge symmetry in a supersymmetric theory, using (in a modern language) a pair of chiral doublet Higgs superfields that would now be called H_1 and H_2. This involves a mixing angle intially called δ but now known as β, defined by

$$\tan\beta = \frac{v_2}{v_1}. \tag{5.7}$$

The fermions of this early supersymmetric model, which are in fact gaugino-higgsino mixtures, should no longer be considered as lepton candidates, but became essentially the "charginos" and "neutralinos" of the Supersymmetric Standard Model [5, 6].

Despite the general lack of similarities between known bosons and fermions, we tried as an exercise to see how far one could go in attempting to relate the spin-1 photon with a spin-$\frac{1}{2}$ neutrino. If we want to attempt to identify the companion of the photon as being a "neutrino," despite the fact that it initially appears as a self-conjugate Majorana fermion, we need to understand how this particle could carry a conserved quantum number that we might interpret as a "lepton" number. This was made possible by the introduction of *a continuous* $U(1)$ *R-symmetry* [4], which also guaranteed the masslessness of this "neutrino" ("ν_L" carrying +1 unit of R), by acting chirally on the Grassmann coordinate θ which appears in the expression of the various gauge and chiral superfields. The supersymmetry generator Q_α carries one unit of the corresponding additive conserved quantum number, called R, so that one has $\Delta R = \pm 1$ between a boson and a fermion related by supersymmetry.

[115] This model is reminiscent of a pre-supersymmetry two-Higgs-doublet model [19], first formulated as an Inert Doublet Model, including a Dirac fermion doublet prefigurating the higgsinos. It was constrained to be invariant under a symmetry Q allowing to rotate independently the two Higgs doublets according to $\varphi' \to e^{-i\alpha}\varphi'$, $\varphi'' \to e^{i\alpha}\varphi''$, that became known later as a $U(1)_{PQ}$ symmetry. This symmetry was redefined (according to $Q = RU$) and extended to quarks and leptons [4, 5], so as to provide both the continuous $U(1)_R$ symmetry in (5.12), and an axial $U(1)_A$ (or $U(1)_{PQ}$) symmetry commuting with supersymmetry, acting as in (5.18).

Attempting to relate the photon with one of the neutrinos could only be an exercise of limited validity. The would-be "neutrino," in particular, while having in this model a $V - A$ coupling to its associated "lepton" and the charged $W^\pm$ boson, was in fact what we would now call a "photino," not directly coupled to the Z boson. Still this first attempt, which essentially became a part of the Supersymmetric Standard Model, illustrated how one can break spontaneously a SU(2) × U(1) gauge symmetry in a supersymmetric theory through an electroweak breaking induced by *a pair of chiral doublet Higgs superfields*, now known as H_1 and H_2. (Using only a single doublet Higgs superfield would have left us with a massless charged chiral fermion, which is, evidently, unacceptable.) Our previous charged "leptons" were in fact what we now call two winos, or charginos, obtained from the mixing of charged gaugino and higgsino components, as given by the mass matrix

$$\mathcal{M} = \begin{pmatrix} (m_2 = 0) & \dfrac{g v_2}{\sqrt{2}} = m_W \sqrt{2} \sin\beta \\ \dfrac{g v_1}{\sqrt{2}} = m_W \sqrt{2} \cos\beta & \mu = 0 \end{pmatrix}, \tag{5.8}$$

in the absence of a direct higgsino mass that would have originated from a $\mu\, H_1 H_2$ mass term in the superpotential.[116] The whole construction showed that one could deal elegantly with elementary spin-0 Higgs fields (not a very popular ingredient at the time), in the framework of spontaneously-broken supersymmetric theories. Quartic Higgs couplings are no longer completely arbitrary, but fixed by the values of the gauge coupling constants — here the electroweak couplings g and g' — through the following "D-terms" (i.e. $\frac{\vec{D}^2}{2} + \frac{D'^2}{2}$) in the scalar potential given in [4],[117]

$$\begin{aligned} V_{\text{Higgs}} &= \frac{g^2}{8}\,(h_1^\dagger\, \vec{\tau}\, h_1 + h_2^\dagger\, \vec{\tau}\, h_2)^2 + \frac{g'^2}{8}\,(h_1^\dagger h_1 - h_2^\dagger h_2)^2 + \cdots \\ &= \frac{g^2 + g'^2}{8}\,(h_1^\dagger h_1 - h_2^\dagger h_2)^2 + \frac{g^2}{2}\,|h_1^\dagger h_2|^2 + \cdots . \end{aligned} \tag{5.9}$$

[116] This $\mu\, H_1 H_2$ term, which would have broken explicitly the continuous $U(1)_R$ symmetry then originally intended to be associated with the "lepton" number conservation law, was already replaced by a $\lambda\, H_1 H_2 S$ trilinear coupling involving an *extra neutral singlet chiral superfield* S. This led, subsequently, to the various N/nMSSM or USSM versions of the Supersymmetric Standard Model, with a $f(S)$ singlet superpotential term or an extra U(1) gauge group.

[117] With a different denomination for the two Higgs doublets, such that $\varphi'' \mapsto h_1$, $(\varphi')^c \mapsto h_2$, $\tan\delta = v'/v'' \mapsto \tan\beta = v_2/v_1$.

This is precisely the quartic Higgs potential of the "minimal" version of the Supersymmetric Standard Model, the so-called MSSM,[118] with its quartic Higgs coupling constants equal to

$$\frac{g^2+g'^2}{8} \quad \text{and} \quad \frac{g^2}{2}. \tag{5.10}$$

Further contributions to this quartic Higgs potential also appear in the presence of additional superfields, such as the neutral singlet chiral superfield S already introduced in this model, which will play an important role in the NMSSM, i.e. in "next-to-minimal" versions of the Supersymmetric Standard Model. Charged Higgs bosons (now called $H^{\pm}$) are present in this framework, as well as several neutral ones. Their mass spectrum depends on the details of the supersymmetry breaking mechanism considered: soft breaking terms, possibly "derived from supergravity," presence or absence of extra-U(1) gauge fields and/or additional chiral superfields, role of radiative corrections, etc.

4 The Supersymmetric Standard Model

These two Higgs doublets are precisely the two doublets used in 1977 to generate the masses of charged leptons and down quarks, and of up quarks, in supersymmetric extensions of the standard model [5]. Note that at the time having to introduce Higgs fields was generally considered as rather unpleasant. While one Higgs doublet was taken as probably unavoidable to get to the standard model or at least simulate the effects of the spontaneous breaking of the electroweak symmetry, having to consider two Higgs doublets, necessitating charged Higgs bosons as well as several neutral ones, was usually considered as too heavy a price, in addition to the "doubling of the number of particles", once considered as an indication of the irrelevance of supersymmetry. As a matter of fact considerable work was devoted for a time on attempts to avoid fundamental spin-0 Higgs fields, before returning to fundamental Higgses, precisely in this framework of supersymmetry.

In the previous SU(2)×U(1) model [4], it was impossible to view seriously for very long "gaugino" and "higgsino" fields as possible building blocks for our familiar lepton fields. This led us to consider that all quarks and leptons ought to be associated with new bosonic partners, the *spin-0 squarks and sleptons.* Gauginos and higgsinos, mixed together by the spontaneous breaking of the electroweak symmetry, correspond to a new class of fermions,

[118]Minimal Supersymmetric Standard Model

now known as "charginos" and "neutralinos". In particular, the partner of the photon under supersymmetry, which cannot be identified with any of the known neutrinos, should be viewed as a new "photonic neutrino," the *photino*; the fermionic partner of the gluon octet is an octet of self-conjugate Majorana fermions called *gluinos*, etc. — although at the time *colored fermions* belonging to *octet* representations of the color SU(3) gauge group were generally believed not to exist (to the point that one could think of using the absence of such particles as a general constraint to select admissible grand-unified theories [20]).

The two doublet Higgs superfields H_1 and H_2 generate quark and lepton masses [5],[119] in the usual way, through the familiar trilinear superpotential

$$\mathcal{W} = \lambda_e \, H_1 \cdot \bar{E} L + \lambda_d \, H_1 \cdot \bar{D} Q - \lambda_u \, H_2 \cdot \bar{U} Q. \tag{5.11}$$

L and Q denote the left-handed doublet lepton and quark superfields, and $\bar{E}$, $\bar{D}$ and $\bar{U}$ left-handed singlet antilepton and antiquark superfields. The vacuum expectation values of the two Higgs doublets described by H_1 and H_2 generate charged-lepton and down-quark masses, and up-quark masses, given by $m_e = \lambda_e v_1/\sqrt{2}$, $m_d = \lambda_d v_1/\sqrt{2}$ and $m_u = \lambda_u v_2/\sqrt{2}$, respectively. This constitutes the basic structure of the **Supersymmetric Standard Model**, which involves, at least, the ingredients shown in Table 1. Other ingredients, such as a $\mu \, H_1 H_2$ direct mass term in the superpotential, or an extra singlet chiral superfield S with a trilinear superpotential coupling $\lambda \, H_1 H_2 \, S + \cdots$ possibly acting as a replacement for a direct $\mu \, H_1 H_2$ mass term, [4] and/or extra U(1) factors in the gauge group (which could have been responsible for spontaneous supersymmetry breaking) may or may not be present, depending on the particular version of the Supersymmetric Standard Model considered.

In any case, independently of the details of the supersymmetry breaking mechanism ultimately considered, we obtain the following minimal particle content of the Supersymmetric Standard Model, given in Table 2. Each

[119] The correspondence between earlier notations for doublet Higgs superfields, and modern ones, is as follows:

$$\underset{\text{(left-handed)}}{S = \begin{pmatrix} S^0 \\ S^- \end{pmatrix}} \quad \text{and} \quad \underset{\text{(right-handed)}}{T = \begin{pmatrix} T^0 \\ T^- \end{pmatrix}} \quad \mapsto \quad \underset{\text{(both left-handed)}}{H_1 = \begin{pmatrix} H_1^0 \\ H_1^- \end{pmatrix} \quad \text{and} \quad H_2 = \begin{pmatrix} H_2^+ \\ H_2^0 \end{pmatrix}}.$$

Furthermore, we originally denoted, generically, by S_i, left-handed, and T_j, right-handed, the chiral superfields describing the left-handed and right-handed spin-$\frac{1}{2}$ quark and lepton fields, together with their spin-0 partners.

Table 1: The basic ingredients of the Supersymmetric Standard Model.

1) the $SU(3) \times SU(2) \times U(1)$ gauge superfields;
2) the chiral quark and lepton superfields corresponding to the three quark and lepton families;
3) the two doublet Higgs superfields H_1 and H_2 responsible for the spontaneous electroweak symmetry breaking, and the generation of quark and lepton masses through
4) the trilinear superpotential (5.11).

Table 2: Minimal particle content of the Supersymmetric Standard Model (MSSM). Neutral gauginos and higgsinos mix into a photino, two zinos and a higgsino, further mixed into four neutralinos. Ordinary particles, including additional Brout-Englert-Higgs bosons, in blue, have R-parity $+1$. Their superpartners, in red, have R-parity -1. The N/nMSSM includes an extra singlet with a trilinear $\lambda\, H_1 H_2\, S$ superpotential coupling, describing a singlino and two additional neutral spin-0 bosons. The USSM also includes an extra neutral gauge boson Z' (or U) and its associated gaugino.

Spin 1	Spin 1/2	Spin 0
gluons photon	gluinos $\tilde{g}$ photino $\tilde{\gamma}$	
$W^\pm$ Z	winos $\widetilde{W}^\pm_{1,2}$ zinos $\widetilde{Z}_{1,2}$ higgsino $\tilde{h}$	$H^\pm$ H h, A } Higgs bosons
	leptons l quarks q	sleptons $\tilde{l}$ squarks $\tilde{q}$

spin-$\frac{1}{2}$ quark q or charged lepton l^- is associated with *two* spin-0 partners collectively denoted by $\tilde{q}$ or $\tilde{l}^-$, while a left-handed neutrino ν_L is associated with a *single* spin-0 sneutrino $\tilde{\nu}$. We have ignored for simplicity further mixings between the various "neutralinos" described by neutral gaugino and higgsino fields, denoted in this table by $\tilde{\gamma}$, $\tilde{Z}_{1,2}$, and $\tilde{h}^0$. More precisely, all such models include four neutral Majorana fermions at least, corresponding to mixings of the fermionic partners of the two neutral $SU(2) \times U(1)$ gauge bosons (usually denoted by $\tilde{Z}$ and $\tilde{\gamma}$, or $\tilde{W}_3$ and $\tilde{B}$) and of the two neutral higgsino components ($\tilde{h}_1^0$ and $\tilde{h}_2^0$). Non-minimal models also involve additional gauginos and/or higgsinos.

5 On supersymmetry breaking, and the way from $U(1)_R$ symmetry to R-parity

Let us now return to the definition of the continuous $U(1)_R$ symmetry and discrete R-parity transformations. R-parity is associated with a Z_2 subgroup of the group of continuous $U(1)_R$ symmetry transformations, acting on the gauge superfields and the two doublet Higgs superfields H_1 and H_2 as in Ref. [4], with their definition extended to quark and lepton superfields so that quarks and leptons carry $R = 0$, and squarks and sleptons, $R = \pm 1$ (more precisely, $R = +1$ for $\tilde{q}_L$, $\tilde{l}_L$, and $R = -1$ for $\tilde{q}_R$, $\tilde{l}_R$). [5] As we shall see later, R-parity appears in fact as the remnant of this continuous $U(1)_R$ symmetry when gravitational interactions are introduced, [6] in the framework of local supersymmetry (supergravity). Either the continuous $U(1)_R$, or simply its discrete version of R-parity, if imposed, naturally forbid the unwanted direct exchanges of the new squarks and sleptons between ordinary quarks and leptons.

These continuous $U(1)_R$ symmetry transformations, which act chirally on the anticommuting Grassmann coordinate θ appearing in the definition of superspace and superfields, act on the gauge and (left-handed) chiral superfields of the Supersymmetric Standard Model as follows:

$$U(1)_R : \begin{cases} V(x,\theta,\bar\theta) \to V(x,\theta e^{-i\alpha},\bar\theta e^{i\alpha}) & \text{for the } SU(3)\times SU(2)\times U(1) \text{ gauge superfields,} \\ H_{1,2}\,(x,\theta) \to H_{1,2}\,(x,\theta e^{-i\alpha}) & \text{for the left-handed doublet Higgs superfields } H_1 \text{ and } H_2, \\ [\,Q,\bar U,\bar D; L,\bar E\,]\,(x,\theta) \to e^{i\alpha}\,[\,Q,\bar U,\bar D; L,\bar E\,]\,(x,\theta e^{-i\alpha}) & \text{for the left-handed (anti)quark and (anti)lepton superfields.} \end{cases} \tag{5.12}$$

They are defined so as not to act on ordinary particles, which have $R = 0$, while their superpartners have, therefore, $R = \pm 1$. They allow us to distinguish between two separate sectors of R-even and R-odd particles. R-even

Table 3: R-parities in the Supersymmetric Standard Model.

Bosons	Fermions
gauge and Higgs bosons graviton $(R=0)$ R-parity +	gauginos and higgsinos gravitino $(R=\pm 1)$ R-parity −
sleptons and squarks $(R=\pm 1)$ R-parity −	leptons and quarks $(R=0)$ R-parity +

particles include the gluons, photon, $W^{\pm}$ and Z gauge bosons, the various Higgs bosons, the quarks and leptons — and the graviton. R-odd particles include their superpartners, i.e. the gluinos and the various neutralinos and charginos, squarks and sleptons — and the gravitino (cf. Table 3). According to this first definition, R-parity simply appears as the parity of the additive quantum number R, as given [7] by the following expression:

$$R\text{-parity } R_p = (-1)^R = \begin{cases} +1 & \text{for ordinary particles,} \\ -1 & \text{for their superpartners.} \end{cases} \tag{5.13}$$

But why should we limit ourselves to the discrete R-parity symmetry, rather than considering its full continuous parent $U(1)_R$ symmetry? This *continuous* $U(1)_R$ symmetry, from which R-parity has emerged, is indeed a symmetry [5] of all four necessary basic building blocks of the Supersymmetric Standard Model:

1) the Lagrangian density for the SU(3) × SU(2) × U(1) gauge superfields;
2) the SU(3) × SU(2) × U(1) gauge interactions of the quark and lepton superfields;
3) the SU(2) × U(1) gauge interactions of the two chiral doublet Higgs superfields H_1 and H_2 responsible for the electroweak symmetry breaking;
4) and the trilinear "super-Yukawa" interactions (11) responsible for quark and lepton masses. Indeed this trilinear superpotential transforms under the continuous $U(1)_R$ symmetry (5.12) with "R-weight" $n_{\mathcal{W}} = \sum_i n_i = 2$, i.e. according to

$$\mathcal{W}(x,\theta) \to e^{2i\alpha}\, \mathcal{W}(x, \theta e^{-i\alpha}). \tag{5.14}$$

Its auxiliary "F-component" (obtained from the coefficient of the bilinear $\theta\theta$ term in the expansion of $\mathcal{W}$), is therefore $U(1)_R$-invariant, generating $U(1)_R$-invariant interaction terms in the Lagrangian density.[120]

However, an unbroken continuous $U(1)_R$ symmetry, which acts chirally on the Majorana octet of gluinos,

$$\tilde{g}_a \to e^{\gamma_5 \alpha}\, \tilde{g}_a, \tag{5.15}$$

would constrain them to remain massless, even after a (spontaneous) breaking of the supersymmetry. We would then expect the existence of relatively light "R-hadrons" [8,9] made of quarks, antiquarks and gluinos, which have not been observed. In fact we know today that gluinos, if they do exist, should be heavy, requiring for Majorana gluinos a significant breaking of the continuous $U(1)_R$ symmetry, in addition to the necessary breaking of the supersymmetry. Once the continuous $U(1)_R$ symmetry is abandoned, and supersymmetry is spontaneously broken, radiative corrections do indeed allow for the generation of gluino masses, [21] a point to which we shall return later.

Furthermore, the necessity of generating a mass for the Majorana spin-$\frac{3}{2}$ *gravitino*, once *local* supersymmetry is spontaneously broken, also forces us to abandon the continuous $U(1)_R$ symmetry, in favor of its discrete R-parity version, thereby also allowing for gluino and other gaugino masses, at the same time as the gravitino mass $m_{3/2}$, as already noted in 1977. [6] A third reason for abandoning the continuous $U(1)_R$ symmetry could now be the non-observation at LEP of a charged *wino* — also called *chargino* — lighter than the $W^\pm$, that would exist in the case of a continuous $U(1)_R$ symmetry [4, 5] as shown by the mass matrix $\mathcal{M}$ of Eq. (5.8); the just-discovered τ^- particle could tentatively be considered, in 1976, as a possible light wino/chargino candidate, before it got clearly identified as a sequential heavy lepton.

Once we drop the continuous $U(1)_R$ symmetry in favor of its discrete R-parity version, we may ask how general is this notion of R-parity, and, correlatively, are we *forced* to have this R-parity conserved? As a matter of fact, there is from the beginning a close connection between R-parity and baryon and lepton number conservation laws, which has its origin in our desire to get supersymmetric theories in which B and L could be conserved,

[120] Note, however, that a direct Higgs superfield mass term $\mu\, H_1 H_2$ in the superpotential, which has R-weight $n = 0$, does *not* lead to interactions invariant under the continuous $U(1)_R$ symmetry; but it gets in general reallowed, as for example in the MSSM, when the continuous $U(1)_R$ symmetry gets reduced to its discrete version of R-parity.

and, at the same time, to avoid unwanted exchanges of spin-0 squarks and sleptons. Actually the superpotential of the theories discussed in Ref. [5] was constrained from the beginning, for that purpose, to be an *even* function of the quark and lepton superfields. *Odd* superpotential terms, which would have violated the "matter-parity" symmetry $(-1)^{(3B+L)}$, were excluded, to be able to recover B and L conservation laws, and avoid direct Yukawa exchanges of spin-0 squarks and sleptons between ordinary quarks and leptons. Tolerating unnecessary superpotential terms which are *odd* functions of the quark and lepton superfields (i.e. R_p-violating terms) does create, in general, immediate problems with baryon and lepton number conservation laws (most notably, a much too fast proton instability, if both B and L violations are simultaneously allowed).

This intimate connection between R-parity and B and L conservation laws can be made quite obvious by noting that for usual particles $(-1)^{2S}$ coincides with $(-1)^{3B+L}$, so that the R-parity (5.13) may be re-expressed [8] in terms of the spin S and the "matter-parity" $(-1)^{3B+L}$, as follows:

$$R\text{-parity} = (-1)^R = (-1)^{2S}(-1)^{3B+L}. \tag{5.16}$$

This may also be written as $(-1)^{2S}(-1)^{3(B-L)}$, showing that this discrete symmetry may still be conserved even if baryon and lepton numbers are separately violated, as long as their difference $(B - L)$ remains conserved, at least modulo 2.

The R-parity symmetry operator may also be viewed as a nontrivial geometrical discrete symmetry associated with a reflection of the anticommuting fermionic Grassmann coordinate, $\theta \to -\theta$, in superspace. [22] This R-parity operator plays an essential role in the discussion of the experimental signatures of the new particles. A conserved R-parity implies that the new spin-0 squarks and sleptons cannot be directly exchanged between ordinary quarks and leptons, and that the new supersymmetric particles must be produced in pairs, before ultimately decaying into the lightest one. This "lightest supersymmetric particle", or LSP, is usually considered to be a neutralino. A conserved R-parity guarantees the absolute stability of the LSP, which became a good candidate for the non-baryonic Dark Matter in the Universe.

Let us come back to the question of supersymmetry breaking, which still has not received a definitive answer yet. The inclusion, in the Lagrangian density, of universal soft supersymmetry breaking terms for all squarks and sleptons,

$$-\sum_{\tilde{q},\tilde{l}} m_0^2\ (\tilde{q}^\dagger\tilde{q} + \tilde{l}^\dagger\tilde{l}), \tag{5.17}$$

was already considered in 1976. But it was also understood that such terms should in fact be generated by a spontaneous supersymmetry breaking mechanism, especially if supersymmetry is to be realized locally. As a matter of fact they were first spontaneously generated with the help of the "D-term" associated with *an extra* $U(1)_A$ *gauge symmetry*, acting *axially* on quark and lepton fields, thereby allowing to lift, by the same positive amount, the mass^2 of *both* "left-handed" and "right-handed" squark and slepton fields. When the new gauge coupling constant g" of this (still unbroken) extra U(1) was taken to be very small, the supersymmetry was spontaneously broken "at a very high scale" $\sqrt{d}$ (or $\Lambda_{\rm SS}$) $\gg m_W$. In the limit $g" \to 0$ (then with a very large parameter ξ for the ξD term associated with the extra $U(1)_A$), the corresponding Goldstone fermion, the gaugino of the extra $U(1)_A$, became decoupled, with supersymmetry still broken with heavy squark and slepton masses. In this limit the breaking became explicit instead of spontaneous, although only softly through the dimension-2 mass terms (5.17).

To get a true spontaneous breaking of the supersymmetry with a physically coupled goldstino (of course to be subsequently "eaten" by the spin-$\frac{3}{2}$ gravitino) rather than an explicit (although soft) one, as well as a spontaneous breaking of the extra $U(1)_A$ symmetry, and also to render, at the same time, the superpotential (5.11) invariant under this extra $U(1)_A$ symmetry so that it can actually be responsible for the generation of quark and lepton masses, we took it to act on the Higgs superfields H_1 and H_2 as well as on quark and lepton superfields, as follows:

$$U(1)_A : \left\{ \begin{array}{rl} V(x,\theta,\bar\theta) \to & V(x,\theta,\bar\theta) \\ & \text{for the } SU(3)\times SU(2)\times U(1) \\ & \text{gauge superfields} \\ \\ H_{1,2}(x,\theta) \to & e^{i\alpha}\, H_{1,2}(x,\theta) \\ & \text{for the left-handed doublet} \\ & \text{Higgs superfields } H_1 \text{ and } H_2 \\ \\ [\,Q,\bar U,\bar D;L,\bar E\,](x,\theta) \to & e^{-i\frac{\alpha}{2}}\,[\,Q,\bar U,\bar D;L,\bar E\,](x,\theta) \\ & \text{for the left-handed (anti)quark} \\ & \text{and (anti)lepton superfields.} \end{array} \right. \tag{5.18}$$

This newly-defined extra $U(1)_A$ (acting on the two Higgs doublets so that it gets spontaneously broken together with the electroweak symmetry) is a new symmetry of the trilinear superpotential interactions (5.11), so that quarks and charged leptons can now acquire masses in a way compatible with the spontaneous supersymmetry breaking mechanism used. This extra U(1) is associated, in the simplest case of Eq. (5.18), with a purely axial current for all quarks and charged leptons.

Gauging such an extra U(1), which must in any case be different from the weak-hypercharge U(1), is in fact necessary, [23] if one intends to generate large *positive* mass^2 for *all* squarks ($\tilde{u}_L$, $\tilde{u}_R$, $\tilde{d}_L$, $\tilde{d}_R$) and sleptons, at the classical level, in a spontaneously-broken globally supersymmetric theory (otherwise we could not avoid some of the squarks having negative or at best very small mass^2). But this method of spontaneous supersymmetry breaking also led to several difficulties. In addition to the question of anomalies and how they may be cancelled, it required new neutral current interactions beyond those of the Standard Model. This was fine at the time, in 1977, but such interactions did not show up while the SU(2)×U(1) neutral current structure of the Standard Model got experimentally confirmed. This mechanism also left us with the question of generating large gluino masses. The gauging of an extra U(1) no longer appears as an appropriate way to generate large superpartner masses. One now uses again, in general, soft supersymmetry-breaking terms [24] generalizing those of Eq. (5.17) — possibly "induced by supergravity" — which essentially serve as a parametrization of our ignorance about the true mechanism of supersymmetry breaking chosen by Nature to make superpartners heavy.

Let us return to gluino masses. As we said, continuous $U(1)_R$ symmetry transformations act *chirally* on gluinos, so that an unbroken $U(1)_R$ symmetry would require them to remain massless, even after a spontaneous breaking of the supersymmetry! Thus the need, once it became experimentally clear that massless or even light gluinos could not be tolerated, to generate a gluino mass either from radiative corrections [21], or from supergravity (see already in Ref. [6]), with, in both cases, the continuous $U(1)_R$ symmetry reduced to its discrete R-parity subgroup.

In the framework of global supersymmetry it is not so easy to generate large gluino masses. Even if global supersymmetry is spontaneously broken, and if the continuous $U(1)_R$ symmetry is not present, it is still in general rather difficult to obtain large masses for gluinos, since: i) no direct gluino mass term is present in the Lagrangian density; and ii) no such term may be generated spontaneously, at the tree approximation, gluino couplings involving colored spin-0 fields. A gluino mass may then be generated by radiative corrections involving a new sector of quarks sensitive to the source of supersymmetry breaking [21], that would now be called "messenger quarks," but iii) this can only be through diagrams which "know" both about: a) the spontaneous breaking of the global supersymmetry, through some appropriately-generated VEV's for auxiliary components, $\langle D\rangle$, $\langle F\rangle$ or $\langle G\rangle$'s; b) the existence of superpotential interactions which do not preserve the continuous $U(1)_R$ symmetry. Such radiatively-generated gluino masses (possibly involving a see-saw mechanism for Dirac gluinos), however, generally tend to be rather small, unless one introduces, in some often rather complicated "hidden sector," very large mass scales $\gg m_W$.

Fortunately gluino masses may also result directly from supergravity, as already observed in 1977 [6]. Gravitational interactions require, within local supersymmetry, that the spin-2 graviton be associated with a spin-$\frac{3}{2}$ partner, [17] the gravitino. Since the gravitino is the fermionic gauge particle of supersymmetry it must acquire a mass as soon as the local supersymmetry gets spontaneously broken. This mass, $m_{3/2} = \kappa d/\sqrt{6}$ or $\kappa F/\sqrt{3}$ with $\kappa = \sqrt{8\pi\, G_N} \simeq 4.11 \times 10^{-19}$ GeV^{-1}, is now usually expressed in terms of the so-called supersymmetry-breaking scale parameter $\sqrt{F} = \Lambda_{SS}$, as

$$m_{3/2} = \frac{\kappa\, F}{\sqrt{3}} \simeq \frac{2.37\ \Lambda_{SS}^2}{10^{19}\ \text{GeV}}. \tag{5.19}$$

Since the gravitino is a self-conjugate Majorana fermion its mass breaks the continuous $U(1)_R$ symmetry which acts chirally on it, just as for the gluinos, forcing us to abandon the continuous $U(1)_R$ in favor of its discrete R-parity subgroup. In particular, in the presence of a spin-$\frac{3}{2}$ gravitino mass term $m_{3/2}$, which corresponds to $\Delta R = \pm 2$, the "left-handed sfermions" $\tilde{f}_L$, which carry $R = +1$, can mix with the "right-handed" ones $\tilde{f}_R$, carrying $R = -1$, through mixing terms having $\Delta R = \pm 2$, which may naturally (but not necessarily) be of order $m_{3/2} m_f$.

Supergravity theories offer, in addition, a natural framework in which to include direct gaugino Majorana mass terms

$$-\frac{i}{2}\, m_3\, \bar{\tilde{g}}_a\, \tilde{g}_a - \frac{i}{2}\, m_2\, \bar{\tilde{W}}_i\, \tilde{W}_i - \frac{i}{2}\, m_1\, \bar{\tilde{B}}\, \tilde{B}, \tag{5.20}$$

which also correspond to $\Delta R = \pm 2$. The mass parameters m_3, m_2 and m_1, for the SU(3) × SU(2) × U(1) gauginos, could naturally (but not necessarily) be of the same order as the gravitino mass $m_{3/2}$. Incidentally, once the continuous $U(1)_R$ symmetry is reduced to its discrete R-parity subgroup, a direct Higgs superfield mass term $\mu\, H_1 H_2$, which was not allowed by the continuous $U(1)_R$ symmetry, gets reallowed in the superpotential, as for example in the MSSM. The size of this supersymmetric μ parameter, which breaks explicitly both the continuous $U(1)_R$ symmetry (5.12) and the (global) extra $U(1)_A$ symmetry (5.18), may then be controlled by considering one or the other of these two symmetries. In general, irrespective of the supersymmetry breaking mechanism considered, one normally expects the various superpartners not to be too heavy, otherwise the corresponding new mass scale would tend to contaminate the electroweak scale, thereby creating a hierarchy problem in the Supersymmetric Standard Model. Superpartner masses are then normally expected to be naturally of the order of m_W, or at most in the $\sim \mathrm{TeV}/c^2$ range.

The Supersymmetric Standard Model ("minimal" or not), with its R-parity symmetry (absolutely conserved, or not), provided the basis for the experimental searches for the new superpartners and Higgs bosons, starting with the first searches for gluinos and photinos, selectrons and smuons, at the end of the seventies. How the supersymmetry should actually be broken, if indeed it is a symmetry of Nature, is not known yet. Many good reasons to work on the Supersymmetric Standard Model and its various extensions have been discussed, dealing with dark matter, supergravity, the high-energy unification of the gauge couplings, extended supersymmetry, new space-time dimensions, superstrings, "M-theory", ... However, despite all the efforts made for more than twenty years, and now forty, to discover the new inos and sparticles, from PETRA to the Tevatron, and LEP and LHC at CERN, we are still waiting for experiments to disclose the missing half of the Super-World. Still supersymmetry may well be, beyond quantum physics and general relativity, the next fundamental symmetry to be discovered in the physics of fundamental particles and interactions, enlarging our vision of space and time to the new anticommuting dimensions of superspace.

Supersymmetric theories also provide spin-0 bosons as extra states for massive spin-1 gauge bosons, the same massive gauge superfields describing the $W^\pm$ and Z bosons responsible for weak interactions, and charged and neutral Brout-Englert-Higgs bosons associated with the electroweak breaking [4, 25]. This property further applies in extended supersymmetric theories, and supersymmetric grand-unified theories with extra dimensions. Even if supersymmetric particles still did not show up at LHC, the 125 GeV spin-0 boson H discovered at CERN in 2012 [26, 27], not so far from the Z mass, finds a natural place within supersymmetric theories. It may be interpreted in this framework, up to supersymmetry-breaking effects, as the spin-0 partner of the Z under two supersymmetry transformations, i.e. as a Z that would be deprived of its spin, then providing the first example of two known fundamental particles that may be related by supersymmetry.

References

1. Yu. A. Golfand and E.P. Likhtman, *ZhETF Pis. Red.* **13**, 452 (1971) [*JETP Lett.* **13**, 323 (1971)].
2. D.V. Volkov and V.P. Akulov, *Phys. Lett.* **B46**, 109 (1973).
3. J. Wess and B. Zumino, *Nucl. Phys.* **B70**, 39 (1974); *Phys. Lett.* **B49**, 52 (1974); *Nucl. Phys.* **B78**, 1 (1974).
4. P. Fayet, *Nucl. Phys.* **B90**, 104 (1975).
5. P. Fayet, *Phys. Lett.* **B64**, 159 (1976); **B69**, 489 (1977).
6. P. Fayet, *Phys. Lett.* **B70**, 461 (1977).
7. P. Fayet, in *New Frontiers in High-Energy Physics*, Proc. Orbis Scientiae, Coral Gables (Florida, USA), 1978, eds. A. Perlmutter and L.F. Scott (Plenum, N.Y., 1978), p. 413, (Stud. Nat. Sci. 14, 413–430 (1978)).
8. G.R. Farrar and P. Fayet, *Phys. Lett.* **B76**, 575 (1978).
9. G.R. Farrar and P. Fayet, *Phys. Lett.* **B79**, 442 (1978).
10. G.R. Farrar and P. Fayet, *Phys. Lett.* **B89**, 191 (1980).
11. P. Ramond, *Phys. Rev.* **D3**, 2415 (1971); A. Neveu and J. Schwarz, *Nucl. Phys.* **B31**, 86 (1971).
12. J.-L. Gervais and B. Sakita, *Nucl. Phys.* **B34**, 632 (1971).
13. J. Iliopoulos and B. Zumino, **B76**, 310 (1974).
14. P. Fayet and J. lliopoulos, *Phys. Lett.* **B51**, 461 (1974).
15. P. Fayet, *Phys. Lett.* **B58**, 67 (1975); L. O'Raifeartaigh, *Nucl. Phys.* **B96**, 331 (1975).
16. E. Cremmer *et al.*, *Phys. Lett.* **B147**, 105 (1979).
17. S. Ferrara, D.Z. Freedman and P. van Nieuwenhuizen, *Phys. Rev.* **D13**, 3214 (1976); S. Deser and B. Zumino, *Phys. Lett.* **B62**, 335 (1976).
18. P. Fayet, *Phys. Lett.* **B175**, 471 (1986).
19. P. Fayet, *Nucl. Phys.* **B78**, 14 (1974).
20. M. Gell-Mann, P. Ramond and R. Slansky, *Rev. Mod. Phys.* **50,** 721 (1978).
21. P. Fayet, *Phys. Lett.* **B78,** 417 (1978).
22. P. Fayet, in *History of Original Ideas and Basic Discoveries in Particle Physics*, eds. H. Newman and T. Ypsilantis, Proc. Erice Conf., *NATO Series B* **352**, 639 (Plenum, N.Y., 1996).
23. P. Fayet, *Phys. Lett.* **B84**, 416 (1979).
24. L. Girardello and M.T. Grisaru, *Nucl. Phys.* **B194,** 65 (1982).
25. P. Fayet, *Nucl. Phys.* **B237**, 367 (1984); *Phys. Lett.* **B142**, 263 (1984); *Phys. Rev.* **D90**, 015033 (2014).
26. ATLAS coll., G. Aad *et al.*, *Phys. Lett.* **B716**, 1 (2012).
27. CMS coll., S. Chatrchyan *et al.*, *Phys. Lett.* **B716**, 30 (2012).

Figure 5.5: John Iliopoulos, 2013.

SUPERSYMMETRIC FIELD THEORIES[121] ⊎ THE LITTLE I UNDERSTAND ABOUT THEM

J. ILIOPOULOS

Laboratoire de Physique de l'Ecole normale supérieure
ENS-PSL, CNRS, Sorbonne Univ., Univ. Paris Cité, Paris, France
ilio@phys.ens.fr

Supersymmetric field theories are beautiful and mysterious. Few are those who have studied them and have not surrendered to their charms. I am not one of them. I am convinced that what we know is only the tip of an iceberg. I always wished to explore all of it and, periodically, I was going diving in icy waters. Every time I was coming up with new wonders and new puzzles. With old age such expeditions became rare but the desire is still alive. In this short note I will narrate mainly the results of my own explorations although they are not among the most significant ones. Naturally, pleasure and excitement come from being part of the whole entreprise, sharing the joy of new discoveries and feeling the disappointment of unfulfilled expectations. I will mention some of them but the selection is arbitrary and I will not follow the chronological order.

The general form of an infinitesimal global transformation acting linearly on a set of fields $\phi^i(x)$, $i = 1, \ldots, m$ can be written as

$$\delta\phi^i(x) = \epsilon^\alpha (T_\alpha)^i_j \phi^j(x) \tag{5.21}$$

where $\alpha = 1, \ldots, n$ with n denoting the number of independent transformations. The ϵ's are infinitesimal parameters and T_α is the matrix of the representation of the fields. Usually the ϵ's are taken to be c-numbers, in which case the transformation mixes only fields with the same spin obeying the same statistics. It is clear that if we want to change the spin of the fields with a transformation (5.21), the corresponding ϵ's must transform non-trivially under rotations. If they have a non-zero integer spin they can mix scalars with vectors, spin-$\frac{1}{2}$ with spin-$\frac{3}{2}$, etc. This was the case with the old $SU(6)$ transformations. If, on the other hand, the ϵ's are anticommuting parameters they will mix fermions with bosons. If they have zero spin, the transformations (5.21) will change the statistics of the fields without

[121]Submitted on February 12, 2024.

changing their spin, which means that they will turn a physical field into a ghost. The BRST transformations[122] which we use in the quantization of gauge theories belong to this class. Supersymmetry however is a symmetry between physical fermions and physical bosons, therefore the infinitesimal parameters must be anticommuting spinors. The first "miracle"' of supersymmetry is that such counterintuitive transformations can be implemented consistently in field theory [1].

Although some among the pioneers of the theories that gave rise to supersymmetry were my colleagues at Orsay [2], my interest in this theory is rather late. In the early seventies I did not feel particularly excited with strange algebraic schemes apparently limited to two dimensions. The *coup de foudre* occurred in the fall of 1973. I was visiting CERN and Bruno Zumino showed me his work with Julius Wess [3] in which the main result was their construction of a peculiar four-dimensional algebra. Apparently it contained the generators of the four-dimensional translations P_μ together with those of four "super-gauge" transformations (I had no idea what was a super-gauge transformation), which Bruno wrote as the four components of a Majorana spinor Q_α. In a more modern notation the algebra can be written as

$$[P_\mu, Q_\alpha] = 0; \quad [Q_\alpha, \bar{Q}_\beta]_+ = -2\gamma^\mu_{\alpha\beta} P_\mu \tag{5.22}$$

where $[A, B]_+ = AB + BA$. I use the notation $[\ldots, \ldots]_+$ as the anticommutator of two operators and, for a Majorana spinor, $\bar{Q} = Q\gamma^0$. I was ignorant of all previous work on this subject [4, 5] and all this sounded like Chinese to me.[123]

In particular, I could not understand how such a scheme could contain in a non-trivial way the four-dimensional translations and I wanted to see a concrete field theory example. Bruno showed me the preliminary calculations which gave rise to their second paper [6] with the well-known Wess-Zumino model which, in modern language, is that of a self interacting chiral multiplet. It was the simplest toy model we can possibly think, invariant under Lorentz transformations, fully massive, obviously renormalizable as it had

[122]Introduced by Carlo Becchi, Alain Rouet, Raymond Stora and Igor Tyutin.

[123]The notation used in the early days was not very transparent and many of us had trouble understanding the physical meaning of transformations with spinorial anticommuting parameters. An anecdote: When Bruno Zumino sent me a draft of the QED paper, he asked me whether it was clear. I answered with a French colloquial expression *clair comme le jus de boudin*. Boudin is a French sausage filled with blood. Obviously, its juice is deep dark!

only Yukawa, ϕ^3- and ϕ^4-type interactions, with an action invariant under these strange "super-gauge" transformations. The Lagrangian density $\mathcal{L}$ is a sum of a kinetic energy term $\mathcal{L}_0$, a mass term $m\mathcal{L}_m$ and an interaction term $g\mathcal{L}_I$:

$$\begin{aligned}\mathcal{L} = &-\frac{1}{2}(\partial A)^2 - \frac{1}{2}(\partial B)^2 - \frac{\mathrm{i}}{2}\bar{\psi}\partial\!\!\!/\psi + \frac{1}{2}F^2 + \frac{1}{2}G^2 + m\left(FA + GB - \frac{\mathrm{i}}{2}\bar{\psi}\psi\right) \\ &+ g(FA^2 - FB^2 + 2GAB) - \mathrm{i}g\bar{\psi}(A - \gamma_5 B)\psi \end{aligned} \quad (5.23)$$

where ψ is a Majorana spinor, A (B) are scalar (pseudoscalar) and F and G are auxiliary fields. Note that (5.23) is NOT invariant under supersymmetry transformations; the Lagrangian density changes, but the variation can be written as a four-divergence of the form $\delta\mathcal{L} \propto \partial_\mu R^\mu$ with R a function of the fields and the parameters of the transformation. It follows that the corresponding action is invariant. Many among the strange features of supersymmetric field theories stem from this property: supersymmetry transformations act like generalized translations. Note also that (5.23) appears to be a counter-example of the Coleman-Mandula no-go theorem [7]. This property by itself was quite intriguing,[124] but Bruno added another strange feature: at the one-loop level this model did not require any mass, or coupling constant, renormalization. I do not believe in miracles, at least not in those concerning quantum field theory, and I bet these cancellations were just an accident of the one-loop order. Just to make sure I would win the bet, the same evening I computed the two-loop divergences and, of course, I found I lost it; the divergences still cancelled, although now in a more subtle way because they involve diagrams with different topologies, see Figure 5.6. Next day I found out that Bruno had also done the two-loop calculation with the same result. I remember I told him look, in physics if a result is true for $n = 1$ and 2, it is true for any n. He looked at me mockingly and asked "is this a theorem?" I said no, but it is a very good guess. So, we decided to sit down and prove it. As a first step we looked for a regularization scheme preserving supersymmetry.[125] This was not difficult because the transformations are global; it follows that the derivative of a field ϕ transforms like ϕ itself. Therefore, I proposed to add to the kinetic energy part $\mathcal{L}_0$ of (5.23) a term $\xi\mathcal{L}_0(\phi \to \Box\phi)$ with ϕ standing for all fields ψ, A, B, F and G. For example,

[124] Today we know the answer, see reference [8].

[125] The terminology we were using at that time was super-gauge symmetry. The term supersymmetry was introduced later.

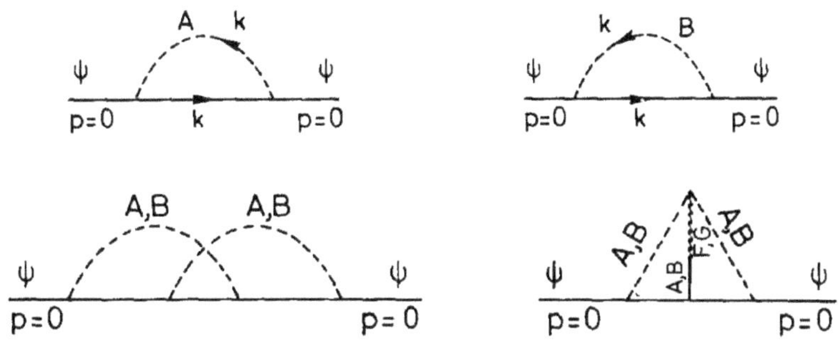

Figure 5.6: Diagrams contributing to the fermion mass renormalization: at one loop (first row) and two loops (second row). We see that at two loops the cancellation involves diagrams with different topologies. Similar results are obtained regarding the coupling constant renormalization.

the fermion kinetic energy term in (5.23) gives rise to a new term

$$-\xi\frac{\mathrm{i}}{2}(\Box\bar{\psi})\not\partial(\Box\psi).$$

The parameter ξ can be taken to zero at the end of the calculation. Supersymmetry is preserved for all values of ξ and for $\xi \neq 0$ all diagrams are finite and calculable. This way we knew that the relations imposed by supersymmetry — equal masses and just one coupling constant — could be preserved by renormalization. The theory would require at most three counterterms: a wave function renormalization common to all fields $\phi_r = Z^{-\frac{1}{2}}\phi$, a common mass counterterm δm for $m_r = Zm + \delta m$ and a coupling constant counterterm Z' for $g_r = Z^{\frac{3}{2}}Z'g$. Our purpose was to prove that we could choose $\delta m = 0$ and $Z' = 1$.

I will not give here the details of the proof [9, 10] (which today presents only a purely historic interest) and I will rather show how its significance has evolved in time. Our first proof was diagrammatic: we looked at the general 1P-I diagrams which could induce mass, or coupling constant counterterms and we showed explicitly that all relevant divergences cancelled to all orders. Although completely straightforward, the proof was rather heavy and hard to explain. Furthermore, it did not seem to offer any new insight in the theory. However, while looking at the various sets of graphs, we discovered that the supersymmetry Ward identities, combined with the particular structure of the Lagrangian (5.23), gave rise to some new relations among Green functions which offered a simpler and potentially more interesting proof of the divergence cancellation property. They were of two sorts: First we proved that the vacuum expectation values of all fields vanish. The only fields for which this does not follow from Lorentz and parity invariance, are A and F, so we proved that $\langle A\rangle_0 = \langle F\rangle_0 = 0$. Second, we proved a relation among the vertex functions — that is the one-particle irreducible Green functions — of the theory, which states that the derivative with respect to the bare mass of every vertex function other than the 1P-I part of $\langle F\rangle_0$, gives an insertion of a zero momentum A field.[126] Combining these two results we proved the absence of mass and coupling constant renormalization counterterms. We also checked that the theory was not super-renormalizable because a common wave-function renormalization counterterm appears to persist to all orders.

This was the first proof of a non-renormalization theorem in quantum field theory and for me it was a revelation. Although its potential physical applications were not obvious, we were puzzled and excited because it appeared to be counter-intuitive. It contradicted the folklore belief according to which in a Lagrangian only terms protected by a symmetry were not subject to renormalization. In our example supersymmetry alone did not forbid the appearance of mass and coupling constant renormalization counterterms; it was the combination of supersymmetry and the particular form of (5.23) that did it. I will show later on how our deeper understanding of these questions has opened new pathways to the study of supersymmetric quantum field theories as well as to possible applications in particle physics phenomenology.

Coming back to the model described in (5.23), we see that, apart from the field content, all that supersymmetry is doing is to impose certain relations among masses and coupling constants. On the other hand, if we did not know that the action was supersymmetric, it would not have been easy to guess the underlying supersymmetry by looking at the Lagrangian (5.23). Therefore, the obvious question is: are there any other relations among masses and coupling constants that would lead to similar, or even more spectacular, divergence cancellations? Together with D.V. Nanopoulos and T.N. Tomaras we found the answer [11]. We studied a general massless model of the form

$$\begin{aligned}\mathcal{L} = &-\frac{1}{2}(\partial A)^2 - \frac{1}{2}(\partial B)^2 - \frac{\mathrm{i}}{2}\bar{\psi}\partial\!\!\!/\psi - \mathrm{i}g\bar{\psi}(A - \gamma_5 B)\psi \\ &- \frac{1}{2}\lambda(A^4 + B^4) - fA^2B^2 \end{aligned} \tag{5.24}$$

[126] This property can be traced to the fact that the derivative with respect to the A field of the interaction part $\mathcal{L}_I$ of the Lagrangian density (5.23), is proportional to the mass term $\mathcal{L}_m$.

with three dimensionless coupling constants[127] g, λ and f. Invariance under supersymmetry is obtained for $\lambda = f = g^2$ which is a fixed point of the renormalization group equations. The possible existence of other relations yielding divergence cancellations can be studied already at the one loop level with the following results: (i) Only under the supersymmetric relation we obtain the non-renormalization theorem. (ii) As expected, this relation is a fixed point of the renormalization group flow. (iii) This fixed point is an infrared attractor, in other words if we start at high energies somewhere in its vicinity, we shall be driven towards it at large distances.

In our paper with Bruno Zumino [9] we studied also various patterns of supersymmetry breaking. We first tried a spontaneous breaking by adding to the Lagrangian (5.23) a term linear in the auxiliary field F. Although F is not invariant under supersymmetry transformations, such term does not give an explicit breaking because F transforms by a total derivative. However it was easy to see that it does not give a spontaneous breaking either, because it can be absorbed by a shift of the field A. Then we tried an explicit breaking and the simplest was to add to the Lagrangian a term linear in the field A: $\mathcal{L}_{\mathrm{Br}} = cA$, with c a constant with dimensions $[\mathrm{mass}]^3$. This term breaks supersymmetry explicitly but softly. The masses of the three physical fields A, B and ψ are split in a very simple way:

$$m_A^2 = m_\psi^2 - 2gc/m_\psi; \quad m_B^2 = m_\psi^2 + 2gc/m_\psi \Rightarrow m_A^2 + m_B^2 = 2m_\psi^2. \tag{5.25}$$

We studied the renormalization properties of the broken theory and found that this relation was robust. We also tried to find other patterns of spontaneous symmetry breaking but we failed; no matter how we tried to induce spontaneous symmetry breaking, there was always a stable supersymmetric solution. Misled by this observation, we formulated a heuristic proof indicating that spontaneous supersymmetry breaking was impossible. I will summarise the argument here because it is instructive to see where it fails. It goes as follows: the fact that the anticommutator of two supersymmetry charges is the operator of four-dimensional translations implies that the hamiltonian H can be expressed as the sum of the squares of the supersymmetry charges, i.e. a sum of positive definite operators. It follows that the spectrum of H is positive semi-definite. Therefore, the zero-energy eigenstate will be annihilated by all the charges, which means that it will be supersymmetric.

---

[127]We can easily check that, for our calculation, this model is stable under renormalization.

We knew that the argument was only heuristic but we believed it could be made rigorous.

We were discussing these results with our colleagues at CERN and Sergio Ferrara pointed out a possible extension. The non-renormalization theorem we had proven shows that only a common wave function renormalization counterterm is needed. This in turn implies that the renormalization group β and γ functions of the theory are proportional to each other. We obtained $\beta(g) = 3g\gamma(g)$. Imagine that β had a non-trivial fixed point $g = g^*$, that is, imagine that $\beta(g^*) = 0$ for $g^* \neq 0$. This would imply also that $\gamma(g^*) = 0$, which means that all fields have canonical dimensions. Since this theory satisfies all properties such as Lorentz invariance and positivity of the metric, it must be a free field theory. This is impossible if $g^* \neq 0$ since g^* is the value of the three-point function. We conclude that $\beta(g)$ has no zero other than the free theory fixed point $g = 0$ and the effective coupling constant $g(q^2)$ must increase indefinitely without bound with q^2 [12]. Furthermore, the particular form of the derivative with respect to the bare mass of the 1P-I Green functions, allowed us to write the corresponding Callan-Symanzik equation in a very simple form in which the right-hand-side is replaced by the insertion of a zero-momentum A field [12].

I went back to Orsay as a supersymmetry zealot trying, through seminars and talks, to spread the good word. The first adept was Pierre Fayet who became soon one of the Great Masters of supersymmetry [13]. At the same time several developments gave a considerable impetus to the subject and opened new horizons.

One of them was the introduction of the superfield formalism [14]. It gave a precise definition of the supersymmetry transformations as generalized translations, it offered well-defined rules for tensor calculus as products of superfields and made possible the systematic construction of supersymmetric field theories. It also shed new light to the non-renormalization theorems. In this formalism a supersymmetric field theory is formulated in the eight-dimensional superspace in which the four extra dimensions can be viewed as the anticommuting components of a Weyl spinor θ_α and $\bar{\theta}_{\dot{\alpha}}$, with $\alpha, \dot{\alpha} = 1, 2$. A superfield is a function of the superspace coordinate,

$$z^M = (x, \theta, \bar{\theta}),$$

namely, $\phi_{...} \to \phi_{...}(x, \theta, \bar{\theta})$, where the dots stand for possible Lorentz indices. The interest of the superfields derives from the fact that, like any function of Grassmann variables, they are polynomials in θ and $\bar{\theta}$.

$$\phi(x, \theta, \bar{\theta}) = A(x) + \theta\psi(x) + \bar{\theta}\bar{\chi}(x) + \cdots + \theta\theta\bar{\theta}\bar{\theta}R(x) \tag{5.26}$$

where the coefficient functions $A(x)$ (complex scalar), $\psi(x)$ (Weyl spinor), ... are ordinary fields. In other words a superfield is a finite multiplet of fields, thus providing for a finite dimensional representation — not necessarily irreducible — of the supersymmetry algebra. It is straightforward [14] to impose covariant constraints to the superfields and obtain the irreducible representations. In particle physics we use mainly two types. The first is the *chiral*, (or *anti-chiral*) multiplet which is a function of θ (or $\bar{\theta}$) only: $\phi(x,\theta)$. The second is the *vector* multiplet which is a function of both θ and $\bar{\theta}$ but on which we have imposed a reality condition. Their respective expansions can be brought to the following form

$$\phi(x,\theta) = A(x) + \theta\psi(x) + \theta\theta F(x), \tag{5.27}$$

$$\begin{aligned} V(x,\theta,\bar{\theta}) &= C(x) + \mathrm{i}\theta\chi(x) - \mathrm{i}\bar{\theta}\bar{\chi}(x) + \cdots - \theta\sigma_\mu\bar{\theta}v^\mu(x) + \cdots \\ &\quad + \frac{1}{2}\theta\theta\bar{\theta}\bar{\theta}D(x). \end{aligned} \tag{5.28}$$

The integration rules on superspace involve the usual integral over the four-dimensional x-space and the integrals over the θ's. The latter are given by the Berezin rules with the result that only the last term in the expansion, the F or D term, survives.

$$\int \delta x\delta\theta\ \phi(x,\theta) = \int \delta x F(x); \quad \int \delta x\delta\theta\delta\bar{\theta}\ V(x,\theta,\bar{\theta}) = \int \delta x D(x). \tag{5.29}$$

Finally, it is possible to introduce covariant derivatives applied on superfields. The tensor calculus, which makes possible the formulation of a Lagrangian field theory, results from the simple fact that the product of two superfields is again a superfield. Supersymmetry transformations act on superfields as generalized translations and, by expansion, we can find the transformation properties of the components fields. Of particular interest are the transformation properties of the last terms in the expansion, the $F(x)$ or $D(x)$ fields in Eqs. (5.27) or (5.28). Since these fields are already proportional to the maximum allowed power of θ, it is easy to see that their transformation under supersymmetry is a space derivative. A Lagrangian density in superspace will be a superfield. If it is chiral it will be of the form $\mathcal{L}(x,\theta)+\bar{\mathcal{L}}(x,\bar{\theta})$ and if it is vector of the form $\mathcal{L}(x,\theta,\bar{\theta})$. Therefore the action will be the integral of either an F or a D term. For example, we can easily check that the kinetic energy part in equation (5.23) is a D term while the mass and the interaction terms are F terms. It is possible to avoid integrating over θ and $\bar{\theta}$ and work directly in the eight-dimensional superspace [15].

We can find the corresponding Feynman rules and formulate the renormalization program of the perturbation expansion. The interesting result is that, for any theory, ALL required counterterms are D terms, i.e. the terms of the form $\int \delta x \delta\theta\delta\bar{\theta}$. In other words, although we can use a chiral superfield as part of a Lagrangian and integrate this piece only over $\mathrm{d}\theta$ to obtain the action, only the integrals over the entire superspace appear as counterterms. This explains the non-renormalization theorem we proved with Zumino and generalizes it to any other field theory model.

Another very important development was the construction of more interesting supersymmetric field theories, beyond the one given in equation (5.23). The first was the supersymmetric extension of quantum electrodynamics [4, 16].[128] It is a straightforward generalization of QED, in the sense that it describes the interaction of a photon with an electron, each one coming with its supersymmetric partners. Thus we have the photon with a massless spin-$\frac{1}{2}$ Majorana fermion — we call it *the photino* — and the electron with a pair of charged spin-0 partners, a scalar and a pseudoscalar. The Lagrangian can be written as

$$\begin{aligned}
\mathcal{L} = &-\frac{1}{4}(\partial_\mu \mathcal{V}_\nu - \partial_\nu \mathcal{V}_\mu)^2 - \frac{\mathrm{i}}{2}\bar{\lambda}\gamma^\mu\partial_\mu\lambda + \frac{1}{2}D^2 \\
&-\frac{1}{2}\big[(\partial A_1)^2 + (\partial A_2)^2 + (\partial B_1)^2 + (\partial B_2)^2 - F_1^2 - F_2^2 - G_1^2 - G_2^2 \\
&+\mathrm{i}\bar{\psi}_1\gamma_\mu\partial^\mu\psi_1 + \mathrm{i}\bar{\psi}_2\gamma_\mu\partial^\mu\psi_2\big] \\
&+m\left[F_1A_1 + F_2A_2 + G_1B_1 + G_2B_2 - \frac{\mathrm{i}}{2}\bar{\psi}_1\psi_1 - \frac{\mathrm{i}}{2}\bar{\psi}_2\psi_2\right] \\
&+e\mathcal{V}_\mu\big[\mathrm{i}\bar{\psi}_1\gamma^\mu\psi_2 - A_1\partial^\mu A_2 + A_2\partial^\mu A_1 - B_1\partial^\mu B_2 + B_2\partial^\mu B_1\big] \\
&-\frac{1}{2}e^2\mathcal{V}_\mu\mathcal{V}^\mu\big[A_1^2 + A_2^2 + B_1^2 + B_2^2\big] + eD(A_1B_2 - A_2B_1) \\
&-\mathrm{i}e\bar{\lambda}\big[(A_1 + \gamma_5 B_1)\psi_2 - (A_2 + \gamma_5 B_2)\psi_1\big].
\end{aligned} \tag{5.30}$$

The interpretation is straightforward. Here $\mathcal{V}_\mu$ and λ are the fields of the photon and the photino respectively. The real Majorana spinors ψ_1 and ψ_2

[128] In Ref. [4] the version with a massive photon and parity breaking was considered, but this work only became known in the West much later.

can be combined together to form a complex Dirac spinor, the field of the electron. A_1, A_2, B_1 and B_2 are the real and imaginary parts of two complex, charged, spin-zero fields, a scalar and a pseudoscalar. They are the supersymmetric partners of the electron, sometimes called *selectrons*. As before, the fields F_1, F_2, G_1, G_2 and D are auxiliary. The Lagrangian (5.30) is invariant under ordinary gauge transformations. In fact, if we eliminate the auxiliary fields, we obtain the usual interaction of a photon with a charged scalar, pseudoscalar and spinor field including the seagull term and the quartic term among the scalar fields. Supersymmetry has introduced only two new elements: (i) The coupling constant in front of the quartic self-interaction of the spin-zero fields is not arbitrary, but it is equal to $e^2/2$ and (ii) new terms, the ones of the last line in (5.30), appear which describe a Yukawa-type interaction between the Majorana spinor (the photino) and the spin 1/2 and zero fields of the electron multiplet. The strength of this new interaction is again equal to the electric charge e. Strictly speaking, (5.30) is not invariant under supersymmetry transformations. However a supersymmetry transformation can be compensated for by a gauge transformation, so all physical results will be supersymmetric.

The theory is renormalizable by power counting and we can easily show that all relations implied by supersymmetry can be maintained. Following our previous discussion, we conclude that there will be no mass counterterm for the electron multiplet because the latter is an F-type term. On the contrary, all other expected counterterms, i.e. the wave function renormalization counterterm for the vector multiplet, that for the chiral one as well as the coupling constant counterterm, will be present. Note also that, although the photon kinetic energy term may be written as an F-term, the fact that it can also be written as a D-term, implies that it is not protected by a non-renormalization theorem and it will require its own special counterterm.

Immediately after we received the preprint of the Wess-Zumino paper [16], Pierre Fayet came and told me that he had found a method to break supersymmetry spontaneously. I told him he was certainly wrong because we had proven with Zumino that such breaking was impossible [9]. He insisted and I checked the calculation. To my great surprise he was right and I could even identify the corresponding massless Goldstone fermion [17]. The method was the analogue of what we had tried with Zumino in the model (5.23), except that here it works. The Lagrangian (5.30) is not the most general we can write compatible with supersymmetry. As in (5.23) we could add a term linear in the field F without any real effect in supersymmetry breaking, here, if we accept to break parity explicitly but softly,[129] we can add a term linear in the auxiliary field D,

$$\mathcal{L} \to \mathcal{L} + \xi D$$

which preserves supersymmetry. Contrary to what happens in (5.23), this term cannot be absorbed by a shift of a physical field. Supersymmetry is spontaneously broken and we obtain a massless Goldstone fermion. Indeed, upon elimination of all the auxiliary fields, the mass terms in the Lagrangian (5.30) with the addition of the ξD term, take the form

$$\mathcal{L}_m = -\frac{im}{2}(\bar{\psi}_1\psi_1 + \bar{\psi}_2\psi_2) - \frac{1}{2}m^2(A_1^2 + A_2^2 + B_1^2 + B_2^2) - \xi e(A_1B_2 - A_2B_1). \tag{5.31}$$

We see that the ξD term leaves the electron mass unaffected but produces a non-diagonal mass term for the selectrons. Since the masses inside a multiplet are split, we conclude that supersymmetry is spontaneously broken. By diagonalising we obtain the fields $\tilde{A}_1 = (A_1 + B_2)/\sqrt{2}, \ldots$ with mass terms

$$-\frac{1}{2}[(m^2 + \xi e)(\tilde{A}_1^2 + \tilde{B}_1^2) + (m^2 - \xi e)(\tilde{A}_2^2 + \tilde{B}_2^2)]. \tag{5.32}$$

What happens next depends on the sign of the square mass terms $m^2 \pm \xi e$. If they are both positive, the story ends here. Supersymmetry is spontaneously broken and λ is the Goldstone spinor. Without loss of generality we can assume that $\xi e > 0$. If $m^2 - \xi e < 0$ the corresponding fields become the Brout-Englert-Higgs fields for the $U(1)$ gauge symmetry, and the photon becomes massive. The Goldstone spinor is a linear combination of λ and the ψ's given by

$$\zeta = \frac{1}{\sqrt{2}}\sin\beta(\psi_1 + \gamma_5\psi_2) + \cos\beta\ \lambda \tag{5.33}$$

with $\beta = \arctan(ev/m)$ and $v^2 = 2(\xi e - m^2)/e^2$. We can verify that ζ is massless and, under an infinitesimal supersymmetry transformation with spinorial parameter α, it transforms as

$$\delta\zeta = -\frac{m^2}{e\cos\beta}\gamma_5\alpha + \cdots \tag{5.34}$$

where the dots stand for terms linear in the other fields. We see that $\delta\zeta$ contains a constant term which shows that ζ is indeed a Goldstone spinor.

---

[129]The term ξD breaks parity but conserves CP. It is easy to see that the operation of charge conjugation on the members of the vector multiplet $\mathcal{V}$, λ and D amounts to a change of sign while it interchanges the members of the two matter multiplets $\phi_1 \leftrightarrow \phi_2$.

It is instructive to see what went wrong with our argument [9] claiming the impossibility of spontaneous supersymmetry breaking. It went through two steps.

(i) The spectrum of the Hamiltonian is positive semi-definite, therefore the ground state has zero energy.
(ii) A zero energy eigenstate is annihilated by all supersymmetry charges, therefore it is supersymmetric.

The implicit assumption is that such a supersymmetric solution exists. It is easy to check in the QED example that it does not. The mistake was to consider supersymmetry as an ordinary internal symmetry where the equations of motion admit always symmetric solutions. In such cases we induce spontaneous symmetry breaking by considering a scalar field with negative square mass. We have no such option in supersymmetry breaking because a negative square mass of a scalar would imply an imaginary mass for its partner fermion. In fact our argument with Zumino would have been exact were it not for this peculiar property of supersymmetric field theories to allow in the Lagrangian density the presence of terms that do not break supersymmetry explicitly, although they are not invariant. They are precisely the terms linear in the auxiliary fields.

Now that we understand the problem we can proceed to a more general analysis. Let us consider a Lagrangian density which is polynomial in the fields and their first derivatives with terms of dimension not higher than 4. It has auxiliary fields F_i and G_i, $i = 1, \ldots, n$ and D_j, $j = 1, \ldots, k$. The potential of the scalar fields in the classical approximation has the form:

$$-V(\phi) = \frac{1}{2}\left[\sum_{i=1}^{n}\left(F_i^2 + F_iF_i(\phi) + G_i^2 + G_iG_i(\phi)\right) + \sum_{j=1}^{k}\left(D_j^2 + D_jD_j(\phi)\right)\right] \tag{5.35}$$

where ϕ denote collectively the physical fields. The equations which eliminate the auxiliary fields are $F_i = F_i(\phi)$, $G_i = G_i(\phi)$ and $D_j = D_j(\phi)$. $F_i(\phi)$, $G_i(\phi)$ and $D_j(\phi)$ are polynomials in the physical fields of degree not higher than second. Eliminating the auxiliary fields we write the potential as

$$V(\phi) = \frac{1}{2}\left[\sum_{i=1}^{n}\left(F_i^2(\phi) + G_i^2(\phi)\right) + \sum_{j=1}^{k} D_j^2(\phi)\right]. \tag{5.36}$$

The positivity property of the potential is the reflexion of the positivity of the Hamiltonian. The important observation is that V is non-negative and vanishes only for:

$$F_i(\phi) = 0; \quad G_i(\phi) = 0; \quad D_j(\phi) = 0. \tag{5.37}$$

We shall assume that neither Lorentz nor translational invariance is broken, consequently only spin zero fields can acquire non-zero, constant, vacuum expectation values. Using the transformation properties of the fields we can easily show that spontaneous symmetry breaking occurs only when one, or more, of the auxiliary fields acquires a non-vanishing vacuum expectation value because they are the only ones which appear without derivatives in the transformation laws of the spinors. When (5.37) are satisfied, all auxiliary fields have zero vacuum expectation values and, as we said above, supersymmetry is unbroken. It follows that the only way to break supersymmetry spontaneously, at least in the classical approximation,[130] is to arrange so that the system of algebraic equations (5.37) has no real solution. In such a case, at least one of the auxiliary fields will have a non-vanishing vacuum expectation value and supersymmetry will be broken. Armed with this result we can construct several examples of spontaneous supersymmetry breaking beyond the QED one [18].

The phenomenon of spontaneous symmetry breaking in the presence of supersymmetry has been a mine of beautiful results which I will not review here [19]. I will only mention one obtained by E. Witten [20] who investigated the possibility of spontaneous supersymmetry breaking by non-perturbative effects. We recall that when supersymmetry is spontaneously broken, the ground state of the Hamiltonian corresponds to strictly positive energy, therefore, the existence of zero energy eigenstates implies that supersymmetry is unbroken. E. Witten introduced the operator $(-)^F$, where F is the fermionic number which is equal to 1 for a fermion and 0 for a boson. We can construct the index of this operator as:

$$W = \text{Tr}(-1)^F. \tag{5.38}$$

W is called the *Witten index*. If we knew the complete spectrum of the Hamiltonian we could easily compute W, but this is not usually the case for a physically interesting system. In general, the result will depend on the values of the various parameters that characterise the system, such as the total volume V as well as the values of masses and coupling constants.

[130] If supersymmetry is exact at the classical level, it means that the equations (5.37) have a real solution. Since these terms are usually protected by non-renormalization theorems, this situation is not likely to change when higher order corrections are taken into account.

At best we could compute W only at some limited region of the parameter space, for example very weak couplings. The interest of the index W is that it is a topological quantity and it is expected to be largely independent of the particular values of the parameters we used to compute it. What Witten obtained can be summarized as follows: If $W \neq 0$ supersymmetry cannot be spontaneously broken, perturbatively, or non-perturbatively. On the other hand, if $W = 0$ we cannot predict the outcome for sure. We can test this criterion in the Wess-Zumino model of equation (5.23) in the semi-classical approximation and we find $W = 2$. We knew that this model had no spontaneous supersymmetry breaking at the classical level and the non-renormalization theorems showed that this remained true at any finite order in perturbation theory, but this argument tells us that it is also true when non-perturbative effects are taken into account. An example of a beautiful non-perturbative result, which we can obtain only in supersymmetric quantum field theories.

The mass spectrum we found in equation (5.32) resembles that of equation (5.25); the square masses in the multiplet are equally split with the fermion mass in the middle. S. Ferrara, L. Girardello and F. Palumbo showed that it is a special case of general mass formula in broken supersymmetry of the form

$$\sum_J (-)^{2J}(2J+1)m_J^2 = 0 \tag{5.39}$$

where m_J is the mass associated with the real field of spin J. It is valid not only in the case of spontaneous supersymmetry breaking, but also in most interesting cases of explicit breaking [21]. It shows a remarkable resilience of supersymmetry and imposes severe constraints in model building. For example, all bosons cannot be made heavy leaving the fermions light. In collaboration with Girardello, I studied the quantum corrections to this general formula with the result that it is very robust [22].

I consider supersymmetry to be the natural extension of gauge invariance. For example, if we write an $SU(N)$ gauge theory with an $N^2 - 1$-multiplet of Majorana fermion fields in the adjoint representation, the theory is automatically supersymmetric [23]. Similarly, the general relativistic invariant theory of a massless, neutral, Rarita-Schwinger field, is supergravity invariant [24]. Furthermore, if we consider theories with extended supersymmetry, we find even more spectacular results. A first example is given by a four dimensional gauge theory with two spinorial charges, known as $\mathcal{N} = 2$ supersymmetry [25,26]. Not only this large supersymmetry can be preserved by renormalization, but also the divergence cancellation mechanisms yield an astonishing result: the renormalization group β-function of the theory has a non-zero term at one loop only; all higher order corrections vanish. N. Seiberg and E. Witten have explored this fact, in combination with a property of electric-magnetic duality, and obtained an exact determination of the low-energy theory [27]. I want to emphasize that, although the $\mathcal{N} = 2$ gauge theory does not resemble anything like the real world, this construction is, nevertheless, an important achievement. It shows that a fully interacting four-dimensional dynamical theory can be exactly determined beyond the perturbation expansion. A second, more astonishing example is the $\mathcal{N} = 4$ super Yang-Mills theory [26] for which the β-function vanishes identically to all orders [28]. It follows that this theory has exact conformal invariance, yet it is not at all equivalent to a free field theory; gauge invariant composite operators have non-vanishing anomalous dimensions. It is quite well established that the $\mathcal{N} = 4$ theory is dual to quantum gravity formulated in a five-dimensional anti-deSitter space, and it has been the subject of intense studies in the last decades. Nevertheless, I am convinced that its most remarkable features belong still to the unexplored part of the iceberg. Similarly, the largest $\mathcal{N} = 8$ supergravity [29] has unexpected hidden symmetries and its divergence properties have not yet been fully explored.

The non-renormalization properties we found in supersymmetry seem to be unique and it is not surprising that this theory is part of practically all attempts to go beyond the Standard Model. In fact, supersymmetry is the only theory that has fascinated theorists over many decades, without having received any experimental support. Before the commissioning of any new accelerator we are convinced that this New Physics will be finally discovered; in vain so far [30]. I remember that around 2010 I was with the late Stephen Hawking in a meeting in Corsica. It was the time we were expecting the first collisions from LHC and we made a bet regarding the great discoveries. I bet that during the first eighteen months of operation LHC would discover both, the Brout-Englert-Higgs scalar boson AND supersymmetry. Steve bet for supersymmetry but against the Higgs. I do not know of any other bet in which both participants succeeded to lose. I do not dare to bet anymore since the next collider is too far away for me, but I secretly believe supersymmetry will be found.

References

1. For an introduction to Supersymmetry, both historical and physical, see the article by G. L. Kane and M. Shifman in this volume, page xv.
2. See, for example, the articles in the first chapters of this book and references therein.

3. J. Wess and B. Zumino, *Nucl. Phys.* **B70**, p. 39 (1974).
4. Y. A. Golfand and E. P. Likhtman, *JETP Lett.* **13**, p. 323 (1971).
5. D. V. Volkov and V. P. Akulov, *Phys. Lett.* **46B**, p. 109 (1973).
6. J. Wess and B. Zumino, *Phys. Lett.* **49B**, p. 52 (1974).
7. S. Coleman and J. Mandula, *Phys. Rev.* **159**, p. 1251 (1967).
8. R. Haag, J. T. Łopuszanski and M. Sohnius, *Nucl. Phys.* **B88**, p. 257 (1975).
9. J. Iliopoulos and B. Zumino, *Nucl. Phys.* **B76**, p. 310 (1974).
10. J. Iliopoulos, Invited talk at the Conference on 30 Years of Supersymmetry, Minneapolis, Minnesota, Oct. 13–15, 2000, *Nucl. Phys. Proc. Suppl.* **B101**, p. 99 (2001).
11. J. Iliopoulos, D.V. Nanopoulos and T. N. Tomaras, *Phys. Lett.* **94B**, p. 141 (1980).
12. S. Ferrara, J. Iliopoulos and B. Zumino, *Nucl. Phys.* **B77**, p. 413 (1974).
13. P. Fayet, Article in this Volume.
14. A. Salam and J. Strathdee, *Nucl. Phys.* **B76**, p. 477 (1974); *Phys. Rev.* **D11**, p. 1521 (1975); S. Ferrara, B. Zumino and J. Wess, *Phys. Lett.* **51B**, p. 239 (1974).
15. M. T. Grisaru, W. Siegel and M. Roček, *Nucl. Phys.* **B159**, p. 429 (1979).
16. J. Wess and B. Zumino, *Nucl. Phys.* **B78**, p. 1 (1974).
17. P. Fayet and J. Iliopoulos, *Phys. Lett.* **51B**, p. 461 (1974).
18. L. O' Raifeartaigh, *Nucl. Phys.* **B96**, p. 331 (1975); P. Fayet, *Phys. Lett.* **58B**, p. 67 (1975).
19. Most of the early results can be found in the reprint collection *Supersymmetry* edited by S. Ferrara and published by North Holland and World Scientific (1987).
20. E. Witten, *Nucl. Phys.* **B185**, p. 513 (1981); E. Witten, *Nucl. Phys.* **B202**, p. 253 (1982).
21. S. Ferrara, L. Girardello and F. Palumbo, *Phys. Rev.* **D20**, p. 403 (1979).
22. L. Girardello and J. Iliopoulos, *Phys. Lett.* **B88**, p. 85 (1979).
23. S. Ferrara and B. Zumino, *Nucl. Phys.* **B79**, p. 413 (1974).
24. D. Z. Freedman, P. van Nieuwenhuizen and S. Ferrara, *Phys. Rev.* **D13**, p. 3214 (1976); S. Deser and B. Zumino, *Phys. Lett.* **62B**, p. 335 (1976).
25. P. Fayet, *Nucl. Phys.* **B113** p. 135 (1976).
26. L. Brink, J. H. Schwarz and J. Scherk, *Nucl. Phys.* **B121**, p. 77 (1977).
27. N. Seiberg and E. Witten, *Nucl. Phys.* **B426**, p. 19 (1994); *Nucl. Phys.* **B430**, p. 485 (1994); *Nucl. Phys.* **B431**, p. 484 (1994). For a review, see K. Intriligator and N. Seiberg, *Nucl. Phys. Proc. Suppl.* **45BC**, p. 1 (1996).
28. S. Mandelstam, *Nucl. Phys.* **B213**, p. 149 (1983).
29. E. Cremmer, B. Julia and J. Scherk, *Phys. Lett.* **76B**, p. 409 (1978); E. Cremmer and B. Julia, *Phys. Lett.* **80B**, p. 48 (1978); *Nucl. Phys.* **B159**, p. 141 (1979).
30. The first detailed study of the supersymmetric extension of the Standard Model is due to P. Fayet, *Phys. Lett.* **69B**, p. 489 (1977); *Phys. Lett.* **84B**, p. 416 (1979). See also several articles in this volume as well as in the reprint collection in Ref. [19].

SUPERSYMMETRIC HIGGS POTENTIALS

LOCHLAINN O'RAIFEARTAIGH
School of Theoretical Physics, Dublin Institute for Advanced Studies, 10 Burlington Road, Dublin 4, Ireland

My first encounter with supersymmetry was at the Warsaw conference on mathematics and physics in March 1974, the forerunner of the present-day M ∩ P conferences. The initial preprint of Wess and Zumino [1] had just appeared and was a subject of informal discussion. Arthur Jaffe, who immediately saw the potential of the idea, and was organizing the field theory section of the 1974 Aspen summer school, invited me to come to the school and lecture on the subject, on the assumption that by June I would be familiar with it. In this way I was precipitated into learning supersymmetry — and fast. In fact, although I immediately embarked on a study of supersymmetry from the few papers available, there was still much to learn when I arrived in Aspen. There I was overawed to find that not only was there a large and distinguished audience of axiomatic field theorists, but also large and distinguished audiences of string theorists and particle physicists eager to learn about this exotic new subject. I had to work extremely hard to prepare the lectures and I recall staying up an entire night at the Institute in preparation for one particular lecture.

It was evident from the beginning that supersymmetry would be phenomenologically acceptable only if it were broken, but an important question that came up at Aspen was whether it should, or could, be spontaneously broken. The initial attraction of a spontaneously broken theory was that it would allow the interpretation of the neutrino as a Goldstone field and, although this possibility was later ruled out, [2] spontaneous supersymmetry breaking still remained an attractive proposition. At the time the only known mechanism for such a breakdown was the Fayet-Iliopoulos mechanism [3], and this depended on the presence of an Abelian gauge field. There was no analogue of a Higgs potential i.e. a supersymmetry breaking potential $V(\Phi)$ that depended on (chiral) scalar superfields Φ alone.

In the two months after Aspen I was busy writing up my lecture notes and did not give the supersymmetric Higgs potential much further thought. However, at a conference in Austria that September, during a short review of supersymmetry, I had occasion to discuss the breakdown of internal gauge symmetry in the presence of supersymmetry. To put that discussion into context I should say that, in an elegant 1968 paper called *The Geometry of the Octet*, Michel and Radicati [4] had shown that, for a field ϕ in the adjoint representation, SU(3) broke naturally to U(2) through an equation of the form

$$m\phi + g\phi \wedge \phi = 0 \quad \text{where } (\phi \wedge \phi)_a \equiv d_{abc}\phi^b\phi^c$$

m and g being constant parameters and d_{abc} the Gell-Mann D-matrices. It was obvious that the same applies to the breaking of SU(N) to U($N-1$) and the point I was making was that supersymmetry produces the Michel-Radicati equation in a natural way. This happens because the only F-type renormalizable supersymmetric potential for chiral scalar superfields is F^*F, where F is bilinear, and for the adjoint of SU(N) the only bilinear is $F = m\Phi + g\Phi \wedge \Phi$. The snag is that the vacuum is degenerate in the sense that the broken and unbroken solutions, $m\Phi = -g\Phi \wedge \Phi$ and $\Phi = 0$ respectively, both occur at the absolute minima of the potential. Thus, while a spontaneous symmetry breakdown can occur, it is not mandatory. Further, it was possible to show that, to the two-loop approximation at least, the effective potential had the same absolute minima as the original one, and the initial non-renormalization theorems suggested that the degeneracy persisted to all orders in perturbation.

After the lectures I was talking to Abraham Pais, who remarked that, although this phenomenon was interesting and probably merited a letter, the real problem was to find a Higgs potential for supersymmetry itself. For the record, a letter duly appeared in *Physics Letters*, [5] but by that time Pais' remark had re-awakened my interest in a supersymmetric Higgs potential, and from September onward I began to concentrate on this problem.

There was some conflicting evidence. On the one hand the existence of the Fayet-Iliopoulos mechanism showed that supersymmetry could be spontaneously broken in principle. On the other hand there was a rumored statement by Zumino that there could be no supersymmetry-breaking Higgs potential for a single chiral scalar superfield and, although I had not seen a proof of this statement, I soon convinced myself that it was true. Did this mean that the presence of a gauge field was necessary for supersymmetry breaking? In those early days anything was possible.

The first step forward came with the discovery that with more than one chiral scalar superfield it was possible to construct a potential with degenerate supersymmetry-breaking, that is, a potential whose absolute minimum had both a supersymmetry-preserving and supersymmetry-breaking phase.

This was analogous to the Michel-Radicati situation except that the broken and unbroken phases were continuously connected. In fact the potential minimum lay along straight lines and planes (nowadays called flat directions) through the origin in the Φ-plain, with the unbroken minimum at the origin. In spite of the non-renormalization theorems [6] for perturbative quantum corrections, I was convinced that the effective potential would choose the unbroken phase. A supersymmetric version of Murphy's law would surely apply! This belief was reinforced by a letter from Fayet and Sibold, who had seen the preprint of the model, and were of the same opinion.

The failure of the model to produce a mandatory spontaneous breakdown suggested going in the opposite direction and looking for a no-go theorem showing that a supersymmetric Higgs potential would always be degenerate. This direction looked promising but somehow the no-go proofs could never be completed. However, the effort turned out to be worthwhile, because, after one particularly promising attempt, I realized that the flaw in the argument pointed the way toward a counter-example. More precisely, it pointed the way toward a three-field potential of the form $V = F^*F$ with a minimum at $F \neq 0$, a clear signal for a spontaneous breakdown of supersymmetry. In terms of the standard superpotential

$$W = \sqrt{\frac{1}{2}}\lambda_a\Phi_a + \frac{1}{2}m_{ab}\Phi_a\Phi_b + \frac{1}{3}\sqrt{\frac{1}{2}}\Phi_a\Phi_b\Phi_c + \text{Herm. conj.}$$

for any number of chiral scalar superfields Φ_a, the parameters chosen were

$$g_{abc} = (g_{ab}\lambda_c + g_{ca}\lambda_b + g_{bc}\lambda_a)/\lambda^2$$

where

$$\lambda_a = \begin{pmatrix} 0 \\ 0 \\ \lambda \end{pmatrix}, \quad g_{ab} = \begin{pmatrix} g & 0 & 0 \\ 0 & 0 & 0 \\ 0 & 0 & 0 \end{pmatrix} \quad \text{and} \quad m_{ab} = \begin{pmatrix} \mu & m & 0 \\ m & 0 & 0 \\ 0 & 0 & 0 \end{pmatrix}.$$

After the elimination of dummy fields, the part containing the ordinary chiral scalar fields ϕ_a in the Grassmann variable expansion $\Phi_a = \phi_a + \theta \cdot \psi_a + \cdots$ of the superfields then became a potential of the form $2V = F_a^* F_a$, where

$$F_1 = m\phi_2 + \mu\phi_1 + 2g\phi_3\phi_1, \quad F_2 = m\phi_1, \quad F_3 = \lambda + g\phi_1^2$$

and led to a potential of the form

$$\begin{aligned} 2V &= \lambda^2 + (m^2 + 2\lambda g)(\operatorname{Re}\phi_1)^2 + (m^2 - 2\lambda g)(\operatorname{Im}\phi_1)^2 \\ &\quad + g^2(\phi_1^*\phi_1)^2 + F^*F_1. \end{aligned}$$

This potential did indeed produce a spontaneous breakdown of supersymmetry [7]. Later, of course, more sophisticated models based on the same principle were constructed.

Thus the sequence of discovery of a supersymmetric Higgs potential was from first trying for positive result for more than one superfield, then trying for a negative result for any number of superfields, and ending up with a positive result for three superfields.

The story does not quite end at this point, because (2) is only the classical potential. To copperfasten the result it had to be shown that the potential was the most general renormalizable potential consistent with all the available symmetries, since otherwise quantum corrections could induce new terms and possibly restore the supersymmetry. What guaranteed the generality of the potential was the existence of R-symmetry (phase-symmetry of the Grassmann variables). In fact the above potential is the most general one of its kind consistent with supersymmetry and R-symmetry. Furthermore, it could be shown that, at the perturbative level at least, the R-symmetry was not broken by the quantum corrections.

In fact the perturbative corrections and the general question of renormalization was soon afterwards investigated in a systematic and clear manner by Piguet, Sibold and their collaborators [8, 9]. A decade later it was shown by Affleck *et al.* [10] that the R-symmetry and the mechanism as a whole were preserved by a class of non-perturbative quantum corrections.

Both the potential V above and the Fayet-Iliopoulos potential break the supersymmetry at tree-level i.e. they break it at the level of the classical action, or, more precisely, at the level of the action in the path-integral. This raises the question as to whether, for a classically unbroken supersymmetry, a spontaneous breakdown could be induced by the quantum corrections. The usual non-renormalization theorems indicate that this does not happen perturbatively (there is no analogue of the Coleman-Weinberg mechanism) but it might happen at the non-perturbative level. In fact it has been shown that, in the presence of gauge fields, a spontaneous breakdown of supersymmetry can be induced in principle by instantons and anomalies. [6, 10]

Whether some modern version of either the tree-level or the non-perturbative models have anything to do with reality is, of course, another question!

References

1. J. Wess and B. Zumino, *Phys. Lett.* **B49**, 39 (1974).
2. B. de Wit and D. Freedman, *Phys. Rev. Lett.* **35**, 827 (1975).

3. P. Fayet and J. Iliopoulos, *Phys. Lett.* **B51**, 461 (1974).
4. L. Michel and L. Radicati, *Ann. Phys.* **66,** 758 (1971).
5. L. O'Raifeartaigh, *Phys. Lett.* **B56**, 41 (1975).
6. S. Weinberg, *The Quantum Theory of Fields III* (Cambridge Univ. Press, 2000).
7. L. O'Raifeartaigh, *Nucl. Phys.* **B96**, 331 (1975).
8. T. Clarke, O. Piguet and K. Sibold, *Nucl. Phys.* **B119**, 292 (1977).
9. O. Piguet, K. Sibold and M. Schweda, *Nucl. Phys.* **B168**, 337 (1980).
10. I. Affleck, M. Dine and N. Seiberg, *Nucl. Phys.* **B256**, 557 (1985).

Figure 5.7: Peter West, 2024.

THE SUPERSYMMETRIC EFFECTIVE POTENTIAL

PETER WEST
King's College, Department of Mathematics, University of London,
Strand, London WC2R 2LS, United Kingdom
peter.west540@gmail.com

I would first like to thank Misha Shifman for his kind invitation to write an article on the paper of Ref. [1] which had the above title. This paper proved that if supersymmetry is preserved at the classical level then the effective potential in a supersymmetric theory had no perturbative quantum corrections. As the paper pointed out, this meant that the degeneracies that were generic in the classical potentials of supersymmetric theories were not removed by perturbative quantum corrections; in modern language supersymmetric theories had moduli. This result, together with earlier work, provided the theoretical basis for the realization that supersymmetry solved the technical hierarchy problem [2]. In other words, the scalar masses in supersymmetric grand unified theories did not receive large corrections and as a result in supersymmetric theories, once one set a small scale for electroweak breaking, quantum corrections did not lead to corrections of order the grand unified scale. It also meant that supersymmetry could not be broken by perturbative quantum corrections if it was not broken at the classical level and as a result it placed considerable constraints on the way realistic supersymmetric models could be constructed.

I would like to concentrate on explaining the developments that lead up to this paper and at the same time try to give some feeling for what it was like to be a a Ph.D. student during the early days of supersymmetry. After I had finished my undergraduate studies at Imperial College in the summer of 1973, I had the good fortune to be accepted as a Ph.D. student of Abdus Salam who was a member of the theoretical physics group of Imperial College and also the director of the International Center for Theoretical Physics in Trieste. Abdus Salam spent most of his time in Trieste, but he did return to London each month. As was normal for students who had not done Part III in Cambridge, I spent most of my time during my first year taking courses at Imperial and King's Colleges. Even in those days there was a large gap between what was in the undergraduate program and the material that was required to do research. Clearly, there was no material on supersymmetry, but there was also no course on string theory, since this was not viewed as an important subject and perhaps more surprisingly no lectures on the spontaneous symmetry breakdown and the developments we now call the Standard Model. However, we did have lectures on quantum field theory, QED, general relativity, group theory and hadronic dual models and Regge poles. The course on field theory was taught by Tom Kibble and among many other topics he explained the problem of infinities in field theories and the renormalization program. I was somewhat surprised to find that the most promising theories of nature required such unaesthetic manipulations. One of the hot topics at that time was the discovery of a new resonance which was eventually identified as the bound state of the charm quark, but at Imperial, which had perhaps the best theory group in England, a debate raged as to whether it might be the fourth color.

Despite the somewhat reduced content compared to today, I was fully occupied studying these courses and so there was little reason to see Abdus Salam. However, once the courses had finished in the summer of 1974, I went to see him to find a suitable topic to work on. Although supersymmetry had been discovered in 1971 by Golfand and Likhtman [3] and independently by those working in string theory [5], there was very little development, except for the paper of Volkov and Akulov [4], until the paper of Wess and Zumino in 1974. [6] This latter paper found a four-dimensional quantum field theory with this symmetry which became known as the Wess-Zumino model. It was with this advancement that small groups of physicists, mainly in France, Germany and England, began to explore the consequences of supersymmetry in earnest. An early result was the discovery of a supersymmetric theory that included a vector particle, the $\mathcal{N} = 1$ Yang–Mills theory. [7] Just before I went to see Abdus Salam, he and John Strathdee had written a paper [8] introducing superspace. This provided a manifestly covariant description of the Wess-Zumino model and later all supersymmetric theories. At the end of this paper was an appendix giving the Feynman rules for this theory in the superspace formalism. These became known as the super-Feynman rules. Abdus Salam suggested I should study this paper. Given my preference to study a subject that did not possess such undesirable features as infinities, I had an initial reluctance. However, once I started reading the paper I became fascinated with these new models and the superspace formalism.

Even today, with the benefit of a number of pedagogical text books, learning supersymmetry is not easy. It involves a systematic level of manipulation of Clifford algebras, such as Fierz rearrangements, that is substantially more involved than previous developments. Hence studying the paper of Strathdee and Salam was not straightforward. However, being at Imperial had another very substantial advantage; Bob Delbourgo was a member of the staff. He explained a number of the technicalities and provided me with some sheets, that contained all the required identities. In effect, he became my supervisor while Abdus Salam was not there. As a result, I was able to start rederiving some of the results in the Salam and Strathdee paper and begin doing research.

Initially, Bob thought he might have a problem I could work on. Just before then it had been shown that the Wess-Zumino model possessed only one infinity when, even if supersymmetry was preserved, one might have expected that it should have three infinities. This result was first shown at one loop by Iliopoulos, Wess and Zumino [9] and to all orders by Iliopoulos and Zumino. [10] The idea of Bob was to use the new super-Feynman rules of Salam and Strathdee to rederive this result. Understandably, he found the problem too tempting and he solved it within a few days. [11] The result was the first paper using the new technique of super-Feynman rules.

Hence at this stage I had yet to find a specific problem to work on. Since in supersymmetric theories the masses of all the particles in the same supermultiplet were equal, one of the earliest problems was how to spontaneously break supersymmetry. It was clear that the potentials in supersymmetric theories were much more constrained than those of a generic quantum field theory. For example, they were positive definite. While breaking internal symmetries was not so problematic, it was far from clear how to break supersymmetry. However, it was found that by a careful choice of field content and interactions one could break supersymmetry spontaneously at the tree level. This was achieved in two different ways; one was discovered by Fayet and Iliopoulos [12] and the other was due separately to Fayet and O'Raifeartaigh. [13] Although the masses of the particles that emerged from these breakings were not the same in each supermultiplet, they did obey a pattern of masses that was rather constrained and not favorable from a model building viewpoint. It had been thought that things would be better in supersymmetric models in which supersymmetry was preserved at the classical level, but spontaneously broken at the quantum level.

We are so familiar with properties of supersymmetric theories today that it is worth recalling the thought process at that time. Most of the quantum field theories that were studied before the advent of supersymmetry had potentials which uniquely fixed the vacuum of the theory at the classical level up to the symmetries of the theory. The effective potential of quantum field theories received quantum corrections at all orders of perturbation theory and the one loop corrections had a particularly elegant form due to Coleman-Weinberg [14]. To calculate these latter effects, one first found a minimum of the classical potential and then computed the masses of the particle in terms of the shifted scalar fields from the minimum. In some exceptional cases, the potentials might have degenerate minimum at the classical level, but these would be removed by quantum corrections.

One very usual feature of the potentials in many supersymmetric theories was that the classical potentials did not uniquely determine the vacua of the theory even up to the symmetries of the theories. This could be viewed as a consequence of the very constrained nature of the superpotentials which meant that they often had a larger symmetry than the theory as a whole. The potentials were said to have flat directions, that is directions along which the value of the potential was a minimum. This was viewed as an undesirable feature and another reason to calculate the effective potential in supersymmetric theories was to see how these degeneracies were removed by the quantum corrections.

Abdus Salam and John Strathdee suggested that Bob Delbourgo and I should compute the one loop effective potential in some supersymmetric models in which supersymmetry was not broken at the classical level to see if it was broken by quantum corrections. They also provided us with a number of promising models. To do this we had to compute the Coleman–Weinberg effective potential as described above. We discovered that the one loop effective potential had a habit of vanishing. To find a model that "worked" we were forced to consider more complicated models with many fields. To compute the effective potential became more and more complicated requiring the diagonalization of large matrices one of which was 26 by 26-dimensional. We displayed these matrices on large sheets of drawing paper and patiently swapped rows and columns until it was diagonal. After taking more than a week to diagonalize such a matrix we found that the result was a complex effective potential. This meant that we had begun from the wrong minimum. It was only after this defeat that it became clear that one had to think about the problem in a more fundamental way.

It was at this point that I began to wonder if the effective potential in a supersymmetric theory might actually vanish. By writing out the part of the Feynman graph that involved the Grassmann odd components of superspace it became clear that this was the case. Initially, I did this loop by loop, but eventually the proof of the all orders result became apparent [1].

Whenever I learnt that Abdus Salam was back on one of his visits to Imperial College, I would knock on his door. He was almost always available and keen to discuss physics in a very friendly way. Since I had not really made much significant progress in my research up until then, I was rather relieved to have found a result which I thought was of some significance. Somewhat to my surprise Abdus Salam did not share my enthusiasm. However, on his return to Trieste he explained the result to John Strathdee who doubted it was true. After some checks they did agree it was correct.

At this point I still had six months left of the three years which it usually took to get a Ph.D. I began working with Ali Chamseddine who was also a Ph.D. student at Imperial. Supergravity had just been discovered [15] and we decided to try to construct supergravity by gauging the super Poincaré group. Although we were aware of the paper by Tom Kibble which gave the first order formalism of general relativity in which the spin-connection was the gauge field for the Lorentz group, we were too naive to realize that gauging of space-time groups, such as the Poincaré group, to gain a complete theory of gravity had not been attempted. We soon found [16] that, with one assumption, gauging the super Poincaré group did lead to the theory of supergravity already proposed. This method had one other advantage, it gave the first algebraic proof of the invariance of supergravity.

After my Ph.D. had finished in the summer of 1976 I had to find a job, which in those days were few and far between. I tried to get a research council fellowship, but in the interview I was asked about the detailed connection of supersymmetry to the known particles. I was not successful since I, and indeed no one else, had much idea of this connection. This attitude was in keeping with the general belief in that part of the research council responsible for theoretical physics that only work rather directly related to experiment should be funded. Fortunately, I was given a Royal Society postdoctoral fellowship to go to the École Normale in Paris for a year and so was able to continue working in supersymmetry.

The following does not constitute a complete list of references, but with the exception of Ref. [2], contains those I found most useful during my Ph.D.

References

1. P. West, *Nucl. Phys.* **B106**, 219 (1976).
2. E. Witten, *Nucl. Phys.* **B186**, 513 (1981); S. Dimopoulos and H. Georgi, *Nucl. Phys.* **B193**, 150 (1982); N. Sakai, *Z. Physik* **C11**, 153 (1982); R. Kaul, *Phys. Lett.* **B109**, 19 (1982).
3. Y.A. Golfand and E.S. Likhtman, *JETP Lett.* **13**, 323 (1971).
4. D.V. Volkov and V.P. Akulov, *Pis'ma Zh. Eksp. Teor. Fiz.* **16**, 621 (1972); *Phys. Lett.* **B46**, 109 (1973).

5. P. Ramond, *Phys. Rev.* **D3**, 2415 (1971), 2415 (1971); A. Neveu and J.H. Schwarz, *Nucl. Phys.* **B31**, 86 (1971); *Phys. Rev.* **D4**, 1109 (1971); J.-L. Gervais and B. Sakita, *Nucl. Phys.* **B34**, 477, 632 (1971); F. Gliozzi, J. Scherk and D.I. Olive, *Nucl. Phys.* **B122**, 253 (1977).
6. J. Wess and B. Zumino, *Nucl. Phys.* **B70**, 139 (1974).
7. S. Ferrara and B. Zumino, *Nucl. Phys.* **B79**, 413 (1974); A. Salam and J. Strathdee, *Phys. Rev.* **D11**, 1521 (1975).
8. A. Salam and J. Strathdee, *Phys. Rev.* **D11**, 1521 (1975); *Nucl. Phys.* **B86**, 142 (1975).
9. J. Wess and B. Zumino, *Phys. Lett.* **B49**, 52 (1974).
10. J. Iliopoulos and B. Zumino, *Nucl. Phys.* **B76**, 310 (1974).
11. R. Delbourgo, *Nuovo Cim.* **25**, 646 (1975). This result was also given in D.M. Capper, *Nuovo Cim.* **25**, 259 (1975).
12. P. Fayet and J. Iliopoulos, *Phys. Lett.* **B51**, 461 (1974).
13. L. O'Raifeartaigh, *Nucl. Phys.* **B96**, 331 (1975); P. Fayet, *Phys. Lett.* **B58**, 67 (1975).
14. S. Coleman and E. Weinberg, *Phys. Rev.* **D7**, 1888 (1973).
15. D. Freedman, P. van Nieuwenhuizen and S. Ferrara, *Phys. Rev.* **D13**, 3214 (1976); *Phys. Rev.* **D14**, 912 (1976); S. Deser and B. Zumino, *Phys. Lett.* **B62**, 335 (1976).
16. A. Chamseddine and P. West, *Nucl. Phys.* **B129**, 39 (1977).

Figure 5.8: Victor Isaakovich Ogievetsky.

AN APPRECIATION OF VICTOR ISAAKOVICH OGIEVETSKY[131] ⊎

K.S. STELLE
The Blackett Laboratory, Imperial College London,
Prince Consort Road, London SW7 2AZ, UK
k.stelle@imperial.ac.uk

I met Victor Ogievetsky during an historic moment in the evolution of quantum gravity, which was the first Moscow Quantum Gravity Seminar, organized by Professor Moisei Alexandrovich Markov in 1978. At that moment, Russia and the West were beginning to open up to each other, and the pioneering work of a number of our Soviet colleagues in the fields of quantum gravity, supersymmetry and supergravity was becoming properly, although belatedly, recognized. Victor and I immediately became good friends, and this friendship continued since. I am indebted to Victor for the many insights he gave me into the secrets of supersymmetric theories, as well as giving me a chance to become acquainted with the intense intellectual world of physics in Russia.

Many memories of times shared come back to mind, from a variety of places in Russia and its neighbors, and also later from meetings in Europe and the UK, once that became possible. I was honoured to host Victor and Rita (his lovely wife, Margarita Mikhailovna) in my flat in London and was happy finally to be able to show them the city which has become my home. Upon Victor's invitation, I visited Dubna for the first time, in 1981. Aside from intense physics discussions with Victor, Emery Sokatchev, Evgeny Ivanov, Alexander Galperin, Sergei Krivonos, Stilyan Kalitzin and other members of the Dubna group, I remember with great pleasure walks in the woods and swimming in the Dubna river near the historic church at Ratmino. Returning from those walks with a selection of mushrooms whose edibility I was completely incompetent to assess, which were then promptly separated into the appropriate categories by Rita in their charming house at the end of Lesnaya Street, the evening would then continue on with a marvelous dinner, lubricated of course by a certain amount of vodka.

[131] From the Ogievetsky Memorial Site http://theor.jinr.ru/people/Ogievetsky/index.html, see http://theor.jinr.ru/people/Ogievetsky/memories.html.

One side trip I remember from that first extended visit to Dubna was to Kimry, on the "Raketa" ferry that used to travel up and down the Volga. Kimry was as far as a foreigner could go in Soviet times, and the town was for me a view into the deeper world of Russia. I remember Victor telling me about how this old town, known since the time of Ivan Grozny as a leather-working center, had developed a surprisingly lively artistic culture, mainly owing to its geographical position more than 101 kilometers from Moscow, and thus allowed as a place of residence for artists and actors in exile during the 1930s. On the way home, Victor spotted a woman selling milk and other products of her farm and gleefully bought cottage cheese — "tvorog," and excitedly told Rita about this immediately upon our return, leading on to yet another marvelous dinner, this time featuring syrniki.

That visit to Kimry stimulated in me a curiosity relating to another introduction Victor made to me: great Russian literature. At first, this was just ornery curiosity: foreigners were allowed to visit the town of Kimry, but the allowed limit stopped right at the bridge across the Volga to the village of Savyolovo. So, of course, I wanted to know what was so interesting in Savyelovo. At the time — frankly — nothing much, except for the end station of a train line leaving Savyolovsky station in Moscow, from which one also departs for Dubna. But Savyolovo stuck in my mind because Victor also introduced me to the writings of the great Russian poets of the 20th century, and in particular to Osip Mandel'shtam. It was later on, upon reading Nadezhda Mandel'stam's Hope against Hope that I realized that Savyelovo was in fact a place where she and Osip had lived in exile in 1937, shortly before his final arrest. Other writers whom Victor introduced me to were Yury Trifonov and Vasily Grossman. It was from such interactions that Victor taught me about the deep connections in Russia between many aspects of culture — Dubna is a scientists' city but it was also the location for important developments in the arts, such as one of the first public concerts of the famous poet and singer Vladimir Vysotsky.

Other crossings of Victor's and my own worldlines took place outside Moscow and Dubna. There was a conference in Erevan, where I remember watching a lightning storm together with Victor and Rita during a warm evening on a balcony outside their hotel room. And there were a couple of meetings in Alushta, at the JINR laboratory's sanatorium. On the first of those visits to Alushta, I remember arriving from Simferopol airport in a taxi. As it happened, Victor and Sasha Galperin were outside and greeted me enthusiastically just as I arrived. Then Victor mischievously informed me that it had been arranged for me to have a "special" room. In fact, Sasha had originally been settled in that room, but when it became known that I was arriving, he was unceremoniously bumped into another room in the complex. However, Victor and Sasha assured me that, to all visible appearances, Sasha's new room looked identical to the one he had just been moved from. So, Victor concluded, it might be best if our discussions, about supersymmetry or about anything else, were to take place during walks along the Black Sea front.

Aside from the pleasures and mysteries of Russia with Victor as my guide, what remains of greatest importance for me is of course the long scientific interaction that I had with the work of Victor and his Dubna colleagues. This began even before we met in person, from struggles together with Peter West to understand the papers of Victor and Emery Sokatchev about the structure of the supercurrent multiplet in minimal $D = 4$ supergravity [1]. The outcome of that struggle led Peter and me to construct an "off-shell" representation of the supergravity multiplet which did not involve the equations of motion in the anticommutator algebra. This then led on, through the work of a number of authors, to a general understanding of the couplings between supergravity and supermatter. Another important intersection of our interests came in connection with the construction I had made, together with Paul Howe and Paul Townsend, of an off-shell representation for $\mathcal{N} = 2$ extended supersymmetry. This then came to full fruition with the beautiful work of Victor, Zhenia Ivanov, Sasha Galperin, Stilyan Kalitzin and Emery Sokatchev on the *harmonic superspace formulation* of $\mathcal{N} = 2$ supersymmetry [2]. Aside from the direct interactions that we had, however, of equal importance to me were the many things that I indirectly learned from Victor through his writings. I think in particular about the beautifully clear lectures that he gave in the 10th Winter School in Karpacz on Nonlinear Realizations of Internal and Space-Time Symmetries [3] or his work with Borisov on gravity as a theory of intersecting affine and conformal symmetries [4].

Victor's passing was a great scientific loss to all of us who have explored this subject, and to me, personally, it was the loss of a great friend. As a friend, I miss him very much. As a physicist, I miss his guidance. In both ways, I am grateful to have known him.

References

1. V. Ogievetsky and E. Sokatchev, *On Vector Superfield Generated by Supercurrent*, Nucl. Phys. B 124 (1977) 309.
2. A. Galperin, E. Ivanov, S. Kalitsyn, V. Ogievetsky and E. Sokatchev, *Unconstrained $\mathcal{N} = 2$ Matter, Yang-Mills and Supergravity Theories in Harmonic Superspace*, Class. Quant. Grav. 1 (1984) 469 Erratum: [Class. Quant. Grav. 2 (1985) 127].
3. V.I. Ogievetsky, *Nonlinear Realizations of Internal and Space-time Symmetries*, in Proc. X-th Winter School of Theoretical Physics in Karpacz, ACTA UNIVERSITATIS WRATISLAVIENSIS, No 207 Vol. 1, (Wroclaw 1974), see https://www.imperial.ac.uk/media/imperial-college/research-centres-and-groups/theoretical-physics/msc/current/smb/other-resources/Ogievetsky–Karpacz-nonlinear-realization-lectures.pdf.
4. A. B. Borisov and V. I. Ogievetsky, *Theory of Dynamical Affine and Conformal Symmetries as Gravity Theory,* Theor. Math. Phys. 21 (1975) 1179 [Teor. Mat. Fiz. 21 (1974) 329]. doi:10.1007/BF01038096.

RECOLLECTIONS OF VICTOR ISAAKOVICH OGIEVETSKY[132]

E. IVANOV[a] and E. SOKATCHEV[b]

[a]*Bogoliubov Laboratory of Theoretical Physics, JINR, 141980 Dubna, Moscow region, Russia*
eivanov@theor.jinr.ru

[b]*LAPTh, Universite Savoie Mont Blanc, CNRS, B.P.110, F-74941 Annecy-le-Vieux, France*
emeri.sokatchev@cern.ch

Evgeny Ivanov

I first heard about Victor Isaakovich Ogievetsky[133] in 1967, being a 4th-year student in the Theory group of the Physics Faculty in the Saratov State University (SSU). At the time an agreement between SSU and the Dubna branch of the Moscow State University (MSU) allowed selected students to spend two years in Dubna and prepare their diploma (Masters) thesis at the Joint Institute for Nuclear Research (JINR). A student (one year senior than me) told me about an extraordinary theoretician at the Laboratory of Theoretical Physics (LTPh), V.I. Ogievetsky who, together with Igor Vasilievich Polubarinov, was very actively working on a new approach to gauge theories and gravity. At the time I knew almost nothing on the subject, still I thought that it would be great to graduate under the supervision of such a distinguished scientist. My friend immediately cooled my ardor by saying that it was very hard to convince V.I. that a student was capable of working with him. He rather quickly lost any interest if he felt that the student was not good enough. I was a very shy youngster and I thought that I was just the right candidate for failure. Nevertheless, I was selected and in February 1968 I arrived in Dubna and started attending remarkable lectures of the leading JINR scientists. At the library I found the recent (now seminal) papers by Ogievetsky and Polubarinov on the spin principle in gauge theories and gravity. They seemed to me rather understandable and very exciting, both because of their beautiful logic (different from the standard textbooks) and the future wide prospects they opened. I frequently met I.V. Polubarinov[134]

[132]Submitted on February 9, 2024.
[133]V.I. in what follows.
[134]I.V. in what follows.

at the library and decided (after some hesitation) to ask him something about his very recent preprint. He answered my query in detail and I asked him if I could come to his office occasionally with more questions. He accepted with a smile. Soon he agreed to be my Diploma advisor and in December 1969 I successfully defended my Diploma Thesis on the tree diagrams in Yang-Mills theory.

Soon after our acquaintance, I.V. introduced me to V.I. Ogievetsky. I was rather surprised to find out that V.I., who seemed to me a kind of "a heavens inhabitant" at the distance, turned out very friendly and amicable in personal contacts. In 1969, by request of I.V. Polubarinov, he wrote a very positive report on my Diploma.

The following year, after graduating from SSU, I was temporally employed at LTPh as a probationer for two years. During this period I came across Boris Zupnik, a PhD student of V.I. from his native city of Dniepropetrovsk (now Dniepr in Ukraine). He was working on the theory of non-linear realizations, the main focus of V.I. in those days. Boris introduced me to some important papers on the subject. This direction seemed to me very interesting and I started studying it with great enthusiasm. Boris and I wrote two joint papers, and in the process I had closer and more frequent contacts with V.I. Then, I realized that he appreciated (above all) independent thinking resulting in novel ideas and well-motivated suggestions in his students and collaborators. Perhaps I possessed this quality a little bit and, as a result, at the end of 1972 I became a PhD student ("aspirant," as they say it in Russian) at the Physics Faculty of MSU under the supervision of V.I. and Dmitry Ivanovich Blokhintsev. At the time V.I. was excited by the idea of applying the nonlinear realizations method to space-time symmetries including Poincaré symmetry. He worked on this subject with his PhD student Alexander Borisov from Sverdlovsk (now Ekaterinburg). They reformulated gravity as the joint nonlinear realization of two well-known space-time symmetries, the conformal and affine ones. This construction was based on the remarkable "Ogievetsky theorem": the infinite-dimensional group of space-time diffeomorphisms is the closure of the two finite-dimensional subgroups above. The graviton field got the beautiful interpretation of a Goldstone field. I was highly impressed by these results and came to V.I. with the suggestion of how a similar idea could be applied to gauge theories, with the Yang-Mills or Maxwell fields as Goldstone fields. We published a joint paper on this subject. My first joint paper with V.I. was "Inverse Higgs phenomenon in nonlinear realizations". In it we showed that the standard Goldstone theorem does not always fully apply to the nonlinearly realized space-time symmetries and some Goldstone fields are not independent. These results, in particular, constituted my PhD Thesis which I defended in 1976. At that time I was already employed as a Junior Researcher at LTPh, in the group headed by Academician Moisei Aleksandrovich Markov (to which Polubarinov and Ogievetsky belonged).

The discovery of supersymmetry in the 1970s deeply influenced the research activity in Markov's group. V.I. quickly realized the potential importance of this new concept in particle and mathematical physics. One of the pioneers of supersymmetry was Dmitry Vasil'evich Volkov (1925–1996) from the Kharkov Institute of Physics and Technology. For a long time he and his group had close scientific and human contacts with Ogievetsky and his collaborators. So it was not surprising that supersymmetry became the main research direction in the group of young researchers around Ogievetsky. The active members of this team were Luca Mezincescu from Bucharest and Emery Sokatchev from Sofia. Boris Zupnik, who defended his PhD earlier and got a position in the Institute of Nuclear Physics near Tashkent, also dwelled deeply into this new area. I also focused on this line of investigation. Later on, this research team constituted the main staff of the sector "Supersymmetry" at LTPh headed by V.I. Ogievetsky. In the beginning of the 1980's the group was enriched by a talented and enthusiastic PhD student Alexander (Sasha) Galperin from Tashkent. Later on, it was further strengthened by D. V. Volkov's student, Anatoly Pashnev. I should also mention Stilyan Kalitzin from Sofia who, like A. Galperin, belonged to the second generation of the supersymmetry adepts at LTPh.

Since the start of the supersymmetry era, the main interests of V.I. and his entourage were concentrated on the superspace approach to supersymmetric theories. The spin principle which inspired the view of gauge invariance as a way to ensure the propagation of a definite spin in an interacting field theory proved very fruitful in supersymmetric gauge theories. This includes, most notoriously, supergravity which is the unique self-consistent theory of interacting gauge fields of the spins 2 (graviton) and 3/2 (gravitino). Actually, Ogievetsky and Polubarinov (in their 1964 lectures) raised the question about a possible gauge theory of an interacting Rarita-Schwinger field carrying spin 3/2. They made a serious efforts to find an answer, but failed because nobody was aware of supersymmetry at the time. The 1971 discovery by Golfand and Likhtman from the Lebedev Institute of Physics (FIAN) of the supersymmetric extension of the Poincaré algebra remained largely unnoticed, including by Ogievetsky.

A non-trivial generalization of the spin principle to superfields was the elegant formulation of $\mathcal{N} = 1$ supergravity by V.I. and E. Sokatchev in 1977–1980 as the theory of an axial-vector gauge superpotential carrying superspin $\mathbf{3/2} = (2, 3/2)$. Supersymmetry led to a new gauge principle, the preservation of $\mathcal{N} = 1$ chirality in interacting theories. $\mathcal{N} = 1$ chirality amounts to the realization of $\mathcal{N} = 1$ supersymmetry in a complex superspace containing only half of the original four Grassmann coordinates. This geometric principle underlies both $\mathcal{N} = 1$ gauge and supergravity theories in superspace. A key remaining problem was finding a geometric formulation of theories with extended $\mathcal{N} \geq 2$ supersymmetry. A decisive step toward this goal was our 1981 paper with V.I. and A. Galperin where we studied $SO(2)$ extended $\mathcal{N} = 2$ supersymmetry and found a generalization of $\mathcal{N} = 1$ chirality. $\mathcal{N} = 2$ superspace contains a subspace with half the original eight Grassmann coordinates, different from the standard chiral subspace. V.I. invented the name "Grassmann analyticity" for this new phenomenon of complex subspaces with a fraction of the Grassmann coordinates. Our hope was that the new concept underlies all gauge theories with extended supersymmetry. The idea about the right type of Grassmann analyticity, in the simplest $\mathcal{N} = 2$ case, came to my mind at the end of 1983. One has to extend the standard $\mathcal{N} = 2$ superspace by pure internal coordinates of the two-dimensional sphere $S^2 \sim SU(2)_R/U(1)_R$. Here $SU(2)_R$ is the group of automorphisms (or R-symmetry group) of the $\mathcal{N} = 2$ Poincaré superalgebra. In this enhanced superspace one can define a new type of the Grassmann analytic subspace containing as usual half of the Grassmann coordinates but also (necessarily!) the set of internal coordinates of S^2. Both the $\mathcal{N} = 2$ matter hypermultiplet and gauge prepotential admit a nice representation as unconstrained superfields living on this analytic subspace. The expansion of such Grassmann analytic superfields in terms of spherical harmonics implies the presence of an infinite number of ordinary component fields. In the $\mathcal{N} = 2$ gauge theory these are pure gauge degrees of freedom. On the other hand, in the hypermultiplet case they are genuine auxiliary fields needed for the off-shell description of the multiplet in the enhanced superspace, something impossible in the standard $\mathcal{N} = 2$ superspace. The new type of superspace was later dubbed "harmonic superspace" (HSS).

During our first discussion of this new approach together with V.I. and Sasha Galperin, V.I. called Emery Sokatchev, who had just returned to Sofia, and pronounced the sacramental phrase: "Emery, come urgently to Dubna, the $\mathcal{N} = 2$ case has been done!" All the basic features of the modern $\mathcal{N} = 2$ HSS (the invariant superfield actions for all $\mathcal{N} = 2$ theories of interest, including supergravity, quantization, etc.) were worked out together with Emery and Stilyan Kalitzin who joined the four of us somewhat later. The name "harmonic" of the new extended superspace was suggested by Boris Zupnik, who also made essential contribution to the field and became its adherent for the rest of his life. Soon after publishing our ground breaking paper on $\mathcal{N} = 2$ HSS (in 1984 in Class. Q. Grav.), we found a relevant $\mathcal{N} = 3$ harmonic superspace. In it we (with the active involvement of Stilyan Kalitzin) constructed for the first time an off-shell formulation of a gauge theory with three manifest supersymmetries. Its main features were the infinite sets of both gauge and auxiliary fields, as well as the action of Chern-Simons type. Further developments and applications of the HSS approach were summarized in the Monograph "Harmonic superspace", (CUP, 2001), by A. Galperin, E. Ivanov, V. Ogievetsky, E. Sokatchev. Recently updated results and relevant references can be found in E. Ivanov, 1604.01379.

It is rather difficult to adequately express my impressions of Victor Isaakovich as a scientist and human being. I admired him. My own scientific style and circle of interests were formed under his great influence. I think that his main character traits were his absolute honesty and integrity, as well as the unlimited devotion to science. He shared the political views of the finest layer of Russian intellectuals in those years. His hero was Academician A.D. Sakharov. I remember how upset he was by the exile of Sakharov to the city of Gorky.[135] He really believed in Gorbachev's "perestroika" only after Sakharov returned to Moscow. V.I. suffered from the fact that, being recognized world-wide for his outstanding work, he was not allowed to travel abroad despite the numerous invitations from top scientists. Only in the 1990s, close to the end of his life, could V.I. undertake a long scientific tour in the USA. He also spent a year in Germany as a Humboldt Prize winner. He was on friendly terms with many people from literature and art, for example the poets Alexander Galich and Yuli Kim. While we were actively working on HSS, I was a frequent guest at V.I.'s home, where occasionally we would drink a little bit of vodka. He taught me how to present talks at seminars according to the so-called "Zermelo rules", two of which were "You can never underestimate the ignorance of the audience" and "Pass by the essential, but stop at the trivial". His most negative judgment of a physicist

[135]Editor's note: Academician Sakharov was detained on his way to work by KGB agents on January 22, 1980, and then, together with his wife Elena Bonner, were exiled to Gorky without trial. –MS

was "he is a scoop-up." We are good friends with his son Oleg who is now a Professor in Luminy (France).

Emery Sokatchev

In 1973, when I was a student at the University of Sofia (Bulgaria), one of my professors proposed to send me to JINR to get my Masters degree at LTPh. At that time students from the member states were not admitted to JINR. With the administrative problem solved, the question was who would supervise my master thesis. I had already met Prof. Ivan Todorov from the Bulgarian Academy of Sciences, who had just returned after many years spent at LTPh. My university professor asked him this question and the answer was: "I will try to convince Victor Ogievetsky, one of the very best people at LTPh, to take care of Emery". He succeeded and so in September 1973 I arrived in Dubna, at the age of 22. I remember wearing a very long sheepskin coat, to cope with the "Siberian winter" in Dubna. In this ridiculous outfit I presented myself to Victor Isaakovich in his office at LTPh. I was greeted by him with his notorious broad and very warm smile. This turned out to be more efficient against the cold Russian winters than my coat!

V.I. gave me a problem for my master thesis. I had to read a paper by a well-known American physicist on the representations of SL(3, R) and try to apply it to the Regge classification. I had a very vague idea of group theory, let alone complicated representations of non-compact groups. I spent my days at the library, reading books and trying to understand the paper. Eventually I reached the conclusion that the paper was wrong and the most important representation did not exist. I could not believe myself and did not dare go to V.I. with such an arrogant claim. So, I simply did not show up in his office for many days. One day we met at the canteen and he said that he was worried that I might be ill. I had no choice but to confess my findings. I saw his smile melting away. He also could not believe that the author had made such a mistake. He asked me to come to his office the next day. I could see myself packing my sheepskin coat and flying back to Sofia, after such a blunder at the very beginning of my work in Dubna. When I entered his office, the smile was back and even broader than usual: "I checked what you said, you are right!". This was the beginning of a very long and fruitful collaboration (43 joint papers!) with Victor Isaakovich Ogievetsky, which laid the main directions of my scientific career.

Throughout the years I learned so many things from V.I. that it is hard to give even a brief summary here. I would like to tell just two stories of how

V.I. failed to make two important discoveries, one with Igor Polubarinov and the other with myself.

The first story is about supersymmetry. V.I. was one of the pioneers of gauge theories in the Soviet Union in the early 1960's. He was one of the first to understand their fundamental role in particle physics. He told me that once he gave a talk on gauge theories in Moscow. After the talk a famous person approached him and said: "Why are you wasting your time with these gauge theories, they have no future!" V.I. did not listen and continued his pioneering studies. Together with Igor Polubarinov he developed a very original (for the time, today it is in the textbooks) approach. The idea is that every gauge theory is the theory of a consistent self-interacting massless field with a given spin. Thus, the theory of a spin 1 particle is Maxwell (or Yang-Mills) theory, that of spin 2 is Einstein's gravity. This approach starts from the first principles of quantum field theory and needs no geometric notions, as these theories are usually presented. In fact, the local symmetry arises as a corollary, not as an input. Having made such a profound observation about the underlying principle of gauge theories, V.I. asked himself: between spins 1 and 2 there is also spin 3/2. Is there a gauge theory that makes a spinor-vector (or Rarita-Schwinger) field consistently self-interact? If so, what is the local symmetry group?

Clearly, since the field of half-integer spin is fermionic, so must be the group parameters. Is there such a group? V.I. went around Moscow asking all of his acquaintances if they knew a group with fermionic parameters. People like I. M. Gelfand and F. A. Berezin did not know. V.I. failed to ask only one person, Yu. A. Golfand, who he met regularly at the theory seminars of the Lebedev Physics Institute (FIAN). He did not know that Golfand and Likhtman had invented the Poincaré supersymmetry algebra in 1971!

The other story, in a way the continuation of the first, is about my work together with V.I. on supersymmetry in 1974–77. I was very lucky in many ways, one was that thanks to Prof. Ivan Todorov I met one of the brightest theorists in Dubna, another was that the beginning of my career coincided with the discovery of supersymmetry. This happened just after my Masters degree in 1974. V.I. and my Bulgarian friends made it possible for me to stay on in Dubna, now as a research assistant. When the first Wess-Zumino paper on supersymmetry appeared in the early 1974, it was immediately clear to V.I. that this was THE symmetry he had been looking for. He and I developed further the Ogievetsky-Polubarinov approach to gauge theories, adapting it to superfields. He thought that his dream of a self-interacting spin 3/2 field will finally be realized. In 1975 we wrote a paper on the Rarita-Schwinger superfield, containing particles with spins 3/2 and 1. We found the free theory equations of motion, but we were unable to make the superfield self-interact. We concluded that we should combine the spin 3/2 field not with spin 1, but rather with spin 2 (graviton). Such a supersymmetry multiplet $(2, 3/2)$ existed and was described by an axial vector superfield. We started working hard on its equation of motion and self-interactions, but we were too slow. We insisted on the beautiful superspace formalism, and this is what slowed us down. Other, more pragmatic physicists worked out the component field formulation in terms of a gravitational and a Rarita-Schwinger fields, and this is how supergravity was born in 1976! A year later we did find an elegant geometric formulation in terms of our vector superfield but it was already too late, the discovery had already taken place.

Apart from physics, I learned a lot about life and human relations from Victor Isaakovich. When I arrived in Dubna at the age of 22, I had no parents alive. In a way, Victor Isaakovich took the role of a step-father for me. But this story is too long and too emotional to be presented here ...

In conclusion, Victor Isaakovich was a representative of the best Russian tradition of creating scientific schools of people with whom the Teacher would maintain good working relations for many decades. More than this, Ogievetsky's school was a kind of a family, and this very human aspect of our collaboration was not the least important one!

Acknowledgements

We thank Misha Shifman for the kind invitation to write our memories about V.I. Ogievetsky and his role in developing supersymmetry.

Figure 5.9: Nathan Seiberg, circa 2010.

CONVERSATION WITH NATHAN SEIBERG[136]

School of Natural Sciences, The Institute for Advanced Study
Princeton, NJ 08540, USA
seiberg@ias.edu

Ooguri: Over the past few decades, there have been remarkable developments in quantum field theory and string theory, and you have made significant contributions to them. There are many ideas and techniques that have been named after you, such as the Seiberg duality in four-dimensional $\mathcal{N} = 1$ theories, the Seiberg-Witten solutions to four-dimensional $\mathcal{N} = 2$ theories, the Seiberg-Witten map of noncommutative gauge theories, the Seiberg bound in the Liouville theory, the Moore-Seiberg equations in conformal field theory, the Affleck-Dine-Seiberg superpotential, the Intriligator-Seiberg-Shih metastable supersymmetry breaking, and many more. Each one of them has marked important steps in our progress. Today, we would like to look back on the history and hear from you how you made these discoveries and your perspective on the future development of quantum field theory and string theory.

Seiberg: Before we start, I would like to thank you very much for your invitation and for your kind words. The hospitality at Kavli IPMU during my visit here has been fantastic.

Ooguri: You started to work on supersymmetry almost immediately or maybe a year after you went to the Institute, is that right?

Seiberg: Almost immediately. I remember studying supersymmetry during the 1982/83 Christmas break.

Ooguri: So, you changed the direction of your research completely after arriving at the Institute. I understand that, at the Weizmann, you were working on model building, something related to technicolor.

Seiberg: Yes, it was not successful. So when I moved to my postdoc position, I thought I should change direction and I thought that supersymmetry was interesting.

Ooguri: What did you find most interesting about supersymmetry at that time?

---

[136]Interview given by Nathan Seiberg to *Hirosi Oouri* and *Yuji Tachikawa*, Kavli IPMU News No. 34 June 2016. Abbreviated. For the full version see http://ow.ly/Ywb63037jHd.

Seiberg: Supersymmetry looked like an interesting intellectual structure, which might also be relevant for phenomenology. Michael Dine had been working on supersymmetry before and he encouraged me to learn it. He recommended that I study the Wess and Bagger book, which was still in preprint form. (Julius Wess had just completed a series of talks at Princeton the year before. And Jonathan Bagger, who was a student at the time, took notes, which eventually became the famous book.)

Ooguri: Was Michael (Dine) also a postdoc there?

Seiberg: Michael was a 5-year member at IAS and I came in as a fresh postdoc. He took me under his wings and became my mentor. He had a huge impact on me. Our first project was about the dynamics of supersymmetric QCD. It's hard to imagine now how confusing it was at the time.

Tachikawa: Then came your paper with Ian Affleck and Dine on supersymmetry breaking.

Seiberg: The first paper we wrote was with Anne Davis, where we wrote the superpotential based on symmetries. Even that was not obvious at that time. Crucial earlier work on the subject had been done by Edward Witten.

Ooguri: Are you talking about Witten's paper on the index?

Seiberg: There were two papers. One about the index and the other was called dynamical supersymmetry breaking. I remember studying these papers very carefully because they were full of wonderful things. In particular, the behavior of supersymmetric Quantum Chromo Dynamics with massive quarks was analyzed. However, at the time there was no coherent picture about the behavior of the theory when the quark masses vanish. We wrote a paper with Anne about the analysis of the flat directions of the theory and we argued that a certain superpotential should be generated there. Our analysis was based on consistency conditions and the symmetries of the problem, but we did not prove that this superpotential is indeed generated.

Seiberg: Then Anne left and we started working with Ian Affleck. That summer we wrote a paper showing that in some circumstances this superpotential is generated by instantons. This was very surprising at that time because people had believed that the non-renormalization theorem was exact and was also true non-perturbatively.

Ooguri: Were you convinced that the standard argument was wrong?

Seiberg: Most of the previous arguments for the non-renormalization theorem were based on Feynman diagrams and were intrinsically perturbative. They did not apply to instantons. I remember that the instanton computation itself took maybe a week. Then it took another two months to go through all the arguments in the literature and to debug them. So we wrote a short letter and then we wrote a longer paper deriving the superpotential...

Tachikawa: This is the Affleck-Dine-Seiberg superpotential.

Seiberg: We showed that, depending on the number of flavors and colors, it is generated by instantons or by gluino condensation. For more flavors, no superpotential is generated and the vacuum degeneracy of the massless theory is not lifted. For me, this was the biggest surprise. These are quantum field theories with several ground states.

Ooguri: Continuously many ground states parameterized by the moduli space.

Seiberg: For me that was shocking.

Ooguri: So, some of the very important concepts in supersymmetric theory appeared within a year or so.

Seiberg: There were precursors of that in previous papers, primarily several papers by Edward Witten and in particular, a paper by Ian Affleck, Jeffrey Harvey and Edward Witten, which had analyzed the same problem in 2+1 dimensions. We also spent some time trying to build models of particle physics phenomenology. So we were primarily interested in breaking supersymmetry. The theories with a moduli space of vacua were a nuisance. They could not be used for that purpose. Then in the summer of 1984 Michael Green and John Schwarz wrote their revolutionary paper and the physics world changed overnight.

Ooguri: When did you hear about it?

Seiberg: I heard about it in late August or early September. Everybody was talking about this fantastic breakthrough, but none of us, the postdocs, had any idea what the breakthrough was about. There was some mysterious theory, string theory that we had never heard about. This was something that the older people had worked on and abandoned. They thought "This was a failed attempt and younger people should not learn about it."

Ian Affleck left for a year in Paris, and Michael Dine and I had to decide, "Are we going to pursue our work on supersymmetric gauge theories, or are we going to study this new thing?" And we decided to study string theory. That fall David Gross gave a fantastic string theory course at Princeton University, mostly in order to educate the young people.

Power of Holomorphy

Ooguri: You went back to the supersymmetric field theory in the early 1990s — maybe a couple of years before the second superstring revolution.

Did you foresee anything coming? I am asking this since what you did in these few years in the early 1990s on supersymmetry turned out to play essential roles in the developments of the second revolution. What motivated you to come back to the supersymmetric field theory?

Seiberg: I was working on conformal field theory and two-dimensional gravity, and I felt that it was time to change. So I worked with Yossi Nir on model building — theories of quark and squark masses. And then I learned of a new paper by Michael Dine and Ann Nelson on gauge mediation of supersymmetry breaking. (Later they collaborated also with Shirman and with Nir.) They decided to take up the same question about supersymmetry breaking that Michael, Ian, and I had dropped when string theory came along.

Ooguri: So, this was a continuation of your work with Affleck and Dine. Was it roughly 10 years after that?

Seiberg: Indeed. They did their work in 1992–1993. Ann visited Rutgers and we talked a lot. At that time, I was tired of supersymmetry and I didn't want to go back to that topic. But she succeeded to push me back into it... I did not want to go back because I thought "Okay, I've already worked on this, I want to do something new." Ann visited us at Rutgers and gave talks about her work with Michael Dine, which were very stimulating for me. We started talking and eventually we wrote a paper about the relation between R-symmetry and supersymmetry breaking. (Some people refer to it as the Nelson-Seiberg theorem.) This followed an observation Ian, Michael, and I had made 10 years earlier, and it turned it into a more concrete and much clearer principle.

Ooguri: This was the beginning of the modern approach to supersymmetric field theory.

Seiberg: This was one element. The second element was influenced by my work with Yossi Nir, where we used spurions and the fact that the superpotential had to be holomorphic in them.

Ooguri: Was that the first time the spurion technique was used in supersymmetric theory?

Seiberg: Spurions had appeared earlier, especially in the context of supersymmetry breaking. I think the new point here was to view all the ordinary supersymmetric coupling constants as spurions by viewing them as background superfields. And the main application was to derive the non-renormalization theorem.

Ooguri: Were you the one who introduced this technique in supersymmetric theory?

Seiberg: I am not sure about that. But perhaps that was the first time all the coupling constants were viewed as background fields in the context of a supersymmetric theory. I remember that within hours or so after I thought of it many things fell into place. I'd like to offer a historical perspective. Before the 1980s there was a clear understanding that the behavior of quantum field theory satisfies some genericity requirement. Murray Gell-Mann described it as the "Totalitarian Principle": Everything not forbidden is compulsory. 't Hooft described the same thing as "naturalness" — parameters take natural generic values unless there is a good reason, e.g. a symmetry, not to do that. But then, supersymmetric theories seemed to violate that principle. The superpotential is not renormalized and can be non-generic. This looked strange and the cancellations behind it seemed miraculous. Then, when it was realized that non-perturbative effects violate the perturbative non-renormalization theorem, it became clear that we need an organizing principle. In other words, to what extent are the genericity properties of quantum field theory true in supersymmetric theories?

The modern point of view, based on holomorphy of the superpotential vindicated Gell-Mann's principle and is consistent with 't Hooft's naturalness. The superpotential is subject to the same genericity properties as every other term, except that when we use this genericity we should also take its holomorphy into account. For this reason, I picked the title of my paper "Naturalness versus supersymmetric non-renormalization theorems." For me that was very satisfying. Many different computations and many different phenomena were understood together using one organizing principle. I felt like I had been circling around this for years and all of a sudden it all came together. It was clear that that was the right way to think about it.

Ooguri: When did you come to appreciate the importance of holomorphy?

Seiberg: The appreciation of the power of holomorphy of the superpotential occurred to me gradually during these years. We have seen many examples of computations of a superpotential, where the real part and the imaginary part were computed independently, and surprisingly the result turned out to be holomorphic. At the time this looked like a miracle and a non-trivial consistency check of the computation.

Using this for a non-renormalization theorem was also not completely new. For example, Edward Witten argued for a perturbative non-renormalization theorem in the string worldsheet using holomorphy and Michael Dine and I used similar reasoning about string perturbation theory. In those cases, the holomorphic dependence was on fields, not parameters (although these were parameters in the worldsheet theory). Holomorphic dependence on parameters in 3+1 dimensional field theory also appeared in papers of Victor Novikov, Mikhail Shifman, Arkady Vainshtein, and Valentin Zakharov and in an unpublished work that Joe Polchinski and I did. And I must be forgetting other examples.

Ooguri: That was in the summer of 1993. But, in the late 1980s, you were essentially using it, weren't you?

Seiberg: Indeed, I used it. But at the time I did not view it as the underlying organizing principle.

Ooguri: That's interesting because that was also the year I wrote this paper with Bershadsky, Ceccotti, and Vafa on topological string theory, which also used the holomorphy and its relation to supersymmetry to derive the recursion relation, the so-called BCOV equation.

Seiberg: Your paper was a milestone. It clarified many issues and it led to many significant consequences.

But this was a separate line of development. It is surprising and interesting that similar ideas came up independently in different contexts at more or less the same time. As we discuss historical developments in science I am reminded of the famous Kurosawa movie Rashomon.

Tachikawa: It is based on a book by Akutagawa that we studied in school...

Seiberg: I am not familiar with the book, but I really like the movie. One lesson from that story is that different people view the same reality differently. This is common in the history of science. There are several lines of development and they are typically motivated by different questions. So if you interview another researcher who worked at the same time, you are likely to hear a completely different version of the events. And it is not that one description is more correct than another. They simply reflect different perspectives. Going back to your question, at the time I was mostly interested in four-dimensional quantum field theory. The idea that you could say anything about the non-perturbative behavior of four-dimensional quantum field theory was totally unimaginable.

Ooguri: Right. You would not have expected that there would be analytic control over any non-perturbative physics in four dimensions.

Seiberg: With the understanding of holomorphy as an organizing tool all of a sudden things became clear and easy.

Ooguri: So, that became the guiding principle.

Seiberg: When you have a new tool you should be maximally ambitious. I remember that I thought, "We should be able to address all the open questions in quantum field theory using holomorphy"

As a first step I looked back at the supersymmetric version of Quantum Chromo Dynamics. With Ian and Michael, we had understood the behavior of the theory for small number of flavors, so I wanted to understand what happens with more flavors. Armed with the new perspective and new tools I realized that although the vacuum degeneracy of the classical theory is not removed in the quantum theory, extremely interesting effects are still present. For example, the complex structure of the moduli space of vacua can be deformed in the quantum theory. In other cases, there are new massless composite particles. That was surprising. Until then it had been believed that with strong dynamics the theory is gapped or has some massless Nambu-Goldstone bosons. Here, on the other hand, there can be massless composites that are not associated with spontaneous symmetry breaking. Later, this observation strongly motivated the understanding of the long distance behavior of $\mathcal{N} = 2$ theories (where there are massless monopoles) and other $\mathcal{N} = 1$ theories (where there are massless composite gauge fields, massless glueballs, massless exotics, etc.). Then I started working with Ken Intriligator and Robert Leigh on increasingly more complicated models and we saw that the new techniques are very powerful leading to many new exact results. Then Ashoke Sen wrote an extremely interesting paper...

Ooguri: Are you talking about his work on monopoles?

Duality and Seiberg-Witten Solutions

Seiberg: Yes. I am referring to Sen's paper establishing the existence of a charge-2 monopole in $\mathcal{N} = 4$ supersymmetric theories. That paper removed a real obstruction to duality. People had thought about duality before that paper. I think it was Witten who emphasized the importance of this charge-2 monopole. And it was believed that such a monopole does not exist and therefore the whole idea of duality had to be wrong. Sen's paper found that monopole and overnight it was clear that duality must be right.

Ooguri: Since Yuji is here, maybe one of us should explain it to him. In early 1990, not too many people believed in the electric-magnetic duality, the S-duality.

Tachikawa: Yes, I came too late to this part of the party.

Ooguri: There was of course the work by, I think, Montonen and Olive and then Olive and Witten, and...

Tachikawa: Goddard, Nuyts and Olive?

Ooguri: Right. But, when I read the Olive-Witten paper in the late 1980s, I had an impression that rather than giving evidence for duality, they were actually explaining why this miraculous formula works without duality — everything comes from supersymmetry. Now we use it to motivate duality. But at that time, it was basically explaining that you don't need duality.

Seiberg: They considered states in small representations of the supersymmetry algebra. Today we refer to such states as BPS states. And they explained many of the special properties of these states as following from this fact. Since these special properties had a rational explanation, everyone thought that this should not be used as evidence for duality. In addition, duality demanded the existence of a certain bound state of two monopoles in $\mathcal{N} = 4$ theories. And it was thought that such a state does not exist...

Ooguri: Yes, but then, Ashoke's demonstration of the bound state was...

Seiberg: ...a phase transition. Before that paper duality was viewed as some technicality. Some things look as if they're dual, but there is no reason to believe this duality was an exact statement. Then, overnight it became obvious that duality is crucial.

Ooguri: Did it convince you that duality is actually a real phenomenon in quantum physics?

Seiberg: I was immediately convinced that it was true in $\mathcal{N} = 4$ theories. But I did not imagine that it will play such a crucial and central role as it later turned out to do. And then I started collaborating with Edward Witten on the $\mathcal{N} = 2$ supersymmetric theory.

Ooguri: Were you interested in $\mathcal{N} = 2$ theory before you started collaborating with Witten?

Seiberg: Yes. I viewed it simply as another theory with $\mathcal{N} = 1$ supersymmetry. And as I was working out examples of increasing complexity with $\mathcal{N} = 1$ supersymmetry, this was a natural member of that list. Also, it was clear that the pure gauge $\mathcal{N} = 2$ theory has massless photons and the pure gauge $\mathcal{N} = 1$ theory was expected to confine. So I was hoping that a better understanding could explain the mechanism for confinement.

Ooguri: And, you did. Did you think that you had a better chance of doing that in the $\mathcal{N} = 2$ case, as opposed to $\mathcal{N} = 1$?

Seiberg: I was viewing it more as a special case of $\mathcal{N} = 1$. The specific properties of $\mathcal{N} = 2$ are nice and allow you to compute additional quantities. But for the question of confinement they are not essential. If you go back some time — I think it was in the fall of 1987 — Witten wrote a paper explaining Donaldson theory as a twisted version $\mathcal{N} = 2$ supersymmetric Yang-Mills theory. It made Donaldson theory accessible to physicists. But it didn't solve that mathematical problem.

Ooguri: It gave a physical interpretation of the mathematical problem.

Seiberg: But I am told by mathematicians that from their viewpoint it didn't solve the problem, because at the end of the day you had to do exactly the same computations that Donaldson did anyway. However, viewed more broadly, this is an amazing paper. In this paper Edward Witten introduced the notion of a Topological Quantum Field Theory. This is an extremely deep idea with far reaching consequences both in mathematics and in physics. When I heard about that work I used my tool kit of instanton technology to write a short note on the $\mathcal{N} = 2$ theory thinking about its moduli space of vacua and showing how the metric is corrected asymptotically. I didn't even want to publish that paper because I did not think it was interesting. Edward encouraged me to publish it. In fact, in his paper on Donaldson theory, he said that the ideas in that paper may well be important for further developments of the theory. He was completely right about that. That was in 1988.

Ooguri: That turned out to be the starting point and the boundary condition of what is now called the Seiberg-Witten solution.

Seiberg: Yes. I wrote the paper on the "Behavior at Infinity" but I didn't pursue it. Then, in 1994 when I came back to supersymmetric theories, I remembered my 1988 paper. I knew that the vacuum degeneracy was not lifted and an infinite series of instantons corrected the metric on the moduli space of vacua. The question was how to evaluate it and how to sum up the series. As I said, at the time Ken Intriligator, Rob Leigh, and I were collaborating. We wrote a paper studying various models in which the superpotential is given by an infinite sum of instantons. We succeeded to sum up the infinite series in an explicit closed-form formula using the knowledge of the singularities of the superpotential and its asymptotic behavior, combined with its holomorphy. So I was optimistic that the same thing could be done also in the $\mathcal{N} = 2$ theory. When I started collaborating with Witten progress came in a stunning rate. Within weeks we had the complete solution of the pure gauge theory and the theory with matter. All sorts of interesting physical phenomena were elucidated including confinement and chiral symmetry breaking. Witten immediately realized that it would help simplifying the problem with four-dimensional topology. But, I was more interested in understanding the dynamics of four-dimensional quantum field theory. Immediately afterwards, I wrote the paper on duality in $\mathcal{N} = 1$. In that paper a dual description of $\mathcal{N} = 1$ supersymmetric gauge theories was presented. This work taught us many lessons. First, it was realized that electric-magnetic duality is ubiquitous. $\mathcal{N} = 1$ theories are more generic than $\mathcal{N} = 4$ and $\mathcal{N} = 2$ theories, and they exhibit similar dualities. Second, here one finds weakly coupled composite gauge fields. This underscores the fact that gauge symmetries are not fundamental. This point had been known before, mostly in the context of Abelian gauge theories, but here it was more dramatic. In the meantime, Witten wrote the paper on the connection of our solution of $\mathcal{N} = 2$ supersymmetric gauge theories to four-dimensional topology.

Ooguri: Yes, which is now called the Seiberg-Witten equation. It turned out to be more powerful than the original Donaldson theory.

Seiberg: Physically, it is very clear because many of the complications of Donaldson theory were associated with small instantons. And every time you have a computation associated with a new manifold you have to control the same small instantons. What the renormalization group allows you to do is to compute the small instantons once and for all in flat space and to find an effective theory without them. Then you can place that effective theory on the curved space of interest. Since this effective theory no longer has these small instantons, many of the complications in the original theory are no longer present. So that's the reason it was...

Ooguri: In the effective theory it's already built-in, so you don't...

Seiberg: It's already built-in and you don't need to worry about that. In fact, the low-energy theory doesn't have small instantons. So that's what made these equations so much more powerful.

*****

Seiberg: Theorists are stimulated by many sources. They listen to seminars, participate in informal discussions, read papers, etc. This impacts their research direction in unexpected ways. It might even lead them to abandon their existing line of research and to start another one. And even within a given project, in most cases the outcome of the research could not be anticipated from the beginning. In our conversation about my research path I shared with you some examples of such unexpected results from my own experience. So there is no way I could outline what I'll be doing in the next five years — this is ridiculous.

The reason research is interesting is because we're surprised by the answers. If we could predict the answers, we would not be surprised by them. Almost by definition, we cannot predict the outcome. So we should not attempt to do that. Pursue what you're interested in, keep working hard, pay attention to what's going on around you and be flexible — these are the rules. Sometimes it works, sometimes it doesn't.

Editor's Note 1: –MS

Below are review articles authored by Seiberg (and Kenneth Intriligator). Collectively, they contain an exhaustive list of references.

- N. Seiberg, *The Power of holomorphy: Exact results in 4-D SUSY field theories,* [arXiv:hep-th/9408013 [hep-th]].
- K. A. Intriligator and N. Seiberg, *Lectures on supersymmetric gauge theories and electric-magnetic duality,* Nucl. Phys. B Proc. Suppl. **45BC**, 1–28 (1996) [arXiv:hep-th/9509066 [hep-th]].
- N. Seiberg, *The Power of duality: Exact results in 4D SUSY field theory,* Prog. Theor. Phys. Suppl. **123**, 337–347 (1996) [arXiv:hep-th/9506077 [hep-th]].
- K. A. Intriligator and N. Seiberg, *Lectures on Supersymmetry Breaking,* Class. Quant. Grav. **24**, S741–S772 (2007) [arXiv:hep-ph/0702069 [hep-ph]].

Editor's Note 2: –MS

In the First Edition of "The Supersymmetric World" released in 2000, the power of supersymmetry as a unique tool for solving dynamical problems at strong coupling was not even mentioned. At that time it was not fully recognized that this pioneering idea eventually would grow into a separate branch which impacted many aspects of theoretical and mathematical physics (see Fig. 1 on page vii). Suffice it to mention two basic achievements: (i) the Seiberg duality [1] and (ii) the first analytic proof of the Nambu-'t Hooft-Mandelstam mechanism [2] of *confinement* (the so-called dual Meißner effect) in $\mathcal{N} = 2$ super-Yang-Mills theory slightly deformed by an $\mathcal{N} = 1$ perturbation by Seibeg and Witten [3]. These breakthrough discoveries are discussed in detail in Seiberg's interview presented in this section. They can be viewed as a culmination of the continuous advances in SUSY-based exact results. The above two papers became world classics.

Advances in this direction started much earlier, however. The first exact result in SUSY — the exact vanishing of the vacuum energy density — is of a purely algebraic nature. It was mentioned in passing by Likhtman in one of his early preprints [4] and discussed in detail by Wess, Zumino, Salam, and others. It is curious that in 1948, long before the discovery of SUSY, Wolfgang Pauli noted [5] that in free field theories containing the same number of the bosonic and fermionic degrees of freedom of the same mass the vacuum energy density vanishes.

In this note I briefly outline the inception/development of the area associated with exact results in supersymmetry. I limit myself to four-dimensional SUSY and some general aspects.

Endowing Yang-Mills theories with supersymmetry created a broad class of the so-called protected quantities which can be calculated exactly at weak and strong coupling regimes. They are also referred to as BPS-protected or short supermultiplets. BPS stands for Bogomol'nyi, Prasad, and Sommerfield [6]. The authors of [6] considered solitons in a nonsupersymmetric setting. They found, however, that under certain conditions solitons can be described by first-order differential equations rather than the second-order equations of motion. Moreover, under these conditions the soliton mass was shown to be proportional to the topological charge. We understand now that the limiting models considered in [6] correspond to the bosonic sectors of supersymmetric models [7].

In 1976 Witten and Olive demonstrated that in supersymmetric theories with solitons, the supersymmetry algebra is modified to include the topological quantum numbers as central charges. As a result, in some well-defined cases the saturation of the Bogomol'nyi bound implies that the quantum corrections to the classical mass spectrum are absent. The mass spectrum formulae become exact.

In 1982 Witten designed the eponymous index I_{W} — an integer counting the number of vacua in supersymmetric theories [8]. He then calculated I_{W} in a number of theories, including supersymmetric Yang-Mills (SYM).

The first analysis of non-perturbative dynamics in Polyakov's compact electrodynamics [9] in $1+2$ dimensions was carried out in Ref. [10]. Affleck, Harvey and Witten supersymmetrized this theory and studied the instanton-monopole contribution with the conclusion that SUSY remains unbroken.

This paper brings in vivid memories of how the problem of superinstantons in four dimensional Yang-Mills theories tortured us at approximately the same time — since summer of 1982.[137]

[137] See Ref. [11] for our first publication on this problem. By "us" I mean here Novikov, Vainshtein, Zakharov, and myself, or a subgroup of the above group.

In 1982–83 we observed [12] that not only in the vacuum, but in *all* backgrounds which conserve a part of supercharges exact results are attainable. This observation was beginning of a long journey... Supersymmetrized BPST instanton [13] was the first object of this type. We realized that the exact superinstanton measure is exhausted by the zero modes which, in turn, led us to a concise all-order expression for the β function. In SYM theory (without matter) it takes the form

$$\beta(\alpha) = -\left(n_b - \frac{n_f}{2}\right)\frac{\alpha^2}{2\pi}\left[1 - \frac{(n_b - n_f)\,\alpha}{4\pi}\right]^{-1}, \tag{5.40}$$

where n_b and n_f are the numbers of the corresponding zero modes. All coefficients in the β function (5.40) are integers and, moreover, of a purely geometric nature. They are in one-to-one correspondence with the number of symmetries nontrivially realized on instantons. Equation (5.40) is valid for $\mathcal{N} = 1, 2$ and 4, ($n_f = \frac{1}{2}n_b$, n_b, and $2n_b$, respectively).

Adding matter we obtained [12, 14] the NSVZ β function — an exact relation between the β function and the anomalous dimensions of the matter fields γ_i,

$$\beta(\alpha) = -\frac{\alpha^2}{2\pi}\left[3\,T_G - \sum_i T(R_i)(1-\gamma_i)\right]\left(1 - \frac{T_G\,\alpha}{2\pi}\right)^{-1}. \tag{5.41}$$

where $T(R_i)$ is $\frac{1}{2}$ of the Dynkin index in the representation R_i and $T_G = T_{\rm adjoint}$. Equation (5.41) was used in [1] for determination of the edges of the conformal window.

The next step was the instanton-based calculation of the gluino condensate $\langle\lambda\lambda\rangle$ in SYM at strong coupling [15]. In SU(2) theory we found the two-point function

$$\langle\lambda\lambda(x)\,\lambda\lambda(0)\rangle \tag{5.42}$$

fully saturated by a single instanton. We discovered that it is x independent. This was the first example of the supersymmetric theorem: If an operator O is Q closed and $\partial_\mu O$ is Q exact the correlation function $\langle O(x)\,O(0)\rangle$ is x independent. In SYM, the operator $\lambda\lambda$ satisfies the condition of this theorem. Our calculation was carried out at small $|x|$; however, since we could use the above theorem to send $|x| \to \infty$ — the limit in which the two-point function (5.42) factorizes — we arrived at (the double-valued) prediction for $\langle\lambda\lambda\rangle$. For a more detailed review the reader is referred to [16, 17].

In 1984 a powerful Affleck-Dine-Seiberg superpotential was obtained in SU(2) SQCD with one flavor [18] (in the low-energy limit). Approximately at the same time we published our paper [19] which dealt with the a different aspect of same model. At weak coupling (in the Higgs regime) the above model was well-defined at all distances, including large, because of complete Higgsing. Under these circumstances we were certain that $\langle\lambda\lambda\rangle$ was saturated by a single instanton. It was natural to think that the size of the saturating instanton would be of the order of $\rho \sim v^{-1}$ where v is the vacuum expectation value of the Higgs field. However, a great surprise expected us *en route*. A remarkable observation was made in [19]: the integral over the instanton size proved to be completely saturated by zero-size instantons, because we represented it as a full derivative of the integrand which reduced to a delta function, $\delta(\rho^2)$. This was the first example of the instanton self-localization which later grew into a powerful construction of the Nekrasov localization [20].

Second, the method of analytic continuation in a holomorphic parameter[138] (the matter field mass term) from weak to strong coupling was first used in this paper in calculating $\langle\lambda\lambda\rangle$. It gave rise to a notorious "4/5 puzzle" [16]. With great surprise we realized that the strong- and weak-coupling calculations of $\langle\lambda\lambda\rangle$ produce numerical coefficients which do not coincide with each other. A long polemics in the literature followed. Although the fact itself can be considered established, its roots are not fully clear [16].

A review of further developments along these lines can be found in [17]. In this publication, among other topics, the reader will find discussions of the domain walls in SYM as *D* branes and non-Abelian flux tubes as SQCD solitons.

References

1. N. Seiberg, *Electric-magnetic duality in supersymmetric non-Abelian gauge theories*, Nucl. Phys. B **435**, 129–146 (1995) [arXiv:hep-th/9411149 [hep-th]].
2. Y. Nambu, *Phys. Rev. D* **10**, 4262 (1974); G. 't Hooft, *Gauge theories with unified weak, electromagnetic and strong interactions*, in *Proc. EPS Int. Conf. on High Energy Physics*, Palermo, June 1975, ed. A. Zichichi (Editrice Compositori, Bologna, 1976); S. Mandelstam, *Phys. Rept.* **23**, 245 (1976).
3. N. Seiberg and E. Witten, *Nucl. Phys. B* **426**, 19 (1994); Erratum, *ibid.* **430**, 485 (1994) [hep-th/9407087]; *Nucl. Phys. B* **431**, 484 (1994) [hep-th/9408099].

[138]This method is one of the cornerstones of the Seiberg-Witten solution [3]. Seiberg coined a special term, "the power of holomorhy."

4. E. Likhtman, Lebedev Physical Institute Preprint # 41, Moscow, 1971 [English Transl. in E. P. Likhtman, Nucl. Phys. B Proc. Suppl. **101**, 5–14 (2001) [arXiv:hep-ph/0101209]]. See page 12 in Proceeding, the discussion following Eq. (7).
5. W. Pauli, *Pauli Lectures on Physics, Selected Topics in Field Quantization* (MIT Press, Cambridge, 1973), Vol. 6, p. 33.
6. E. B. Bogomol'nyi, *Yad. Fiz.* **24**, 861 (1976) [*Sov. J. Nucl. Phys.* **24**, 449 (1976)] [reprinted in C. Rebbi and G. Soliani (eds.), *Solitons and Particles* (World Scientific, Singapore, 1984) p. 389]; M. K. Prasad and C. M. Sommerfield, Phys. Rev. Lett. **35**, 760 (1975), reprinted in C. Rebbi and G. Soliani (eds.), *Solitons and Particles* (World Scientific, Singapore, 1984) p. 530.
7. H. J. de Vega and F. A. Schaposnik, Phys. Rev. D **14**, 1100 (1976), reprinted in C. Rebbi and G. Soliani (eds.), *Solitons and Particles* (World Scientific, Singapore, 1984) p. 382.
8. E. Witten, *Constraints on Supersymmetry Breaking*, *Nucl. Phys. B* **202**, 253 (1982) [Received July 19, 1982], [reprinted in S. Ferrara (ed.), *Supersymmetry* (North Holland/World Scientific, Amsterdam-Singapore, 1987), Vol. 1, p. 490].
9. A. Polyakov, *Quark confinement and topology of gauge theories*, Nuclear Physics B **120**, 429–458 (1977).
10. I. Affleck, J. A. Harvey and E. Witten, *Instantons and (Super)Symmetry Breaking in (2+1)-Dimensions,* Nucl. Phys. B **206**, 413–439 (1982) [Received October 18, 1982].
11. V. A. Novikov, M. A. Shifman, A. I. Vainshtein and V. I. Zakharov, *Instanton in Supersymmetric Theories*, Preprint ITEP-93, July 1, 1982; Nucl. Phys. B **223**, 445–473 (1983).
12. V. A. Novikov, M. A. Shifman, A. I. Vainshtein and V. I. Zakharov, *Exact Gell-Mann-Low Function of Supersymmetric Yang-Mills Theories from Instanton Calculus,* Preprint ITEP-89, June 6, 1983; Nucl. Phys. B **229**, 381 (1983); *Instantons Versus Traditional Approach*, Phys. Lett. B **166**, 329 (1986).
13. A. Belavin, A. Polyakov, A. Schwartz, and Yu. Tyupkin, *Pseudoparticle Solutions of the Yang-Mills Equations*, Phys. Let. B, **59**, 85–87 (1975).
14. M. A. Shifman, A. I. Vainshtein and V. I. Zakharov, *SUSY Non-Abelian gauge models: Exact beta function from one loop of perturbation theory*, Preprint ITEP-137, August 15, 1985. Sov. J. Nucl. Phys. **43**, 1028 (1986).
15. V. A. Novikov, M. A. Shifman, A. I. Vainshtein and V. I. Zakharov, *Instanton Effects in Supersymmetric Theories,* Preprint ITEP-77, May 13, 1983; Nucl. Phys. B **229**, 407 (1983).
16. M. A. Shifman, *Exact results in gauge theories: Putting supersymmetry to work. The 1999 Sakurai Prize Lecture,* Int. J. Mod. Phys. A **14**, 5017–5048 (1999) [arXiv:hep-th/9906049 [hep-th]].
17. M. Shifman, *Supersymmetric tools in Yang–Mills theories at strong coupling: The beginning of a long journey,* Int. J. Mod. Phys. A **33**, no.12, 1830009 (2018) [arXiv:1804.01191 [hep-th]].
18. I. Affleck, M. Dine and N. Seiberg, *Dynamical Supersymmetry Breaking in Supersymmetric QCD,* Nucl. Phys. B **241**, 493–534 (1984).
19. V. A. Novikov, M. A. Shifman, A. I. Vainshtein and V. I. Zakharov, *Supersymmetric Instanton Calculus (Gauge Theories with Matter),* Preprint ITEP-31, January 24, 1985; Nucl. Phys. B **260**, 157 (1985).
20. N. A. Nekrasov, *Seiberg-Witten Prepotential from Instanton Counting*, Adv. Theor. Math. Phys. **7**, no. 5, 831 (2003) [hep-th/0206161].

6

The Historical Perspective

REVEALING THE PATH TO THE SUPERWORLD[139]

M. S. MARINOV
Physics Department, Technion — Israel Institute of Technology, Haifa, 32000 Israel

Below I present a brief history of ideas which paved the way to supersymmetry.

One of the most impressive discoveries of the 20th century is the understanding that neither real nor complex numbers are sufficient to describe physical systems. It was found that, besides the usual coordinates, electrons and other elementary particles have spin degrees of freedom, which have no intuitive classical analogues, and *anti-commuting variables* must be employed to produce a proper and elegant representation of dynamics at the microscopical level. *The Superworld* has commuting and anti-commuting dimensions, owing to (or giving rise to) two kinds of particles, bosons and fermions.

The way to this discovery was paved by the mathematics of the 19th century. In 1843 William R. Hamilton discovered quaternions, an associative algebra of number quadruples with three *anticommuting* imaginary units, including division (each nonzero element has its inverse). The discovery was accepted as a miraculous revelation — "the light dawned upon Hamilton on a certain October day in Dublin," as some of his adepts recalled [1]. The complete theory, including elements of the vector analysis, was published in *Lectures on Quaternions* (1853) and in a more exhaustive way in *Elements of Quaternions* (1866) [2]. About three decades after the work on quaternions was first published, William K. Clifford introduced a more general associative algebra (though with no division in general, unlike that of quaternions) of *bi-quaternions* with n auxiliary anticommuting units. The bi-quaternions were presented in the form $c = a + \sum_{k=1}^{n} b_k\epsilon_k$, where a and b_k are real and ϵ_k^2 may be $+1$, -1 or 0. The latter is the case used in functional formalism for Fermi-type variables and in the supersymmetry in particular. In general, the Clifford algebra is given by the anti-commutation relations for its generators,

[139] This is a revised version of the article prepared by M. Marinov for the Yuri Golfand Memorial Volume *The Many Faces of the Superworld* (World Scientific, Singapore, 2000), page 32. M. Marinov worked on it in the summer of 1999, and then started revising it to include a part devoted to supergravity. The work was not completed: he died of cancer on January 17, 2000.

$\epsilon_k\epsilon_l + \epsilon_l\epsilon_l = g_{kl}$, where the quadratic form given by g_{kl} may be degenerate. The matrices introduced in physics by Pauli and Dirac are two examples of the Clifford algebras with non-degenerate g_{kl}.

The introduction of quaternions by the Royal Astronomer of Ireland was a spectacular achievement, but at about the same time a less celebrated mathematician, Hermann G. Grassmann, a high school teacher in Stettin, Prussia, developed a more general, profound and abstract "theory of extension." Grassmann was an outstanding person of encyclopedic knowledge in science, mathematics and philology (one of his famous works was a dictionary of Sanskrit) [3]. The treatise describing Grassmann's theory was published [4] in 1844, and the most complete second edition appeared in 1878. Both Hamilton and Grassmann discovered vector analysis, obtained new associative algebras, and found a number of other beautiful things. A struggle for priority (not the first one in the history of mathematics) between the British and German mathematical communities continued for decades. In fact, the "theory of extension" provided a basis for modern differential geometry and tensor analysis. It was very much ahead of its time, but now it is studied in some form at most departments of mathematics. Referring to the Clifford algebra with $\epsilon_k^2 = 0$, Berezin introduced the term "Grassmann algebra with anti-commuting generators," which is widely used now in theoretical physics.

Until the beginning of the 20th century Grassmann's theory was available to only a few initiated persons, since Grassmann's book was hardly easy reading. However, we have evidence that a great physicist studied from it as a boy; that was 16-year-old Enrico Fermi. Emilio Segré published a letter by Adolfo Amidei, a colleague and friend of Enrico's father, in his scientific biography of Fermi [5]. We read in the letter:

> I also deemed it appropriate for him [Enrico] to study the *Ausdehnungslehre* by H. Grassmann, which has an introduction on the operations to deductive logic by Giuseppe Peano. These books were lent to him in 1918. I thought it appropriate because it was my opinion that the *Ausdehnungslehre* [similar to vector analysis] was the most suitable tool for the study of different branches of geometry and theoretical mechanics, because in this system of calculus the operations are not performed on numbers [coordinates] which define the geometrical objects but on the objects themselves, thus obtaining formulae which are simple and of easy interpretation. Furthermore, this calculus contains, as a particular case, analytical geometry, and is specially suited to the study of theoretical mechanics and graphical statics... Enrico found vector analysis very interesting, useful and not difficult...

Now one may only speculate if Fermi had in mind Grassmann's construction when he elaborated the theory of degenerate "fermion gas" [6]. No reference to anti-commutation relations was given. It was in Göttingen that the canonical anti-commutation relations (CAR) were consistently introduced by Pascual Jordan and Eugene Wigner [7] as a necessary addition to the canonical commutation relations (CCR) of the Heisenberg-Weyl algebra. Thus, commuting and anti-commuting canonical operators were put on an equal footing. Boson and fermion fields appeared in conventional quantum field theory.

The operator formalism, including the standard Fock space in field theory, is based consistently upon CCR and CAR, but in order to employ the functional formalism of Schwinger and Feynman one needs a further development of the corresponding classical theory. In the boson sector, where CCR correspond to the standard Poisson brackets, the classical limit leads to commuting quantities, so the operators (q-numbers) are reduced to complex numbers (c-numbers), where the standard calculus is known. In the fermion sector, the classical limit leads to "anti-commuting c-numbers," while the Clifford algebra with $\epsilon_k^2 = +1$ is reduced to the degenerate case where $\epsilon_k^2 = 0$ (since CAR are proportional to $\hbar$), and the necessary elements of calculus need an extension. First, Julian Schwinger advocated the universality of his "quantum action principle" in a reply [8] to opponents who doubted its applicability to fermion fields. According to Schwinger's arguments, the canonical variables may be of two kinds; first-kind variables with anti-symmetrical Poisson brackets leading to CCR upon quantization, and second-kind variables with symmetrical Poisson brackets leading to CAR. The general principle may be formulated in the following way. The canonical variables of the boson type x_a (the usual phase-space vector) and the fermion type (anti-commuting) ξ_α are united in one vector $X_A \equiv (x, \xi)$, where the index is $A \equiv (a, \alpha)$, of the extended phase space (which may be called "phase superspace"). For any pair of the dynamical functions $\mathcal{A}(X)$ and $\mathcal{B}(X)$, the Poisson brackets were defined

$$\{\mathcal{A}, \mathcal{B}\}_{\text{P.B.}} \equiv \frac{\partial \mathcal{A}}{\partial X_A}\frac{\partial \mathcal{B}}{\partial X_A}\Omega_{AB}. \tag{MM.1}$$

The (X-independent) bilinear form Ω must be nondegenerate and skew-Hermitian, i.e. $\Omega^\dagger = -\Omega$. Its canonical real form is $\Omega_{ab} = \omega_{ab} = -\omega_{ba}$, $\Omega_{\alpha\beta} = i\delta_{\alpha\beta} = i\delta_{\beta\alpha}$, $\Omega_{a\alpha} = 0 = \Omega_{\alpha a}$. The Poisson brackets defined in this way constitute a Z_2-graded Lie algebra of the simplest type, provided the partial derivatives in ξ are defined properly. In agreement with Grassmann's concepts, Schwinger considered the second-kind variables as elements of the *exterior algebra* (another name for Grassmann's algebra of differential forms). The formalism was developed in a series of papers published in *Proc. Natl. Acad. Sci.*, which were later reprinted and commented on in Schwinger's book [9]. In these terms, the formulation of a linear realization of supersymmetry looked straightforward. Let us assume that the system is described by a quadratic Hamiltonian $\mathcal{H}(X) = \frac{1}{2}(X, \check{H}X)$, where $\check{H}$ is a linear operator in the extended phase space. The symmetry of the system under linear transformations $X = \check{L}Y$ is given by the Poisson-bracket-generated Lie algebra of *all* quadratic functions $\mathcal{C}(X) = \frac{1}{2}(X, \check{C}X)$ which are in involution with $\mathcal{H}(X)$, namely

$$\{\mathcal{H}, \mathcal{C}\}_{\text{P.B.}} = 0. \tag{MM.2}$$

Evidently, besides the blocks leaving the boson and fermion sectors invariant, the generic operator $\check{C}$ satisfying the above condition contains a part mixing x and ξ and proportional to anti-commuting parameters. Actually, all the quadratic functions $\mathcal{C}(X)$, which are the integrals of motion for the system of harmonic oscillators of the boson and fermion type, constitute a simple Z_2-graded Lie algebra with respect to the Poisson brackets. Evidently, the foundation of supersymmetry was elaborated by Schwinger, but further development came from another side.

Feynman's functional integral in QED was restricted to electromagnetic (boson) field A_μ. For instance, integrating out A_μ, which appears quadratically in the Lagrangian, one gets an effective (nonlocal) action functional for the electron (fermion) field ψ. The original QED Lagrangian was bilinear also in ψ, so why not start from the integral in ψ? This was the question asked by Isaak Khalatnikov, and he found [10] that it may be done, indeed — the result would be consistent with the canonical operator formalism, provided that the Gaussian integral in the anti-commuting variables ψ results in the determinant in the numerator, and not in the denominator, as in the usual calculus. Khalatnikov presented his work at the seminar steered by Landau, which was attended by Felix Berezin, then a student of mathematics under the supervision of Israel Gelfand. The job performed by Berezin was to make Khalatnikov's conjecture consistent mathematically. The arguments were quite simple. A way to evaluate the usual Gaussian integral

$$I(B) \equiv \int_{\mathsf{R}_n} \exp[-(x, Bx)] d^n x \tag{MM.3}$$

in the n-dimensional Euclidean space R_n with the standard measure $d^n x$ and a positive definite linear operator $B = \tilde{B}$ is to employ its invariance. After any linear change of variables $x = L_y$, where $\det L \neq 0$, one gets the integral of the same type, so

$$I(B) = I(\tilde{L}BL)\det L, \quad \forall L, \tag{MM.4}$$

where the second factor is due to the Jacobian, as $d^n x = d^n y \det L$. Hence, one concludes that

$$I(B) = [\det B]^{-1/2} \times \text{const.} \tag{MM.5}$$

Here the constant is known, of course, but it is practically irrelevant for the functional calculations. Now let us consider the integral in the anti-commuting variables and *define* it in such a way as to get the desired result,

$$J(F) \equiv \int_{\mathsf{G}_n} \exp[-(\xi, F\xi)]d^n\xi = [\det F]^{1/2} \times \text{const}, \tag{MM.6}$$

where $F = -\tilde{F}$ is a linear operator in the anti-commuting space G_n, and the integration measure should be defined properly. Apparently, in order to get the result one has to *assume* that under the linear substitution $\xi = L\eta$ one must have $d^n\xi = d^n\eta/\det L$. Therefore, as soon as for the anti-commuting variables $\prod_{k=1}^n \xi_k = \det L \prod_{k=1}^n \eta_k$, the integral $\int(\prod \xi)d^n\xi$ is an invariant of the linear transformations. Consequently, Berezin *postulated* the basic properties of the the elementary integrals,

$$\int \xi\, d\xi = 1, \quad \int d\xi = 0. \tag{MM.7}$$

These consistent rules opened the way to complete calculus in the anti-commuting variables, since the partial derivatives in ξ were defined straightforwardly, and in applications to quantum field theory. The results appeared in a series of Berezin's papers and in the book [11] published in Russian in 1965, which was translated immediately. A work on the extension of the Lie groups to the groups with commuting and anti-commuting parameters (now called the super-Lie groups) was also published by Berezin and Katz [12]. The ground for supersymmetry was prepared; further progress was prompted by a new concept.

About 1969, the dual reggeon-resonance model of strong interactions, found by Veneziano and developed by a number of people, was formulated in terms of four-dimensional harmonic oscillators. In constructing the Fock space, the problem was to decouple some ghost states of the negative norm, owing to the time-like components of the oscillators. Constraints had to be imposed on the physical states. As was shown by Virasoro, the consistency of the approach resulted from the structure of an infinite-dimensional Lie algebra. After Nambu advanced the idea that the oscillators represent relativistic string, the scheme was reformulated as a field theory in the (τ, σ) space, where the field was the Minkowski coordinates $x_\mu(\tau, \sigma)$ of the points on the world sheet of the string. Respectively, the Virasoro algebra was found to be the Lie algebra of the group of the conformal transformations on the string world sheet. In QED the gauge symmetry is known to be necessary for consistent quantization of the four-vector potential, and therefore the conformal symmetry was also called the *gauge symmetry* in the dual models. In order to include fermions and pion-like negative-parity bosons, Neveu, Schwarz [13] and Ramond [14] introduced fermion oscillators besides the boson oscillators, and the problem of decoupling of the ghost states arose again. Neveu, Schwarz and Thorn [15] showed that the proper Fock space is constructed by means of anti-commuting operators $\hat{G}_r$ satisfying the following algebra, together with the previously known Virasoro operators $\hat{L}_n$:

$$[\hat{G}_r, \hat{G}_s]_+ = -2\hat{L}_{r+s} + a_1 \hat{I}\delta(r+s), \tag{MM.8}$$

$$[\hat{G}_r, \hat{L}_m]_- = (r - m/2)\hat{G}_{r+m}, \tag{MM.9}$$

$$[\hat{L}_m, \hat{L}_n]_- = (m-n)\hat{L}_{n+m} + a_0 \hat{I}\delta(n+m). \tag{MM.10}$$

Here (m, n) are integers, (r, s) are both integers or half-integers, $\hat{I}$ is the unit operator, and $\delta(k) = 1$ for $k = 0$ and 0 for $k \neq 0$. The additional terms with constants a_0 and a_1 (their values are irrelevant now) were discovered later. They are due to quantum anomalies and do not appear in the classical Poisson brackets. In Ref. [15] the operators $\hat{L}_m$ and $\hat{G}_r$ were represented as bilinear combinations of the oscillator creation–annihilation operators, so that Eqs. (8) result from CCR and CAR. In order to account for the fermion degrees of freedom in the framework of string theory, the anti-commuting fields $\psi_\mu(\tau, \sigma)$ were introduced to represent the fermion oscillators. As in the purely bosonic case, the symmetry providing consistent constraints decoupling the ghost states had a geometrical meaning discovered by Gervais and Sakita [16]. In an appropriate (conformal) gauge, it was just an infinite-dimensional group with commuting and anti-commuting parameters acting *linearly* upon x_μ and ψ_μ. Thus, the Z_2-graded Lie algebra (8) was interpreted by Gervais and Sakita as the algebra of generators of the field transformations. This property of the boson and fermion fields on the string world sheet was called by the authors the *supergauge* symmetry, which was probably the first application of the word "super" in this context. It is noteworthy that

Gervais and Sakita referred to the textbook [17] by Jan Rzewuski, where Berezin's functional formalism was described in full detail. Thus, Berezin's ideas influenced the work on supersymmetry.

In two-dimensional space-time the group of conformal (and superconformal) transformations is infinite-dimensional, but it has a very simple realization. It is just the group of diffeomorphisms of the light-cone coordinates. In the four-dimensional Minkowski space the dimensionality of the conformal group is 15, and its geometrical action is more intricate. Given the conformal group, can one define the generators of the superconformal transformations so that their anticommutators would be the generators of the conformal group in four dimensions? This was the question asked by Julius Wess and Bruno Zumino [18], and, miraculously, they found the positive and complete answer. The four-dimensional superconformal transformations were represented linearly on multiplets having equal numbers of boson and fermion fields. It looked like a miracle that the Lie algebra of the generators was closed and included the conformal generators as a subalgebra. Only later, when the formalism of superspace and superfields was developed, was the miracle explained properly. As soon as the transformations were realized linearly in terms of the local fields, the generators were represented as integrals of the corresponding supercurrents over a space-like surface, like the four-momentum, which is the integral of the energy–momentum tensor. Examples of the invariant Lagrangians were also constructed, and it appeared that the key to the solution was the introduction of auxiliary fields, which enter the theory without derivatives and have trivial equations of motion. That was an unusual and a very important feature, which enabled the linear realization of the symmetry. It was soon found that the Poincaré group, as a subgroup of the conformal group, has also a nontrivial superextension. The use of the geometrical ideas enabled the construction of the beautiful theory of supersymmetry and supergravity, which is presented now in a number of books (e.g. by Wess and Bagger [19]).

Yuri Golfand discovered the super-Poincaré algebra in his own way, at least three years before the wonderful work by Wess and Zumino was published, and even before the works [15, 16] on supersymmetry in dual models were done. It is known that Golfand discussed the new symmetry with his colleagues in the late 1960's, trying to solve the puzzle of weak interactions, before the electroweak theory did the job. That is why the parity violation was incorporated as well in the algebra. Though Golfand lived in Moscow and belonged to the same community as Khalatnikov and Berezin, his approach was quite independent and the arguments were more aesthetical than pragmatical. The known basic fields were those representing vector bosons (like photons) and fermion spinors (like electrons). By the standard gauge invariance, the vector fields are intimately related to the four-momentum operators. Are there operators associated with the fermion fields? If such operators exist, they must be spinorial and satisfy anti-commutation relations, naturally involving the momentum operators in the right-hand side. The crucial question was whether one could construct a super-Lie algebra, so that the anticommutators would satisfy the Jacobi identities. This was checked by Golfand and Likhtman, and their first note was published [20]. As soon as the algebra was constructed, the next problem was to realize it in terms of the local fields and to construct an invariant theory. This program was formulated and the first supersymmetric model was constructed [20]. The general scheme of supersymmetric model-building was also proposed, in a recursive way [21], since Golfand and Likhtman had no linear realization for supersymmetry. This was done with auxiliary fields by Wess and Zumino. They started from the local fields and constructed supersymmetric field theory independently. Their work [18] then triggered the subsequent explosive development. Wess and Zumino were inspired by Neveu, Schwarz and Ramond. At the time of their work they knew neither of Golfand nor of any other research on four-dimensional supersymmetry.

In fact, an independent research was carried out in 1972 by D.V. Volkov and V.P. Akulov who posed the question whether Goldstone particles with spin 1/2 could exist. They formulated [22] a nonlinear realization of the super-Poincaré algebra and suggested a Lagrangian for the self-interacting Goldstino field, the Goldstone fermion of (the spontaneously broken) supersymmetry. The idea of presenting neutrino as a Goldstino was then abandoned because of phenomenological difficulties. Thus, unfortunately, this work went relatively unnoticed, producing no impact in the outside world. It was revived only after Wess and Zumino.

Gauging the super-Poincaré group (i.e. making supersymmetry local) leads to supergravity. A description of gravity based on the minimal supersymmetry and including fields of spin 3/2, 1 and 1/2 was suggested as early as in 1973 by Volkov and Soroka [23]. A comprehensive theory of supergravity was worked out by[140] [Freedman *et al.* [24] and by Deser and Zumino [25]].

Almost three decades after its discovery, supersymmetry remains one of the most beautiful theories in physics. Is it just a theoretical option, or

[140]This paragraph was not completed in the manuscript.

a symmetry really existing in Nature? This is probably one of the most important questions in modern particle physics.

Acknowledgments

I am grateful to Misha Shifman for encouragement and helpful advice. Support for the work from the Fund for Promotion of Research at the Technion is acknowledged.

References

1. D. J. Struik, *A Concise History of Mathematics* (Dover, New York, 1948).
2. W.R. Hamilton, *Elements of Quaternions*, 2 vols. (London, 1866); 3rd edition (Chelsea Pub. Co. New York, 1969).
3. *Hermann Günther Graßmann (1809–1877): Visionary Mathematician, Scientist and Neohumanist Scholar*, Ed. G. Schubring (Kluwer, Dordrecht — Boston — London, 1996).
4. H.G. Grassmann, *Die lineale Ausdehnungslehre, ein neuer Zweig der Mathematik* (O. Wigand, Leipzig, 1844); 2nd rev. edition (1878).
5. E. Segré, *Enrico Fermi — Physicist* (University of Chicago Press, Chicago — London, 1970), p. 10.
6. E. Fermi, *Rend. Lincei* **3**, 145 (1926); *Z. Physik* **36**, 478 (1926).
7. P. Jordan and E. Wigner, *Z. Physik* **28**, 631 (1928).
8. J. Schwinger, *Phil. Mag.* **44**, 1171 (1953).
9. J. Schwinger, *Quantum Kinematics and Dynamics* (W. A. Benjamin, New York, 1970).
10. I.M. Khalatnikov, *JETP* **28**, 635 (1954).
11. F.A. Berezin, *Method of Second Quantization* (Academic, NY, 1966).
12. F.A. Berezin and G. I Katz, *Mat. Sbornik* **82**, 343 (1970) [*Math. USSR — Sbornik* **11**, 311 (1970)].
13. A. Neveu and J.H. Schwarz, *Nucl. Phys.* **B31**, 86 (1971).
14. P. Ramond, *Phys. Rev.* **D3**, 2415 (1971).
15. A. Neveu, J.H. Schwarz, and C.B. Thorn, *Phys. Lett.* **35B**, 529 (1971).
16. J.-L. Gervais and B. Sakita, *Nucl. Phys.* **B34**, 632 (1971).
17. J. Rzewuski, *Field Theory II* (Iliffe Books, London, 1969).
18. J. Wess and B. Zumino, *Nucl. Phys.* **B70**, 39 (1974).
19. J. Wess and J. Bagger, *Supersymmetry and Supergravity*, 2nd edition (Princeton University Press, Princeton, 1992).
20. Yu.A. Golfand and E.P. Likhtman, *JETP Letters* **13**, 323 (1971) [Reprinted in *Supersymmetry*, Ed. S. Ferrara (North-Holland/World Scientific, Amsterdam-Singapore, 1987), Vol. 1, p. 7.] See also page 478 in this volume.
21. Yu.A. Golfand and E.P. Likhtman, in I.E. Tamm Memorial Volume, *Problems of Theoretical Physics* (Moscow, Nauka, 1972), p. 37 [Engl. translation in *The Many Faces of the Superworld*, Ed. M. Shifman (World Scientific, Singapore, 2000)].
22. D.V. Volkov and V.P. Akulov, *Phys. Lett.* **B46**, 109 (1973).
23. D.V. Volkov and V. Soroka, *JETP Lett.* **18**, 312 (1973).
24. D.Z. Freedman, P. van Nieuwenhuizen and S. Ferrara, *Phys. Rev.* **D13**, 3214 (1976).
25. S. Deser and B. Zumino, *Phys. Lett.* **62B**, 335 (1976).

COMMENT ON CHRONOLOGY AND NUMEROLOGY OF RESEARCH ON SUPERSYMMETRY — 2023

M. SHIFMAN

Theoretical Physics Institute, University of Minnesota,
Minneapolis, MN 55455
shifman@umn.edu

I will refer to the "Prehistoric Era" all works relevant to supersymmetry which were done before the first publication of Wess and Zumino on this subject.

These papers are:

(1) **Yu.A. Golfand and E.P. Likhtman**, *Extension of the Algebra of Poincaré Group Operators and Violation of P-Invariance*, JETP Lett. **13**, 323–326 (1971). Received by the Editorial Office on March 10, 1971; [Reprinted below on page 478];

(2) **E. Likhtman**, *On Possible Extensions of the Generator algebra of the Poincaré Group by Bispinor Generators*, "*Short Communications in Physics*", vol. 5, p. 33 (1971) (in Russian), see https://ksf.lebedev.ru/contents.php?post=1&year=1971&number=05&z=0;

(3) **Yu.A. Golfand and E.P. Likhtman**, *On the Extensions of the Algebra of the Generators of the Poincaré Group by the Bispinor Generators*, in I. E. Tamm Memorial Volume *Problems of Theoretical Physics*, Eds. V.L. Ginzburg *et al.*, (Nauka, Moscow 1972), pp. 37–44.[141] "Signed (authorized) for print" on March 20, 1972;

(4) **E.P. Likhtman**, *Irreducible Representations of the Extension of the Algebra of the Poincaré Group Generators by the Bispinor Generators*, Report of the Lebedev Physics Institute # 41, 1971, pp. 1–15.[142] "Signed (authorized) for print" on April 12, 1971;

(5) **A. Neveu and J.H. Schwarz,** *Factorizable Dual Model of Pions*, Nucl. Phys. **B31**, 86–112 (1971). Received by the Editorial Office on March 25, 1971;

[141]English translation is published in *The Many Faces of the Superworld*, Ed. M. Shifman (World Scientific, Singapore, 2000), p. 45.

[142]English translation of this preprint is published in *SUSY-30*, Proc. Int. Symposium Celebrating 30 Years of Supersymmetry, Eds. K. Olive, S. Rudas, M. Shifman, Nucl. Phys. B (Proc. Suppl.) 101 (2001), p. 9.

(6) **P. Ramond**, *Dual Theory for Free Fermions*, Phys. Rev. **D3**, 2415–2418 (1971). Received by the Editorial Office on January 4, 1971;

(7) **A. Neveu, J.H. Schwarz, and C.B. Thorn,** *Reformulation of the Dual Pion Model*, Phys. Lett. **35B**, 529–533 (1971). Received by the Editorial Office on May 7, 1971;

(8) **J.-L. Gervais and B. Sakita,** *Field Theory Interpretation of Supergauges in Dual Models*, Nucl. Phys. **B34**, 632–639 (1971). Received by the Editorial Office on August 13, 1971;

(9) **D.V. Volkov and V.P. Akulov,** *Possible Universal Neutrino Interaction*, JETP Lett. **16**, 438–440 (1972). Received by the Editorial Office on October 13, 1972;

(10) **D.V. Volkov and V.P. Akulov,** *Is the Neutrino a Goldstone Particle?* Phys. Lett. **B46**, 109–110 (1973). Received by the Editorial Office on March 5, 1973;

(11) **D.V. Volkov and V.A. Soroka,** *Higgs Effect for Goldstone Particles with Spin 1/2*, JETP Lett. **18**, 312–314 (1973). Received by the Editorial Office on September 3, 1973;

(12) **J. Wess and B. Zumino,** *Supergauge Transformations in Four Dimensions*, Nucl. Phys. **B70**, 39–50 (1974). Received by the Editorial Office on October 5, 1973. The dates quoted for the papers (1)–(4) and (9)–(11) must be considered as upper bounds rather than the actual submission dates. This is due to the fact that all materials intended for publication in the USSR had to be pre-cleared by various censoring offices, including the omnipresent GLAVLIT, the secret governement agency in charge of the final clearance. A latent period of a few months was unavoidable. During this time the paper prepared for publication was officially non-existent.[143] In the works (6)–(8) authored by the string-theory pioneers two-dimensional supersymmetry was introduced on the string world sheet. To this end two-dimensional fermion fields were added on the world sheet. They became building blocks for supersymmetries of various types in string theory.

At this point it is worth quoting John Schwarz's clarification (see his interview on page 73),

> The Ramond-Neveu-Schwarz string theory actually contains local space-time supersymmetry, but we were slow to realize that. The realization came after supersymmetry in four dimensions was thoroughly studied by others.

[143] According to recollections of Likhtman (page 98) and Marinov (page 362) Golfand's superalgebra (8.1) on p. 479 was found in 1969.

The last paper (12) in the above list presents an independent discovery of supersymmetric field theory in four dimensions. It opens the "Historic Era" of supersymmetry.

The subsequent explosive proliferation of publications devoted to supersymmetry is presented in Figs. 6.1–6.5. The growth continued until 1987 and then, after a brief decline, resumed in 1993 reaching the absolute peak around 1998. For the subsequent 14 years (until 2012) the rate of

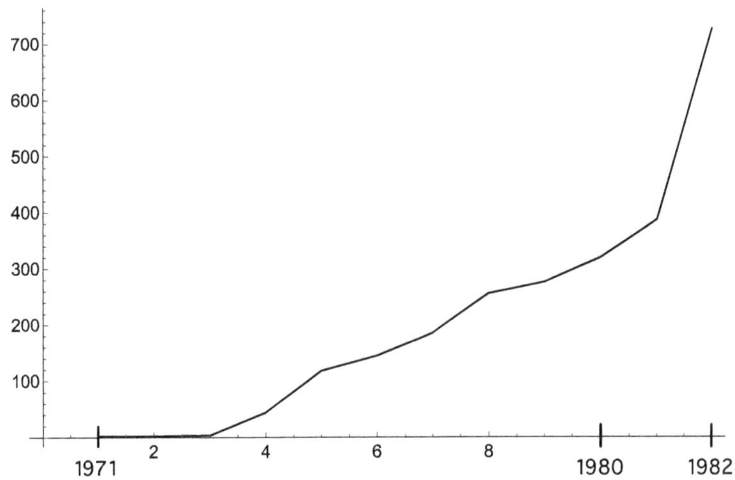

Figure 6.1: The number of papers on supersymmetry published between 1971 and 1982. The beginnings of the accelerated growth in 1973 and even more accelerated growth in 1982 are visible.

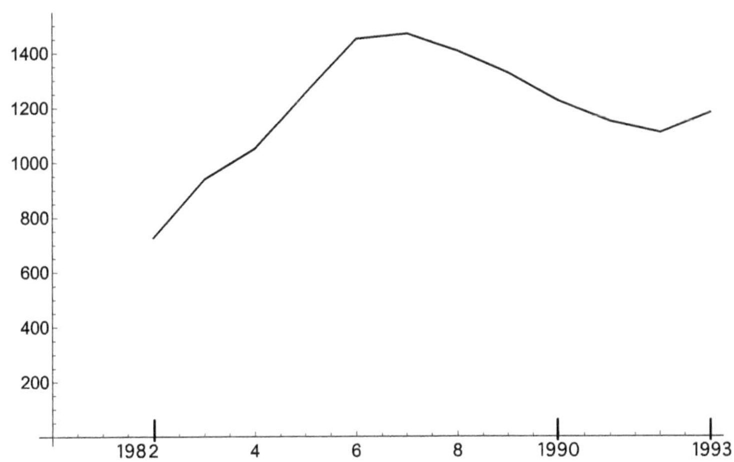

Figure 6.2: The number of papers on supersymmetry published between 1982 and 1993. Note the maximum at the level of 1400 papers in 1987 and a local minimum in 1992–1993.

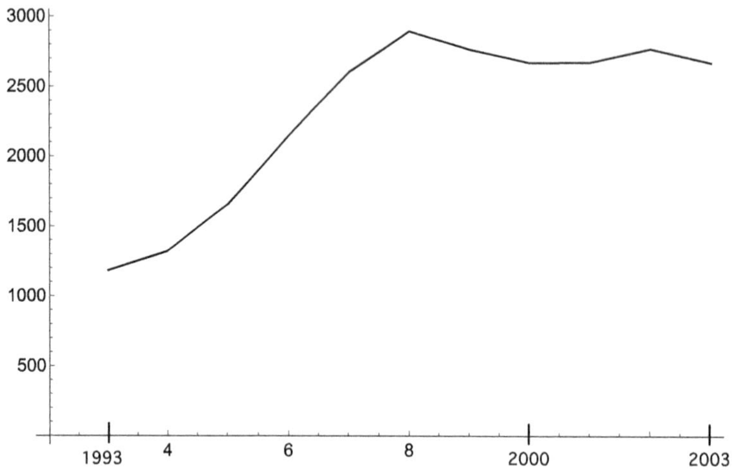

Figure 6.3: The number of papers on supersymmetry published in 1993–2003. The absolute maximum at the level of 2900 papers per year is achieved in 1998–1999.

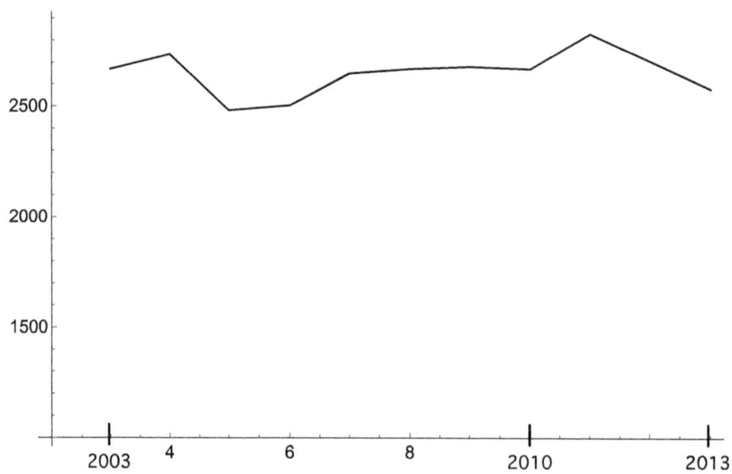

Figure 6.4: The number of papers on supersymmetry published in 2003–2013.

supersymmetry-related papers stabilized at roughly 2700 papers per year. Since 2012 we observe a continuous decline which exhibits a tendency of acceleration in the last few years. The current rate is around 40% of that at the maximum.

Some features of the plots in Figs. 6.1–6.5 deserve comments. The steep rise starting in 1982 is apparently correlated with the appearance of two papers written by E. Witten [1], which caused a strong resonance in the community. In 1987 we see the beginning of a moderate fall which lasted

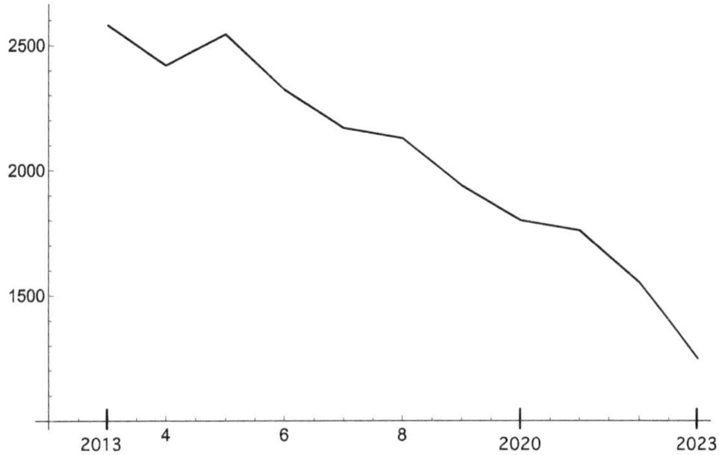

Figure 6.5: The number of papers on supersymmetry published in 2013–2023.

until 1992. Presumably, it can be explained by the release of the famous work by work of P. Candelas *et al.* [2] which diverted an active part of the community to string theory. The papers resulting from this rearrangement of interests were classified under the topic "strings" in the iNSPIRE-hep data base.

The downturn tendency was overcome by the end of 1992. A quite rapid growth rapid and continuous growth in the publication rate becomes evident in 1994, after the breakthrough work of N. Seiberg and E. Witten [3], which appeared in July of 1994. For four years fascinating features of supersymmetric gauge dynamics at strong coupling [3] captured the imagination of many theorists who seemed to had left the field. The peak in 1998 is the absolute maximum in the number of publications (at the level of 2900 publications/year). In 1998/99 we observe a slight decline with the subsequent stabilization at the level of 2700. An approximate plateau lasted up to 2011. Then, a continuous decline started (Fig. 6.5), with a tendency to acceleration in the last few yeas. In my mind, the reason lies in the negative results in the searches for superpartners at LHC. For decades young theorists were investing their efforts to the fullest into phenomenology-based SUSY, improving MSSM and adding extra features which gave birth to NMSSM, NNMSSM and so on. Now, being deprived of any hints from nature, they got disheartened.

One can say that the phenomenological branch of SUSY has almost dried out at the time being (See Fig. 1, p. vii). I do not think this statement to be an exaggeration. We will have to wait for the experimental discovery of supersymmetry (or perhaps some clear-cut hints from nature) to revive the phenomenological studies of SUSY. The other branch in Fig. 1 — SUSY at strong coupling — is alive and growing.

The iNSPIRE-hep date base tells us that the overall number of publication on supersymmetry during the 52 years that elapsed since its inception is 58,600. Supergravity adds 16,600.

References

1. E. Witten, *Dynamical Breaking of Supersymmetry*, Nucl. Phys. **B185**, 513–554 (1981); *Constraints on Supersymmetry Breaking*, Nucl. Phys. **B202**, 253–316 (1982).
2. P. Candelas, G. T. Horowitz, A. Strominger and E. Witten, *Vacuum Configurations for Superstrings,* Nucl. Phys. B **258**, 46–74 (1985).
3. N. Seiberg and E. Witten, *Electric-magnetic duality, monopole condensation, and confinement in N=2 supersymmetric Yang-Mills theory,* Nucl. Phys. B **426**, 19–52 (1994) [Erratum: Nucl. Phys. B **430**, 485–486 (1994)].

SUPERGRAVITY WAS DISCOVERED BY D.V. VOLKOV AND V.A. SOROKA IN 1973[144]

STEVEN DUPLIJ

Center for Information Technology
University of Münster, Münster, Germany
Email: douplii@uni-muenster.de

Abstract: Supergravity is a remarkable theory combining supersymmetry and general relativity. While the theory has a number of creators from across the globe, we wish to address the question of the origins of this fantastic idea and relevant chronology.

The idea of supergravity — SUGRA in short — (without mentioning this name per se) was first formulated on December 5, 1972, in the last paragraph of the paper by Prof. Dmitry V. Volkov (47 years old at that time), and his PhD student Vladimir P. Akulov (28 years old at that time) entitled *Possible universal neutrino interaction*" [1] followed by the paper *"Is the neutrino a Goldstone particle?"* received by the PLB office on March 5, 1973 [2]. I quote verbatim

> ...[T]he gravitational interaction may be included by means of introduction the gauge fields for the Poincaré group. [...] if the gauge field for the transformation (3) is also introduced; then, as a result of the Higgs effect the massive gauge field with spin 3/2 appears — the Goldstone spin-$\frac{1}{2}$ particle combines with it and disappears.

The outlined program was materialized in a concrete model in 1973 by Volkov and his PhD student Vyacheslav A. Soroka (29 years old at that time) and reported in the paper *"Higgs effect for Goldstone particles with spin 1/2"* [3] (Received by the JETP Lett. Editorial Office on October 20, 1973. JETP Letters is the English translation of the Russian journal Pisma ZhETF, published with a one-month delay or so.)

The above date can be viewed as a day of the discovery of nonlinearly realized supergravity in a physical model. Unfortunately, this publication went largely unknown to young researchers on the other side of the Iron Curtain.

---

[144]East Eur. J. Phys. 3. 81–82 (2019), https://arxiv.org/pdf/1910.03259, with abbreviations.

The development and thorough studies of SUGRA in the West — at first, pure gravity without matter, with the subsequent inclusion of matter — were carried out three years later in 1976. Only then a new "magic" word "supergravity" which was absent in the Volkov-Soroka paper was coined [4,5]. The Freedman-van Nieuwenhuizen-Ferrara (FNF) paper [4] was titled *"Progress toward a theory of supergravity"* [4] and it cites the Volkov-Soroka article [3].

In the same month (June, 1976), the Deser-Zumino (DS) paper *"Consistent supergravity"* was published [5], which cited the FNF preprint and also Akulov, Volkov, and Soroka's paper of 1975 *"Gauge fields on Superspaces with Different Holonomy Groups"* [6].

The history of supergravity is given in the *SUSY story* (narrated by its founders) ([7], pp. 1–28) and in the article *Supergravity* [8]. The connection between the Volkov-Soroka and FNF approaches was explained in a clear-cut manner several times, e.g. (i) by D.V. Volkov in [9,10] and (ii) by V.A. Soroka in *The Sources of Supergravity* [11,12] or in *The Starting point of supergravity,* [13].

I feel sorry that in most of the mass media articles about prizes given for the "discovery of supergravity" no mention is made of the 1973 discoverers of supergravity, Dmitriy Vasilievich Volkov and Vyacheslav Soroka. I refer the reader to their Memorial Page [14].

References

1. D.V Volkov and V.P. Akulov, *Possible Universal Neutrino Interaction*, JETP Letters, 16(11), 621–624 (1972), in: http://www.jetpletters.ac.ru/ps/1766/article_26864.shtml.
2. D.V. Volkov and V.P. Akulov, *Is The Neutrino A Goldstone Particle?* Physics Letters **B**, 46(1), 109–110 (1973).
3. D.V. Volkov and V.A. Soroka, *Higgs effect for Goldstone particles with spin 1/2*, JETP Lett. **18**, 312–314), in: http://jetpletters.ru/ps/1568/article_24038.pdf. Page 489 below.
4. D.Z. Freedman, P. van Nieuwenhuizen, and S. Ferrara, Progress toward a theory of supergravity, Phys. Rev. D, 13, 3214–3218 (1976).
5. S. Deser and B. Zumino, Consistent supergravity, Physics Letters B, 62(3), 335–337 (1976).
6. V. Akulov, D.V. Volkov, and V.A. Soroka, *Gauge fields on Superspaces with Different Holonomy Groups*, JETP Letters, 22, 187–188 (1975), http://jetpletters.ru/ps/1526/article_23351.pdf.
7. E. Likhtman, D. Volkov, V. Akulov, H. Miyazawa, G. Stavraki, V. Kac, V. Pakhomov, J. Lopuszanski, R. Haag, D. Leites, *SUSY Story (narrated by its founders)*, in: S. Duplij, W. Siegel, J. Bagger, editors, *Concise Encyclopedia Of Supersymmetry and Noncommutative Structures in Mathematics and Physics*, (Springer, Berlin, New York, 2005), *Concise Encyclopedia* in what follows, see https://www.uni-muenster.de/IT.StepanDouplii/old/susy/SUSYEnc_Story.pdf.

8. S. Duplij, *Supergravity*, in *Concise Encyclopedia* [7], page 420.
9. D.V. Volkov, *Supergravity Before and After 1976*, NATO Sci. Ser. B **352**, 663–675 (1996) [arXiv:hep-th/9404153 [hep-th]], 1994.
10. D.V. Volkov, in *Consise Encyclopedia*, pp. 6–9.
11. V.A. Soroka, The Sources of Supergravity, in: G. Kane and M. Shifman, editors, The Supersymmetric World. The Beginnings of the Theory, 1st Edition, (World Scientific, 2000), pp. 88–92. See also pp. 171–174 and 489–492 in the present second edition.
12. V.A. Soroka, The Sources of Supergravity, (2002), in: https://arxiv.org/abs/hep-th/0203171.
13. V. A. Soroka, *Starting point of supergravity,* [arXiv:hep-th/0111271 [hep-th]], presented at *Supergravity at 25*, December 1–2, 2001. Stony Brook, NY, United States.
14. D.V. Volkov, Memorial page, https://www.uni-muenster.de/IT.StepanDouplii/old/volkov/

    V.A. Soroka Memorial page is at https://www.uni-muenster.de/IT.StepanDouplii/old/soroka/.

NOTES ON THE CONCEPTUAL DEVELOPMENT OF SUPERSYMMETRY[145]

ROSANNE DI STEFANO
Center for Astrophysics — Harvard & Smithsonian
Cambridge, MA 02138
rdistefano@cfa.harvard.edu

A short history of the conceptual development that led to the notion of supersymmetry is presented, and some of the developments that arose from the growing understanding of supersymmetry are sketched. Particular emphasis is placed on the role that both ideas about symmetry and symmetry breaking (this latter particularly in the Soviet Union), and ideas born of hadronic string theories (particularly in the West) played in pointing the way to supersymmetry. The related role that supersymmetry later played in reviving interest in string theories is also discussed. Although this brief account is necessarily somewhat narrowly focused, an attempt is made to view these developments from a long-term perspective. A study of the history of supersymmetry provides an ideal opportunity to obtain such a view, because supersymmetry has been at the center of intense interest during a time when both the theoretical and experimental study of high energy physics have changed in significant ways.

In 1987 a conference on the history of gauge field theories was held in Logan, Utah. The paper that follows is the paper I wrote as a contribution to the conference proceedings. The reason I agreed to give the talk, and then spent a good deal of time writing the paper, was that the seeds of interest in the history of science had been planted while I was a student at the Institute for Theoretical Physics at Stony Brook during the years after supergravity was invented. We were all excited about supersymmetry and supergravity in those days, and students had a chance to attend talks by just about every major figure in the West. In addition, because so many of these researchers made extended visits, we had a chance to talk with them,

[145]This paper was prepared in 1987 for a conference proceedings which have never been published. The paper circulated as SUNY (Stony Brook) preprint ITP-SB-8878.

sometimes at length. Thus, we had a wonderful opportunity to witness the flow of work in supersymmetry, the rise of some ideas and the fall of others. This is the story I have tried to tell in the paper. I focused on the years up to 1985 simply because I did not have more time to devote to the project.

I benefited tremendously from contact with some historians of science and technology while I wrote the paper and afterward. Sam Schweber was a mentor to me, Spencer Weart was a source of support and encouragement. I learned a good deal from faculty and students at MIT, where I was conducting my own research in astrophysics, and where I also decided to expand my horizons by learning more about how to study the history of science. Perhaps the most precious resources came from the supersymmetry researchers themselves, many of whom read various versions of the paper for me, and made comments. The fact that this rather tough and demanding group of people reacted in a way that was almost uniformly positive reinforced my impression that it had been a good idea to rely solely on the published literature for my historical paper. The interesting conversations we had made me feel, however, that it would also be useful to future historians of science to capture an oral history of these conceptual developments. My goal was to write a book that would combine a systematic oral history with the bare facts of the physics as written in the literature and summarized here. Many of the key contributors to the conceptual developments have already contributed to this larger project by being interviewed. The book is "in progress," although in an arrested state of development, because the demands of my own research have not left enough time to pursue much work on the history of supersymmetry, supergravity, and superstrings.

Since the proceedings of the Logan conference were apparently never published, I was pleased to hear from Gordy and Misha that they wanted to include my paper in this volume. I had not thought that anyone would remember this old work that had circulated as an ITP preprint some 12 years ago. As I read the other contributions I once again feel of the fire and excitement in this field and become aware of the need for an even more systematic and comprehensive history.

Rosanne Di Stefano
September 25, 2000

Contents

Introduction

Supersymmetry is a symmetry which relates fermions to bosons. In the years since its discovery it has proved to be a powerful symmetry which has provided:

- internal symmetries which allow particles of different spin to sit in the same irreducible multiplet;
- quantum theories with "miraculous" ultraviolet cancelations;
- theories of supergravity with convergence properties that are better than those of standard general relativity;
- candidates for theories in which all of the known forces are unified;
- consistently coupled spin-$\frac{3}{2}$fields;
- insights into the positive energy theorems of general relativity;
- improved string theories;
- a possible connection between gauge and Higgs particles;
- a possible solution to the hierarchy problem.

This partial list of its successes makes it easy to understand why supersymmetry has played such an important role in theoretical physics during the last fifteen or so years. Figure 6.6,[146] in which the number of published papers classified under the heading of supersymmetry is plotted as a function of time, serves to illustrate the great deal of attention that has been focused on this area. An examination of the papers themselves reveals that the majority of them have been written by theorists. The enthusiasm of experimentalists has been more restrained. This is undoubtedly related to the fact that, as far as I know, the only data available to date on the existence of supersymmetry is data of the type plotted in Fig. 6.6.

In other words, one of the features that has characterized the development of supersymmetry so far has been the absence of any experimental evidence of its existence. The lack of experimental verification is not due to predictions of supersymmetry which have failed, but rather due to the fact that the crucial confrontations between supersymmetry and experiment have

[146]The data plotted in figures one through three are the number of papers listed under these categories in the subject index of *Physics Abstracts*. Each number represents the sum of the January–June and July–December figures. (These numbers have been estimated.) In July–December 1986 there were no entries under the heading of *string theory*, but the listings under *duality and dual models* were dominated by papers on superstrings and related topics. In January–June 1987 there were approximately 300 papers listed under the heading of *string theory*, and *duality and dual models* was still dominated by superstring related topics.

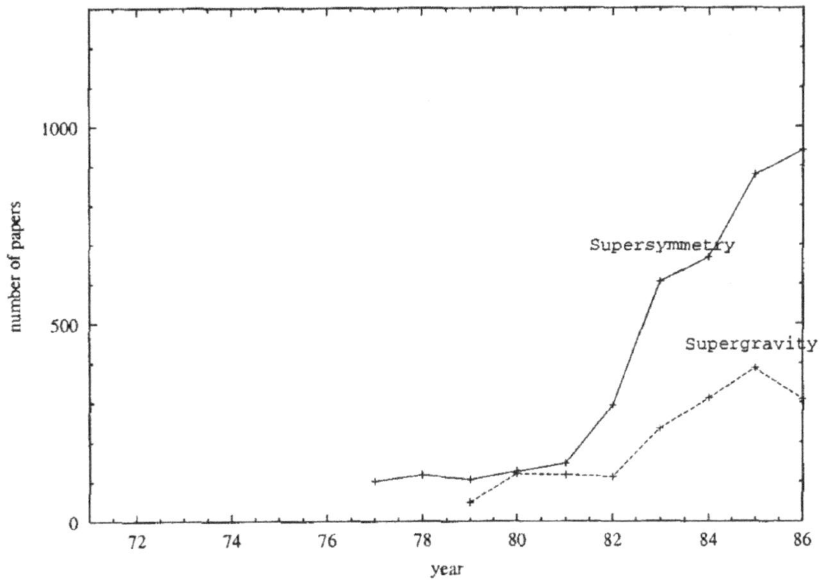

Figure 6.6: Supersymmetry and Supergravity.

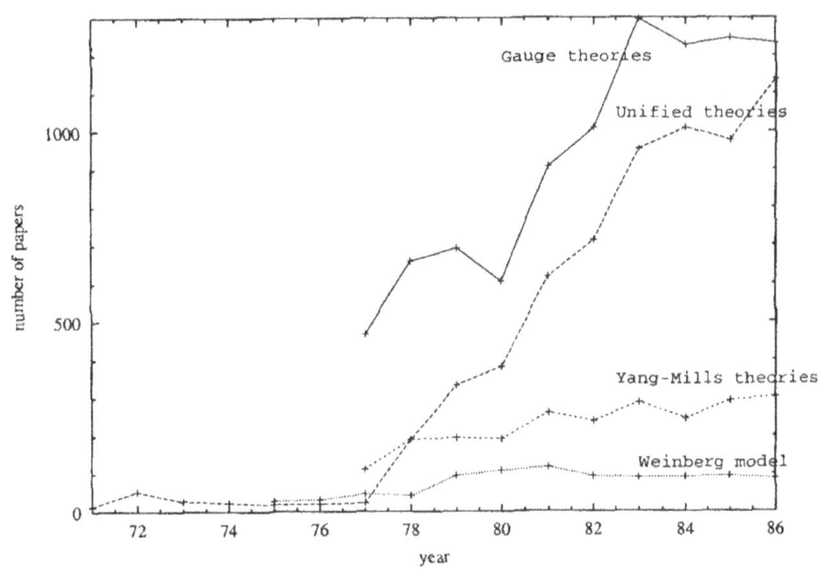

Figure 6.7: Field Theories.

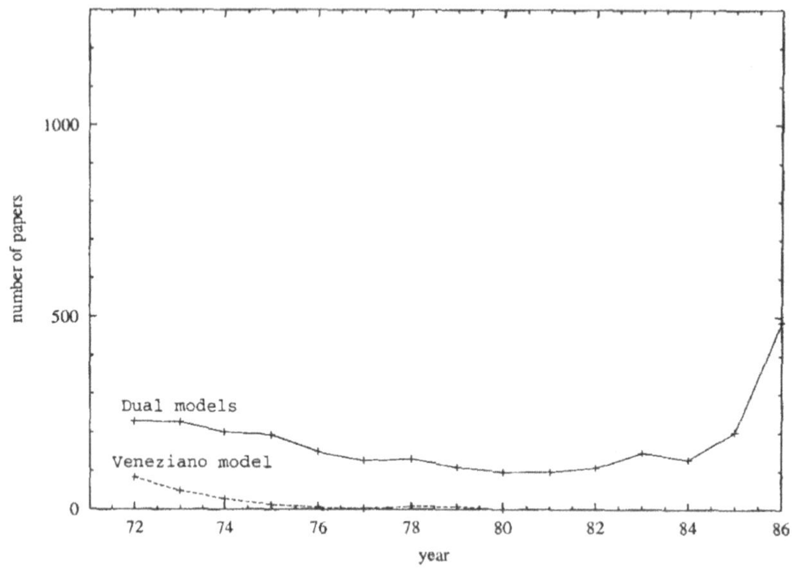

Figure 6.8: String Theories.

yet to take place. Nevertheless, this situation does put supersymmetry, at this point of time, on a different footing from many of the other symmetries which have been discussed at this conference.

To some extent, the idea of the possible existence of these other symmetries was inspired by physical observations. For example, the similarity between the proton and the neutron when electromagnetic effects can be ignored inspired the development of non-Abelian gauge field theories. Or, the observed equivalence of inertial and gravitational mass was one of the elements that led Einstein to construct a theory invariant under general coordinate transformations. Theories with such invariances led to new predictions that were then verified by further experiments. In contrast, the history of supersymmetry thus far is the history of an idea.

One might imagine that the history of an idea would be rather tame. Nevertheless the story of supersymmetry contains all of the elements of high drama: an existence which defies all expectations, a double discovery, the birth of a new idea from an old and dying one which is then rejuvenated through the success of its offspring. The telling of this story will also provide an opportunity to explore what factors influence the generally perceived importance of an area of research, and to compare the research environments within which physicists have worked at different times during this century.

Outline

The sheer volume and range of work that has been done on supersymmetry makes it impossible to include a comprehensive list of significant developments in a presentation of this length. Even if such a list, with a few descriptive remarks and the relevant attributions and dates, could be tailored to fit in this limited space, what purpose would it serve? Redundant to the specialist, not particularly enlightening to those uninitiated in its various subspecialities, it would not even begin to do justice to the dynamic nature of its subject. (Although the dynamic nature of the supersymmetry researchers themselves could not fail to be noticed by anyone presented with such a list.)

On the other hand, if one were to attempt only an overview, there would be the danger of obscuring the details which make the field so rich. This presentation is an attempt to strike a balance. An overview will place supersymmetry within the larger tapestry of ideas that have been important in the evolution of theoretical physics over the past 65 years. On this tapestry, certain threads that are central to the overall pattern will be brought into sharper focus. Since some of the most interesting features of the pattern began to be studied in the earliest work on supersymmetry, some of this work will be discussed in detail. We will then follow a few of these individual threads, although in somewhat less detail, through subsequent developments. Because only a few threads can be studied in a review of this size, the reader who is not already familiar with the literature should be warned that many topics are not discussed here at all. For example superparticles and supersymmetric nonlinear σ-models have not been included, even though both of these areas have provided fertile ground for research. However, references to these, and to most of the other areas that are not touched upon here, can be found in the sources mentioned at the end of this section.

Section 1, on pre-superhistory, and Section 9, which briefly addresses the question of the future of supersymmetry, form a sort of historical frame which places supersymmetry within the broad context of high energy physics.

Section 2, on the dawn of supersymmetry, discusses the early work on supersymmetry which was done in the context of string theories of the strong interactions. The focus will be on the work of Ramond, and of Neveu and Schwarz, on dual models with graded algebras, and also on the discovery by Gervais and Sakita of the world-sheet supersymmetry of the Ramond-Neveu-Schwarz theory. This section, together with Section 8, which is a short section on the early work on superstrings, form a somewhat narrower frame. This more narrow frame will emphasize how the idea of supersymmetry was formulated in the context of dual string models of the hadrons, and how later, the study of supersymmetry led to consistent string theories as candidates for "theories of everything."

In the West, supersymmetry first started to be studied in 1971, in work on hadronic string theories. At that time it was hoped that such models might provide a good and useful description of the hadrons. The hadronic string theories were soon eclipsed by quantum chromodynamics. However the idea of supersymmetry lived on, largely because it was applied to local field theories in four dimensions by Wess and Zumino. The work of Wess and Zumino inspired a great deal of further work on supersymmetric field theories. Actually, supersymmetric field theories had already been constructed, and some of their properties had been studied, in the Soviet Union. Golfand and Likhtman, and Volkov, Akulov and Soroka had studied supersymmetric local field theories, but their work does not seem to have attracted much attention in the West. Thus, it was not until the independent work of Wess and Zumino, that supersymmetry came to be widely studied.

The question of what makes the time right for the study of any particular type of theory is an interesting one. Because of the way in which its development has been linked to other developments in string and field theory, and because it was discovered twice — once creating only a few ripples, and once creating a veritable tidal wave — supersymmetry provides a unique opportunity to explore possible answers to this question. Section 3, on the changing fortunes of string theories and local field theories, is an attempt to understand what made the time so right for the study of supersymmetric field theories in 1974.

Section 4 is on the dual discovery of supersymmetric field theories. Here the early papers by Golfand and Likhtman, by Volkov, Akulov and Soroka, and by Wess and Zumino are studied. The major themes that began to be explored after the work of Wess and Zumino are discussed. Of these, the unfolding understanding of the unique quantum properties of supersymmetric theories is studied in more detail in Section 5. Several sources are now available which provide detailed chronologies of the results which have been achieved. (See for example Ref. [132] and the contributions of Stelle and of West to Ref. [15].) The goal of Section 5 is to highlight the evolution of some of the important ideas and techniques.

One of the most important developments in the history of supersymmetry was the development of supergravity. The possibility that gravitation could be naturally formulated in the context of supersymmetry had been noticed, both by Volkov and Akulov, and by Wess and Zumino, in their earliest work on supersymmetry. Since the requirement of supersymmetry would naturally give the graviton at least one companion (either a spin-$\frac{3}{2}$ or a spin-$\frac{5}{2}$ field) and possibly other matter couplings as well, it was clear that supergravity might provide a natural path toward the unification of all of the know forces. As the cancelation of divergences which occurs in supersymmetric theories began to be studied, it appeared that supersymmetry might be able to overcome a barrier that had prevented the construction of a unified quantum theory — the barrier of the non-renormalizability of gravitation. It was hoped that supergravity theories (or at least one such theory) might be finite. This double possibility of unification and finiteness generated a great deal of excitement, and a great deal of creative work. The development of supergravity is outlined in Section 6.

One of the ideas that has played an important role in the progress of supergravity has been the Kaluza-Klein idea of achieving unification in higher dimensions. Of course, the notion of studying physics in higher dimensions had also of necessity been considered in the dual models, and we will discuss this connection as well. Section 7 is devoted to the study of supersymmetry in higher dimensions.

A Note on 'Conceptual Development'

It is difficult to define the notion of the 'conceptual development' of an area of research in a precise way. One is interested in knowing when key ideas were first developed and introduced, and also in understanding when they gained acceptance and became an established part of the foundation of the subject.

A survey of this kind cannot hope to record exactly when a particular idea might have occurred to, or might have been suggested by, a particular individual. Nor can it identify a point in time at which it became generally accepted that a particular approach might be important to the further evolvement of the subject. What it can do instead is to follow certain trends and broad ideas in the development of the subject, from the rough time that papers began to be published on these ideas. This survey is therefore only a step towards a more complete history of the subject. Because there has not been time to interview a broad enough cross section of the people who have participated in these developments, the published literature alone (in particular the dates on which manuscripts have been received by a journal) has been used to date the introduction of new ideas. Since much of the true infusion of these ideas occurred through lectures, discussions, and the distribution of preprints, the published record provides at best a first approximation to the timing of this process, but at least it is a well-defined approximation. To give a sense of the general perception of the progress of the field, these notes have relied on statements that were published at the time (for example in review articles) and also on statistics drawn from *Physics Abstracts* and *Current Contents.*

Because it was necessary to limit the space devoted to the list of references, this list surveys only a small portion of the relevant literature. Fortunately, many reviews, collections of reviews, and collections of pivotal papers are available. These would be useful to the reader who wants to go beyond the presentation sketched here. A collection of *Physics Reports* edited by Jacob [81] surveys the work on hadronic strings, while a collection of papers edited by Schwarz [115] deals with the first fifteen years of work on superstrings. Another useful source of information on string theories is the text by Green, Schwarz and Witten [69], which also includes a comprehensive bibliography. Ferrara has recently edited a collection of papers on supersymmetry and supergravity [47], while a collection of *Physics Reports* on supersymmetry has been edited by Jacob [82]. Texts include contributions by Wess and Bagger [131], Gates, Grisaru, Roček, and Siegel [57] and West [132]. A commentary on the history of Kaluza-Klein theories is included in the collection of papers on that subject which has been put together by Appelquist, Chodos and Freund [6]. Finally, the book by Abraham Pais, *Inward Bound* [101], is an excellent source of information on the pre-supersymmetry background that is touched upon in Sec. 1.

1 Pre-Superhistory

1.1 *Spin and Statistics*

The idea of a fermi-bose symmetry presupposes a clear knowledge of quantum spin and of the relation between spin and statistics. These fundamental notions were developed during the 1920's. The research environment of the physicists who invented these ideas was quite different from that of the pioneers of supersymmetry. The differences are clear to physicists who work in today's environment, and who have studied something of the history of quantum theory. The contrast is highlighted by some of the facts and remarks which have been gathered together by Pais in his book. The quotations mentioned in this section, as well as more details about developments that occurred before 1970 or so, can be found there.

One of the most important differences is the relation between experiment and theory. In the early days of quantum theory, experimental results forced the development of theoretical constructs. Commenting on the history of the Zeeman effect during the twenty fifth anniversary of its discovery, Lorentz said in 1921 that "theory could not keep pace with experiment." Indeed it was his study of the anomalous Zeeman effect which led Pauli to his 1925 formulation of the exclusion principle. When discussing his and Uhlenbeck's discovery of spin, Goudsmit remarked that the pair had been "saturated with a thorough knowledge of the structure of atomic spectra."

These comments serve to emphasize the mutual feedback and interplay between theory and experiment. There were unexplained experimental results which required the development of new ideas. And the consequences of those ideas, and further conjectures based on them, were then amenable to quick experimental tests.

This meant that the theories of those days were vulnerable to the discomfort of being proven wrong. Pais has compiled a list of conjectures which include: Einstein's suggestion in January of 1925 to apply Bose-Einstein statistics to electrons and to helium; Fermi's use, in March of 1926, of helium to illustrate the Fermi-Dirac equation of state; Dirac's conjecture of August 1926 that "the solution with antisymmetrical eigenfunctions... is probably the correct one for electrons in an atom and one would expect molecules to resemble electrons more closely than light quanta." Before the end of the decade, analysis of data on the specific heats of molecular hydrogen, and Wigner's proof that a gas of systems, each containing N fermions, obeys Fermi-Dirac statistics if N is odd, and Bose-Einstein statistics if N is even, showed that at least some of these conjectures were wrong.

On the other hand, access to experimental results gave theorists the ability to correct course quickly. This meant that they could afford to be wrong, without necessarily devoting a significant portion of their careers to work which was not immediately relevant to forming an understanding of the physical world. Furthermore, even if the results contradicted their guesses or predictions, they provided valuable guidance for the direction of further work.

The relation between physics and mathematics was also simpler in the 1920's. Although the analysis of quantum systems introduced many physicists to mathematics that was new to them, at least some of this mathematics was relatively accessible and was related to concepts that had already been used in the study of other physical systems. For example when, in 1926, Wigner introduced group theory to quantum theory during the course of his study of a system of N identical particles, he was using mathematics that he had worked with before. He had already used classical space groups while working on a chemical engineering problem on the lattice structure of rhombic sulfur.

1.2 *The Role of Symmetry in High Energy Physics*

From the early days of quantum mechanics and the work of Wigner, group theory played an important role in the study of atoms. Internal symmetry groups started to become important in the study of nuclei from the early 1930's when Heisenberg introduced the concept of isospin. (A brief history and references can be found in Ref. [135].) In 1937 Wigner introduced SU(4) symmetry as a way to extend the SU(2) symmetry group associated with isospin. He wrote: "To the extent that the strong forces inside the nucleus depend on neither orientation of spin nor isospin, they obey a stronger symmetry, SU(4)."

As the domain of experimental high energy physics expanded to include ever higher energies, more particles were discovered. With the discovery of strange particles, the search for a hadronic symmetry which could embrace more than isospin, became more urgent. By 1964, when the discovery of the Ω^- confirmed a prediction of Gell-Mann which had been based on SU(3), SU(3) was widely recognized to be such a symmetry. At this point SU(6) was introduced into high energy physics, to play the same role for SU(3) that Wigner's SU(4) had played for isospin. This work attracted a great deal of interest, and was successful in describing the hadronic spectrum. It also helped lead to the development of the no-go theorems. These will be discussed in the next subsection.

The symmetries discussed above were understood as global symmetries. In 1954 Yang and Mills suggested that the notion of symmetry (specifically the isospin symmetry) be expanded to include internal symmetry transformations which can be made independently at each space-time point [135]. The non-Abelian gauge theories that were born of their idea have been central to the role that symmetry has played in high energy physics. These Yang–Mills theories have been discussed in more detail at this conference; we will come back to them here in the section on field theory and, of course, when discussing supergravity.

1.3 *No-go Theorems*

The work on SU(6) inspired an interesting line of questioning: to what extent was it possible to combine internal and space-time symmetries in a non-trivial way? For example, could different particles in the same multiplet have different mass, spin, and internal quantum numbers? The consideration that seems to have been most cited during the mid 1960's had to do with the mass, and was motivated by the observed mass differences between particles in the same multiplet of a symmetry group I, like the hadronic SU(3). Were these mass differences necessarily due to symmetry breaking? Or was it possible that there was a larger group G, with respect to which the world was exactly symmetric, and which could encompass both I and the Poincaré group, P, in a non-trivial way? If so, then the generators associated with I and P would not necessarily commute, $[I, P] \neq 0$, and masses of particles within the multiplets associated with I could therefore be different.

In 1964 W.D. McGlinn studied this question for a special case in which I is a semisimple Lie group [89]. McGlinn showed that some general assumptions (specifically that the structure constants are antisymmetric, that the Jacobi identity holds, and that $[I, L] = 0$, where L represents the generators of the Lorentz group) are sufficient to ensure that $[I, P] = 0$. This was the first of a series of so-called no-go theorems. (See Refs. [16], [98–100] and references therein.) The last and most powerful of these was a theorem that was presented by Coleman and Mandula in 1967 [17]. In a paper titled '*All possible symmetries of the S-matrix*', they were able to extend the no-go result to the case in which the internal symmetry group is an infinite parameter group. In order to delineate those conditions under which a non-trivial combination of space-time and internal symmetries is impossible, they made a set of well-defined assumptions about the one-particle spectrum. These included the assumptions that there are a finite number of particle types with mass less than any given finite mass, that the elastic scattering amplitudes are analytic functions, and that there is scattering.

There is no evidence that these no-go theorems prevented the exploration of useful avenues of research. In fact they played a positive role in that they gave a set of well defined situations in which the known symmetry groups could not be extended in the desired way.

Eventually these theorems were circumvented by considering systems with symmetry generators which participate in a (graded) algebra which includes anticommutators as well as commutators. An interesting side note is that in 1971 Lopuszański did consider the possibility of including fermionic generators among the symmetry generators of a theory [88]. He was working in the context of axiomatic field theory and his work seemed to indicate that such generators could not be included in a consistent way. In 1975, when supersymmetric theories had begun to be studied, it was Lopuszański who, with Haag and Sohnius, worked to determine the most general algebraic structures which include fermionic generators. The paper in which they presented their results was titled *All possible generators of supersymmetries of the S-matrix* [76].

2 The Dawn of Supersymmetry

In the late 1960's, data about the hadronic spectrum suggested a theoretical hypothesis about hadronic scattering amplitudes which was referred to as duality. In 1968, Veneziano was able to construct a scattering amplitude which satisfied the duality hypothesis. This inspired a great deal of further theoretical work on dual models. In 1970 it was discovered that the dual models could be viewed as models of hadronic strings. These hadronic string theories were found to have a number of interesting features. Some of these features are so attractive and compelling that they are even now drawing a new generation of physicists into the study of string theories as possible 'theories of everything'. There were also some problems. Some of these have since been worked out. Others are not problems when the arena for the string theories is on the order of the Planck length, rather than on the order of the size of a nucleon. One of the problems with the early theories was the fact that they could describe only bosons. Other problems included the presence of tachyons in the spectrum and a consistent formulation in only 26 space-time dimensions.

It was in the study of hadronic string theories that algebras which include both anticommutators and commutators were first used as symmetry algebras for high energy physical systems. (Graded algebras had already been studied by mathematicians. See Ref. [11] and references therein.) Such algebras were used by Ramond in an attempt to extend the dual theory formalism in such a way as to include fermions. They played an important role in the further development of string theories. Theories associated with such algebras proved capable of describing both fermionic and bosonic states. Such theories were found to be consistent in 10 space-time dimensions which, if not a particularly popular number for working physicists during that period, at least seemed preferable to the 26 dimensions necessary for such string theories to have a space-time supersymmetry and to be free of tachyons. (Here the word space-time refers to the full space-time arena in which the physics is taking place, as opposed to the two-dimensional world-sheet that is swept out by the string.)

The first paper to derive a graded algebra for dual models was written by Pierre Ramond. It was almost immediately followed by important work by Neveu and Schwarz.

2.1 *Graded Algebras*

In 1971 Pierre Ramond extended the reach of dual models by inventing a way to incorporate fermionic states [103]. (Actually the work must have been done in 1970, for the paper was received by *Physical Review* on the 4th of January, 1971 and was published in the May 15th issue.)

His approach is remarkable, both for its ingenuity and for its simplicity. Ramond described the known bosonic states in terms of an internal time coordinate τ. He noted that this description had already been used by others. However he went on to define a physical limit in which the observables of the system can be written as averages taken over the period of the internal motion. For the free bosonic states this internal motion was assumed to be generated by the Nambu Hamiltonian,

$$H_B = \frac{1}{2}\sum_{n=0}^{\infty}[p^{(n)} + \omega_n^2 q^{(n)} \cdot q^{(n)}], \tag{RS2.1}$$

where $\omega_{n+1} - \omega_n = \omega$, for $n = 0, 1, 2, \ldots$, and where the mode variables $q_\alpha^{(n)}$ and $p_\alpha^{(n)}$ are four vectors which satisfy bosonic bracket relations.

Ramond considered that the physical limit is achieved by taking the limit $\omega_0 \to \infty$. In this limit the observables were expressed as averages over the internal time coordinate. For example he defined the momentum of a bosonic state as

$$p_\mu = \langle P_\mu(\tau) \rangle. \tag{RS2.2}$$

In (2.2), $\langle P_\mu(\tau) \rangle$ represents the average

$$\langle P_\mu(\tau) \rangle = \frac{\omega}{2\pi}\int_{-\frac{\pi}{\omega}}^{\frac{\pi}{\omega}} P_\mu(\tau) d\tau, \tag{RS2.3a}$$

and P_μ is defined to be

$$P_\mu = \sum_{n=0}^{\infty} p_\mu^{(n)}. \tag{RS2.3b}$$

The Klein-Gordon equation was generalized: $p^2 - m^2 \to \langle P \cdot P\rangle - m^2$. Solutions of this generalized equation were identified as the states of the free bosonic system.

To identify the fermionic states, he used the same procedure, but developed an equation which was analogous to the Dirac equation rather than the Klein-Gordon equation. Since the Dirac equation involves the Dirac matrices, Ramond first found generalization of these. He assumed that they should also be viewed as functions of the internal time coordinate, τ, with

$$\langle \Gamma_\mu(\tau)\rangle = \gamma_\mu. \tag{RS2.4}$$

By noting that the algebraic relations satisfied by the Γ's must reduce to the appropriate relations between the γ's when the averaging is performed, and by making the simplest assumptions which accomplish this, he was able to derive an explicit form for the generalized Dirac matrices:

$$\Gamma_\mu(\tau) = \gamma_\mu + i\omega_0\tau\delta_\mu + i\sqrt{2}\gamma_5\sum_{n=1}^{\infty}[b_\mu^{(n)\dagger}e^{i\omega_n\tau} + b_\mu^{(n)}e^{-i\omega_n\tau}]. \tag{RS2.5}$$

The important thing about this expression from the point of view of developing a fermi-bose symmetry is the presence of the b_μ^n's. These obey anticommutation relations.

$$\begin{aligned}\{b_\mu^{(n)}, b_\nu^{(m)}\} &= \{b_\mu^{(n)\dagger}, b_\nu^{(m)\dagger}\} = 0,\\ \{b_\mu^{(n)}, b_\nu^{(m)\dagger}\} &= -g_{\mu\nu}\delta^{n,m}, \quad n, m = 1, 2\ldots\end{aligned} \tag{RS2.6}$$

It is through these fermi-like coefficients, that anticommutators enter into the algebra of the dual models. To see how this works it is appropriate to pause in this development to see how an algebra is associated with a string theory in the absence of fermions. The treatment given below largely follows the development of Ref. [69] where the interested reader will find a detailed account.

Constraints in a theory often play two roles. They may serve as generators of symmetry transformations, and they naturally restrict the space of possible physical states. In bosonic string theories the energy-momentum tensor is constrained to be zero. Define L_m to be appropriately chosen Fourier components of the energy–momentum tensor. The L_m are naturally also zero, i.e. they are themselves constraints. It is possible to define them in such a way that they satisfy the following classical algebra:

$$[L_m, L_n]_{P.B.} = (m-n)L_{m+n}. \tag{RS2.7}$$

This has been referred to as a classical algebra because the bracket relations are Poisson Brackets and not commutators. This algebra is the Virasoro algebra and the L_m are the Virasoro generators. They are symmetry generators because the transformations they generate are reparametrizations of the world sheet of the string, and such reparametrizations do not affect any physical results.

For the quantum system, the physical states $|\phi\rangle$ are those for which

$$L_m|\phi\rangle = 0. \tag{RS2.8}$$

We note that the L_n can be expanded in terms of harmonic oscillator-like creation and annihilation operators. Because of ordering ambiguities for $m = 0$ the condition on the physical states must be permitted to be more general: $(L_o - a)|\phi\rangle = 0$. Related to this is the fact that the quantum (commutator) algebra is slightly different from the classical one. It contains an anomaly term,

$$[L_m, L_n] = (m - n)L_{m+n} + \frac{D}{12}(m^3 - m)\delta_{m+n}. \tag{RS2.9}$$

This is a central extension of the Virasoro algebra.

The development above applies to the bosonic string. What Ramond did is to invent a generalization of the Dirac equation, using his generalized matrices,

$$[\langle \Gamma_\mu(\tau)P_\mu\rangle - m]|\psi\rangle = 0, \tag{RS2.10}$$

where $|\psi\rangle$ represents a fermionic state. He then defined Fourier components of the operator $\langle \Gamma \cdot P\rangle$,

$$F_{\pm n} = \langle e^{\pm i\omega_n \tau}\Gamma_\mu(\tau)P_\mu(\tau)\rangle. \tag{RS2.11}$$

With this definition, Ramond found that the algebra

$$\begin{aligned} &= \frac{1}{2}\omega(2m - n)F_{n+m}, \\ \{F_n, F_m\} &= 2L_{n+m} \end{aligned} \tag{RS2.12}$$

is satisfied, where the L_n are generalizations of the Virasoro generators for the bosonic theory. Thus a graded algebra has emerged.

It was also possible to show, by using (2.10) and (2.12), that, in analogy to (2.8),

$$F_n|\psi\rangle = 0. \tag{RS2.13}$$

Important to Ramond's investigation is that the spectrum of physical states was fermionic.

2.2 *Dual Pion Models*

Just a few months after Ramond's work was submitted, Neveu and Schwarz presented a paper in which they introduced a new dual model for pions [94]. The motivation and approach of their paper was quite different from that of Ramond, but their work makes contact with his in a very interesting way.

The goal of Neveu and Schwarz was to create a bosonic string theory with a more realistic spectrum. Toward that end they augmented the usual bosonic creation and annihilation operators with fermionic ones. The theory that they introduced was an interacting theory while, as (2.10) suggests, Ramond's work had applied to a free theory. In the course of their work they constructed generators G_r, where r is a half integer index, which participate in an algebra almost identical to the one derived by Ramond. This suggested that perhaps the fermi states of Ramond and the new bosonic states constructed by Neveu and Schwarz could be viewed as different states in a single model.

How this came to be understood takes us to the modern understanding of the space-time supersymmetry of superstrings, and so we will continue this discussion in Sec. 8. However at this point it may be appropriate to emphasize that the symmetries that were first understood to be present in the work of Ramond and of Neveu and Schwarz, were symmetries of the two-dimensional world-sheet of the strings. The nature of these world sheet symmetries was illuminated by the work of Gervais and Sakita [58].

2.3 *Strings with World-Sheet Supersymmetry*

In string theories one can define fields which are functions of the coordinates of the two-dimensional world-sheet. Since the world-sheet is itself defined in a space-time of higher dimension (for example 26 for the old bosonic strings and 10 for the theories discovered by Ramond, Neveu and Schwarz), these fields also have transformation properties on the larger space. For example the field Φ considered below is a space-time vector. However, in this section we will be concerned not with space-time, but rather with the world-sheet. For it was world-sheet supergauge symmetry which provided the most direct link between the work on string theories with a graded algebra, and the work of Wess and Zumino on supersymmetric field theory.

In their 1971 paper, '*Field theory interpretation of supergauges in dual models*' [58], Gervais and Sakita proposed to interpret the operators G_r, which had been introduced by Neveu and Schwarz, and later used by Neveu, Schwarz and Thorn [95], as a generator of supergauge transformations. These transformations linked the field Φ, which is quantized using commutators, to fields ψ_i which were considered to be quantized using anticommutators. In order to write the transformations, Gervais and Sakita introduced an anticommuting c-number, χ

$$\begin{aligned} \delta\psi_1 &= \epsilon g_1(\partial_x\Phi)\chi, \\ \delta\bar{\psi}_2 &= \epsilon g_2(\partial_{\bar{x}}\Phi)\chi, \\ \delta\Phi &= -\frac{i}{2}\epsilon(g_2\psi_1 + g_1\bar{\psi}_2 + \bar{g}_2\bar{\psi}_1 + \bar{g}_1\psi_2)\chi. \end{aligned} \tag{RS2.14}$$

Under these supergauge transformations, and other (non-linear) supergauge transformations also introduced in Ref. [58], the Lagrangian associated with the functional integral of the theory transforms as a total derivative, indicating that the G_r can indeed be viewed as symmetry generators. These supergauge transformations are strikingly similar to the supergauge transformations later introduced by Wess and Zumino. However, the symmetry transformations that they generate are not space-time symmetries that present a non-trivial extension of the Poincaré algebra, and so do not themselves directly tread on the toes of the no-go theorems. It was left to the generalizations of these new symmetries to four-dimensional space-time to do that.

2.4 *The Interplay Between High Energy Theory, Experiment, and Mathematics, circa 1974*

This is a good point to make contact with the broader theme of how physics and the way it is done have evolved during the period around (and especially during) the era of supersymmetry.

There are both many similarities and many differences between the way that physics was done in the 1920's and 30's and the way which it was done by 1974. Perhaps the most obvious difference is the greater degree of specialization that existed in 1974, even among theorists. This specialization is responsible for the fact that most of the concerns that have been addressed by supersymmetry are the concerns of high energy physicists. And perhaps the most obvious similarity, one which may have been taken for granted at that time because it had been so central to the advance of science until then, was the relatively tight connection between theory and experiment. This relationship is perhaps most vividly illustrated by the discovery of the J/ψ in the autumn of 1974, the year that Wess and Zumino's first paper on supersymmetry was published. Its discovery was announced on he 11th of November, and ten days later the discovery of the ψ' was announced. Pais writes, "The general pandemonium following these discoveries compares only, in my experience, to what happened during the parity days of late 1956... theorists delved in their grab bag of exotica. I recall early discussions of three options: it was a weakly interacting Z-like boson; or a resonance carrying free color, or charmonium" [101].

Although the discovery of what turned out to be charmonium was particularly exciting, at that time there was nothing unusual about this sort of close contact between theory and experiment. The ISR (intersecting storage rings) had come on line at CERN in 1971, and Fermilab had begun operations at 200 GeV in 1972 and quickly achieved higher energies. Neutral currents were discovered in neutrino reactions in 1973.

It is clear that things were beginning to change though, because very high energies were needed to test some of the key predictions of theories like the electroweak theory and the new grand unified theories. As a result, the size and difficulty of the necessary experiments were increasing. One gets a feeling for the longer time delays between prediction and possible experiment from the example of the experimental discovery of the intermediate vector bosons of the electroweak theory, which did not occur until 1983, approximately 16 years after the formulation of the now standard electroweak model.

As for the sort of mathematics that high energy theorists were using, it had begun to change in many ways. In particular the more sophisticated mathematics that is associated with Yang–Mills theories was being mastered and used by physicists, and a great deal of interaction between mathematicians and physicists was associated with this process. Nevertheless, many of the notions that were still most useful to high energy theorists were related to the basic group theoretic concepts that were introduced to many physicists during the era of the development of quantum mechanics.

3 The Changing Fortunes of String Theories and Local Field Theories

Developments in local field theory and in string theory have generally not occurred in phase. When work on the hadronic string was cresting, interest in field-theoretic studies (having been buoyed by work on symmetry breaking, by the possibility of constructing a unified theory of the weak and electromagnetic forces, and by new results on the renormalization of Yang–Mills theories), was still rising. When quantum chromodynamics was generally accepted as a fundamental theory of the strong interactions, interest in hadronic strings plunged. Much later, when it began to seem unlikely that field theories with local supersymmetry would be found to give finite results, new results in string theory produced a surge of interest which seems to still be growing.

This section will be devoted to studying the relationship between the fortunes of local field and string theories over the period from 1970 or so through the present time. This evolving relationship has played a major role in the development of supersymmetric theories. And supersymmetry has, in turn, provided an important and much used link between field and string theories.

The story begins in the late 1960's with efforts to understand the strong interactions. At that time there were concerns that the very strength of the interaction might make sensible application of a field theoretic perturbation scheme impossible. And not much was understood about non-perturbative approaches. Furthermore, hadronic resonances of high spin had been discovered, and it was known that there were problems of consistency in field theories that describe interactions of particles of higher spin, like spin-$\frac{3}{2}$ for example. The duality concept grew out of an attempt to understand hadronic data, and it incorporated the higher spin resonances in a natural way, It was discovered in 1970 that the dynamics of the theories that satisfy the duality hypothesis is the dynamics of a relativistic string.

The explorations which had led to the formulation of hadronic string theory had occurred at energies available to experimentalists at the machines that operated during the mid-1960's. By the end of that decade, scattering experiments at higher energies, which might have been capable of actually exploring the structure of the strings, had found evidence for point-like behavior instead. This behavior was inconsistent with the predictions of the string models, but could be described by a local field theory, quantum chromodynamics.

In the meantime, during the decade that spanned the early 1960's and the early 1970's, several important discoveries had been made about local field theories. In 1974 Wess and Zumino brought the idea of supersymmetry to the study of field theories in four dimensions. Their work, which was submitted to *Nuclear Physics* on the 5th of October 1973, triggered an avalanche of further work on supersymmetric field theories. Fayet and Ferrara [39] listed papers on supersymmetry that had been written by March 1977, when they updated the list of references in their *Physics Reports* during proof. They included approximately 135 papers on supersymmetry, which had been written by about 52 separate authors. In 1977 itself, 104 papers were listed under supersymmetry in *Physics Abstracts.*

Much of the excitement that was generated by these theories had to do with their remarkable ultraviolet properties, which were due to cancelations of divergences contributed by the fermi sector against divergences from the bose sector. These cancelations were of particular interest to theorists who had been struggling with the problem of quantizing gravity using perturbation schemes. Einstein's theory of gravitation had long been known to be non-renormalizable. This is to say that, if infinities appear order-by-order in the quantum perturbation expansion, then it is impossible to eliminate them while retaining the predictive power that a physical theory should have. Non-renormalizability need not be fatal if either the quantum expansion gives finite results in each order (so that renormalization is not needed), or if it can be shown that the sum of the terms in the perturbation expansion turns out to be finite. To test whether or not this latter possibility is realized would require the ability to do non-perturbative calculations for quantum gravity. To develop a workable approach to such calculations is, to say the least, extremely difficult. Although work has been done to develop non-perturbative methods, it would seem that we still have a long way to go toward a non-perturbative evaluation of the status of Einstein's theory as a predictive quantum theory. Most high energy theorists have preferred to study the perturbation expansion to see if the first hope, term-by-term finiteness, is realized. This is also a difficult enterprise. Although, in 1974, 't Hooft and Veltman were able to show that there are no physical problems for Einstein's theory in one-loop order, the calculations are so difficult that the two-loop result was not obtained until 1985. As many had long suspected, the two-loop result for pure gravity was not finite. Of course the calculations that would seem to be more physically relevant are those in which gravity is coupled to matter. Hopes that matter couplings might help to eliminate order-by-order divergences were dashed when, starting in 1974, possible couplings were shown to yield to divergent results even in one-loop order. This made the idea of incorporating supersymmetry into theories which include gravitation seem very attractive.

Fortunately, it proved possible to incorporate supersymmetry into gravitational theories by gauging the symmetry. For, when supersymmetry is a local symmetry, gravitation is automatically and necessarily present.

Although the task of constructing and working with these theories of supergravity was not an easy one, such theories were first constructed in 1976 and a great deal had been understood about them by 1984. In particular the possible cancelation of divergences had been studied, and it had indeed been found that supersymmetry improved the situation vis-a-vis the divergences. The reason for the improvement had also come to be understood. However, although there was no direct evidence to suggest that supergravity theories could not be finite, it was not clear why they *necessarily* should be. In surveying the literature, one senses that by the early 1980's many of the experts had begun to doubt that even the maximally symmetric ($N = 8$) theory of supergravity, which had provided the greatest hope, would be finite.

Even if the hope that supersymmetry would provide a predictive field theory of quantum gravity has apparently not so far been realized, supersymmetry has nevertheless made it easier to discuss quantum predictions of gravitational theories. This comes about because the presence of the symmetry makes the necessary calculations more manageable, and is illustrated by the fact that Grisaru was able to discover that pure supergravity is finite up to two-loop order, approximately eight years before a two-loop result was obtained for pure gravity. This tremendous increase in calculational power makes the incorporation of supersymmetry into theories of gravity still seem attractive.

If field theory was not to provide a predictive theory of quantum gravity, perhaps there was another way. An alternative was suggested by Scherk and Schwarz, way back in 1974. Their suggestion was based on the fact that field theories of gravity are necessarily plagued by ultraviolet divergences. The source of these divergences can be understood as follows. Although the Newtonian constant, G_N, has the dimensions of $(mass)^{-2}$, the terms in a probability amplitude must be dimensionless. Thus, those terms which are of order $(G_N)^n$ must include $2n$ factors of the momentum, and will therefore diverge at high energy. Because strings have finite extent, Scherk and Schwarz postulated that a string theory which includes gravitation should be able to avoid the ultraviolet divergences of the gravitational field theories. Although they and others did pursue this idea, only a few papers on it were published during the 1970's. One of the things that had begun to be understood, however, was the way that string theories could incorporate space-time supersymmetry [59, 60]. This line of investigation was then systematically pursued by Green and Schwarz during the early 1980's, and by 1984 had begun to attract a great deal of attention. These more recent developments will be discussed in Sec. 8. The following subsections, on duality and local field theory respectively, give something of the flavor of the research that was being done in each area during the early 1970's. The role that this state of affairs played in the development of supersymmetry, as well as the role eventually played by supersymmetry in both field and string theories, is also discussed.

3.1 *Duality and Hadronic String Theories*

The goal of the dual resonance models was an ambitious one. As stated by Fubini in his 1974 introduction to a collection of *Physics Reports* reprints [81], "The final aim is to collect all particles with the same baryonic number in a single supermultiplet." Furthermore the conceptual and mathematical structure of these models was extremely rich. It was discovered that dual theories can be formulated as theories of relativistic strings, indicating that the dual theories were describing the physics of extended objects.

In his 1974 *Physics Reports* article, Veneziano paints a picture of how the dual theories were regarded. (This review is reprinted in Ref. [81].) "The impression one gains through discussion of this subject in the scientific community is that of considerable polarization in the attitude towards it: there is a group of enthusiastic theorists who consider the problem of strong interactions as practically solved (up to some 'technical' points) and a group of more sceptical physicists who can hardly follow what is going on, and, basically, do not see compelling physical motivation behind the dual approach to strong interactions."

As mentioned in the previous section, there were three main weaknesses of the dual resonance models: the early models described only bosonic states; a consistent quantum formulation seemed to require 26 space-time dimensions; and the presence of tachyons plagued the spectrum. Supersymmetry proved capable of affecting all of these. The introduction of graded algebras enabled the string theories to describe both bosonic and fermionic states. These improved string theories with world-sheet supersymmetry had a consistent quantum formulation in 10 dimensions, which must have at least seemed like a step in the right direction. And the projection of these theories which turned out to have space-time supersymmetry eliminated the tachyons.

In spite of the benefits that supersymmetry bestowed upon hadronic strings, for most physicists interested in understanding more about hadronic systems, the focus of attention shifted from string theories to QCD during the early 1970's. Since the reason for the shift did not have to do with any of the problems of string theories, to most people the supersymmetric 'cure' seemed irrelevant. However ideas drawn from duality did not fade from hadronic physics altogether. The bag model, which was also developed in 1974 (see Johnson in Ref. [116]) was an attempt to create a predictive scheme which had some of the features of the dual models, but also features of the quark and parton models. The bag, like the string, was an extended, finite object. The potential energy associated with it was taken to be a constant times the volume of the bag, analogous to the situation with the string. The development of QCD also included string like structures, in the tubes of SU(3) color. In fact, a significant number of papers on duality continued to be published through the 1970's and into the early 1980's. The low point occurred in 1980 and 1981, in each of which slightly more than 90 papers were published. Most of these papers, at least up until the early 1980's, were focused on the role of strings in hadronic physics.

3.2 *Strings as 'Theories of Everything,' and the Connection Between String and Field Theories*

Even as interest in string theory seemed to dry up, the seeds of its renaissance in a more powerful form were planted. In a paper which was submitted to *Nuclear Physics* on the 14th of May 1974, Joel Scherk and John Schwarz suggested that string theories could describe both the hadronic and the non-hadronic world [111]. (They noted that this had also been suggested by Bardakçi and Halpern [10].) One of the key elements of their proposal was that it is possible to derive field theories as the low energy limit of string theories. The discovery that this is so had been initiated by Scherk in a paper which was published in 1971 [110].

Scherk had noted that the so-called Feynman-like diagrams (FLDs) of the dual models had been derived as generalizations of the Feynman diagrams of the ϕ^3 field theory, and had studied the questions of whether and how the ϕ^3 theory could be extracted from the dual resonance model. As in the later superfield formalism, each line in a dual FLD represents more than one field. In the FLDs, each internal line represents the exchange of an infinite number of excited states lying along a Regge trajectory. Scherk expanded the amplitudes of the theory in powers of the slope parameter α'; this corresponds to an expansion in terms of the fundamental length scale of the theory. He then considered the $\alpha' \to 0$ limit, which should correspond to a low energy limit. Since in this limit the mass of the excited states becomes infinite, Scherk reasoned that the ϕ^3 theory should be recovered.

Because the slope is not the only parameter in the theory — for example, there is also the coupling constant g — there is more than one way to take the $\alpha' \to 0$ limit. Neveu and Scherk took the limit differently and found that the on-shell amplitudes in the tree approximation are those of the 10-dimensional Yang–Mills theory [96]. Yoneya considered the $\alpha' \to 0$ limit of the Virasoro-Shapiro model (with $g\sqrt{\alpha'}$ fixed), and found that it corresponds to Einstein's theory of gravity [136, 137].

The massless particles associated with the gauge fields of these theories had no obvious role in a theory of hadrons. However, in their pivotal 1974 paper, Scherk and Schwarz noted that if string theories were considered as theories of non-hadrons, as well as of the hadrons, then the "presence of massless particles becomes a virtue" [111]. They proposed to take this view seriously and to consider the consequences. They wrote, "... it is tempting to speculate as to whether a dual scheme might also be appropriate for the non-hadronic world (including leptons, photons, gravitons, and other gauge fields). Obviously there is no empirical evidence of duality or Regge behavior in non-hadronic interactions. However, the idea of having string-like rather than point-like particles is so general that it might extend to non-hadrons as well, provided that the fundamental length (or, equivalently, slope) characterizing the non-hadronic world is much smaller that the hadronic value, so as not to conflict with the limits imposed by the successes of quantum electrodynamics." Indeed they found that α' had to be quite small (approximately 10^{-34} GeV^{-2}), in order for a non-hadronic string theory to yield the correct value of the electric charge and of Newton's constant.

Scherk and Schwarz pursued this idea and its implications. In 1975 they won an honorable mention in the contest sponsored by the Gravity Research Foundation, for their essay, '*Dual model approach to a renormalizable theory of gravitation.*' They observed that the good ultraviolet behavior of string theories might make a consistent and predictive theory of quantum gravity possible. The Einstein field theory limit might then be to the full string theory, as the V-A point interaction scheme is to the full electroweak gauge theory. (This essay is reprinted as paper #18 in Ref. [115].)

One aspect of the string theories which needed to be reconsidered if they were to be thought of as fundamental theories within which all of the forces, including gravity, could be unified, was the meaning of their definition in higher dimensions. The possibility of taking the extra dimension seriously was studied by Cremmer, Scherk, and Schwarz [11, 18, 112], who, along with Englert, were among the first to introduce the notion of using higher dimensions to study supersymmetric theories.

Even though this early conjecture about string theories as theories of everything did not attract very many active converts, it did seem to attract a fair amount of attention and favorable comment. For example, Olive considered it important enough to include in his plenary report on dual models at the XVII International Conference on High Energy Physics, which was held in 1974. After remarking that "almost all the classical field theories can be obtained as special cases" of the dual models, he went on to say that "we have the tantalizing possibility that there may emerge a unified theory of all the interactions [116]".

3.3 *Local Field Theory*

The years between 1964 and 1974 had been years of tremendous development for field theory. Work on spontaneous symmetry breaking showed that observations in which a symmetry is not obvious, might nevertheless be predicted by a Lagrangian with a "secret symmetry." The discovery of the Higgs mechanism for such theories showed that the particle associated with a Yang–Mills gauge field could have nonzero mass. This established that the associated force could have finite range, and thus that the weak and strong interactions might be described by a Yang–Mills field theory. It also made it possible to see how, even in theories which can be described by a Lagrangian which respects an internal symmetry group, it is possible for different particles in the same multiplet of the group to acquire different masses. A gauge theory (now the "standard model") was constructed for the unified electroweak interaction. The renormalization of Yang–Mills theories was understood and shown to work even in theories in which the symmetry is spontaneously broken. A Yang–Mills theory was developed for the strong interactions. This theory, QCD, was shown to exhibit asymptotic freedom. Grand unification of the strong and electroweak forces was on the horizon.

The list of some of the accomplishments that had been achieved through the study of local field theory within a ten year period, gives some idea of the excitement that must have been generated by the study of field theory. Abers and Lee commented that, after the 1971 work of 't Hooft on the renormalization of Yang–Mills theories with spontaneously broken symmetry, there was "an explosion of interest in the subject, and the study of spontaneously broken gauge theories has become a major industry among theorists" [1]. The advent of supersymmetric field theories provided an opportunity to explore these themes in a different setting. And indeed this is exactly what was done by a fairly large and growing number of field theorists. Spontaneous symmetry breaking, both for supersymmetry and for other symmetries which may exist in a supersymmetric field theory, was studied and attempts were made to construct realistic theories (particularly for the electroweak interactions) which were globally supersymmetric. The question of whether the neutrino was a Goldstone fermion was studied in detail, and it was found to be highly improbable. It was found that supersymmetric theories had special ultraviolet properties, and that there typically were cancelations among divergences in quantum calculations. A new formalism, the superfield formalism, which was specifically suited to the study of supersymmetric theories was developed. By 1976 supergravity had been discovered and a whole new area began to be explored.

3.4 *Mapping the Changing Fortunes*

The graphs on Figs. 6.6, 6.7, and 6.8 are plots of the number of papers published in several relevant areas during the years 1971 through 1986. The data was taken directly from *Physics Abstracts*. Naturally such information can give only a general idea of the amount of activity in each area and of general trends. [Among its weaknesses are these: there is a time lag in that it registers when the publication was listed and not when the paper was actually written; it relies on a classification scheme which must wait to confirm sustained interest before creating new categories; it does not give an idea of the proportion of physicists working in these areas as the number of physicists changes.] Nevertheless it does give a sense of what areas are being actively explored, and a way of comparing activity in different areas during any given year.

The graphs clearly indicate that the rise of supersymmetry and supergravity coincides with a time of increasing interest in field theories, particularly theories with a local internal symmetry and theories in which different forces are unified. It also indicates that interest in dual models never completely died out.

Unfortunately these graphs cannot by themselves give a clear idea of the situation in the years 1973–1975, which form the central focus of the discussion above. Yang–Mills theories and gauge theories did not even have their own listings until later. Furthermore, the number of papers on unified theories, which seems to have been more or less constant until about 1977, is not necessarily a good indicator. This is because in the early 1970's this heading was dominated by papers that discussed unification in the sense of Einstein, i.e. a unification of gravity with the other forces, in particular with electromagnetism, a unification that most high energy physicists did not dare to approach until somewhat later.

Perhaps the clearest indication that something special had happened in the early to mid 1970's was the creation of new headings by *Physics Abstracts* from 1975 through 1977. 'Weinberg models' started to be listed separately in 1975. This means that there must have been a steady or increasing number of publications per year on electroweak unification schemes during at least the few years prior to this. Yang–Mills theories and gauge theories were each given a heading in 1977. In that year there were approximately 119 papers listed under Yang–Mills theories, and a whopping 465 papers listed under gauge theories. Interest in these theories did not spring suddenly from nothing, and so it is clear that the intensity of interest in these areas must have been building by 1974. In the meantime, papers listed under the heading of 'duality' were already in the midst of a gentle decline by 1974, with a more pronounced decline evident for the number of papers listed under the *Veneziano model.*

4 The Dual Discovery of Supersymmetric Field Theories

Physics is an international venture. It was therefore with some hesitation that this section was divided into two parts, one to deal with the discovery of supersymmetric theories in the Soviet Union, and one to deal with its discovery in the 'West'.

Field theories invariant with respect to fermionic symmetry transformations which participate in a non-trivial extension of the Poincaré algebra, were first considered in the Soviet Union. A paper on this topic was written by Golfand and Likhtman in 1971 [62], about two and a half years before Wess and Zumino submitted their first paper on supersymmetry. Golfand and Likhtman explicitly emphasized the possible fermionic extension of the Poincaré algebra. They presented an algebra, symmetry generators for two multiplets of free fields, and a method for constructing interacting Hamiltonians which respect their fermionic symmetry, along with an example of an interacting theory. Although the algebra that they presented is somewhat different from what is now called the supersymmetry algebra[147]; and

[147]Editors' note: The superalgebra found by Golfand and Likhtman was in fact the standard $\mathcal{N}=1$ super-Poincaré algebra, as we know it now, written in a somewhat non-standard notation. Instead of using the Weyl Spinorial generators of the superalgebra which is customary at present, Golfand and Likhtman worked with the Dirac spinors. This is the reason why they had to introduce $(1 \pm \gamma_5)$ projectors. –MS.

although their approach was not used in the later development of the field, it is clear that key elements of later work were there.

Volkov, Akulov, and Soroka also did work based on the existence of a fermionic symmetry. Their interest was primarily in symmetry breaking. A paper by Volkov and Akulov, with the intriguing title, *Is the neutrino a Goldstone particle?*, was published in *Physics Letters* in 1973 [126].

None of this work seemed to attract much attention outside the Soviet Union. Nor does the subject seem to have caught fire within the Soviet Union, although discussions with other physicists were mentioned by Volkov and collaborators, and although a great deal of later work on supersymmetry was to be done by Soviet physicists. As a matter of fact, the *Physics Letters* article of Volkov and Akulov did not refer to the paper by Golfand and Likhtman.

If there is a puzzle, it is not only to understand why this work did not attract more attention, but also why it did not do so at a time which was so right for the development of supersymmetric field theories. One may also wonder what a history of supersymmetry written today would have been like had it not been for the work of Wess and Zumino, which did catch fire. Would there have been follow-up on the work of the Soviets which would have eventually excited worldwide interest in the subject? Would someone else have discovered or invented supersymmetric field theories? If so, what circumstances would have ensured that this work would have been widely noticed and built on?

These are questions without answers. Yet to consider them seriously is not a frivolous exercise. For although there is no way to directly explore the past (much less a hypothetical past), it is important to understand what factors led to one body of work being more or less ignored, while another body of work based on the same concepts, drew the attention and further work of physicists around the world.

The only aspects of these issues that can even begin to be addressed, are the specific ones of why the early Soviet work was not widely studied, and why the work of Wess and Zumino was. And since it is easier to explore the reasons why a positive action was taken than the reasons why an action did not occur, most of our efforts here will be directed toward understanding the follow-up to the work of Wess and Zumino.

It seems obvious that the fact that much of the early Soviet work was first published in Russian, and that only one paper of that entire body of work was published in a journal that is widely read by members of the international physics community, must have been at least partially responsible for the lack of international response. That is why this section has been split into two subsections, of 'East' and 'West'. There may have been other factors, such as the slight difference in timing, and the ease with which papers can get lost among the huge number of papers published every year, but these will not be speculated on here. As for the success of Wess and Zumino, it seems clear from the considerations of the previous section that timing did play a crucial role. There are also other factors which may have been important. These involve both the style and content of their early work — specifically that they explored the central ideas one by one, that they started to study the special ultraviolet properties of supersymmetric theories, and that they almost immediately began collaborating with other physicists on the study of supersymmetry.

4.1 *Supersymmetry Rises in the East*

Golfand and Likhtman

In 1971 Golfand and Likhtman wrote down the first field theory which was built to be invariant with respect to a fermionic symmetry [62]. Their motivation is clearly stated: they noted that "only a fraction of the [possible Poincaré invariant interactions] is realized in nature. It is possible that these [physical] interactions, unlike others, have a higher degree of symmetry ... we propose ... a special algebra $\mathcal{R}$, which is an extension of the algebra $\mathcal{P}$ of the Poincaré group generators."

They introduced fermionic generators W_α and $\bar{W}_\beta$, which they called generators of spinor translations. ($\bar{W} = W^\dagger \gamma_0$.) The commutator of W with the generators $M_{\mu\nu}$ of Lorentz transformations are such that W transforms as a spinor. The usual relations of the Poincaré algebra are assumed to hold for the $M_{\mu\nu}$ and for the P_μ, the generators of translations. The new relations are:

$$\begin{aligned} \{W, \bar{W}\} &= \frac{(1+\gamma_5)}{2}\gamma^\mu P_\mu, \\ \{W, W\} &= 0, \\ [P_\mu, W] &= 0, \end{aligned} \tag{RS4.1}$$

where the square brackets represent commutators and the curly brackets represent anticommutators. Having presented this algebra, they then constructed two separate generators which satisfy it:

$$W_1^0 = \frac{1}{i}\frac{1+\gamma_5}{2}\int d^3x(\phi^*(x)\overleftrightarrow{\partial}_0\psi_1(x) + \omega(x)\overleftrightarrow{\partial}_0\psi_1^c(x)), \tag{RS4.2}$$

and

$$W_2^0 = \frac{1}{i\sqrt{2}} \frac{1+\gamma_5}{2} \int d^3x (\chi^*(x) \overleftrightarrow{\partial}_0 \psi_2(x) + A^\mu(x) \overleftrightarrow{\partial}_0 \gamma_\mu \psi_2(x)). \quad \text{(RS4.3)}$$

The superscript zero on each generator indicates that, when this generator is substituted into the equations (4.1) of the algebra, the P_μ that appears on the right hand side of the first equation corresponds to the free theory of the fields which are included in each multiplet. Thus W_1^0 and W_2^0 are each supersymmetry generators, as is their sum.

There are two features of this work thus far which form an interesting contrast to the work which was later done in the West. The first is that the algebra above is not really the algebra that is usually referred to as the supersymmetry algebra.[148] Because of the presence of the $\frac{(1+\gamma_5)}{2}$ which multiplies P_μ in the algebra, it is a sort of chiral version of the supersymmetry algebra. In fact Golfand and Likhtman comment that the theories associated with their algebra cannot be expected to conserve parity. Indeed the title of their paper was '*Extension of the Algebra of Poincaré Group Generators and Violation of P Invariance*'.

The other feature that immediately strikes someone who is already familiar with the later and now more standard approach initiated by Wess and Zumino, is the absence of auxiliary fields. As will be discussed in subsection two, the multiplets that Wess and Zumino constructed contained both dynamical fields and other (auxiliary) fields that have no dynamics associated with them. These auxiliary fields entered into the Lagrangian in such a way that a subset of the Euler-Lagrange equations of motion are just algebraic equations for the auxiliary fields, and can be used to eliminate the auxiliaries from the action without loss of generality. When the auxiliary fields are eliminated however, two things happen. The first is that the supersymmetry transformations become non-linear, theory-dependent functions of the dynamical fields. Secondly, the supersymmetry algebra closes only when the equations of motion are used, that is only 'on-shell'.

Golfand and Likhtman did in fact comment that the equation of motion must be used in order to reproduce the equations of the algebra by substituting the particular forms W_1^0 and W_2^0 into (4.1).

As for the theory dependence of the generators, it was just this that they proposed to use in order to construct interacting theories which are invariant with respect to the transformations generated by W. In fact, in this paper, Golfand and Likhtman presented a method to construct invariant theories.

[148]See footnote on p. 403.

Their method was based on expanding the generators of the interacting theory in powers of a coupling constant g. In particular, W and the Hamiltonian H are so expanded, and in this form, they are substituted into the relation $[W, H] = 0$. Terms proportional to equal powers of g are equated. After this it seems that the equations of motion, $\dot{W} = [W, H]$, are used.

Although one example of an interacting theory which has been constructed this way is given in their paper, it does not seem as if this method was used by others. (A somewhat different Hamiltonian-oriented method for the construction of supersymmetric theories was used later [34–36].) In fact, given what was later understood to be the importance of the idea presented by Golfand and Likhtman in this paper, i.e. the idea of expanding the notion of symmetry of a field theory to include transformations generated by fermionic generators, the lack of response to their work is mystifying.

Volkov, Akulov, and Soroka

On the fifth of December 1972, Volkov and Akulov submitted a paper, '*Possible Universal Neutrino Interaction*', for publication. In it they entertained the hypothesis that the neutrino is a Goldstone particle [125]. In a paper that was submitted for publication a little more than a month later, they discussed the attraction of considering Goldstone and gauge particles with half integral spin [3]. "In the theory of elementary particles, a particularly important role is played by particles whose existence is made necessary and whose interaction is virtually uniquely determined by the presence of symmetry. ... [these are] all those associated with gauge fields ... and also Goldstone particles that arise in systems with degenerate vacuum." In order to study gauge and Goldstone fields with half-integral spin, it was necessary to study groups that are non-trivial extensions of the Poincaré group. This requires an extension of the group parameters, which must now contain spinor parameters.

The space of the coordinates must likewise be extended. So, in addition to the space-time coordinates x_μ, spinor coordinates $\psi(x)$ are introduced. The transformations of these coordinates that are expected to be symmetry transformations can be viewed as an extension of a symmetry transformation of the equations of motion of a free neutrino:

$$\begin{aligned} \psi &\longrightarrow \psi + \xi, \\ x_\mu &\longrightarrow x_\mu + \frac{1}{4i}(\xi^\dagger \sigma_\mu \psi - \psi^\dagger \sigma_\mu \xi), \end{aligned} \tag{RS4.4}$$

where the σ_μ represent the relativistic Pauli matrices.

This is both strikingly similar to and strikingly different from the superfield approach which was later introduced by Salam and Strathdee, and which is described in more detail in the next section. It is similar, in that in both approaches the coordinates of the four-dimensional space-time are augmented by a set of spinor coordinates. It is different, in that Volkov and Akulov give a physical interpretation to the spinor coordinate, which is assumed to be a function of the space-time coordinate $x_\mu \cdot \psi(x)$ plays the role of the Goldstone field. In the superfield as first introduced by Salam and Strathdee [105], the spinor coordinates are multiplied by functions of x_μ, and it is these functions that are identified as the physical fields.

There are other differences between this work and most of the later work on supersymmetry, and this is related to the point made above. In the later work on supersymmetry, a supersymmetry transformation generally related fermi fields to bose fields. On the other hand, in this early work of Volkov and Akulov, the only physical field was the spinor field ψ. Furthermore the transformation of each of the fields was later generally expressed as a linear transformation. Because of the x dependence of Volkov and Akulov's ψ, its full transformation is inherently nonlinear. Although nonlinear transformations have been considered in later work, this was the exception and not the rule, especially for the early work on supersymmetry in the West.

In their work on Goldstone fermions, Volkov and Akulov introduced differential forms:

$$\omega_\mu = dx_\mu + \frac{a}{2i}(\psi^\dagger \sigma_\mu \psi - d\psi^\dagger \sigma_\mu \psi). \tag{RS4.5}$$

These were used to construct an invariant action,

$$S = \int |V| d^4x,$$

where

$$\begin{aligned} V_{\mu\nu} &= \delta_{\mu\nu} + aT_{\mu\nu}, \\ T_{\mu\nu} &= \frac{1}{2i}(\psi^\dagger \sigma_\mu \partial_\nu - \partial_\nu \psi^\dagger \sigma_\mu \psi). \end{aligned} \tag{RS4.6}$$

The symmetry determined the possible interactions of the Goldstone particle.

There were formal considerations which seem to have led Volkov and Akulov to their idea of a Goldstone fermion. These can be understood from the text of their papers, and also from a paper by Volkov on phenomenological Lagrangians [124]. This latter, seems to have been published close to the time that Volkov's work with Akulov was done. Once the idea proved to be theoretically feasible, they began to consider its phenomenological implications. For example in Ref. [4] they discussed the possible role of Goldstone fermions in astrophysics. "At a definite stage of stellar evolution, when there is a high density of high energy photons, the universal interaction of the Goldstone neutrinos can ... play the fundamental role in the cooling of the star. It would be if interest to consider the influence of this interaction during the early stage of expansion in the hot universe model."

Volkov and Akulov ended their first paper on supersymmetry [125] by noting that: "The gravitational interaction can be introduced into the scheme in analogous fashion, by introducing gauge fields corresponding to the Poincaré group. ...If we introduce also gauge fields corresponding to the [fermionic] transformations, then, as a consequence of the Higgs effect, a massive gauge field with spin-$\frac{3}{2}$ arises, and the Goldstone particles with spin-$\frac{1}{2}$ vanish." This seems to be the first suggestion to consider a supersymmetric theory of gravity. It is also the first mention of the possibility of the phenomenon which is now referred to as the super-Higgs effect.

The generalization to a local symmetry, and the super-Higgs effect, was studied by Volkov and Soroka in Ref. [127]. This was accomplished by generalizing the differential forms to include the associated gauge fields, and then constructing invariants. The theory which results from such a procedure includes gravitation, a spin-$\frac{3}{2}$ field, a spin-1 Yang–Mills field, and, in cases in which the super-Higgs effect doesn't hold, a spin-$\frac{1}{2}$ fermion.

Although the work of Soviet authors does not seem to have been widely known in the West until after the independent discovery of supersymmetry had occurred there, it did eventually receive attention, and does seem to have influenced some of the later work on supersymmetry. For example, the question of whether or not the hypothesis that the neutrino is a Goldstone fermion is consistent with what is known about the phenomenology of neutrinos, was later studied in the West [49]. (This will be briefly discussed in Sec. 9.) The work of Volkov and Soroka on the inclusion of gravity was cited by Zumino in his talk at the *XVII Conference on High Energy Physics* [116]. Zumino later went on to try to construct a physical theory of supergravity in the arena of what he called the "curved superspace" introduced by Volkov and Soroka. Arnowitt and Nath also did a great deal of work on this. This work will be sketched in Sec. 6.

The Western work on curved superspace seems to have then, in combination with the early Soviet results, sparked further efforts in this direction by Volkov and his collaborators [5]. Likhtman also followed up on the work that he had done with Golfand, in a way that indicated that he was familiar with the Western literature on supersymmetry. He considered questions of renormalizability and also used superfield techniques [86, 87]. And so, as these examples demonstrate, a genuine scientific exchange did begin between 'East' and 'West', thus obviating the need for further separate discussions after the following subsection on the early work of Wess and Zumino.

4.2 *Supersymmetry Rises in the West*

Wess and Zumino

In 1973 Wess and Zumino made the step of transplanting supergauge transformations from the seemingly dying body of dual models to the quickly growing body of field theories in four dimensions. Clearly this was the right step, taken at the right time. Their first paper on the subject was received by *Nuclear Physics* on the 5th of October 1973. Within two years tremendous progress had been made, both by Wess and Zumino and by others, toward understanding the remarkable features of supersymmetric theories.

In less than a year, three papers were published by the Wess-Zumino duo. Each one focused on a particular aspect of supersymmetric field theories.

The first paper, '*Supergauge Transformations in Four Dimensions*', was devoted to studying the supersymmetry algebra [128]. Two multiplets of fields were introduced and a set of supergauge transformations was given for each multiplet.

The scalar multiplet consisted of two scalar fields $A(x)$ and $F(x)$, two pseudoscalar fields $B(x)$ and $G(x)$, and a Majorana spinor field $\psi(x)$. As did the supergauge transformations of Gervais and Sakita, these transformations connected bose and fermi fields,

$$\begin{aligned}
\delta A(x) &= i\bar{\alpha}\psi(x),\\
\delta B(x) &= i\bar{\alpha}\gamma_5\psi(x),\\
\delta\psi(x) &= \partial_\mu(A(x)-\gamma_5 B(x))\gamma_\mu\alpha + (F(x)+\gamma_5 G(x))\alpha\\
&\quad + n(A-\gamma_5 B)\gamma^\mu\partial_\mu\alpha,\\
\delta F(x) &= i\bar{\alpha}\gamma^\mu\partial_\mu\psi(x) + i\left(n-\frac{1}{2}\right)\partial_\mu\bar{\alpha}\gamma^\mu\psi,\\
\delta G(x) &= i\bar{\alpha}\gamma_5\gamma^\mu\partial_\mu\psi(x) + i\left(n-\frac{1}{2}\right)\partial_\mu\bar{\alpha}\gamma_5\gamma^\mu\psi(x),
\end{aligned} \tag{RS4.7}$$

where the α_i are totally anticommuting spinors, and where n is arbitrary.

The canonical dimensions of F and G can be read off from these transformations, and they are clearly not those appropriate for conventional dynamical scalar or pseudoscalar fields. F and G are auxiliary fields.

When supersymmetric Lagrangians were constructed, these fields appeared in such a way that they could be eliminated. However, their presence ensured that the supergauge transformations were linear, theory-independent and, as remarked upon by Wess and Zumino in this paper, was essential for closing the algebra (at least when the equations of motion are not satisfied).

The closure of the algebra was studied by computing the commutator of two successive transformations. If the α_i are not constant, then the commutator of two supergauge transformations involves a conformal transformation, a mixing of the spin-0 fields, and a γ_5 transformation of the spinor field. Because of the problems of anomalies associated with these symmetries, Wess and Zumino limited their attention to constant spinor parameters in their later work; however in this paper they did work out the full algebra, and also verified the Jacobi identities. The algebra that was later referred to as the supersymmetry algebra is the one that is obtained by requiring the spinor transformation parameters to be constant. In this case the result is

$$[\delta_1, \delta_2]\mathcal{F} = 2i\xi^\mu \partial_\mu \mathcal{F} \tag{RS4.8}$$

for all the fields $\mathcal{F}$, where $\xi^\mu = 2i\bar{\alpha}_i\gamma^\mu\alpha_i$. This relation indicates that the generators of translations play a non-trivial role in the supersymmetry algebra, and thus that the algebra is a non-trivial extension of the Poincaré algebra. Equation (4.8), together with the conditions that the Hamiltonian and translation generators be supersymmetric, and the condition that Q transforms as a spinor under Lorentz transformations, defines the algebra of simple supersymmetry. The later work of Haag, Lopuszański, and Sohnius studied the possible extensions of this algebra [76].[149]

Equation (4.8) also holds for the vector multiplet, which Wess and Zumino introduced in this first paper as well. The vector multiplet consists of the fields v_μ, λ, D, C, M, and N, which are subject to the supergauge transformations,

$$\begin{aligned}
\delta v_\mu &= i\bar{\alpha}\gamma_\mu\lambda + i\bar{\alpha}\partial_\mu\chi, \\
\delta\lambda &= \left[-\frac{1}{2}F_{\mu\nu}\gamma^\mu\gamma^\nu + D\gamma_5\right]\alpha, \\
\delta\chi &= [\gamma^\mu v_\mu - \partial_\mu C\gamma_5\gamma^\mu + (M + \gamma_5 N i)]\alpha, \\
\delta D &= i\bar{\alpha}\gamma_5\gamma^\mu\partial_\mu\lambda, \quad \delta C = i\bar{\alpha}\gamma_5\chi, \\
\delta M &= i\bar{\alpha}\lambda, \quad \delta N = i\bar{\alpha}\gamma_5\lambda,
\end{aligned} \tag{RS4.9}$$

[149] Extended superalgebras were first discussed by A. Salam and J. Strathdee, *Nucl. Phys.* **B80**, 499 (1974). — Editors' note.

where $F_{\mu\nu} = \partial_\mu v_\nu - \partial_\nu v_\nu$, and we have assumed that α is constant. Another important part of this paper was the discovery of a way to construct scalar Lagrangians by combining multiplets. This method uses the form of the transformations of the auxiliary fields, for example δF in (4.7) or δD in (4.9), which are simply total derivatives. By combining multiplets in such a way that the combinations themselves transform as the fields of a multiplet do, and by inventing combinations for which the expressions corresponding to auxiliary fields are interesting enough to themselves be Lagrangians, one can construct theories for which the action is invariant up to a surface term. Wess and Zumino used this method to find that the Lagrangian of the free fields of the scalar multiplet corresponds to the auxiliary field D of a composite vector multiplet,

$$\mathcal{L}_s^0 = \frac{1}{2}[\partial_\mu A \partial^\mu A + \partial_\mu B \partial^\mu B + \bar{\psi}\gamma^\mu \partial_\mu \psi] + \frac{1}{2}[F^2 + G^2]. \tag{RS4.10}$$

Furthermore, the free Lagrangian of the vector multiplet is proportional to the auxiliary field F of a scalar multiplet,

$$\mathcal{L}_v^0 = -\frac{1}{4}F_{\mu\nu}F^{\mu\nu} - \frac{i}{2}\bar{\lambda}\gamma^\mu \partial_\mu \lambda + \frac{1}{2}D^2. \tag{RS4.11}$$

In this first paper Wess and Zumino particularly stressed the significance of the [complete] algebraic structure that they had developed. "The fact that the conformal algebra in four dimensions can be extended to the algebraic structure described above...and that the entire algebraic structure is generalized by the supergauge transformations, does not appear to have been realized before. It is interesting in itself, irrespective of the field-theoretic applications that we have in mind." They also noted that one may expect, by generalizing their supergauge transformations, to obtain an algebra which includes general coordinate transformations.

In their second paper, Wess and Zumino presented an interacting theory of the fields of the scalar multiplet [129]. After the auxiliary fields are eliminated, the Lagrangian is

$$\begin{aligned}\mathcal{L} = \mathcal{L}_s^0 &- \frac{1}{2}[m^2(A^2 + B^2) + im\bar{\psi}\psi] - gmA(A^2 + B^2)\\ &- \frac{g^2}{2}(A^2 + B^2)^2 - ig\bar{\psi}(A - \gamma_5 B)\psi.\end{aligned} \tag{RS4.12}$$

In principle, there could have been 10 independent parameters in such a theory (assuming that only terms consistent with parity conservation and renormalization are included). The fact that the requirement of supersymmetry allows only two independent parameters illustrates one of the most striking features of supersymmetric theories.

Wess and Zumino did a one loop calculation and found that renormalization did not force the introduction of new parameters, and later work by Iliopoulos and Zumino showed that the relationship between the parameters holds for the fully renormalized theory [80]. This stability of the relations between parameters plays an important role in phenomenological applications of supersymmetry. For example, it helps to maintain the huge separation between hierarchies in those grand unified theories which incorporate supersymmetry. Once they are set, they are not subject to radiative corrections in super-GUTS theories.

Related to this is the fact that, in this second investigation, Wess and Zumino discovered the special renormalization properties of supersymmetric theories. They especially noted the cancelations of divergences among different diagrams, cancelations which made the theory easier to renormalize than the generic theory of this form that can be built from this multiplet of fields. These properties will be the focus of the next section. They ended the paper by looking toward the future. "The present investigation suggests two interesting lines of research. The first is the study of higher order corrections. The second is the construction of more complex and hopefully more realistic models, invariant under a combination of supergauge and internal symmetries." Within a year both lines had been followed, both by Wess and Zumino and by others. [See Ref. [39] for a more complete list of references on this further work that the one compiled at the end of this paper.]

In their third paper, Wess and Zumino studied the interplay between gauge symmetry and supersymmetry. They constructed a supersymmetric gauge theory, a "supergauge invariant extension of electromagnetism" [130].[150] Two symmetry transformations, performed successively, produced a sort of generalized gauge transformation.

First, a Lagrangian symmetrical with respect to both the gauge and supergauge transformations, and hence also under the generalized transformation, was constructed. Because this form was so unwieldy (an infinite power series in the coupling constant), a special gauge was chosen, so that in its final form only the gauge, and not the supergauge symmetry was manifest. In this form, the only fields from the vector multiplet that survived

[150]Editors' note: In fact, supersymmetric QED was first constructed by Golfand and Likhtman, Ref. [62]. We suggest to call this model the Golfand-Likhtman model. –MS.

were v_μ, λ, and D. For this theory too, there was just one mass parameter and one coupling constant. Again this property was preserved in the one loop approximation, and again there were cancelling divergences.

Although these few paragraphs cannot completely summarize their work, it is clear that the early works of Wess and Zumino formed a step by step approach to understanding some of the important features of supersymmetric theories. In particular they developed the algebraic structure, successfully constructed interacting theories, and had begun to study the special ultraviolet properties of supersymmetric theories. They also constructed a gauge theory and explored the relationship between supersymmetry and gauge symmetry. It is also clear that by this point they had discussed their work with a number of active physicists, some of whom had already begun to work on supersymmetry themselves. In these papers they mentioned discussions with J.S. Bell, S. Coleman, J. Iliopoulos, and E. Remiddi, in addition to B.W. Lee. In their third paper they were able to refer to other work on supersymmetry by Fayet, Iliopoulos, Salam and Strathdee.

Supersymmetry Takes off

The structure that Wess and Zumino had begun to uncover was interesting and highly relevant to other work which had been done in the study of field theory during the previous decade. Supersymmetry gave field theorists an opportunity to explore gauge theories, symmetry breaking, and renormalization in a new and attractive setting. The bypass of the no-go theorems also raised questions of how far one could go toward including generators of other symmetry transformations in the supersymmetry algebra. The cancelations of divergences presented the possibility of a finite quantum theory of gravity. The construction of supergravity raised new questions whose answers were eagerly pursued.

In 1977, a *Physics Report* about supersymmetry was published by Fayet and Ferrara [39]. In studying the references in that report, including the ones that were added in proof in March of 1977, one can see that all of these questions had been or were being considered. In many cases one can trace interest in these questions back to the pre-supersymmetry explorations of the authors. Some of the contributions to that early body of work on supersymmetry (although not a very large fraction) were from people who had worked on hadronic strings.

Certainly the most important reason for the rapid involvement of so many physicists in supersymmetry is the intrinsic interest of the subject.

However, the fact that so many people were prepared to embark on the study of supersymmetry on such short notice is also related to the particular time at which the work by Wess and Zumino appeared. Had hadronic string theories proved capable of explaining the high-energy data, had the problems involving tachyons and higher dimensions been solved, had the calculations of gravity coupled to some form of matter given finite results, and had the electroweak theory already achieved the level of experimental verification that it now enjoys, it seems unlikely that the call of supersymmetry would have been so quickly and enthusiastically answered. It even seems possible that part of the reason that the earlier work of the Soviets was overlooked may have had to do with the small difference in timing.

Wess and Zumino continued to work on supersymmetry and published further work, both alone and in collaboration with others. If one looks at their collaborators over the years from 1974 through 1976, and the collaborators of their collaborators, one finds that they account for a large percentage of the early papers on supersymmetry.

5 Quantum Behavior of Supersymmetric Theories

Long before the invention of supersymmetry, it was realized that divergences associated with fermions might be expected to at least partially cancel some divergences associated with bosons. Given this, one can imagine that it would have been possible to discover supersymmetry in a way that is different from what actually happened. One can imagine that some physicist studying field theories which included both fermi and bose fields, some physicist who was perhaps tired of encountering divergences, would try to see if things could be arranged so as to let the divergences associated with different fields cancel among each other. This mythical physicist might have studied how the divergence structure of the theory depended upon the number of fields considered and upon their masses and coupling constants, and might then have seen how to minimize the divergences by allowing these parameters to be interrelated. Had such a path been taken, it might not have been obvious that the theory had an underlying supersymmetry. However the theories so derived would almost undoubtedly have been considered to be important, both because of their divergence properties, and because of the decreased number of arbitrary parameters.

Of course this is not the way things actually happened. But it is an interesting fact that something like this sort of Bose-Fermi cancelation seems to have been anticipated by Bohr. Bohr alluded to the possibility of such a cancelation in his lecture at the 1947 Cavendish Conference. And, in his lecture at the 1947 Summer Symposium at the University of Michigan, Pais mentioned Bohr's conjecture. "Bohr has suggested that ... one might try to cancel the infinite negative energies of the Dirac seas (for electrons, protons, neutrons, and neutrinos) by the positive infinities of the zero point energies (of photons, and other Bose-Einstein fields, like mesons etc.) Thus, certain infinities would only be apparent because we are only considering one kind of elementary particle. Thus a deeper understanding might yield a connection between the two problems: the problem of the multitude of particles and the paradoxes of the infinities." (This is a quote from the notes of C.S.W. Chang.)

When Wess and Zumino introduced their first interacting theory, the first aspect of it that they explored was its renormalizability. Their early studies of the quantum properties of supersymmetric theories were followed by other investigations that continued to find cancelations among divergences associated with different Feynman graphs. Indeed, one of the questions that was considered in the year or so after Wess and Zumino's first paper, was the Vacuum energy in supersymmetric theories. In his paper, '*Supersymmetry and the Vacuum*' [138], Zumino found that "... the sum of all vacuum diagrams vanishes identically as a consequence of compensations among contributions among different fields of the supermultiplet. This means that the zero-point energy-momentum density is zero to all orders in the supersymmetric interaction."[151]

With Salam and Strathdee's introduction of superfields [105], it became possible to not only construct supersymmetric theories with more ease, but also to perform quantum calculations in a way in which these cancelations could occur automatically. Supergraph techniques to perform these quantum calculations were developed and refined over a number of years. Their power, especially when formulated in an improved form, was demonstrated by a series of so-called non-renormalization theorems.

5.1 *The Early Work Which Was Done Without Superfields*

In their first paper on an interacting supersymmetric theory, Wess and Zumino commented: "It is remarkable that this Lagrangian appears to be renormalizable even when the masses and coupling constants are not

[151]In fact, this property was first noted in the supersymmetric context by E. Likhtman in 1971 (Lebedev Physics Institute Report #41, in Russian). — Editors' note.

independent and that the relations among them are preserved by renormalization. Furthermore the theory turns out to be less divergent when the [relations between the masses and the coupling constants] are satisfied." (At this point the authors included a note to thank B.W. Lee for "pointing out first the occurrence of the divergences and also the fact that relations among couplings are preserved in the one-loop approximation.")

In fact it was the relationships between the mass and coupling constants which were responsible for the good ultraviolet behavior. For these relations were such that divergent contributions from different Feynman graphs were able to cancel among themselves. The authors noted that: "The quadratic divergence of the self-energy cancels among the various diagrams..." They also observed that possible corrections to off-diagonal mass terms, and to the couplings as well, add up to zero. "All these cancelations among different diagrams are due to the supergauge invariance of the Lagrangian."

The result was that, at least in the one-loop approximation, only one wave function renormalization common to all of the fields, was needed.

In a paper which was submitted to *Nuclear Physics* about two and a half months later, Iliopoulos and Zumino extended this result to all orders, and proved that just a single wave function renormalization was needed to renormalize the theory [80]. Their paper places special emphasis on the mutual cancelations, and includes specific diagrams to illustrate the point. In studying the reason for the cancelations they observed that it was not merely the symmetry under supergauge transformations alone which was responsible for them, but also the specific form of the Lagrangian, which had been dictated by both the requirement of supersymmetry and the inclusion of only terms which could be expected to be present in a renormalizable theory.

Lang and Wess later tried to relax this latter condition to test whether the power of supersymmetry to soften divergences would be great enough to allow a theory which by naive power-counting would seem to be non-renormalizable, to be renormalized [84]. They considered a specific Lagrangian, constructed with the fields of the scalar multiplet. The Lagrangian that they considered includes terms which should have spoiled the renormalizability. Although supersymmetry did not prove to be capable of completely saving the situation, Lang and Wess did find that the requirement of supersymmetry softened the divergences of the theory.

The cancelations which made these results possible were sometimes referred to as 'miraculous'. It turned out to be possible to understand the reasons that supersymmetry forces such cancelations, and to more readily predict their occurrence by using the superfield formalism. In that formalism, a single entity, a superfield, describes a set of fields defined on the four-dimensional space-time. These fields are referred to as the component fields of the superfield. They might correspond to the fields of the Wess and Zumino scalar multiplet for example, or to those of the vector multiplet. They generally include both the dynamical fields of the multiplet and the auxiliary fields. By considering diagrams which are associated with the propagation of its components, the cancelation of divergences can happen automatically. That is to say that, if the diagrams associated with the component fields have divergences which cancel among themselves, the superfield diagrams would simply give finite results.

These observations were used in early work which developed and used the superfield Feynman rules to do quantum calculations. Superfield methods have since evolved into even more powerful tools.

5.2 *Superfields*

Within months after the publication of Wess and Zumino's first paper on supersymmetry, Salam and Strathdee introduced the superfield, as a way to "construct at least some of the representations of this symmetry [105]". The superfield turned out to be a powerful tool for the construction of multiplets, supersymmetric theories, and for quantum calculations.

The idea of the superfield used the totally anticommuting property of the elements of a Grassmann algebra in an ingenious way. The mathematics associated with Grassmann algebras was being used in the study of quantum systems by the late 1960's. The reference that was probably most widely quoted at that time was Berezin's book on second quantization which was published in the West in 1966 [11]. Berezin worked in the Soviet Union, and his early work was first published in Russian. One can wonder whether the key role played by Berezin and his colleagues, both in the development of the mathematics and in its application to quantum theory, may have been a factor in the earlier sprouting of some supersymmetric ideas in the Soviet Union.[152]

The work of Berezin also received attention from physicists in the West, especially after the publication of his book in English [11]. Citing this book as a good introduction to the subject, Montonen introduced totally anticommuting quantities to string theory in 1973 [91], and this work was soon

[152]The work of Berezin and Katz served as an important impetus for Volkov and Akulov, see Akulov's note in this volume. — Editors' note.

followed by a paper by Fairlie and Martin which also used Grassmann variables. Whether or not the original work of Salam and Strathdee has roots that go back to the work of Berezin and the work in string theories, it will be useful to briefly review some of this earlier work. This review will provide an opportunity to define the basic concepts and also to get a sense of the ideas that were being discussed in 1974. To avoid unnecessary confusion, the Grassmann variables will be denoted in the same way for all of the work discussed below; they will be represented by θ.

Grassmann Variables

In his book, Berezin considers a finite-dimensional Grassmann algebra $\mathcal{G}_n$, with n generators $\theta_1, \theta_2, \ldots, \theta_n$. The generators are totally anticommuting,

$$\{\theta_i, \theta_j\} = \theta_i\theta_j + \theta_j\theta_i = 0. \tag{RS5.1}$$

An important consequence of this relation is that $\theta_i^2 = 0$. This means that every element of $\mathcal{G}_n$ can be expressed as a sum,

$$f(\theta) = f^{(0)} + f_k^{(1)}\theta_k + f_{k_1k_2}^{(2)}\theta_{k_1}\theta_{k_2} + \cdots + f_{12\ldots n}^{(n)}\theta_1\theta_2\ldots\theta_n, \tag{RS5.2}$$

where the highest order monomial is of degree n, and where the summation convention has been assumed.

Berezin also defined left and right derivatives, and integration. For example the right derivative was given by $\frac{\partial\theta_i}{\partial\theta_j} = \delta_i^j$. The rule for products can be obtained by generalizing from:

$$\frac{\partial}{\partial\theta_i}\theta_j\theta_k = \delta_{ij}\theta_k - \theta_j\delta_{ik}. \tag{RS5.3}$$

Integration was defined as follows:

$$\int d\theta_i = 0, \tag{RS5.4}$$

$$\int d\theta_i\theta_i = 1. \tag{RS5.5}$$

An interesting feature of these Berezin integrals is that the operation of integration is like that of differentiation. δ-functions can also be defined,

$$\int d\theta\, f(\theta)\delta(\theta - \theta') = f(\theta'). \tag{RS5.6}$$

Grassmann Algebra in the Dual Models

In 1973 Montonen used Grassmann variables in the study of dual resonance models. The goal of this work was to invent a way to perform multiloop calculations involving fermi operators. It was noted that, in the Veneziano model, general multiloop amplitudes could be constructed by a procedure which involves taking a multiple trace. The difficult combinatorial problem of calculating the trace could be converted to a more manageable functional integral by introducing coherent states, i.e. eigenstates of the destruction operator. Since such states do not exist for fermi operators, Montonen introduced a "coherent-like state" in terms of anticommuting operators b and Grassmann variables θ [91].

Soon after this, Fairlie and Martin discovered that it was possible to use Grassmann variables to see that the Neveu-Schwarz model is an extension of the Veneziano model [37].

The Introduction of Superfields

Salam and Strathdee suggested that the study of supersymmetry might be done most conveniently in a space in which the usual space-time coordinates x_μ are augmented by a set of Grassmann variables θ_α. The most general form for a function of the superspace variables is that given by (5.2) for $f(\theta)$. A function which can be expanded in this way is referred to as a superfield, and will be denoted by Φ. In their first paper on the subject Salam and Strathdee considered Grassmann variables which were Majorana spinors, and they chose to impose a reality condition on the superfield

$$\Phi(x,\theta)^* = \Phi(x,\theta).$$

Salam and Strathdee noted that "the truly remarkable and exciting feature of the Wess-Zumino group is that the superfield $\Phi(x,\theta)$ in eight dimensions is exactly equivalent to a 16-component set of ordinary fields in four dimensions [105]". The functions of the space-time variables x_μ which multiply each term of the superfield expansion are referred to as the component fields. When the reality requirement is placed on the superfield, the component fields are A, B, ψ, F, G of the scalar multiplet, along with v_μ, λ, D of the vector multiplet. Soon after the paper by Salam and Strathdee, Ferrara, Zumino and Wess showed that it was possible to consider superfields which describe only the scalar multiplet [41]. To do this they introduced two features which have been widely used in work on the superfield formalism ever since. They used two-component spinors, and they constructed superfields (the so-called chiral superfields) which were functions of θ and not of $\bar{\theta}$. A chiral superfield can be expressed in terms of a general superfield which is subject to a constraint. The constraint can be written in terms of a covariant derivative $\bar{D}_{\dot{A}}$. In the two-component notation

$$\begin{aligned} D_A &= \frac{\partial}{\partial\theta^A} - i(\sigma^\mu)_{A\dot{A}}\bar{\theta}^{\dot{A}}\partial_\mu, \\ \bar{D}_{\dot{A}} &= -\frac{\partial}{\partial\bar{\theta}^{\dot{A}}} + i(\sigma^\mu)_{A\dot{A}}\bar{\theta}^{A}\partial_\mu. \end{aligned} \tag{RS5.7}$$

A chiral superfield satisfies

$$D_{\dot{A}}\Phi = 0. \tag{RS5.8}$$

In Ref. [105] Salam and Strathdee showed that supergauge transformations of the component fields could be derived by considering the transformations of the superfield under the superspace transformations: $\theta \to \theta + \epsilon$, $x_\mu \to x_\mu - \frac{i}{2}\bar{\epsilon}\gamma_\mu\theta$, In particular these transformations imply that the variation of the highest component of the superfield is a derivative, so that its "space-time integral ... would be an invariant if surface effects can be neglected."

The highest component can of course be isolated by performing successive superspace differentiations. However, because of the similarity between superspace differentiation and integration, these invariants can also be expressed as integrals over the superspace variables. Thus, it is possible to express superinvariant Lagrangians and possible counterterms as such integrals. For example the mass and interaction terms of the first interacting theory constructed by Wess and Zumino are of the form:

$$\int d^4x\, d^2\theta[\alpha\Phi^2 + \beta\Phi^3], \tag{RS5.9}$$

where Φ satisfies (5.8). Because Φ is chiral, there can be no integral over $\bar{\theta}$.

The elements described above were introduced in the early papers on superfields. They sufficed to develop Feynman diagram methods. References to the early work which developed these methods and demonstrated their usefulness through specific applications can be found in Refs. [108], [109], [40], [132], and in many other sources.

In 1979, Grisaru, Roček, and Siegel proposed a method designed to simplify supergraph calculations by "[making] greater use of supersymmetry" [73]. After discussing the advantages that had been associated with the older super-Feynman rules, they noted that "relatively few superfield calculations were carried out... Although simpler than conventional calculations, they were nonetheless cumbersome because the methods did not fully exploit the inherent advantages of supergraphs as a manifestly supersymmetric formalism. It was necessary to manipulate expressions involving explicit spinor variables, and to rely on complicated algebraic identities to obtain simple final results." The improved rules led directly to what have been called '*non-renormalization theorems*'. In Ref. [73] itself, Grisaru, Roček, and Siegel proved the theorem that "each term in the effective action can be expressed as an integral over a single $d^4\theta$." The consequences of this sort of result can be appreciated by looking at the superfield form of the mass and interaction terms of the Wess-Zumino multiplet. Since these cannot be expressed as such an integral, no terms of this type can pop up in the effective action, and so no renormalization of the mass and coupling constant is needed.

The superfield techniques were used (for example to show that the β-function for $N = 4$ super-Yang–Mills is zero to three-loop order [74]), and further refined (see for example Ref. [75]). Eventually they were used to show that the $N = 4$ super-Yang–Mills theory is finite. (See Refs. [15] and [132] for references.)

The superfield approach was not the only path toward the study of the quantum properties of supersymmetric theories. As a matter of fact, the finiteness of the $N = 4$ Yang–Mills theory was first proved via another path — the use of an 'anomalies' argument. This had been done by Sohnius and West by 1981 [117]. There are also references in the literature to an unpublished work by Ferrara and Zumino. Actually the foundation for the anomalies argument was laid in 1974 in a paper by Ferrara and Zumino [43]. In '*Transformation properties of the supercurrent*', they showed that "the spinor current, if correctly defined, belongs to a supermultiplet, together with the energy-momentum tensor and the axial-vector current." This meant that the anomalies associated with these currents also form a multiplet, so that if some of the associated internal symmetries are preserved, then θ^{μ}_{μ}, the trace of the energy-momentum tensor must vanish. However since θ^{μ}_{μ} is proportional to the β-function, its vanishing implies that the β-function is also zero, and thus that the theory is finite.

Thus, by 1981, supersymmetry had been understood to have given the world a finite quantum field theory. The progress of hopes for a finite theory of supergravity will be outlined in the next section.

6 Supergravity

6.1 *The Discovery and Development of Supergravity*

Motivation and Early Developments

Quantum gravity is inherently non-renormalizable. This problem, which was noticed by Heisenberg in the late 1930's [78], arises because the coupling constant κ, which is proportional to the square root of the Newtonian constant G_N, is not dimensionless. This was discussed briefly in Sec. 3, as was the possible string solution suggested by Scherk and Schwarz. There is another way in which the difficulty can be phrased. Because κ has the dimensions of $(mass)^{-1}$, the number of derivatives in possible counterterms is different from the number of derivatives in the original Lagrangian. Therefore the counterterms cannot be reabsorbed through renormalization. This need not be a problem if such counterterms are outlawed by a symmetry principle. A great deal of literature exists on the problem and on possible solutions. (See for example the contributions to Ref. [15].)

It must be stressed that these problems of renormalization are problems of a perturbation expansion, and may not indicate that there are infinities in the non-perturbative predictions of quantum gravity. However most of the work which incorporates supersymmetry into theories of gravity has focused on properties of a quantum perturbation expansion.

In a paper which was published in 1974, 't Hooft and Veltman put the calculation of perturbative effects in quantum gravity on a firm footing [119]. In their paper they calculated all of the one-loop divergences of pure gravity, and of gravity coupled to a scalar field. Although they showed that the divergences that were encountered in their calculation of pure gravity were not physically relevant, there were non-trivial divergences when a scalar field was coupled to gravity. The length and difficulty of the calculations were indicated in the ending of their paper. "We do not feel that this is the last word on this subject, because the situation... is so complicated that we feel less than sure that there is no way out. A certain exhaustion however prevents us from further investigation, for the time being."

This work by 't Hooft and Veltman inspired other (not-yet-exhausted) physicists to explore the situation when gravity is coupled to other matter systems. Gravity coupled to a charged scalar field [97], to the Maxwell field [25, 26], to spin-$\frac{1}{2}$ fermions [28], to the Yang–Mills system [27], and to quantum electrodynamics [29], were studied. In all cases there turned out to be non-trivial divergences. It is interesting to note that among the most active participants in this search for a finite matter-gravity coupling were Deser and van Nieuwenhuizen. Freedman had also studied gravity-matter couplings, taking a different approach which led to somewhat more hopeful results, by allowing only external gravitons (Ref. [48] and references therein).

Many of the negative results mentioned above had been obtained very soon after the work by 't Hooft and Veltman. By that time tremendous progress had been made toward the goal of a unified theory of the fundamental interactions. The Weinberg–Salam-Glashow model had begun to be experimentally tested, and there was a growing feeling that it might indeed describe a physically realized unification of electromagnetism and the weak force. Grand unification — a unification of the strong, weak, and electromagnetic forces — had been proposed. The force that was still excluded was gravity. Since there was not much hope that gravity could be included unless a consistent and predictive quantum theory of gravity coupled to matter could be formulated, the search for such theory must have seemed particularly urgent.

One idea that was suggested at that time as a possible way to include gravity was that a new symmetry principle should be introduced which might forbid the unwanted counterterms made possible by the dimensionality of κ. It seemed possible that supersymmetry might be just the symmetry to cure the ills of quantum gravity. It had already proved to be capable of softening ultraviolet divergences like those that plagued gravity, and so naturally there was hope that the requirement of supersymmetry would mandate a coupling of the gravitational field to matter for which the ultraviolet divergences would cancel. And there was the further hope that such a coupling would yield a theory in which all of the known forces were naturally unified.

It turned out that even constructing such a supersymmetric theory as a step toward testing these hopes was a Herculean task. The difficulties can begin to be appreciated without even looking at the eventual form of the theories of supergravity. All one needs to do is to note that almost two and one half years elapsed between the first paper of Wess and Zumino and the paper of Freedman, van Nieuwenhuizen, and Ferrara [51].[153]

This time lag was certainly not due to a late-dawning realization of the possibilities of supergravity. In their very first paper on supersymmetry

[153] It is generally accepted that supergravity was independently worked out by two groups: D. Freedman *et al.*, Ref. [51], and S. Deser and B. Zumino, Ref. [30], see e.g. P. West, *Introduction to Supersymmetry and Supergravity* (World Scientific, Singapore, 1990). See also footnote on p. 429. — Editors' note.

Wess and Zumino had mentioned that it would be natural to consider supersymmetric theories with general coordinate invariance. The reason given there was a formal one, based on the structure of the algebra of transformations. However, once the possibility of canceling divergences in supersymmetric theories had been realized, the potential benefits of combining supersymmetry and gravitation began to be discussed. At the *XVII Conference on High Energy Physics* which took place in London in 1974, Zumino gave a talk on supersymmetry [116]. There he asked, "Can a theory of this kind, because of the compensation of divergences due to supersymmetry, provide a renormalizable description of gravitational interactions?" The possible role of supersymmetry in eliminating the divergences of theories of gravity coupled to matter was later mentioned by others in discussions of the negative results described above. (See for example Deser's contribution to Ref. [116], in which he refers to Zumino's conjecture, and van Nieuwenhuizen's contribution to Ref. [104].)

In other areas a time lag of two or three years might not mean much, but in the fast-moving area of supersymmetry and what was to become supergravity, a great deal would generally happen in two years, and a lot did happen in the time between Wess and Zumino's paper on supergravity. Actually a lot happened in the study of local supersymmetry during those years. Arnowitt, Nath, and Zumino all made contributions to this study, for example Ref. [93] and Ref. [7], that were somewhat in the spirit of the work of Volkov and Soroka on curved superspace.

Gauge Supersymmetry

Richard Arnowitt, Pran Nath, and Bruno Zumino used the superfield formalism to attempt to construct a supersymmetric theory which included gravitation. The basic idea behind this approach was to extend the use of general coordinate invariance to superspace. Thus, the coordinate space that they worked with was the space which included Grassmann coordinates in addition to the four space-time coordinates used in standard general relativity. The general coordinate transformations of Einstein's theory were extended to include general coordinate transformations in superspace. Although Arnowitt, Nath, and Zumino collaborated on at least one paper [7], there were differences between the approach taken by Nath and Arnowitt, and the one taken by Zumino. These differences had to do with their approach to the geometry of superspace. While Arnowitt and Nath considered a Riemannian geometry of curved superspace, Zumino considered a non-Riemannian geometry. (The differences between the two approaches were discussed at the conference: *Gauge Theories and Modern Fields Theories*, which was held at Northeastern University in the fall of 1975 [8].)

The approach of expanding the space of the coordinates was also taken in both the earlier and later work on Kaluza-Klein theories, although there the added coordinates had the same (bosonic) character as the original spatial coordinates. The fermionic character of the extra dimensions of superspace makes them quite different from the space-time coordinates x_μ. Because of this difference it seemed clear that the new coordinates should not be regarded as physical. For the Kaluza-Klein theories, the question of the physical interpretation of the extra dimensions, which are not obviously different from the other spatial dimensions, cannot be ignored. Zumino noted that "the fact that the additional space-time dimensions in the present picture are fermionic and spinorial, instead of bosonic as in Klein-Kaluza, presents the combined advantages that physical space-time is obviously still four-dimensional and that half-integral spin fields arise very naturally [8]". It is interesting that the addition of either the usual bosonic coordinates, or of fermionic coordinates, leads to theories in which the fundamental forces are unified.

Nath and Arnowitt began their study of gauge supersymmetry, with a paper which was submitted to *Physics Letters* in February of 1975 [93]. In their approach, supersymmetry and gravity were combined in an interesting way. The metric of superspace, $g_{AB}(z)$, occupied center stage. The subscripts A and B could denote either indices of the usual four-dimensional space-time, or spinor indices. This metric was expanded as a superfield, and the physical fields were taken to be a subset of the components of the superfield.

Nath and Arnowitt noted that, "the usual supersymmetry transformation is ... a linear transformation in the 8-dimensional superspace of $z^A = \{x^\mu, \theta^\alpha\}$. It is natural then to consider a generalization of this to arbitrary coordinate transformations in superspace, $z^{A'} = z^A(z)$ [restricted to maintain the bose (fermi) nature of $x^\mu(\theta^\alpha)$]. We take this to be our basic gauge group" [89]. They then went on to show that this gauge group includes not only the group of general coordinate transformations of general relativity, but also the other gauge transformations of elementary particle physics, including the Yang–Mills transformations of SU(n).

Furthermore, Nath and Arnowitt showed that the component fields of the metric played the role of gauge fields. They inferred that "if *one extends supersymmetry to be a local gauge invariance then it is possible to construct a theory where all fields are gauge fields*, and hence all interactions are determined by a common non-Abelian gauge invariance."

In comparison to the later theories of supergravity, the conceptually rich gauge supersymmetry did not generate the enthusiasm and active participation of many researchers. It would seem that the reason for this was related to the very richness of the formalism. For, the large number of fields that were automatically included in the metric superfields made the identification of the physics described by the theory difficult. This point is mentioned in the first paragraphs of the work by Freedman, van Nieuwenhuizen, and Ferrara [51], where they explain the reason for their resolve to work with only physical fields.

The Discovery of Supergravity

Freedman, van Nieuwenhuizen, and Ferrara chose to start with the simplest possible multiplet of physical fields which could be expected to respect a local supersymmetry. They wrote "we commit ourselves from the start to a formulation without superspace in which the only fields in the gravitational supermultiplet are the metric tensor $g_{\mu\nu}(x)$ [or, equivalently, the vierbein field $V_{a\mu}(x)$] and a Rarita-Schwinger field $\psi_\mu(x)$."

If the gravitational field is to couple to a spinor field, as is to be expected in a supersymmetric theory, then it is convenient to use the vierbein formalism. The vierbein is related to the metric by: $g_{\mu\nu} = V_{\mu a}V_{\nu b}\eta^{ab}$. Thus the vierbein can be taken to be the fundamental field, with the variations of the metric derived through its relation to the vierbein, and not considered to be independent. This formalism, in which the vierbein (or, equivalently, the metric) is the only independent field variable associated with gravitation, is referred to as the second order formalism.

The approach taken by Freedman, van Nieuwenhuizen, and Ferrara was to make an ansatz for the form of the Lagrangian and also for the form of the supersymmetry transformations. They then checked to see whether the chosen Lagrangian was symmetric under the chosen transformations. When it turned out that their original choices did not quite work (although they came close), they modified both the Lagrangians and the transformations. In this next step they completed the construction of a theory of supergravity.

The Lagrangian that they started with was the simplest one possible for gravitation coupled to a Rarita-Schwinger field,

$$I = \int d^4x \left[\frac{\kappa^{-2}}{4}\sqrt{-g}\,R - \frac{1}{2}\epsilon^{\lambda\rho\mu\nu}\bar{\psi}_\lambda(x)\gamma_5\gamma_\mu D_\nu\psi_\rho(x) \right]. \qquad \text{(RS6.1)}$$

The coupling of the spin-$\frac{3}{2}$ field to the gravitational field is just the minimal coupling, and is accomplished through the presence of the covariant derivative in the Lagrangian,

$$D_\nu \psi_\rho(x) = \partial_\nu \psi_\rho(x) - \Gamma^\sigma_{\nu\rho}\psi_\sigma + \frac{1}{2}\omega_{\nu ab}\sigma^{ab}\psi_\rho, \qquad \text{(RS6.2)}$$

where the form of $\Gamma^\sigma_{\nu\rho}$ in terms of the vierbein is the one given by the Christoffel symbol, and where $\omega_{\nu ab}$ represents the so-called spin connection, which, like the metric, can be expressed in terms of the vierbein.

$$\omega_{\nu ab} = \frac{1}{2}\{[V_a^\mu(\partial_\nu V_{b\mu} - \partial_\mu V_{b\nu}) + V_a^\rho V_b^\sigma(\partial_\sigma V_{c\rho})V_\nu^c] - [a \leftrightarrow b]\}. \qquad \text{(RS6.3)}$$

The ansatz for the supersymmetry transformations was

$$\begin{aligned} \delta\psi_\mu(x) &= \kappa^{-1}D_\mu\epsilon(x), \\ \delta V_\mu^a(x) &= i\kappa\bar{\epsilon}(x)\gamma^a\psi_\mu(x). \end{aligned} \qquad \text{(RS6.4)}$$

The trial expression for the variation of the Rarita-Schwinger fields is a generalization of the symmetry transformation $\delta\psi_\mu = \partial_\mu\epsilon$ of the free Rarita-Schwinger field. The authors noted that this is an attempt "to interpret supersymmetry as the curved-space generalization of the old Rarita-Schwinger gauge transformation." This is analogous to the procedure used by Akulov and Volkov to generalize the symmetry transformations of the Dirac field.

Because the chosen action turned out not to be invariant with respect to these transformations, both the action and the transformations were modified, with a term bilinear in the fermi fields added to the transformations and a corresponding quadratic term added to the Lagrangian. $\delta\psi_\mu$ was augmented by $\frac{i\kappa}{4}(2\bar{\psi}_\mu\gamma_a\psi_b + \bar{\psi}_a\gamma_\mu\psi_b)\sigma^{ab}\epsilon$.

With the action and the supersymmetry transformations so modified, the action was indeed invariant under the local transformations; Freedman, van Nieuwenhuizen, and Ferrara had succeeded in constructing a theory of supergravity. As can be seen from the form of the action and of the transformations, these authors had completed a formidable calculational task. Indeed the complete result required a computer computation.

Another important component of this first paper on supergravity was the study of the algebra of the proposed transformations, to verify that it was indeed a supersymmetry algebra. Of course, because there were no auxiliary fields, closure of the algebra could only be accomplished if the equations of motion were used. The analysis of the algebra was completed in a second paper by Freedman and van Nieuwenhuizen [52]. They found that "the algebra of local supersymmetry transformations, when restricted to fields satisfying the equations of motion, consists of general coordinate transformations plus field-dependent local Lorentz rotations and further field-dependent supersymmetry transformations." This situation is analogous to the situation encountered in other supersymmetric gauge theories, in that in those theories as well, the algebra naturally included field-dependent gauge transformations, as described in an earlier work by Freedman and de Wit [50].

In discussing the physics associated with their theory, Freedman, van Nieuwenhuizen, and Ferrara speculated that the super-Higgs effect which had been discussed by Volkov and Soroka might allow the massless Rarita-Schwinger field to acquire a mass. This would solve two of the potential phenomenological problems of supersymmetry at once: the mass acquired by the Rarita-Schwinger field would insure that it was not associated with any long range effects which might have been inconsistent with established observations; and the eating of the spin-$\frac{1}{2}$ Goldstone fermion would have eliminated the goldstino. This was especially desirable in light of earlier work of Freedman and de Wit which had indicated that it was highly unlikely that neutrinos, the only known apparently massless fermions, could correspond to the massless Goldstone fermions associated with the spontaneous breaking of supersymmetry [49].

The construction of supergravity theories suggested or renewed interest in many fields of inquiry. One of the questions that was raised was the consistency of a theory with an interacting spin-$\frac{3}{2}$ field. This question was addressed in a paper which was submitted for publication about a month after[154] the paper by Freedman, van Nieuwenhuizen, and Ferrara. This paper, which was written by Deser and Zumino, also intrinsically raised another more technical question. This question was, what is the best way to approach the study of a theory of gravity which requires long and intricate calculations? That is, is there a formalism which might make such calculations more tractable? The paper of Deser and Zumino will be reviewed below during the discussion of the development of supergravity.

[154]This paragraph as well as the previous discussion of the Freedman *et al.* vs. Deser and Zumino issue suggests that Deser and Zumino's work [30] was secondary with regards to that of Freedman *et al.* [51]. The editors believe that it would be more appropriate to consider both works, Refs. [30] and [51], on equal footing. They present independent constructions of supergravity, as is generally presented in the world literature. — Editors' note; see also footnote on p. 424.

The Study of Supergravity

Immediately after the construction of the first theory of supergravity, the issues that needed to be clarified in order to smooth the way for further progress were addressed. Because the derivation of results about supergravity presented substantial technical difficulties, it was necessary and not at all trivial to think through questions about the most appropriate formalism. For example: what was the best way to handle the gravitational field in a theory which necessarily includes spinors? How best to approach the development of the superfield formalism? Was the component formalism superior to the superfield formalism or vice versa? There was also the question of whether fields of spin even higher than 2 could be consistently included. And perhaps most important, given the primary reason for the development of supergravity in the first place: could theories of supergravity be well defined at the quantum level?

There were other important questions as well. Perhaps most central to understanding whether or not theories of supergravity could be physical theories was the study of symmetry breaking and the construction of phenomenological models. Because of limits of time and space, these other issues will receive only glancing attention in this presentation. However, one can gain insight into the type and pace of developments in these areas, as well as in the areas which will be somewhat more extensively covered here, from examining the proceedings of the many conferences and schools that have been held on supersymmetry and supergravity. The schools which were held at Trieste in 1981 [45], 1982 [46], and 1984 [33], are particularly useful in this regard, especially concerning the increasing use of formulations in higher-dimensional space-times. These will be discussed in Sec. 7.

The conference that was held at Stony Brook in 1979 provides some insight as to the state of affairs when supergravity was about two and one half years old [122]. The study of supergravity in superspace was well underway. Almost one third of all of the papers that were included in the proceedings were about the progress that had been made toward understanding how best to explore supergravity in superspace (including the derivation of superspace constraints), or about results that had been obtained through use of the superfield formalism, or dealt with a comparison between the superfield and component field approaches. Another fairly large group of papers (about 13% of the total), were devoted to the study of coupling among fields which include fields of higher spin. Papers that dealt with the problem of supersymmetry breaking and of the phenomenology of theories of supergravity comprised about 15% of the total. Naturally a significant number of papers (not necessarily distinct from the papers on superfields) dealt with the study of quantum supersymmetry. There were also a number of papers (over 10%) which dealt with the topological features of supergravity theories. It is clear that a lot of creative thought about supergravity had been going on, for there were a number of papers with provocative titles. For example: '*Local Supersymmetry without Gravity?*' (Townsend), '*From Supergravity to Antigravity*' (Scherk), '*Difficulties in Constructing Bose Analogs to Supergravity*' (Milton and Urrutia), and '*Complex and Quaternionic Supergeometry*' (Lukierski).

The Consistency of Theories Which Include Fields with Spin Greater than One

Long before the invention of supersymmetry, studies of interacting spin-$\frac{3}{2}$ fields had encountered problems of consistency. Since the requirement of supersymmetry demanded the existence of at least either one spin-$\frac{3}{2}$ field or one spin-$\frac{5}{2}$ field as a partner for the graviton, the consistency of theories of supergravity required that at least those interactions mandated by the requirement of local supersymmetry be consistent.

This question was first addressed in the paper by Deser and Zumino on supergravity [30]. As a matter of fact it was this consideration that led them to their form for the variation of the fields. They observed that in the theories for which the coupling of the Rarita-Schwinger fields to other fields had been shown to lead to inconsistencies, the requirement of consistency led to too restrictive constraints on the space of solutions to the equations of motion. They discovered that the requirement of local supersymmetry ensured that these would-be constraint equations were automatically satisfied when the equations of motion were, thereby eliminating the problem.

Although the consistency of the (spin-2)-(spin-$\frac{3}{2}$) interaction had been ensured through the requirement of local supersymmetry, it remained to be discovered whether or not a consistent coupling to spin-$\frac{5}{2}$ fields was also possible. Supergauge theories which included these higher spin fields were referred to as theories of hypergravity. In the years just after the discovery of supergravity, this was an active area of research. However the results were discouraging, and it came to be widely felt that, because of problems with interactions, fields with spin greater than two could not be consistently incorporated into theories of supergravity. This had an important consequence, because it placed a limit on the amount of internal symmetry which a theory of supergravity could accommodate. If the highest spin allowed in the multiplet was to be equal to two, group-theoretic considerations limited the number of supersymmetry generators, N, to 8. As will be discussed below, this made it seem less likely that theories of supergravity could be shown to give finite quantum results.

Questions of Formalism

Although this paper deals primarily with the conceptual development of supersymmetry, the complexity and flavor of some of the calculations which led to a greater understanding of the subject is such that the conceptual development cannot be fully appreciated without making contact with the formalism. This was true for the superfield techniques which have played such a crucial role in the construction of supersymmetric theories and in the study of their quantum properties. It is also true of supergravity.

Second order, first order, and 1.5 order formalism: The theory of gravity is difficult to work with, and the ability to understand its quantum predictions can be enhanced by choosing or inventing a formalism that is well suited to the study of the question at hand. When the gravitational action is expressed in terms of the vierbein, its functional form is quite complicated. A substantially simpler form is achieved by introducing the spin connection. When one goes a step further, and considers $\omega_{\mu ab}$ as a field which can be varied independently of the vierbein, then the work is said to be done in the first order formalism. Deser and Zumino chose to work in the first order formalism [30].

The physics discussed in the paper of Deser and Zumino was the same as that in the paper of Freedman, van Nieuwenhuizen, and Ferrara. However, in the first order formalism the functional form of the spin connection in terms of the vierbein and the other fields of the theory, is given by the Euler-Lagrange equations. It turned out that the expression that was so derived by Deser and Zumino was different from that given by (6.3). Nevertheless, when their expression for $\omega_{\mu ab}$ was substituted back into the Lagrangian, the complete Lagrangian of Ref. [51] was obtained. Thus, even those interaction terms which were quadratic in the Rarita-Schwinger field, the terms that had to be added in the second step of the calculations in Ref. [51], were automatically included.

It turned out to be possible to simplify the calculation of the variation of the Lagrangian still further, by using a formalism which has been called the 1.5 order formalism. This simplification is based on the fact that, because of the form of the equation of motion associated with the spin connection, $\left(\frac{\delta\mathcal{L}}{\delta\omega_{\mu ab}} = 0\right)$, it is possible to neglect those terms in $\delta\mathcal{L}$ which are proportional to the variation of the spin connection [123].

The importance of an appropriate formalism has recently been underscored by the introduction of a set of new variables to describe Einstein's gravity. These new variables, which were introduced by Abhay Ashtekar, seem to be well suited to the study of quantum gravity. (The reason for this has to do with the simplification of the equations of constraint which occurs when the new variables are used.) Ashtekar's formalism has already been applied to supergravity by Jacobsen. (See Ref. [9] and references therein.) However, the extent of the impact that the new variables may eventually have on the study of supergravity, is not yet known.

Supergravity in superspace: Another concern that needed to be addressed was the question of the superspace formulation of supergravity. In order to construct a theory whose physical content was clear, it had been necessary to leave superspace. This journey had necessitated giving up the advantages that had been associated with superspace. These include: a set of linear transformations whose algebra is closed without the need to invoke the equations of motion, a systematic method for the construction of further theories, and simplified quantum calculations.

A natural step to take after the successful construction of supergravity, was to try to find a set of auxiliary fields to complement the minimal set of dynamical fields used in that construction. This step was taken by Ferrara and van Nieuwenhuizen [44], and by Stelle and West [118]. This work was followed by a great deal of further work on the study of supergravity in superspace. (For references see Refs. [57] and [77].) As is indicated by the large number of contributions devoted to this subject at the Stony Brook conference, this work and its comparison to work without superspace attracted a great deal of attention in the years just after the discovery of supergravity. Eventually it would prove possible to study even the promising $N = 8$ supergravity theory in the superfield formalism. This study would, by 1981, lead many experts to doubt that even this most symmetric supergravity theory could be shown to yield finite quantum results.

Supergravity in higher dimensions: A third question of formalism arose with the use of higher-dimensional space-times to study theories of supergravity. This will be the focus of Section 7.

The Study of Quantum Supergravity

Since one of the primary motivations for the introduction of supergravity was the possibility of discovering a gravity-matter coupling which led to a well-defined quantum theory, the study of quantum supergravity began immediately after the first construction of a theory of supergravity. Two approaches to this work were taken. One concentrated on the study of the S-matrix, and the other concentrated on the study of possible counterterms. The fact that the results derived in both ways agreed, gave confidence in the validity of the results.

Grisaru, van Nieuwenhuizen, and Vermaseren computed the one-loop corrections to photon-photon scattering and found that the divergences canceled against each other [70]. In the same paper they examined pure supergravity and found it to be one-loop finite. Grisaru went on to use S-matrix-oriented techniques to show that pure supergravity is also two-loop finite [72]. (See Ref. [71] for a discussion of the S-matrix approach.)

The philosophy behind the search for possible counterterms was that, if such terms could be shown to be impossible, then one would know that there were no problems with non-renormalizability. However, if such terms were found to exist, then their very existence is a signal of the possibility of trouble for the quantum theory. Although it would not necessarily mean that there are problems — since the coefficient of such terms in the quantum perturbation expansion could turn out to be zero, especially it there **are** other symmetries at work — it would nevertheless be a discouraging sign. Deser, Kay, and Stelle found that there were possible invariants for pure supergravity from three loops onward [31].

Since it was conceivable that other symmetries would constrain the coefficients of these bothersome counterterms to be zero, it seemed advantageous to study supergravity theories which had more symmetry, i.e. the theories with values of $N > 1$. However, Deser and Kay found a possible three loop counterterm for $N = 2$ supergravity [32].

Using superfields methods, Howe and Lindström studied the theory with maximal symmetry, the $N = 8$ theory, and found that the theory seemed to have non-vanishing invariants from seven loops onward. (See their contribution to Ref. [77].)

Although they indicated that it might be possible to circumvent theory result, they concluded that "On the face of it, it would therefore seem as if supergravity is only an improvement on quantum gravity in the sense that the divergences start at higher loop order."

This result was presented at the Nuffield workshop on superspace and supergravity, which was held in the early summer of 1980 [77]. It seems to have brought the curtain down on most hopes that supergravity could itself cure the ultraviolet problems of quantum gravity coupled to matter. In his *Physics Reports* review on the situation [123], Peter van Nieuwenhuizen recapped the results on quantum supergravity, and commented that the "results seem to leave little hope short of a miracle ... Of course, miracles do sometimes happen, but they would have to happen at every loop order. It seems better to accept a fundamental property of gravitons: that no nearby massless theory exist (the van Dam-Veltman theorem) and to use nonperturbative methods. Perhaps this is the way quantum supergravity should go in the future, but it is easier to say that one should do nonperturbative calculations than to do them."

These discouraging results set the stage for the next act, in which the possible role of strings in eliminating the ultraviolet divergences was (and is being) explored.

7 Supersymmetry in Higher Dimensions

A great deal of work in supersymmetry has been done in the context of space-times that have more than four dimensions.

Higher-dimensional space-times had been used in the study of gravitation before as a possible means of unifying gravitation with the other forces. Kaluza's original idea was written up in 1919 (although it was not published until 1921), and so applied only to gravitation and electromagnetism. This idea was also studied extensively by Klein and many others. (A history of Kaluza-Klein theories has been given by Appelquist, Chodos, and Freund [6]. Those references which are not specifically cited below may be found there.) Although there had not been a great deal of activity centered around the Kaluza-Klein approach to unification in the recent past, Bryce De Witt assigned the problem of unifying gravitation and Yang–Mills forces in the Kaluza-Klein geometrical framework as a problem (no. 77) in his 1963 Les Houches lectures. The effect that this assignment had on the students who attended the lectures and on the many more who read the lecture notes is difficult to gauge. Nevertheless, by 1968 Kerner had presented work on this project. Cho and Freund later added to this, and another important piece of work, which will be discussed further below, was done by Cremmer and Scherk [19].

Given the past link between the Kaluza-Klein approach to gravitation and the possibility of unification, it was perhaps inevitable that higher dimensions would play an important role in the study of theories of supergravity. There was, however, another road that had led to the study of theories in higher dimensions, and this was the road marked out by the study of dual string theories.

It was primarily through the work of theorists who came to the study of higher-dimensional theories via work on hadronic strings, that supersymmetry first came to be studied in higher dimensions. For example the work by Cremmer and Scherk on spontaneous compactification was motivated by considerations that came out of the study of dual models. It is therefore necessary to look at some of this earlier work, in which the role of supersymmetry was largely irrelevant, in order to get a sense of how and why supersymmetric theories first came to be formulated in space-times of more than four dimensions.

7.1 *Higher Dimensions of Dual Models*

One of the problems of the dual resonance model was the presence of ghost states. A great deal of study was devoted to understanding how these non-physical states could be made to decouple from the physical states. One suggestion in this regard had been to add a fifth dimension. Some work in a 1970 paper by Lovelace indicated that the problem might be solved if the dimensionality of the space-time was increased to 26. This was verified by later work, which also indicated that for the models that were associated with graded algebras, the critical dimension was 10. (References can be found in Ref. [69].)

The existence of a critical dimension seems to have been viewed as something of a mixed curse. On the one hand, since the world does not appear to be 26 (or even 10)-dimensional in any obvious way, the known consistent string theories seemed to be unphysical. It seems to have been assumed by most that when a more realistic theory could be constructed, the number of space-time dimensions predicted by it would be four. The positive aspect was that there *was* any prediction of the number of space-time dimensions at all. Even general relativity, which describes the geometry of space-time, had always been done with the number of dimensions as input.

Although the prevalent attitude seemed to be that the extra dimensions themselves did not have any physical significance, there were individuals who explored the possibility that the extra dimensions might indeed have a physical interpretation. Notable among these were Scherk, Schwarz, and Cremmer [18, 111]. This view of the possible physical interpretation of the extra dimensions was part of the larger picture in which the possible role of the dual models had been expanded to include, in addition to the hadronic interactions, at least gravitation as well. Although the themes of string theories as potential 'theories of everything' and the role of the higher dimensionality of string theories are intimately related, in this section we will focus on the aspect of higher dimensions.

7.2 *The Physical Interpretation of Extra Compact Spatial Dimensions: An Extrapolation in the Context of Dual Models*

In a paper which was submitted to *Physics Letters* in March of 1975 [112], Scherk and Schwarz asserted that "it is possible to make sense of a theory formulated in $4 + N$-dimensional space-time, provided that the extra N variables span a compact and spacelike N-dimensional domain. ... The existence of the N extra spatial dimensions is unobservable so long as one probes with energies less than $1/R$ [where the size of the compact space is of order R]. This is a simple consequence of the fact that fields with non-zero values for the quantum numbers associated with the eigenfunctions of [the compact space] have mass of order $1/R$."

This theme was explored in more detail in a paper by Cremmer and Scherk which appeared soon after this [18]. In '*Dual Models in Four Dimensions with Internal Symmetry*', Cremmer and Scherk studied the implications of interpreting the extra dimensions of the dual theories as spatial dimensions of a compact space. They found that, if the total number of dimensions was equal to the critical number, then there are still no ghosts. Although their central interest was in showing that dual models defined on such spaces are consistent, Cremmer and Scherk included a long and illuminating discussion on field theories in which some of the dimensions are unbounded and some are compact. Their approach included a number of features that were important to the later work that was done on higher-dimensional supersymmetric theories.

The particular space that they considered was a four-dimensional space, with two unbounded dimensions and two compact dimensions. x^0 and x^1 corresponded to the unbounded time and space coordinates respectively. x^2 and x^3 will be taken to represent the coordinates of the compact internal space, which was chosen to be a torus of radii R_2 and R_3. Thus, fields with x^2 and x^3 dependence are necessarily periodic. Any field $\Phi(x^0, x^1, x^2, x^3)$ can therefore be written as

$$\Phi[\hat{x}] = \frac{1}{\sqrt{R_2 R_3}} \sum_{n,m} \Phi_{n,m}(x) \exp\left[2i\pi\left(\frac{nx_2}{R_2} + \frac{mx_3}{R_3}\right)\right], \tag{RS7.1}$$

where n and m are integers which run from 0 to ∞; hatted variables and indices denote four-dimensional quantities.

The authors choose an action:

$$S = \int d^2x \int_0^{R_2} dx^2 \int_0^{R_3} dx^3 \left(\frac{1}{2}\eta^{\hat{\mu},\hat{\nu}}\partial_{\hat{\mu}}\Phi\partial_{\hat{\nu}}\Phi - \frac{1}{2}\mu^2\Phi^2 - \frac{1}{4!}\lambda_0\Phi^4\right). \tag{RS7.2}$$

When Φ is substituted into the action, and the integrals over x^2 and x^3 are performed, the result is an action with kinetic energy and mass terms given by

$$S_{ke} + S_m = \int d^2x \left(\sum_{n,m} \frac{1}{2}\partial_\mu\Phi^*_{n,m}\partial_\mu\Phi_{n,m} - \frac{1}{2}\mu^2_{n,m}\Phi^*_{n,m}\Phi_{n,m}\right). \tag{RS7.3}$$

The mass parameter $\mu^2_{n,m}$ can be expressed as

$$\mu^2_{n,m} = \mu^2_0 + \left(\frac{n^2}{R_2^2} + \frac{m^2}{R_3^2}\right). \tag{RS7.4}$$

Cremmer and Scherk went on to interpret the physical meaning of this: "Suppose we would live in such a world. ... Then in high energy experiments one would produce heavy particles in pairs, having two additive conserved quantum numbers n and m... A particle moving in the x_2, x_3 direction would not at all disappear from the two-dimensional space-time, but would simply be seen as a particle with n, m not zero. So in this universe we would have no classical feeling at all that there are four dimensions..."

The ideas presented in this paper were important in two ways. Of course, first of all, the idea of how the existence of extra physical dimensions might be exhibited through the existence of high mass particles was important. But, in addition to this, it turned out that the method of dimensional reduction presented there was very useful for supersymmetric theories. The reason for this is that it provides a mechanism for the breaking of supersymmetry. One can get a sense that this might be so from the fact that the procedure outlined above changes the masses of the particles associated with the fields of the theory. Since the masses of all of the particles in a supermultiplet are the same, if the mechanism can change the masses of particles in the same multiplet by different amounts, then the supersymmetry would no longer be manifest.

7.3 *Precedents to This Early Work*

The paper by Scherk and Schwarz referenced the earlier Kaluza-Klein work which had been done in the context of gravitation. In addition to noting the work of Kaluza, Scherk and Schwarz referenced a work by Thirring which had just appeared in 1972; this particular paper was also referenced in a good deal of the early work on supersymmetric Kaluza-Klein theories.

Work by Ne'eman, and by Ne'eman and Rosen was also cited. This work was not done in the context of gravitation, but was entirely motivated by considerations central to particle physics. However the idea put forward was related to the Kaluza-Klein idea, in that the suggestion made was to view the internal symmetries as being related to transformations of additional internal compact space-time coordinates.

Scherk and Schwarz also mentioned that when Fubini and Veneziano had introduced a fifth dimension to try to deal with the problem of ghost states in the Veneziano model, they had taken the momenta associated with the extra spatial dimension to be discrete.

7.4 *Spontaneous Compactification in Kaluza-Klein Theories*

Although the Kaluza-Klein idea had been applied to Yang–Mills theories, there was a problem associated with this application that had not occurred when electromagnetism alone had been linked to gravitation. The problem had to do with the identification of the vacuum state. It was felt that the vacuum state should be a direct product of four-dimensional Minkowski space and an $(N-4)$-dimensional compact space. This form for the vacuum solution could not be achieved with a pure Kaluza-Klein theory in more than five dimensions. [By a 'pure' Kaluza-Klein theory is meant one in which the only field considered in the higher-dimensional space is the gravitational field.] And, in order to incorporate interactions which went beyond electromagnetism, more than five dimensions were needed.

In 1976 Cremmer and Scherk demonstrated that a product vacuum solution of the desired form could exist if fields other than gravity were included in the higher-dimensional space itself [19]. The solution that was exhibited by Scherk and Cremmer was an example of a 'spontaneous compactification'. The compactification is spontaneous in the same way that the breaking of a spontaneously broken internal symmetry is spontaneous. That is, the action and the equations of motion are completely symmetric (in the Kaluza-Klein case they are symmetric in all of the space-time coordinates), but the ground state solution necessarily breaks the symmetry.

Although the inclusion of fields other than the gravitational field in the higher-dimensional space seemed to be a violation of the spirit of the Kaluza-Klein approach, it turned out to be ideally suited to the study of supergravity theories, since the existence of fields other than the gravitational field is mandated by the requirement of supersymmetry anyway.

7.5 *Supersymmetry in Higher Dimensions*

It turned out to be advantageous to study even theories with only global supersymmetry in higher dimensions. One reason for this was that it was possible to derive supersymmetric theories in four dimensions from theories originally formulated in higher dimensions, by a dimensional reduction procedure. As a matter of fact, theories with sufficient complexity to be thought of as serious candidates for physical theories with unified interactions were originally derived in higher dimensions. This happened for both supersymmetric Yang–Mills theories and for supergravity.

The number of independent fields in the higher-dimensional space is smaller than the number of independent fields in the lower-dimensional space, and so there are fewer fields and indices to keep track of. It is therefore often easier to construct and to work with the theory in its higher-dimensional form. For example, through dimensional reduction to four space-time dimensions, the 10-dimensional $N = 1$ Yang–Mills theory is transformed into a Yang–Mills theory with $N = 4$, while the $N = 8$ supergravity theory in 4 dimensions was derived from the $N = 1$ theory in 11 dimensions.

Just as it proved to be easier to derive some supersymmetric theories in higher dimensions, it also proved to be easier to explore some of their properties in these higher-dimensional space-times. Four issues which were important in the study of supersymmetric theories are discussed in the subsections just below. There are other interesting issues however, and a vast literature exists. References can be found in many places, for example see Refs. [6] and [23].

Is 11 the Maximum Number of Dimensions?

The 11-dimensional theory from which the maximally symmetric, $N = 8$, supergravity was derived is of particular importance. This is so because, as was shown by Nahm in 1978, any theory of supergravity which is formulated in more than 11 dimensions will, upon dimensional reduction to 4 dimensions, necessarily have N greater than 8, and will, therefore, include fields with spin greater than 2 [139]. Since, despite a great deal of study, it seemed as if such theories were inconsistent, this was taken to mean that 11 was the maximum number of dimensions appropriate for the study of supergravity. Furthermore, because the 11-dimensional theory corresponds to the maximally symmetric theory, it was also considered to have the best chance of ultimately being shown to be finite. 11 was a favored number of dimensions for another reason as well, because in 11 dimensions there are no matter multiplets, i.e. multiplets which do not include gravitation. Thus, the requirement that there be just one graviton uniquely specifies the multiplet.

Spontaneous Compactification of 11-Dimensional Supergravity: $11 = 4 + 7 = 7 + 4$

In a paper which was written in 1980, Freund and Rubin commented on the need for additional bosonic degrees of freedom in order for spontaneous compactification to work: "Such a complication of the original theory is contrary to the spirit of unification. Here we remark that bosonic fields other than gravity, automatically appear in higher-dimensional supergravity theories, and that they can induce spontaneous compactification. The number of compactified dimensions can then take only two preferential values. For 11-dimensional supergravity we find that either 7 or 4 space-like dimensions compactify. In the first case ordinary "large" space-time has therefore 1 time and 3 space dimensions, a pleasing result."

Supersymmetry played a key role in the derivation of this result. The total number of dimensions, d, was assumed to be 11, because of the considerations mentioned in Sec. 7.5 above. One of these dimensions was taken to correspond to a time coordinate, and the remaining 10 dimensions were assumed to be spatial. The additional bose fields needed for the compactification were associated with an antisymmetric tensor of rank $s - 1$, with corresponding field strength $F^{\mu_1 \cdots mu_s}$ of rank s. Without the requirement of supersymmetry, s is an arbitrary parameter. However the requirement of supersymmetry, specifically the associated equality of the number of Fermi and Bose degrees of freedom, mandates the existence of a field with rank $s = 4$. Freund and Rubin assumed the existence of a vacuum solution of the product form, and then derived expressions for the scalar curvatures in each of the factor spaces in terms of d and s,

$$R_{d-s} = \frac{(s-1)(d-s)}{d-2}\lambda, \quad R_s = -\frac{s(d-s-1)}{d-2}\lambda. \tag{RS7.5}$$

For $d = 11$ and $s = 4$, the signs of these expressions differ. This indicates that either the four-dimensional space-time is not compact, and the 7-dimensional space is, or a 7-dimensional space-time is not compact, and a 4-dimensional spaces is. Unfortunately, it proved possible to construct an example of the latter type of split [102], so that it could not be argued that supersymmetry unambiguously leads to a prediction of our observed 4-dimensional space-time.

Higher Dimensions for the Spontaneous Breaking of Supersymmetry

Another way in which the extra dimensions turned out to be useful was in the breaking of supersymmetry. Scherk and Schwarz and Cremmer used an idea similar to that outlined above from the paper of Scherk and Cremmer, to devise a way to affect supersymmetry breaking through dimensional reduction. (References can be found in Schwarz's contribution to Ref. [122].) The phrase 'dimensional reduction' can be used in several ways. In their paper titled '*How to Get Masses From Extra Dimensions*', Scherk and Schwarz note that: "In 'ordinary' dimensional reduction, the coordinates of a $D+E$-dimensional theory are divided into D space-time coordinates (x^μ) and E internal coordinates (y^α) that form a compact space. The fields and symmetry transformation laws that the theory possesses in $D+E$ dimensions are taken to be y independent [...]. A limitation of this approach is that starting from a massless theory (as is necessarily the case for supersymmetry theories in 10 or 11 dimensions) the resulting reduced theory is also massless (aside from possible masses due to the Higgs mechanism or topological excitations). Also it is exactly invariant under the supersymmetry algebra, implying Fermi-Bose degeneracy that needs to be broken in order to describe the real world." They go on to discuss their generalization of this approach, "The new ingredient is to allow the fields and transformation laws to depend upon the internal y coordinates in a well-defined fashion determined by a symmetry in $D+E$ dimensions."

The fields are expressed in a form that automatically respects the chosen symmetry, and then this form is substituted into the action. Work by Scherk, Schwarz, and Cremmer showed that it is possible, by following this procedure to induce a spontaneous breaking of supersymmetry. In his contribution to the 1979 conference at Stony Brook, Schwarz illustrated how this sort of supersymmetry breaking works. He gave two examples. In each, one starts with an action that has two or more symmetries, two of which are played off against each other. In his first example, he considered an action which describes only bosonic states. The action is like (4.12), but without the mass and interaction terms. The theory is defined on a 5-dimensional space-time, with one of the coordinates, y, assumed to parameterized a circle. This abbreviated action is symmetric under both global phase transformations

$(\Phi \to e^{i\alpha}\Phi)$ and constant translations $(\Phi \to \Phi + c)$. The former symmetry is used to write a specific form of a solution which is periodic in y,

$$\Phi = e^{imy} \sum_{-\infty}^{\infty} \Phi_n(x) e^{\frac{2\pi i n y}{L}}. \tag{RS7.6}$$

When, in a procedure which parallels the one followed by Cremmer and Scherk [18], this form is substituted into the action, mass terms are derived, with the mass associated with Φ_n given by:

$$m_n = \left| m + \frac{2\pi n}{L} \right|. \tag{RS7.7}$$

The mass terms in the action indicate that the translational symmetry has been lost. This result holds even after the $L \to 0$ limit is taken and the dimensional reduction is complete.

The procedure for the spontaneous breaking of supersymmetry in a theory of supergravity is similar. The major complication is that fermions need to be included, and that the dimensional reduction will therefore necessarily involve a dimensional reduction of the appropriate Dirac matrices as well. (Actually Cremmer and Scherk had already considered fermions in their paper of 1975.)

In order to illustrate how things work out for a theory of supergravity, Schwarz considered a 4-dimensional theory, whose symmetries include both local supersymmetry and a global chiral symmetry: $\psi_\mu \to (e^{im\Gamma_s})\psi_\mu$, where the indices are considered to be 4-dimensional. This chiral symmetry was used to stimulate a spontaneous breaking of the local supersymmetry which allowed the gravitino to become massive (while the graviton remained massless), when the theory was reduced to a three-dimensional one.

Evolving Attitudes Toward Work in Higher Dimensions

One can study the evolving attitude toward work in higher dimensions by looking back at the proceedings of the schools and conferences that have been held on supersymmetry. Higher-dimensional theories were discussed at the Stony Brook conference in 1979 [122], but the work on higher dimensions formed a small percentage of the work that was presented there. It is also significant that, at least in their written form, the papers that did use higher dimensions, did not discuss the physical significance of the extra dimensions. Rather, the extra dimensions seem to be included only as a means of learning something about supersymmetric field theories in four dimensions. For example, the contribution of John Schwarz which discussed supersymmetry breaking through dimensional reduction is entitled '*How to Break Supersymmetry*'. There was a contribution by Ulf Lindström, describing work which had been done in collaboration with Deser, entitled '*Use of Dimensional Reduction in the Search for Supergravity Invariants*'. In explaining their motivation for working with a higher-dimensional theory, Lindström wrote: "Since the simple supergravity action in 11D is far easier to handle than that of the reduced theory in 4D, the search for a current multiplet (generalizing the Bel-Robinson tensor) and its corresponding invariant should be considerably easier." A paper by D'Adda, A'Auria, Fré, and Regge discussed work that was to become important for the study of supergravity in higher dimensions. Although the paper, '*Geometrical Formulation of Supergravity as a Theory on a Supergroup Manifold*', focused on four-dimensional theories, the authors noted that the question of how their formalism worked in any number of dimensions had already been addressed.

The proceedings of the schools on supersymmetry which were held at Trieste provide an opportunity to see which ideas had made it into the mainstream of work on supersymmetry. The topics presented there were chosen so as to educate graduate students and others who were new to the study of supersymmetry, in what were considered to be its fundamentals, and to guide them toward a working knowledge of the basic tools of the trade. Each of these volumes contains an introduction by Salam which provides an additional guide. Studying the topics of the lectures, one does find an increase in emphasis on Kaluza-Klein ideas. In the proceedings of the 1981 school, there is very little mention of higher dimensions, except in the context of dimensional reduction, particularly the derivation of $N = 8$ supergravity in four dimensions from $N = 1$ theory in 11 dimensions. This is also the only context in which Salam mentioned higher dimensions in his introduction. By 1982 things were different. Salam wrote: "The new ground (relative to 1981) lies in the Kaluza-Klein domain and its compactification discussed by J. Strathdee and M. Duff, with related discussions by H. Nicolai on $N = 8$ supergravity, and a contribution from P. Fré on supergravity in 11 dimensions in the context of Cartan integrable systems." The emphasis on Kaluza-Klein theories was even more pronounced in the proceedings of the 1984 school. Salam wrote, "Far from being a peripheral interest, these theories have now come to occupy the center of the stage among supergravity models. ...This shift has been discernible ever since 1979, when Cremmer, Julia and Scherk showed that the maximal $N = 8$ supergravity massless particles in four dimensions could be obtained as a truncation of the $N = 1$ supergravity theory in eleven space-time dimensions. This was the extended super-Kaluza-Klein miracle. The change since 1979 is the deeper study of the compactification mechanism from the higher to the four-dimensional anti-de Sitter and, in some cases, Minkowski space-time, an appreciation of their topological significance, the spectra of the massive (super)-Kaluza-Klein excitations, investigations of their stability properties as well as of the possible supersymmetry breaking. During the Workshop, other aspects of the Kaluza-Klein theories were considered, such as the problems of inducing the compactifications radiatively, the possible relevance of (super)-Kaluza-Klein theories to phenomenology and cosmology — and, among the other riches, the super-string models [...] which after all provided of the original reasons for the higher (ten-)dimensional theories."

Is 11 the Minimum Number of Space-time Dimensions?

One of the papers which was responsible for the increased interest in the Kaluza-Klein theories, which was evident by 1982, is a paper by Witten [134], '*Search for a Realistic Kaluza-Klein Theory*'. In this paper Witten made an observation that gave hope that the 11-dimensional supergravity theory might be chosen by phenomenological considerations. He argued that 11 was the lower bound on the number of space-time dimensions, since it is the minimum dimension of a compact space which can have an $\mathrm{SU}(3) \times \mathrm{SU}(2) \times \mathrm{U}(1)$ symmetry. He commented that, "It is certainly a very intriguing numerical coincidence that [the maximum number of dimensions] for supergravity, is the minimum number in which one can obtain $\mathrm{SU}(3) \times \mathrm{SU}(2) \times \mathrm{U}(1)$ gauge fields by the Kaluza-Klein procedure. This coincidence suggests that the approach is worth serious consideration." This paper was submitted to *Nuclear Physics* in January of 1981. The approach it advocated received a great deal of attention during the following three or so years.

Chiral Fermions

There was however a problem with 11 dimensions, which was noticed in Witten's paper itself. That problem had to do with the possibility of constructing a theory in which the non-compact 4-dimensional world would be inhabited by the chiral fermions that are observed in our own apparently 4-dimensional world. In considerations which were studied in both Ref. [134] and in later work, Witten found indications that it might be impossible to construct such a theory by starting in 11 dimensions. This problem also received a great deal of attention from others and several possible loopholes were found. (For a review of these see Ref. [6].) Work by Wetterich [133], indicated that the problem with chiral fermions could be solved in 2 mod 8 dimensions.

Back to the Higher Dimensions of String Theories

The considerations above provide evidence of a sort of positive feedback loop between string theories and supersymmetric field theories. For, the study of supersymmetry in higher dimensions, which was often initiated by researchers who had worked with hadronic strings, eventually led to a more ready acceptance of the higher dimensions which appeared to be necessary for a consistent quantum supersymmetric field theories. As mentioned above, considerations of chirality in supersymmetric filed theories even led to an independent argument that there might be something special about 10 dimensions, the critical number of dimensions for supersymmetric string theories.

The theme of mutual reinforcement between supersymmetric string and field theory will also be reinforced in the discussion of superstrings in the next section.

8 Supersymmetry and Superstrings: Back to the Future

As a graph of work on dual theories [Fig. 6.8] demonstrates, interest in these theories never completely dried up. However most of the papers that were published about strings in the late 1970's and in the first years of the 1980's were still not about strings as possible fundamental theories. Instead, they considered the role that strings might play as effective theories of the strong interactions. There were some notable exceptions however, and the fruit that these bore was the modern superstring. Supersymmetry played an important role in this development. These notes will outline that role, but will necessarily stop at about 1984.

When we spoke of strings in Sec. 3, Scherk and others had begun to explore the relation of string theories to local field theories. As discussed in that section, the understanding that it was possible to obtain both Yang–Mills theories and gravitation as low energy limits of string theories was largely responsible for the proposal by Scherk and Schwarz to view string theories as possible theories of everything.

It would seem that, beyond its conceptual significance, the implication that string theories contain local field theories, was important to the eventual development of superstrings in another more practical way. For it meant that those who still maintained a strong interest in exploring string theory did not have to part ways with their fellow theorists who were primarily interested in studying fields, since studies of field theory could yield information that would be important to the study of strings, and vice versa. In fact, the study of supersymmetric field theories did provide valuable leads for the investigation of strings, while taking the low energy limit of certain string theories led to the derivation of new supersymmetric field theories. The possibility of such a symbiotic relationship was of course made more likely by the string enthusiasts, like Scherk and Schwarz, who also worked on supersymmetric field theories. For it allowed them to help to set an agenda for the study of supersymmetry in which some of the ideas from string theory became a familiar part of the repertoire of many physicists whose interest was primarily in field theory. When the work of Green, Schwarz, and others indicated the there were important reasons to study strings, the necessary tools were within the reach of a relatively large number of active physicists, and some of the ideas (like the possible need to work in higher dimensions) no longer seemed terribly strange.

8.1 *Strings with Space-time Supersymmetry: The Elimination of Tachyons*

It was strings with world-sheet supersymmetry which provided the point from which supersymmetry jumped into the realm of field theory and was realized as a space-time symmetry. Later it was an understanding of how string theories can themselves incorporate space-time supersymmetry, that led to the renaissance of strings.

In 1976, Gliozzi, Scherk, and Olive did work that proved to be pivotal in both the study of supersymmetric field theories and in the later study of superstrings [59, 60]. As far as the latter is concerned, their work gave a clear indication that a suitable projection of the states of the Ramond-Neveu-Schwarz (RNS) model might respect a space-time supersymmetry. An important characteristic of their projection was that it is free of tachyons. The signal of the latent supersymmetry was that there were an equal number of Bose and Fermi states at each mass level. Oddly enough, this aspect of their result was not followed up for some time. It seems strange that this important lead was not followed up sooner, especially because in their introduction, the authors put particular emphasis on the possible role of string theories as fundamental theories. They said that "[One] line of thinking is to consider supergravity and its refinements as only a first step. ... Then there would be an obvious parallelism between the $V - A$ with coupling constant G_F and the gravitational theory with coupling constant $G_N \sim \kappa^2$ since both constants have the same dimensions. In the renormalizable gauge theories G_F is a phenomenological parameter, expressed as $G_F \sim g^2/M_W^2$ where g is the real dimensionless coupling constant of the theory and M_W^2 plays the role of a cut-off. So one may ask whether one could find a similar expression for $G_N \sim (g^2)^n/\Lambda^2$ and identify g and Λ.... Dual models provide an answer to this kind of question."

The aspect of their work that was immediately followed up was the method that they used to derive supersymmetric field theories. By considering the zero-slope limit while also letting the size of the compact 6-dimensional space go to zero, they derived supersymmetric field theories, some of which were already known and some of which had not yet been constructed. For example a string theory with both open and closed strings had the $N = 4$ supersymmetric Yang–Mills coupled to $N = 4$ supergravity as its field theory limit, while a theory with only closed strings yielded $N = 8$ supergravity.

Gliozzi, Scherk, and Olive emphasized the relation to field theory: "... We would like to show that dual models are in accordance with the program of supergravity" [60]. This relation was apparently also important to the physicists who took up the challenge of exploring the space-time supersymmetry of the RNS model. In 1982 Green wrote, "The suggestion that a truly supersymmetrical string theory might exist motivated John Schwarz and me to construct the theory explicitly. The ... connection with $N = 4$ Yang–Mills theory and $N = 8$ supergravity provided incentive in addition to the intrinsic interest in string theories as a possible framework for a fundamental theory of nature" [66].

8.2 *Explicit Supersymmetry and Anomaly and Infinity Cancelations: Superstrings Take Off*

In 1980, approximately four years after the work by Gliozzi, Scherk, and Olive, Green and Schwarz began a systematic program to understand the space-time supersymmetry of 10-dimensional string theory. They described their first paper on the subject as "another step towards a proof of the conjectured supersymmetry of the ten-dimensional theory" [63]. In this first paper they actually constructed a supersymmetry generator for the on-shell states and showed that this generator satisfied the supersymmetry algebra and that it also transformed as a Lorentz spinor.

In a series of papers they worked their way toward a covariant description of superstrings. The story of this work, and of the development of a superstring formalism, and of the eventual discovery of anomaly and infinity cancelations, is amply documented in Refs. [115], [69], and in several excellent reviews which are referenced in these. Perhaps the result which caused the most excitement was the discovery that the requirement of anomaly cancelation actually picked out a symmetry group, either SO(32) or $E_8 \times E_8$ — this before an $E_8 \times E_8$ superstring theory had been constructed.

The work of Green and Schwarz set off an explosion of further work on superstrings. Although supersymmetry plays an important role in much of this work, it cannot be considered here because of limitations of space. Nevertheless it seems appropriate to at least assess the unusual strength of the trend from a historical perspective. For it does seem that, for a type of theory whose study has not been motivated by experimental observations, string theory has inspired an unprecedented amount of excitement among theorists. (Compare this to the discovery of superconductivity in substances that are maintained at relatively high temperatures. The wave of excitement which has followed this discovery was motivated, and is being constantly fed by experimental results.)

The fact that something special has indeed happened is documented in the November 9, 1987 issue of *Current Contents* [56]. This issue contains a list of the 1985 physical sciences articles that were most-cited in 1985 and 1986.

In his commentary on the list Eugene Garfield notes the rise of superstrings. "The emergence of superstring theory in this year's study represents an important new development in the field. An indication is the very prominence of the term 'superstring'. In our study of the 1983 papers most-cited in 1983 and 1984, not a single ... paper contained the words 'string' or 'superstring' in its title. However by the following year... 'superstrings' appeared in the titles of two papers. This year, 'string' or 'superstring' appear in 24 article titles." In order for a paper to be included on this list of most-cited papers it must have received at least 32 citations. Only 104 papers satisfied this criterion, so the string contingent represents about 23 percent of the total. This is far more than the percentage of papers in any other area. For example there were 12 articles on quasicrystals and 4 on astrophysics. (Papers from all of the "physical sciences" were considered, except for chemistry, which was studied separately.)

Garfield goes on to say that " ... this year's study is dominated by papers concerning string theory. And perhaps the most dramatic indication that something important has occurred in the superstring field is that the five top-cited papers are all related and that three have received an unprecedented number of citations for a two-year study in the physical sciences. The most cited paper on this year's list received 304 citations. But more typical of past studies were the 170 citations collected in two years by the paper announcing the discovery of the W^+ and W^- particles."

It seems fitting, that supersymmetry, which (at least in the West) had its origins in the study of string theory, should have had such a profound effect on the study of strings.

9 Prospects for a Superfuture

9.1 *Supersymmetry and Experiment*

No matter how aesthetically pleasing the concept of supersymmetry is, or how powerful a tool it provides for the exploration of possible theories, its role as a physical symmetry can be assured only through experimental verification. Verification could come in one of two ways. The first might be experimental discovery of a phenomenon that is predicted uniquely by supersymmetry. An example would be the observation of superpartners of the known particles. Verification could also come in another guise, if the theoretical study of supersymmetry can lead to an understanding of aspects of the physical world which at present must be simply taken as given. If, for example, the masses and coupling constants of the known particles, and perhaps even the dimensionality of space-time and of its fundamental inhabitants could, albeit after the fact, be predicted by supersymmetry, then its value as a physical principle would probably be widely accepted.

A great deal of work has been done to understand the phenomenology of supersymmetry. However there does not seem to be any one phenomenological scenario which would be compelled to occur by the existence of supersymmetry. Therefore, although it might be possible for a series of observations during the next few years to confirm the existence of supersymmetry, there does not seem to be any comparable series of observations which could rule it out. This state of ambiguity is cited by Fayet in his contribution to the 1986 Nobel Symposium [12]. "The observation of superpartners at present or future colliders would probably be considered now as a largely expected discovery. But one should also keep in mind that superpartners might only appear at extremely high energies; or, conversely, that both superpartners and extra dimensions might reveal themselves at future colliders." The situation with regard to possible experimental tests of superstring theories is, if anything, even more ambiguous. Nevertheless work is proceeding to try to determine the observational consequences of superstrings, both for collider experiments and for astrophysics. The motivation for these studies is twofold. Beyond the idea of experimentally testing the validity of the hypothesis of superstring theories, is the question of whether or not (or perhaps which of) these hypotheses are consistent with observations which have already been made.

Given this state of affairs, it is difficult to pick out those themes of the history of the study of supersymmetric phenomenology which will ultimately prove to be most significant. Because of this, and because of the limited space available, the discussion of this theory will be brief. More complete discussions and references can be found elsewhere, for example in Refs. [92], [85], [24], [47].

The earliest phenomenological idea that was related to supersymmetry was the one suggested by Volkov and Akulov, that perhaps one might be able to understand the existence and couplings of the neutrino by viewing it as the Goldstone particle associated with the spontaneous breaking of a fermionic symmetry. The fermionic analogue of the Goldstone theorem was proved by Salam and Strathdee in 1974. This work was referenced by de Wit and Freedman when, in 1975, they studied the phenomenology of Goldstone neutrinos [49]. (They also referenced some unpublished work of W.A. Bardeen.) B. de Wit and D. Freedman studied the validity and implications of low energy theorems of spontaneously broken supersymmetry and concluded that, at least for the broad class of theories that they had considered, it was likely that the hypothesis that the neutrino was a Goldstone fermion was inconsistent with experimental results.

The 1974 work of Wess and Zumino illuminated the ability of supersymmetry transformations to connect particles of different spin. In their first paper on supersymmetry they conjectured that "supersymmetry may ... provide a natural way for the formulation of higher internal symmetries linking mesons and baryons." After their work, attempts were made to relate the known elementary particles of different spin through supersymmetry transformations. Unfortunately, it soon began to seem unlikely that such a program could be successfully carried out. It is now widely agreed that supersymmetry requires the introduction of so-called superpartners of the known particles. The half integral spin partners of the known bosons are graced with the suffix "ino". Thus the graviton has the gravitino, the photon has the photino, while the W's and Z's pal around with the more dubious-sounding winos and zinos. (Before this 'ino' convention was adopted, Bruno Zumino referred to what is now generally called a goldstino as a germion" [116].) The spin zero partners of the leptons and quarks are the sleptons and squarks. Most attempts to detect traces of supersymmetry in the physical world have involved searching for a distinctive signal of the superpartners, and have placed limits on their masses based on the absence of such a signal. (See for example the contribution of Savoy-Navarro to Ref. [92].)

Although it does not seem that supersymmetry can relate all of the fundamental particles to each other within supermultiplets, supersymmetric theories can incorporate multiplets which contain both the spin-0 Higgs bosons and the spin-1 gauge fields. (See for example the contribution of Fayet to Ref. [92].)

Another nice feature of supersymmetric theories is related to the lack of radiative corrections, which was discussed in Sec. V. In the context of supersymmetric grand unified theories, this feature ensures that the widely different mass scales (of the mass of the W and the mass of the X for example) stay widely separated, even when radiative corrections are considered; i.e. hierarchies are preserved in supersymmetric quantum theories.

9.2 *A Superfuture?*

The flavor of the above discussion reveals that high energy theory and experiment are not coupled in the way that theory and experiment were coupled during the 1920's or even during the 1960's and early 1970's. There are tests of some of the predictions of supersymmetry that either have been done, are ongoing, or are planned. And the continuing interaction between experimentalists and theoreticians is illustrated by the 1982 meeting that took place at CERN, and by the contributions to that meeting which are collected in the *Physics Reports* issue 'Supersymmetry Confronting Experiment' [92]. However, a clear-cut and decisive confrontation between supersymmetric field theory and experiment has yet to occur.

The development of superstrings continues the general trend of a *widening separation between high energy theory and experiment.* The reasons for the trend are not negative. On the contrary, it reflects the successes of our theories to date, which have been capable of enveloping most of the observed high energy phenomena, and our optimism about the ability of the human mind to explore new territory. It is difficult, however, to know where this exploration will take us, or how we will know that we are there. During his talk at the Nobel Symposium, Ellis referred to the "lone experimentalist at this meeting." Glancing through the proceedings of that meeting one also gets a sense of another trend-the increasingly important role that frontier mathematics is playing in the further development of string theory.

Since at any time a new and perhaps unexpected discovery either experimental or theoretical-could change things dramatically, it is impossible to know definitely what the future holds in store for the further development of supersymmetry. But it is clear that, should a further history of supersymmetry be written ten years from now, it will either include the story of the experimental discovery of supersymmetry, or a more detailed discussion of the ways in which supersymmetry will be known not to manifest itself. A good deal of information of this latter type should be obtainable from observations that are possible within the next decade. *If, at that time, supersymmetry is still a purely theoretical phenomenon, it is possible that it may nevertheless continue to be vigorously studied by theorists. The reason for this is that, beyond its aesthetic appeal, it provides ways of calculating quantities that might otherwise be poorly defined, or at least so much more difficult to compute, that computation is not viable. The lack of experimental verification would, however, present a formidable challenge to theory. For it will probably mean, at least, that either our first attempts at understanding how nature may be supersymmetric have been too naive, or that supersymmetry is a property of the world which can only be understood if one knows about phenomena that occur at much higher energies than those we have thus far been able to observe.*[155]

In considering the possibility that supersymmetry, if it is indeed realized in the physical world, is an ultra high energy phenomenon, we can wonder about how far we can go toward understanding its role through the power of pure thought. One may wonder for example what would have happened if the principle of non-Abelian gauge symmetry had been developed during the nineteenth century. Although the Abelian case might have been understood to correspond to electromagnetism, what use would non-Abelian symmetry, beautiful though it is, have been to physicists who were trying to understand the world at much smaller distances than those they were able to probe? Without any knowledge of atomic, much less nuclear structure, and without knowing that the laws of classical physics could not be applied to nature in the small, how much progress would they have been able to make by constructing gauge theories? At this early stage in the development of

[155]We allowed ourselves to italicize this part of the paragraph (written by the author approximately 12 years ago) which turned out to be prophetic. — Editors' note.

high energy physics it seems possible that we are in an analogous situation, although we may fervently hope not.

Acknowledgments

This paper was written at the Institute for Theoretical Physics at Stony Brook where I have greatly benefited from the presence of several active practitioners of supersymmetry, in particular P. van Nieuwenhuizen. I am also grateful to S. Deser, P. Ramond, A. Salam, D.V. Volkov, and J. Wess for providing helpful information and comments. I would like to thank Max Dresden for telling me about the early conjecture of Bohr, for copies of lecture notes from the 1940's, for his comments on the manuscript, and for his encouragement. I would also like to thank Bruce Gittleman for his help in compiling the data plotted in the graphs, Arvind Borde for useful comments of the manuscript, and Betty Gasparino for typing the list of references.

References

1. E.S. Abers and B.W. Lee, "Gauge theories", *Phys. Rep.* **C9**, 1 (1973).
2. Y. Aharonov, A. Casher and L. Susskind, "Spin $\frac{1}{2}$ partons in a dual model of hadrons", *Phys. Rev.* **D5**, 988 (1972).
3. V.P. Akulov and D.V. Volkov, "Goldstone fields with spin 1/2", *Teor. Mat. Fiz.* **18**, 39 (1974).
4. V.P. Akulov and D.V. Volkov, "Interaction of Goldstone neutrino with electromagnetic field", *ZhETF Pis. Red.* **17**, 367 (1973).
5. V.P. Akulov, D.V. Volkov and V.A. Soroka, "Gauge fields on superspace with different holonomy groups", *JETP Lett.* **22**, 187 (1975) [*Pis'ma Zh. Eksp. Teor. Fiz.* **22**, 396 (1975); V.P. Akulov, D.V. Volkov and V.A. Soroka, "Generally covariant theories of gauge fields on superspace", *Teor. Mat. Fiz.* **31**, 12 (1977); V.P. Akulov and D.V. Volkov, "Reimannian superspace of minimal dimensionality", *Teor. Mat. Fiz.* **41**, 147 (1979)].
6. T. Appelquist, A. Chodos, P.G.O. Freund (Eds.), *Modern Kaluza- Klein Theories* (Addison-Wesley, 1987).
7. R. Arnowitt, P. Nath, and B. Zumino, "Superfield densities and action principle in curved superspace", *Phys. Lett.* **B56**, 81 (1975).
8. R. Arnowitt, P. Nath (Eds.), *Gauge Theories and Modern Field Theory* (MIT Press, 1976).
9. A. Ashtekar, *New Perspectives in Canonical Gravity* (Bibliopolous (Naples) 1988).
10. K. Bardakçi and M.B. Halpern, "Dual M-models", *Nucl. Phys.* **B73**, 295 (1974).
11. F.A. Berezin, *The Method of Second Quantization* (Academic Press, 1966).
12. L. Brink, R. Marnelius, J.S. Nilsson, P. Salomonson and B.-S. Skagerstam, Unification of fundamental interactions. Proceedings of Nobel Symposium 67, *Physica Scripta* **T15** (1987).

13. L. Brink, J. Schwarz and J. Scherk, "Supersymmetric Yang–Mills theories", *Nucl. Phys.* **B121**, 77 (1977).
14. L. Brink and P. Howe, "The $N = 8$ supergravity in superspace", *Phys. Lett.* **B88**, 268 (1979).
15. S.M. Christensen (Ed.), *Quantum Theory of Gravity* (Adam Hilger Ltd., 1984).
16. F. Coester, M. Hammermesh and W.D. McGlinn, "Internal symmetry and Lorentz invariance", *Phys. Rev.* **B135**, 451 (1964).
17. S. Coleman and J. Mandula, "All possible symmetries of the S-matrix", *Phys. Rev.* **D159**, 1251 (1967).
18. E. Cremmer and J. Scherk, "Dual models in four dimensions with internal symmetries", *Nucl. Phys.* **B103**, 399 (1975).
19. E. Cremmer and J. Scherk, "Spontaneous compactification of space in an Einstein-Yang–Mills-Riggs model", *Nucl. Phys.* **B108**, 409 (1976).
20. E. Cremmer, B. Julia and J. Scherk, "Supergravity theory in 11 dimensions", *Phys. Lett.* **B76**, 409 (1978).
21. E. Cremmer and B. Julia, "The $N = 8$ supergravity theory I. The Lagrangian", *Phys. Lett.* **B80**, 48–51 (1978).
22. E. Cremmer, J. Scherk and J. Schwarz, "Spontaneously broken $N = 8$ supergravity", *Phys. Lett.* **B84**, 83 (1979).
23. R. D'Auria and P. Fré, *Superunification and Extra Dimensions* (World Scientific, 1986).
24. Del Agulia, J.A. de Azcárraga, and L.E. Ibáñez (Eds.), *Supersymmetry, Supergravity and Related Topics* (World Scientific, 1985).
25. S. Deser and P. van Nieuwenhuizen, "One-loop divergences of quantized Einstein-Maxwell fields", *Phys. Rev.* **D10**, 401 (1974).
26. S. Deser and P. van Nieuwenhuizen, "Nonrenormalizability of the quantized Einstein-Maxwell system", *Phys. Rev. Lett.* **32**, 245 (1974).
27. S. Deser, H.-S. Tsao and P. van Nieuwenhuizen, "Nonrenormalizability of Einstein-Yang–Mills interactions at the one-loop level", *Phys. Lett.* **B50**, 491 (1974).
28. S. Deser and P. van Nieuwenhuizen, "Nonrenormalizability of quantized fermion-graviton interactions", *Nuovo Cim. Lett.* **11**, 218 (1974).
29. S. Deser and P. van Nieuwenhuizen, "Nonrenormalizability of the quantized Dirac-Einstein system", *Phys. Rev.* **D10**, 411 (1974).
30. S. Deser and B. Zumino, "Consistent supergravity", *Phys. Lett.* **B62**, 335 (1976).
31. S. Deser, J. Kay and K. Stelle, "Renormalizability properties of supergravity", *Phys. Rev. Lett.* **38**, 527 (1977).
32. S. Deser and J. Kay, "Three-loop counterterms for extended supergravity", *Phys. Lett.* **B76**, 400 (1987).
33. B. de Wit, P. Fayet and P. van Nieuwenhuizen (Eds.), *Supersymmetry and Supergravity '84* (World Scientific, 1984).
34. R. DiStefano, Ph.D. Dissertation, SUNY at Stony Brook, 1982.
35. R. Di Stefano, "Disappearance of the axiliary fields in a canonical formulation of supersymmetry", *Phys. Lett.* **B192**, 130 (1987).
36. R. Di Stefano, "Canonical Construction of Symmetry Generators and Algebras: Application to Simple Supersymmetry (I) Systems without Constraints", ITP:SB:87/81; (II) "Systems with Constraints", ITP:SB:88/4.
37. D.B. Fairlie and D. Martin, "New light on the Neveu-Schwarz model", *Nuovo Cim.* **18**, 373 (1973).

38. P. Fayet and J. Iliopoulos, "Spontaneously broken supergauge symmetries and Goldstone spinors", *Phys. Lett.* **B51**, 461 (1974).
39. P. Fayet and S. Ferrara, "Supersymmetry", *Phys. Rep.* **32**, 249 (1977).
40. S. Ferrara and O. Piguet, "Perturbation theory and renormalization of supersymmetric Yang–Mills theories", *Nucl. Phys.* **B93**, 261 (1975).
41. S. Ferrara, B. Zumino and J. Wess, "Supergauge multiplets and superfields", *Phys. Lett.* **B51**, 239 (1974).
42. S. Ferrara and B. Zumino, "Supergauge invariant Yang–Mills theories", *Nucl. Phys.* **B79**, 413 (1974).
43. S. Ferrara and B. Zumino, "Transformation properties of the supercurrent", *Nucl. Phys.* **B87**, 207 (1975).
44. S. Ferrara and P. van Nieuwenhuizen, "The auxiliary fields of supergravity", *Phys. Lett.* **B74**, 330 (1978).
45. S. Ferrara and J.G. Taylor (Eds.), *Supergravity '81* (Cambridge University Press, 1982).
46. S. Ferrara, J.G. Taylor, and P. van Nieuwenhuizen (Eds.), *Supersymmetry and Supergravity '82: Proceedings of the Trieste September 1982 School* (World Scientific, 1983).
47. S. Ferrara, *Supersymmetry* (North-Holland/World Scientific, 1987).
48. D.Z. Freedman and So-Young Pi, "External gravitational interactions in quantum field theory", *Ann. of Phys.* **91**, 442 (1975).
49. D.Z. Freedman and B. de Wit, "Phenomenology of Goldstone neutrinos", *Phys. Rev. Lett.* **35**, 827 (1975).
50. D.Z. Freedman and B. de Wit, "Combined supersymmetric and gauge invariant field theories", *Phys. Rev.* **D12**, 2286 (1975).
51. D.Z. Freedman, P. van Nieuwenhuizen and S. Ferrara, "Progress toward a theory of supergravity", *Phys. Rev. D* **13**, 3214 (1976).
52. D.Z. Freedman and P. van Nieuwenhuizen, "Properties of supergravity theory", *Phys. Rev.* **D14**, 912 (1976).
53. P.G.O. Freund, Mark A. Rubin, "Dynamics of dimensional reduction", *Phys. Lett.* **B97**, 233 (1980).
54. S. Fubini, D. Gordon and G. Veneziano, "A general treatmet of factorization in dual resonance", *Phys. Lett.* **B29**, 679 (1969).
55. S. Fubini and G. Veneziano, "Algebraic treatment of subsidiary conditions in dual resonance models", *Ann. of Phys.* **63**, 12 (1971).
56. E. Garfield, "The most cited 1985 Physical Science articles", *Current Contents* **27**, 3 (1987).
57. S.J. Gates, M.T. Grisaru, M. Roček and W. Siegel, *Superspace or One Thousand and One Lessons in Supersymmetry* (Benjamin/Cummings, 1983).
58. J.L. Gervais and B. Sakita, "Field theory interpretation of supergauges in dual models", *Nucl. Phys.* **B34**, 632 (1971).
59. F. Gliozzi, J. Scherk and D. Olive, "Supergravity and the spinor dual model", *Phys. Lett.* **B65**, 282 (1976).
60. F. Gliozzi, J. Scherk and D. Olive, "Supersymmetry, supergravity theories and the dual spinor model", *Nucl. Phys.* **B122**, 253 (1977).
61. P. Goddard and C.B. Thorn, "Compatibility of the dual pomeron with unitarity and the absence of ghosts in the dual resonance model", *Phys. Lett.* **B40**, 325 (1972).
62. Y.A. Golfand and E.P. Likhtman, "Extension of the algebra of Poincaré group generators and violation of P-invariance", *JETP Lett.* **13**, 323 (1971).

63. M.B. Green and J.H. Schwarz, "Supersymmetrical dual string theory", *Nucl. Phys.* **B181**, 502 (1981).
64. M.B. Green, J.H. Schwarz and L. Brink, "$N = 4$ Yang–Mills and $N = 8$ supergravity as limits of string theories", *Nucl. Phys.* **B198**, 474 (1982).
65. M.B. Green and J.H. Schwarz, "Supersymmetrical string theories", *Phys. Lett.* **B109**, 444 (1982).
66. M.B. Green, "Supersymmetrical dual string theories and their field theory limits — A review", *Surveys in High Energy Physics* **3**, 127 (1983).
67. M.B. Green and J.H. Schwarz, "Covariant description of superstrings", *Phys. Lett.* **B136**, 367 (1984).
68. M.B. Green and J.H. Schwarz, "Anomaly cancelations in superpersymmetric $D = 10$ gauge theory and superstring theory", *Phys. Lett.* **B149**, 17 (1984).
69. M.B. Green, J.H. Schwarz and E. Witten, *Superstring Theory* (Cambridge University Press, 1986).
70. M. Grisaru, P. van Nieuwenhuizen and J.A.M. Vermaseren, "One-loop renormalizability of supergravity and of the Maxwell-Einstein theory in extended supergravity", *Phys. Rev. Lett.* **37**, 1662 (1976).
71. M. Grisaru, H.N. Pendleton and P. van Nieuwenhuizen, "Supergravity and the S-matrix", *Phys. Rev.* **D15**, 996 (1977).
72. M. Grisaru, "Two-loop renormalizability of supergravity", *Phys. Lett.* **B66**, 75 (1977).
73. M. Grisaru, W. Siegel and M. Roček, "Improved methods for supergraphs", *Nucl. Phys.* **B139**, 429 (1979).
74. M. Grisaru, M. Roček and W. Siegel, "Zero value for the three-loop β-function in $N = 4$ supersymmetric Yang–Mills theory", *Phys. Rev. Lett.* **45**, 1063 (1980).
75. M. Grisaru and W. Siegel, "Supergraphity (II). Manifestly covariant rules and higher-loop finiteness", *Nucl. Phys.* **B201**, 292 (1982).
76. R. Haag, J.T. Lopuszański and M. Sohnius, "All possible generators of supersymmetries of the S-matrix", *Nucl. Phys.* **B88**, 257 (1975).
77. S.W. Hawking and M. Roček (Eds.), *Superspace and Supergravity* (Cambridge University Press, 1981).
78. W. Heisenberg, *Solvay Ber. Kap.* III, IV (1939).
79. P. Howe, K. Stelle and P. Townsend, "Miraculous ultraviolet cancelations in supersymmetry made manifest", *Nucl. Phys.* **B236**, 125 (1984).
80. J. Iliopoulos and B. Zumino, "Broken supergauge symmetry and renormalization", *Nucl. Phys.* **B76**, 310 (1974).
81. M. Jacob (Ed.), *Dual Theory* (North-Holland, 1974).
82. M. Jacob (Ed.), *Supersymmetry and Supergravity* (North-Holland/World Scientific, 1986).
83. B. Kursonoglu, A. Perlmutter and L. Scott (Eds.), *Deeper Pathways in High Energy Physics* (Plenum, 1977).
84. W. Lang and J. Wess, "Investigation of a nonrenormalizable Lagrangian model invariant under supertransformation", *Nucl. Phys.* **B81**, 249 (1974).
85. J. León, J. Pérez-Mercader and M. Quirós (Eds.), *Third CSIC Workshop on SUSY and Grand Unification: From Strings to Collider Phenomenology* (World Scientific, 1986).
86. E.P. Likhtman, "Supergauge renormalizable theory of a massive vector field", *JETP Lett.* **21**, 109 (1975).

87. E.P. Likhtman, "A supersymmetric renormalized theory of a massive non-abelian vector field", *JETP Lett.* **22**, 57 (1975).
88. J.T. Lopuszański, "On some properties of physical symmetries", *J. Math. Phys.* **12**, 2401 (1971).
89. W.D. McGlinn, "Problem of combining interaction symmetries and relativistic invariance", *Phys. Rev. Lett.* **12**, 467 (1964).
90. B. Milewski (Ed.), *Supersymmetry and Supergravity 1983* (World Scientific, 1983).
91. C. Montonen, "Multiloop amplitudes in additive dual-resonance models", *Nuovo Cim.* **A19**, 69 (1073).
92. D.V. Nanopoulos and A. Savoy-Navarro, "Supersymmetry confronting experiment", *Phys. Rep.* **105**, 1 (1984).
93. P. Nath and R. Arnowitt, "Generalized super-gauge symmetry as a new framework for unified gauge theories", *Phys. Lett.* **B56**, 177 (1975).
94. A. Neveu and J.H. Schwarz, "Factorizable dual models of pions", *Nucl. Phys.* **B31**, 86 (1971).
95. A. Neveu, J.H. Schwarz and C.B. Thorn, "Reformulation of the dual pion model", *Phys. Lett.* **B35**, 529 (1971).
96. A. Neveu and J. Scherk, "Connection between Yang–Mills fields and dual models", *Nucl. Phys.* **B336**, 155 (1972).
97. M. Nouri-Moghadam and J.G. Taylor, "One-loop divergences for the Einstein-charged meson system", *Proc. Roy. Soc.* **A344**, 87 (1975).
98. L. O'Raifeartaigh, "Internal symmetry and Lorentz invariance", *Phys. Rev. Lett.* **14**, 332 (1965).
99. L. O'Raifeartaigh, "Mass differences and Lie algebras of finite order", *Phys. Rev. Lett.* **14**, 575 (1965).
100. L. O'Raifeartaigh, "Lorentz invariance and internal symmetry", *Phys. Rev.* **B139**, 1052 (1964).
101. A. Pais, *Inward Bound: Of Matter and Forces in the Physical World* (Oxford University Press, 1986).
102. K. Pilch, P. Townsend and P. van Nieuwenhuizen, "Compactification of $d = 11$ supergravity on S^4 (or $11 = 7 + 4$, too)", *Nucl. Phys.* **B242**, 377 (1984).
103. P. Ramond, "Dual theory for free fermions", *Phys. Rev.* **D3**, 2415 (1971).
104. R. Ruffini (Ed.), *Proceedings of the First Marcel Grossman Meeting on General Relativity* (North Holland, 1975).
105. A. Salam and J. Strathdee, "Super-gauge transformations", *Nucl. Phys.* **B76**, 477 (1974).
106. A. Salam and J. Strathdee, "Super-symmetry and non-Abelian gauges", *Phys. Lett.* **B51**, 353 (1974).
107. A. Salam and J. Strathdee, "SU(6) and supersymmetry", *Nucl. Phys.* **B84**, 127 (1975).
108. A. Salam and J. Strathdee, "Superfields and Bose-Fermi symmetry", *Phys. Rev.* **D11**, 1521 (1975).
109. A. Salam and J. Strathdee, "Feynman rules for superfield", *Nucl. Phys.* **B86**, 142 (1975).
110. J. Scherk, "Zero-slope limit of the dual resonance model", *Nucl. Phys.* **B31**, 222 (1971).
111. J. Scherk and J.H. Schwarz, "Dual models for non-hadrons", *Nucl. Phys.* **B81**, 118 (1974).

112. J. Scherk and J.H. Schwarz, "Dual field theory of quarks and gluons", *Phys. Lett.* **B57**, 463 (1975).
113. J. Scherk and J.H. Schwarz, "Spontaneous breaking of supersymmetry through dimensional reduction", *Phys. Lett.* **B82**, 60 (1979); *Nucl. Phys.* **B153**, 60 (1979).
114. J.H. Schwarz, "Dual resonance theory", *Phys. Rep.* **8C**, 269 (1973).
115. J.H. Schwarz (Ed.), *Superstrings: The First Fifteen Years* (World Scientific, 1985).
116. J.R. Smith (Ed.), *Proceeding of the XVII International Conference on High Energy Physics* (Science Research Council, Rutherford Laboratory, 1974).
117. M. Sohnius and P. West, "Conformal invariance in $N = 4$ supersymmetry Yang–Mills theory", *Phys. Lett.* **B100**, 241 (1981).
118. K. Stelle and P.C. West, "Minimal auxiliary fields for supergravity", *Phys. Lett.* **B74**, 333 (1978).
119. G.'t Hooft and M. Veltman, "One-loop divergences in the theory of gravitation", *Ann. Inst. Henri Poincaré* **20**, 69 (1974).
120. P. van Nieuwenhuizen, M.T. Grisaru and J.A.M. Vermaseren, "One-loop renormalizability of pure supergravity and the Maxwell-Einstein theory in extended supergravity", *Phys. Rev. Lett.* **37**, 1662 (1976).
121. P. van Nieuwenhuizen and P.K. Townsend, "Geometrical interpretation of extended supergravity", *Phys. Lett.* **B67**, 439 (1977).
122. P. van Nieuwenhuizen and D.Z. Freedman, *Supergravity: Proceedings of the Supergravity Workshop at Stony Brook* (North-Holland, 1979).
123. P. van Nieuwenhuizen, "Supergravity", *Phys. Rep.* **68**, 189 (1981).
124. D.V. Volkov, "Phenomenological Lagrangians", *Sov. J. Particles Nucl.* **4**, 1 (1973).
125. D.V. Volkov and V.P. Akulov, "Possible universal neutrino interaction", *JETP Lett.* **16**, 438 (1972).
126. D.V. Volkov and V.P. Akulov, "Is the neutrino a Goldstone particle?", *Phys. Lett.* **B46**, 109 (1973).
127. D.V. Volkov and V.A. Soroka, "Higgs effect for Goldstone particles with spin $\frac{1}{2}$", *Pis'ma Zh. Eksp. Teor. Fiz.* **18**, 529 (1973); D.V. Volkov and V.P. Akulov, "Is the neutrino a Goldstone particle?", *Phys. Lett.* **B46**, 109 (1973); D.V. Volkov and V.A. Soroka, "Gauge fields for symmetry group with spinor parameters", *Teor. Mat. Fiz.* **20**, 291 (1974).
128. J. Wess and B. Zumino, "Supergauge transformations in four dimensions", *Nucl. Phys.* **B70**, 39 (1974).
129. J. Wess and B. Zumino, "A Lagrangian model invariant under supergauge transformations", *Phys. Lett.* **B49**, 52 (1974).
130. J. Wess and B. Zumino, "A supergauge invariant extension of quantum electrodynamics", *Nucl. Phys.* **B78**, 1 (1974).
131. J. Wess and J. Bagger, *Supersymmetry and Supergravity* (Princeton University Press, 1983).
132. P. West, *Introduction to Supersymmetry and Supergravity* (World Scientific, 1986).
133. C. Wetterich, "Dimensional reduction of Weyl, Majorana and Majorana-Weyl spinors", *Nucl. Phys.* **B222**, 20 (1983).
134. E. Witten, "Search for a realistic Kaluza-Klein theory", *Nucl. Phys.* **B186**, 412 (1981).
135. C.N. Yang and R.L. Mills, "Conservation of isotopic spin and isotopic gauge invariance", *Phys. Rev.* **96**, 191 (1954).

136. T. Yoneya, "Quantum gravity and the zero-slope limit of the generalized Virasoro model", *Nuovo Cim. Lett.* **8**, 951 (1973).
137. T. Yoneya, "Connection of dual models to electrodynamics and gravidynamics", *Prog. Theor. Phys.* **51**, 1907 (1974).
138. B. Zumino, "Supersymmetry and the vacuum", *Nucl. Phys.* **B89**, 535 (1975).
139. W. Nahm, "Supersymmetries and their representations", *Nucl. Phys.* **B135**, 149 (1978).

7

SUSY-50 Conference, May 18–20, 2013

William I. Fine Theoretical Physics Institute, University of Minnesota

The Conference 50 Years of Supersymmetry (SUSY 50) sponsored by the Simons Foundation and the William I. Fine Theoretical Physics Institute was organized by FTPI on May 18–20,[156] to commemorate the 50th anniversary of theoretical supersymmetry. The total number of participants was around 45, with 27 talks delivered during three days of the conference. The conference was divided into three sections. Section 1 (on Thursday, May 18) was devoted to intersection of black holes and supersymmetry and superconformal field theories. The following 9 speakers presented their talks: Francesco Benini, SISSA, Italy; Nick Dorey, Univ. of Cambridge, UK; Philip Argyres, Univ. of Cincinnati; Luca Iliesku, Stanford University; Simeon Hellerman, Univ. of Tokyo, Japan; Ying-Hsuan Lin, Harvard University; Hirosi Ooguri, Caltech; Nikita Nekrasov, SCGP; and Hitoshi Murayama, UC Berkeley.

Section 2 (on Friday, May 19) was focused of applications of supersymmetry in solving so far unanswered issues in Yang-Mills theory at strong coupling and the use of recently discovered global anomalies. The following 9 speakers presented their talks: Csaba Csaki, Cornell; Adi Armani, Swansea Univ. UK; Clay Cordova, Univ. of Chicago; Ken Konishi, Pisa Univ. Italy; Michael Dine, Yale; Emily Nardoni, Univ. of Tokyo, Japan; Alexey Cherman, Univ. of Minnesota; and Erch Poppitz, Univ. of Toronto, Canada.

Section 3 (on Saturday, May 20) covered mathematical (quantum groups, strings) and topological aspects of supersymmetry as well as maximally extended supersymmetries. The following 9 speakers presented their talks: Elli Pomoni, Hamburg University, Germany; Peter West, King's College, UK; Sara Pasquetti, Univ. of Milan-Bicocca, Italy; Emilio Trevisiani, Institut de Physique Théorique, Saclay, France; Muneto Nitta, Keio University, Japan; Peter Koroteev, UC Berkeley, Shlomo Razamat, Technion, Israel, Silviu Pufu, Princeton University.

[156]The organizers were: M. Shifman, FTPI, Z. Komargodki, SCGP, and T. Dumitresku, UCLA.

Selected talks. Summaries

• *Francesco Benini*
Quantum Black Hole Entropy in AdS/CFT

Formulating and understanding theories of quantum gravity is one of the longstanding problems in theoretical physics. Black holes constitute an excellent theoretical (and hopefully — in the future — also experimental) laboratory where ideas can be developed and tested. One of the striking properties of black holes is their large entropy.

The leading contribution is captured by the Bekenstein-Hawking formula: the entropy is a quarter of the horizon area in units of the Planck length. Can a theory of quantum gravity reproduce this result, and add quantum corrections to it, from a microscopic counting of the quantum states?

Exploiting AdS/CFT, this task can be mapped into a computation in a strongly-coupled boundary QFT, which, in the presence of supersymmetry, can often be exactly performed with supersymmetric localization techniques [1]. In the last decade these computations have become extremely precise. For instance, for BPS black holes in AdS_5, from the superconformal index of 4D $\mathcal{N} = 4$ SYM theory, we are able to extract: (i) the Bekenstein-Hawking entropy as a function of the black-hole charges and angular momenta [2]; (ii) perturbative corrections ascribable to higher-derivative corrections in 5D supergravity; (iii) perturbative corrections ascribable to one-loop effects in 5D supergravity; (iv) non-perturbative corrections reproducing a semi-classical expansion of the Euclidean supergravity path integral [3]; (v) non-perturbative corrections ascribable to Euclidean D3-brane instantons [3]. From the QFT index one can compute other types of corrections, whose meaning in supergravity has not been yet identified.

Going beyond supersymmetry remains a challenge. We made some progress [4] by performing a KK reduction of the boundary QFT on a spatial manifold with the same topology as the black-hole horizon (as suggested by AdS/CFT). This produces a peculiar quantum mechanical system with a large number of modes, in which the couplings are pseudo-randomly distributed. This has traits in common with the SYK models, which are already known to reproduce physical properties of near-extremal (low-temperature) black holes. Detailed studies of these quantum-mechanical systems is under way.

References

1. F. Benini, K. Hristov, A. Zaffaroni, *Black hole microstates in AdS_4 from supersymmetric localization*, JHEP 1605, 054 (2016), arXiv:1511.04085 [hep-th].
2. F. Benini, E. Milan, *Black Holes in 4D $\mathcal{N} = 4$ Super-Yang-Mills Field Theory*, Phys. Rev. X 10, 021037 (2020), arXiv:1812.09613 [hep-th].
3. O. Aharony, F. Benini, O. Mamroud, E. Milan, *A gravity interpretation for the Bethe Ansatz expansion of the $\mathcal{N} = 4$ SYM index*, Phys. Rev. D 104, 086026 (2021), arXiv:2104.13932 [hep-th].
4. F. Benini, S. Soltani, Z. Zhang, *A quantum mechanics for magnetic horizons*, JHEP 2305, 070 (2023), arXiv:2212.00672 [hep-th].

• *Luca Iliesiu*
Constructing all BPS black hole micro-states from the gravitational path integral

Supersymmetry has long played a critical role in quantum gravity. Supersymmetric indices have been used in string theory to provide the first microscopic counts of BPS black hole entropies. More recently, the presence of supersymmetry was exploited to perform the supergravity path integral exactly using the technique of supersymmetric localization. While these techniques have been critical in computing BPS black hole degeneracies or indices, they do not explain how to prepare the individual BPS states. Understanding how to prepare and count black hole micro-states by using the gravitational path integral is one of the most important problems in quantum gravity. There have been many attempts to address this problem, the most famous of which is the "fuzzball program," which aims to find BPS solutions in supergravity that can be associated with single black hole micro-states. Nevertheless, a state-by-state count of such black hole micro-states is difficult. If one includes only some of the supergravity degrees of freedom, the obtained number of micro-states is often exponentially smaller than the correct string theory answer. If one includes all the supergravity degrees of freedom, the apparent number of degrees of freedom available in the gravitational effective theory can vastly exceed the entropy of the black hole, even in the special case of the BPS black holes. Such a discrepancy can be viewed as a supersymmetric version of the black hole information paradox. The focus of the talk was to address this discrepancy and explain that we can indeed use the supergravity path integral to prepare a complete basis of states for the Hilbert space of all BPS black hole micro-states. Even though the presented computation used the statistics of inner products between different BPS black hole micro-states rather than their exact value, a surprise was that the final answer for the dimension of the Hilbert space spanned by these states is exact. The dimension of the resulting Hilbert space is in complete agreement with the degeneracy obtained from the Gibbons-Hawking prescription used in the localization procedure and with the string theory result. Specifically, this match includes all non-perturbative corrections in $1/G_N$.

Reference

Jan Boruch, Luca V. Iliesiu, and Cynthia Yan, arXiv 2307.13051.

• *Erich Poppitz*
Anomalies, tori, and new twists in the gaugino condensate

One of the the rare cases where a nonpertubative quantity in four dimensional quantum field theory can be exactly determined is the bilinear gaugino condensate in minimally supersymmetric Yang-Mills theory (SYM). This is made possible by the nonrenormalization theorem and the holomorphic dependence of the condensate on the strong coupling scale in SYM. It was realized 40 years ago that a semiclassical calculation of the exact gaugino condensate should be possible on a small 4-torus with 't Hooft's twisted boundary conditions. However, previously this calculation was neither performed in detail nor properly interpreted.

This talk uses the striking post-2017 advances in our understanding of generalized symmetries (here: center symmetry) and their mixed anomalies to understand the gaugino condensate. These new anomalies were cast in the Hamiltonian formalism in [1], which, combined with a better understanding of the fractional instanton backgrounds allowed [2] to calculate the gaugino condensate on a small T^4 for $SU(2)$ gauge group. The result differs by a curious factor of 2 from the weak-coupling calculations on R^4 and $R^3 \times S^1$. We believe that this new puzzle points towards the need to better understand the global structure of the moduli space of fractional instantons on T^4, presenting an interesting arena for current and future studies.

References

1. A. Cox, E. Poppitz and F.D. Wandler, The mixed 0-form/1-form anomaly in Hilbert space: pouring the new wine into old bottles JHEP10(2021)069, arXiv 2106.11442.
2. M. Anber and E. Poppitz, The gaugino condensate from asymmetric four-torus with twists JHEP01(2023)118, arXiv 2210.13568.

• *Sara Pasquetti*
Is there a fundamental duality? The local dualization algorithm at work

Infra-red (IR) dualities among supersymmetric gauge theories are interesting phenomena extensively studied over the years. These correspond to the situation in which distinct microscopic theories flow to the same fixed point in the IR. After the discovery of the Seiberg duality, a plethora of new IR dualities have been found in various dimensions. For example in 3D we have various Seiberg-like dualities as well as mirror dualities for theories with eight supercharges. The latter can be neatly explained as a consequence of string theory S-duality thanks to the Hanany-Witten brane setup realization. Given this proliferation of the IR dualities, there are various fundamental questions that one can ask purely from the field theory perspective.

For example, one may ask if we can derive all 3D and lower dimensional IR dualities from 4D dualities. This idea, in the context of a more general program of studying dualities across dimensions, has proven to be a useful perspective and for many lower-dimensional dualities it has been found a common higher dimensional ancestor. 3D mirror dualities, considered an inherently three-dimensional phenomenon, for a long time appeared to evade this pattern. However, recently, we discovered 4D avatars of 3D mirror dualities [1]. These new 4D dualities reduce to 3D mirror dualities after circle reduction and suitable deformations, strengthening the expectation that indeed physics of 3D dualities can be derived from 4D.

Another intriguing question is whether there is a fundamental set of dualities in terms of which all other dualities can be derived. This second question was addressed in various dimensions by using the notion of sequential deconfinement, which can be used to derive dualities for theories with matter fields in tensor representations by only assuming a small set of the basic Seiberg-like dualities [2].

Encouraged by these results we asked whether mirror dualities can be derived assuming only the same set of the fundamental Seiberg-like dualities. This led us to the construction of a local dualization algorithm [3]. This is an algorithmic procedure based on the iteration of a single fundamental duality which dualizes a theory into its mirror dual. This construction provides a purely field-theoretic proof of mirror symmetry; we recently extended it to the more general SL(2, Z) duality group [4] and to theories with only four supercharges [5].

References

1. 4D mirror-like dualities, Hwang-Pasquetti-Sacchi, JHEP 09 (2020) 047.
2. Dualities from dualities: the sequential deconfinement technique, Bottini-Hwang-Pasquetti-Sacchi [JHEP 05 (2022) 069].
3. Rethinking mirror symmetry as a local duality on fields, Hwang-Pasquetti-Sacchi [Phys.Rev.D 106 (2022) 10, 105014].
4. The SL(2, Z) dualization algorithm at work, Comi-Hwang-Marino-Pasquetti-Sacchi [JHEP 06 (2023) 119].
5. Mirror dualities with four supercharges, Benvenuti-Comi-Pasquetti [to appear].

• *Emilio Trevisiani*
Parisi-Sourlas Supersymmetry in Random Field Models

A long-standing problem in statistical physics is to understand the critical behavior of models with random fields (RF). One of the most important milestones of the subject is a conjecture proposed by Parisi and Sourlas which is divided in two parts: (i) Emergence of SUSY: the RF fixed point is described by a special supersymmetric CFT which we call Parisi-Sourlas CFT (P-S CFT) and (ii) Dimensional reduction: the P-S CFT is related to a CFT in two less spacetime dimensions without supersymmetry.

This conjecture however was known to fail in some simple cases, but there was no universal consensus on why this happens. We consider the part 2) of the conjecture and we confirmed it using an axiomatic CFT toolkit, borrowed from the bootstrap. Next we consider part 1) of the conjecture.

We describe an RG framework — which makes use of the replica trick combined with a clever change of variable first introduced by Cardy — in which the RF models can be written as theories with P-S SUSY deformed by extra SUSY-breaking operators. If these operators are relevant then Parisi-Sourlas supersymmetry is lost in the IR.

We then apply this framework to the two most important instances of RF theories, defined by a single scalar field ϕ Lagrangian (with ϕ coupled to the random field) perturbed by either cubic or quartic potentials ϕ^3, ϕ^4. For RF ϕ^4 theory, we show that SUSY-breaking perturbations become relevant below a critical dimension $d_c \approx 4.2 \sim 4.7$. This supports the scenario that the SUSY fixed point exists for all $3 < D \leq 6$, but becomes unstable for $D < D_c$. On the other hand for RF ϕ^3 theory all SUSY-breaking perturbations are irrelevant and thus the fixed point is always a P-S CFT. Our results are in complete accordance with the latest numerical predictions and give a new RG explanation for when/why the Parisi-Sourlas conjecture holds.

References

1. Apratim Kaviraj and Emilio Trevisani, JHEP **08**, 290 (2022).
2. A. Kaviraj, S. Rychkov and E. Trevisani, JHEP **04**, 090 (2020); JHEP **03**, 219 (2021); Phys. Rev. Lett. 129, 045701 (2022).

• *Nick Dorey*
Superconformal Quantum Mechanics and Ultra-Spinning Black Holes

The AdS/CFT correspondence maps conformal field theories in D spacetime dimensions to theories of gravity in $D+1$ dimensions. As such it allows us to reformulate hard questions in gravitational physics, such as understanding the thermodynamic properties of black holes, as potentially more tractable ones in the dual field theory. The case $D=1$ where field theory reduces to ordinary quantum mechanics is particularly interesting but has remained elusive. Although, quantum mechanical systems dual to black hole geometries have been proposed previously, evidence in favour of the corresponding duality depends on the the results of numerical simulations. In the talk, based on the work [1], I present a superconformal quantum mechanical model which provides a dual description of a certain ultra-spinning supersymmetric black hole in AdS_7. In this case, it is possible to check the duality explicitly by computing the asymptotics of the superconformal index of the quantum mechanics and comparing it with the Bekenstein-Hawking entropy of the black hole. In the talk, I described the resulting agreement and prospects for further progress in the given direction.

Reference

1. N. Dorey, R. Mouland and B. Zhao, JHEP **06** (2023), 166.

• *Michael Dine*
Possibility of Demonstrating Confinement in Non-Supersymmetric Theories by Deforming Confining Supersymmetric Theories

Understanding confinement is considered one of the great challenges of theoretical physics. From numerical studies of quantum chromodynamics, there is evidence for this hypothesis, but a more conceptual explanation is desirable. In theories with supersymmetry — where there is a symmetry between fermions and bosons — Seiberg and Witten demonstrated that quarks are confined, by methods which are completely analytic. These theories are not realistic, but one can deform theses theories by giving masses to the extra particles implied by supersymmetry. I argue that it may be possible to establish that confinement survives in the deformed theory. Ideed, for a small deformation, this is straightforward to show. For large deformations, the possibility of establishing confinement is suggested by the fact that confinement is a phenomenon of the far infrared, and seems to have universal features.

Reference

1. Michael Dine, arXiv:2211.17134.

List of Participants[157]

Philip Argyres, Adi Armoni, Francesco Benini, Aleksey Cherman, Clay Cordova, Csaba Csaki, Michael Dine, Nick Dorey, Thomas Dumitrescu, Simeon Hellerman, Luca Iliesiu, Alex Kamenev, Zohar Komargodski, Kenichi Konishi, Peter Koroteev, Zhen Liu, Hitoshi Murayama, Emily Nardoni, Nikita Nekrasov, Muneto Nitta, Hirosi Ooguri, Sara Pasquetti, Elli Pomoni, Erich Poppitz, Silviu Pufu, Shlomo Razamat, Serge Rudaz, Misha Shifman, Emilio Trevisiani, Arkady Vainshtein, Scott Watson, Peter West.

Postdocs and graduate students

Ravneet Bedi, Saif Ullah Baig, Raymond Co, Yohei Ema, Sami Farrag, Yasha Gindikin, Theo Jacobson, Saarik Kalia, Ying-Hsuan Lin, Kunfeng Lyu, Khilav Majmudar, Emily Nardoni, Maria Neuzil, Arpon Paul, Chao-Hsiang Sheu, Andrey Shkerin, Emilio Trevisani.

*****

Figure 7.1: Nick Dorey.

[157]Editor's note: Below photographs of some of the participants are shown. All photographs are mine. –MS

Figure 7.2: Shlomo Razamat and Zohar Komargodsky.

Figure 7.3: Sara Pasquetti.

Figure 7.4: Peter West.

Figure 7.5: Elli Pomoni's talk.

Figure 7.6: Thomas Dumitrescu.

Figure 7.7: Kenichi Konishi and Muneto Nitta.

Figure 7.8: Arkady Vainshtein, Adi Armoni, Alicia Canfield, Misha Shifman, and Meghan Murray.

Figure 7.9: Hitoshi Murayama.

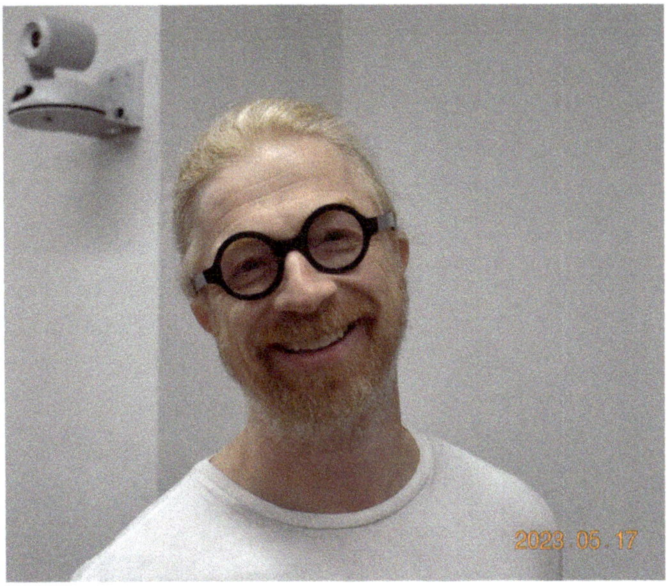

Figure 7.10: Nikita Nekrasov.

Figure 7.11: Maria Neuzil and Chao-Hsiang Sheu.

Figure 7.12: Hirosi Ooguri.

8

Appendices

English translation of articles from Pisma ZhETF

- **Golfand and Likhtman**
- **Volkov and Akulov**
- **Volkov and Soroka**

EXTENSION OF THE ALGEBRA OF POINCARÉ GROUP GENERATORS AND VIOLATION OF P INVARIANCE

Yu.A. GOLFAND AND E.P. LIKHTMAN

Lebedev Physics Institute, USSR Academy of Sciences

Submitted 10 March 1971

ZhETF Pis. Red. 13, No. 8, 452–455 (20 April 1971)

Eng. trans. JETP Lett. **13**, No. 8, pp. 323–326

One of the main requirements imposed on quantum field theory is invariance of the theory under the Poincaré group [1]. However, only fraction of the interactions satisfying this requirement is realized in nature. It is possible that these interactions, unlike others, have a higher degree of symmetry. It is therefore of interest to study different algebras and groups, the invariance with respect to which imposes limitations on the form of the elementary particle interactions. In the present paper we propose, in constructing the Hamiltonian formulation of the quantum field theory, to use as the basis a special algebra $\mathcal{R}$, which is an extension of the algebra $\mathcal{P}$ of the Poincaré group generators. The purpose of the paper is to find such a realization of the algebra $\mathcal{R}$, in which the Hamiltonian operator describes the interaction of quantized fields.

The extension of the algebra $\mathcal{P}$ is carried out in the following manner: we add to the generators P_μ and $M_{\mu\nu}$ the bispinor generators W_α and $\bar{W}_\beta$, which we shall call the generators of spinor translations.[158] In order to obtain the algebra $\mathcal{R}$, it is necessary to find the Lorentz-invariant form of the permutation relations between the translation generators. In order not to violate subsequently the connection between the spin and statistics, we shall consider anticommutators of the operators W_α and $\bar{W}_\beta$. A generalization of the Jacobi identities imposes stringent limitations on the form of possible commutations relations between the algebra operators. We confine ourselves to consideration of only those algebras $\mathcal{R}$, in which there are no subalgebras $\mathcal{Q}$ such that $\mathcal{P} \subset \mathcal{Q}$ and $\mathcal{P} \neq \mathcal{Q}$. This choice is governed by the fact that the remaining algebras $\mathcal{R}$ are obtained by further extending the algebras $\mathcal{R}$,

[158]Currently these generators are referred to as "supercharges" Q_α and $\bar{Q}_\beta$. –MS

and the field theories corresponding to them will have a still higher degree of symmetry.

An investigation of the algebras $\mathcal{R}$ has shown that upon spatial inversion they do not go over into themselves for any choice of the structure constants of the algebra. As a result, in a field theory that is invariant under such an algebra, the parity should not be conserved,[159] and the form of the non-conservation is completely determined by the algebra itself.

We shall now discuss one of the algebras $\mathcal{R}$,

$$\begin{aligned}
[M_{\mu\nu}, M_{\sigma\lambda}]_- &= i(\delta_{\mu\sigma}M_{\nu\lambda} + \delta_{\nu\lambda}M_{\mu\sigma} - \delta_{\mu\lambda}M_{\nu\sigma} - \delta_{\nu\sigma}M_{\mu\lambda}),\\
[P_\mu, P_\nu]_- &= 0,\\
[M_{\mu\nu}, P_\lambda]_- &= i(\delta_{\mu\lambda}P_\nu - \delta_{\nu\lambda}P_\mu),\\
[M_{\mu\nu}, W]_- &= \frac{i}{4}[\gamma_\mu, \gamma_\nu]_-\\
\bar{W} &= W^+\gamma_0;
\end{aligned} \tag{8.1a}$$

$$\begin{aligned}
&\boxed{[W_\pm, \bar{W}]_+ = \gamma_\mu^+ P_\mu}\\
&[W, W]_+ = 0,\\
&[P_\mu, W]_- = 0,
\end{aligned} \tag{8.1b}$$

where $\gamma_\mu^\pm = s^\pm\gamma_\mu$; $s^\pm = \frac{1}{2}(1 \pm \gamma_5)$; $\gamma_5^2 = 1$, the spinor indices have been omitted, and an expression of the type $d_\mu d^\mu$ is defined here and below as $d_0 d_0 - d_1 d_1 - d_2 d_2 - d_3 d_3$. [In the above expressions W is a *bispinor* in the *spinorial* representation and so are the γ matrices. Moreover, $[\ldots]_-$ and $[\ldots]_+$ denote commutators and anticommutators, respectively. –MS]

The irreducible representations of the algebra (8.1) can be reduced in terms of the irreducible representations of its subalgebra $\mathcal{P}$. A case of physical interest is one in which the reduction is carried out in accordance with the representations $\mathcal{P}$ characterized by a mass and spin. In the simplest irreducible representation of this type, the operator W^0 (the index "0" denotes

[159] A more detailed analysis of this question will be the subject of a separate paper. [See Igor Tamm Memorial Volume *Problems of Theoretical Physics*, Eds. V.L. Ginzburg *et al.*, Nauka, Moscow, 1972, page 37 (in Russian); Eng. Transl. in Yuri Golfand Memorial Volume *The Many Faces of the Superworld*, Ed. M. Shifman, World Scientific, Singapore, 2000, page 45. The superalgebra discovered by Golfand and Likhtman is now known as $\mathcal{N} = 1$. In fact, this superalgebra does not necessarily lead to parity violation, e.g. pure super-Yang-Mills theory conserves prity. As far as $\mathcal{N} = 1$ SQED is concerned see Eqs. (8.7), (8.8) below and my addendum. –MS].

henceforth that the operator is bilinear in the field operators) has the following form:

$$W^0 = s^+ W^0 = \frac{1}{i} \int d^3x \left(\phi^*(x) \overleftrightarrow{\partial_0} s^+ \psi_1(x) + \omega(x) \overleftrightarrow{\partial_0} s^+ \psi_1^c(x) \right), \qquad (8.2)$$

where $\phi(x)$ and $\omega(x)$ are scalar Hermitian fields, $\psi_1(x)$ is a spinor field, and the index "c" denotes charge conjugation.[160] The mass of all the fields is the same and equals m. The operators P_μ and $M_{\mu\nu}$ have the same usual form they have in the absence of interaction. Another irreducible representation contains the Hermitian scalar field $\chi(x)$, the transverse Hermitian vector field $A_\mu(x)$, and also the spinor field $\psi_2(x)$. The mass μ of these fields is also the same. The generator of the spinor translations is written in this representation as follows:

$$W^0 = s^+ W^0 = \frac{1}{i\sqrt{2}} \int d^3x \left(\chi(x) \overleftrightarrow{\partial_0} s^+ \psi_2(x) + A_\mu(x) \overleftrightarrow{\partial_0} \gamma_\mu^+ \psi_2(x) \right). \qquad (8.3)$$

To verify the validity of the formulas (8.2) and (8.3) it suffices to substitute them in the algebra (1) and carry out the appropriate calculations, during the course of which it is necessary to use the equations of motion and the commutation relations for the free fields.

We proceed now to describe the interaction of the field on the basis of the algebra (8.1). As seen from (8.1b), the operator P_μ not only commutes with W, but stands also in the right side of the commutation relations. Therefore if the operator W is equal to W^0, i.e., if it is bilinear in the field operators, then the operator $P_\mu = P_\mu^0$ is also bilinear in the field operators. In other words, the operators $W^\pm = W - W^0$ and $H^I = H - H^0$ $(H = P_0)$ either vanish simultaneously (and we obtain the theory of free fields), or else do not vanish (and we deal with the theory of interacting fields). Thus, in describing the interaction of the fields it is necessary to use representations (8.1) such that the operator W^I, like the operator H^I, is expressed in terms of a product of fields of higher degree than the operators W^0 and H^0. Such representations (8.1) are best sought by using perturbation theory in the interaction representation. To this end we expand the operators $W^0 + W^I(t)$

[160] Editor's note: One can combine two real fields ϕ and ω into one complex field Φ. Observing that the Dirac fermion filed in (8.2) actually appears only through its right-handed components, we can replace $s^+\psi_1$ by a Weyl spinor $\bar{\psi}_{\dot{\alpha}}$. Then one observes that the supercharge in (8.2) generates the Wess-Zumino supermultiplet $\{\Phi, \psi_\alpha\}$. –MS.

and $H^0 + H^I(t)$ in powers of the coupling constant g,

$$\begin{aligned} W^0 + W^I(t) &= W^0 + gW_1(t) + g^2W_2(t) + g^3W_3(t) + \cdots \\ H^0 + H^I(t) &= H^0 + gH_1(t) + g^2H_2(t) + g^3H_3(t) + \cdots \end{aligned} \tag{8.4}$$

and substitute (8.4) in the last relation of (8.1b):

$$\begin{aligned} &[W^0, H_1(t)]_- + [W_1(t), H^0]_- = 0 \\ &[W^0, H_2(t)]_- + [W_1(t), H_1]_- + [W_2(t), H^0]_- = 0 \\ &\quad\cdots \end{aligned} \tag{8.5}$$

Solving the i-th equation of (8.5) with respect to the operator $W_1(t)$, we obtain

$$\begin{aligned} W_1(t) &= i\int_{-\infty}^{t} [W^0, H_1(t)]_- \, dr \\ W_2(t) &= i\int_{-\infty}^{t} [W^0, H_2(t)]_- + [W_1(t), H_1]_- \, dr \\ &\cdots \end{aligned} \tag{8.6}$$

The increments in the orbital part of the operator $M_{\mu\nu}$ can be calculated in similar fashion. By direct substitution of (8.6) in (8.1) it is easy to verify that relations (8.6) determine the representations (8.1) for an arbitrary interaction Hamiltonian. However, at an arbitrary local density of the Hamiltonian $H^I(t)$ the density of the operator $W^I(t)$ will generally speaking not be local, owing to the interaction in (8.6) with respect to the time. Therefore, in order to obtain physical consequences, we generalize the concepts of invariance of the theory with respect to the algebra (or group) to a case when the algebra [of] operators themselves can depend strongly on the interaction, We shall stipulate that the spatial density of the operators of the algebra be a local function of the field operators.[161] As a result of this requirement, the integrand in (8.6) should be total derivatives with respect to time, and relations (8.6) are transformed into equations for the determination of the operators $H^I(t)$ and $W^I(t)$.

[161] A separate paper will be devoted to a justification of this postulate, and also to its comparison with the ordinary formulation of the requirement that the theory be invariant against a transformation group.

Equations (8.6) reduce to a system of linear homogeneous equations for constant coefficients, which are introduced as indeterminate coupling constants in the most general form of the interaction Hamiltonian. It was possible to solve this system of equations for the case when $H_1(t)$ is a product of three fields, two of which are transformed in accordance with representation (8.2) and the complex conjugate of (8.2), and the third in accordance with representation (8.3). The system (8.6) has in this case unique solutions, and only the operators $W_1(t)$, $H_1(t)$, $H_2(t)$ differ from zero. Knowing the exact form of the Hamiltonian in the interaction representation, we can reconstruct from it the Lagrangian in the Heisenberg representation,[162]

$$\begin{aligned}\mathcal{L}(x) = {} & (\partial_\alpha\phi^* - igA_\alpha\phi^*)(\partial_\alpha\phi + igA_\alpha\phi) - m^2\phi^*\phi + (\partial_\alpha\omega^* - igA_\alpha\omega^*) \\ & \times(\partial_\alpha\omega + igA_\alpha\omega) - m^2\omega^*\omega + \frac{i}{2}\bar{\psi}_1\gamma_\alpha\overleftrightarrow{\partial}_\alpha\psi_1 - m\bar{\psi}_1\psi_1 - g\bar{\psi}_1\gamma_\alpha\psi_1A_\alpha \\ & + \frac{i}{2}\bar{\psi}_2\gamma_\alpha\overleftrightarrow{\partial}_\alpha\psi_2 - \mu\bar{\psi}_2\psi_2 - \frac{1}{2}(\partial_\beta A_\alpha)^2 + \frac{\mu^2}{2}A_\alpha A_\alpha + \frac{1}{2}(\partial_\alpha\chi)^2 - \frac{\mu^2}{2}\chi^2 \\ & + g\mu(\phi^*\phi - \omega\omega^*)\chi - \frac{g^2}{2}(\phi^*\phi - \omega^*\omega)^2 + \sqrt{2}g(\bar{\psi}_1 s^-\psi_2\phi + \bar{\psi}_2 s^+\psi_1\phi^*) \\ & - \sqrt{2}g(\overline{\psi_1^c}s^+\psi_2\omega^* + \bar{\psi}_2 s^-\psi_1^c\omega). \end{aligned} \tag{8.7}$$

Thus, we have obtained a model for the interaction of quantized fields with parity nonconservation, invariant under the algebra (8.1).

Reference

1. S. Schweber, *Introduction to Relativistic Quantum Field Theory*, Harper, 1961.

[162]Editor's note: In modern terminology m is the mass of the electron/selectrons while μ is the photon/photino mass. In the last line in (8.7) (which is not clearly visible in the original JETP publication) I corrected presumable typos in $s^\pm$. Golfand-Likhtman's ψ_2 is a four-component spinor which is a superpartner to $\{A_\alpha,\ \chi\}$. Their Dirac spinor ψ_1 is a superpartner to the fields ω and ϕ both of which are considered complex in (8.7). Thus, the matter sector in (8.7) presents a reducible representation containing two Wess-Zumino supefmultiplets. –MS

Editor's Addendum (December 29, 2023)

From the modern perspective, the title of the Golfand-Likhtman paper is somewhat misleading. The superalgebra discovered by Golfand and Likhtman is currently known as $\mathcal{N}=1$. Although this suberalgebra can generate P noninvariant models this is *not* its unavoidable feature. The simplest and most well-known example of supersymmetric P-even theory is $\mathcal{N}=1$ Yang-Mills theory without matter. The Golfand-Likhtman model presented in Eq. (8.7) above *might seem to be P violating*[163] *— but only at first sight.* In fact, *it is not!*

The easiest way to see that $\mathcal{N}=1$ SQED conserves parity is to use the superfield formalism with the matter sector composed of two chiral superfieds Q and $\tilde{Q}$ with the electric charges ± 1 respectively. Under the P transformation $\theta \leftrightarrow \bar{\theta}$ and $Q \leftrightarrow \bar{\tilde{Q}}$, $\bar{Q} \leftrightarrow \tilde{Q}$. The invariance of the action is obvious.

However, the Golfand-Likhtman Lagrangian was obtained in components — the superfield formalism came almost four years later. For convenience of my component demonstration I will rewrite the Lagrangian (8.7) of supersymmetric QED in today's "standard" notation, *omitting* for simplicity the photon/photino mass μ from (8.7); it is irrelevant for the issue of P invariance. I will use two Weyl (left-handed) fermions ψ and $\tilde{\psi}$ with the opposite charges (and their Hermitean conjugated) instead of one Dirac field ψ_1 of Golfand-Likhtman.[164] Then Eq. (8.7) takes the form

$$\begin{aligned}\mathcal{L} = &\frac{1}{e^2}\left(-\tfrac{1}{4}F_{\mu\nu}F^{\mu\nu} + \bar{\lambda}_{\dot{\alpha}} i\partial^{\dot{\alpha}\alpha}\lambda_\alpha\right)\\ &+ \left(\mathcal{D}^\mu \bar{q}\,\mathcal{D}_\mu q + \bar{\psi}_{\dot{\alpha}} i\mathcal{D}^{\dot{\alpha}\alpha}\psi_\alpha\right) + \left(\mathcal{D}^\mu \bar{\tilde{q}}\,\mathcal{D}_\mu \tilde{q} + \bar{\tilde{\psi}}_{\dot{\alpha}} i\mathcal{D}^{\dot{\alpha}\alpha}\tilde{\psi}_\alpha\right)\\ &+ \left[i\sqrt{2}(\lambda\psi)\bar{q} + \text{H.c.}\right] + \left[-i\sqrt{2}\left(\lambda\tilde{\psi}\right)\bar{\tilde{q}} + \text{H.c.}\right]\\ &- m\left(\psi\tilde{\psi} + \text{H.c.}\right) - V(q,\tilde{q}),\end{aligned} \tag{8.8}$$

[163] Because of the charged matter fields in the last two lines of (8.7).

[164] Alternatively, it may me convenient to to introduce a Dirac spinor Ψ,

$$\Psi = \begin{pmatrix} \psi \\ \bar{\tilde{\psi}}, \end{pmatrix}$$

so that $\lambda\psi \to \frac{1}{2}\lambda(1-\gamma^5)\Psi\,\bar{q}$ and so on.

where the scalar potential $V(q, \tilde{q})$ is

$$V(q, \tilde{q}) = \frac{e^2}{2}(\bar{q}q - \bar{\tilde{q}}\tilde{q})^2 + |mq|^2 + |m\tilde{q}|^2. \tag{8.9}$$

Above, λ is the photino (Majorana) field, q and $\tilde{q}$ are the scalar fields (selectrons), i.e. the lowest components of the superfields Q and $\tilde{Q}$, respectively, and ψ and $\tilde{\psi}$ are the fermion components of Q and $\tilde{Q}$. The electron mass parameter m is assumed to be real. (This can be always achieved.) One should not forget that the electric charges of Q and $\tilde{Q}$ are opposite; therefore,

$$i\mathcal{D}_\mu q = (i\partial_\mu + A_\mu)\, q, \quad i\mathcal{D}_\mu \tilde{q} = (i\partial_\mu - A_\mu)\, \tilde{q},$$
$$i\mathcal{D}_\mu \psi = (i\partial_\mu + A_\mu)\, \psi, \quad i\mathcal{D}_\mu \tilde{\psi} = (i\partial_\mu - A_\mu)\, \tilde{\psi}. \tag{8.10}$$

The Lagrangian (8.8) *exactly* coincides with that derived by Wess and Zumino (Nucl. Phys. **B78** (1974) 1, received by the Editorial Office on April 19, 1974.) Equation (1) in the above Wess-Zumino paper is written in a slightly more detailed form than our Eq. (8.8),[165] which makes the statement of its P invariance obvious. It is not surprising that Wess and Zumino emphasize — in the very first paragraph of their paper — that their SQED Lagrangian preserves parity. –MS

[165]Wess and Zumino split the boson and fermion fields into real and imaginary parts — a procedure exhibiting the parities of their scalar fields A and B: A is scalar while B is pseudoscalar.

POSSIBLE UNIVERSAL NEUTRINO INTERACTION

D.V. VOLKOV AND V.P. AKULOV

Physical-Technical Institute, Ukrainian Academy of Sciences

Submitted 13 October 1972, ZhETF Pis. Red. 16, No. 11, 621–624 (5 December 1972)

Eng. Trans. JETP Lett. 16, 438–440 (1972)

Much attention is being paid of late in elementary-particle physics to the possible degeneracy of vacuum, and the ensuing spontaneous breaking of one symmetry or another. The most direct consequence of vacuum degeneracy is the occurrence of the zero-mass particles called *Goldstone* particles [1].

Of all the presently known elementary particles, only the neutrino, photon, and graviton have zero mass. The last two, however, correspond to gauge fields and apparently do not require vacuum degeneracy for their description. The neutrino is therefore the only particle whose existence may be directly connected with vacuum degeneracy.

We wish to point out here that the hypothesis that the neutrino is a Goldstone particle leads to a definite type of neutrino interaction both with other neutrinos and with all other particles. The interaction is fully defined by a single phenomenological coupling constant and is in this sense universal. To determine the type of symmetry whose spontaneous violation causes the degeneracy of vacuum and the corresponding properties of the neutrino as a Goldstone particle, let us consider the symmetry properties of the equation for the free neutrino

$$i\sigma_\mu \frac{\partial}{\partial x_\mu}\psi = 0. \tag{8.11}$$

This equation is invariant both with respect to the Poincaré group in chiral transformations , and with respect to shifts in spinor space, i.e., with respect to transformations of the type

$$\psi \to \psi' = \psi + \zeta, \quad x \to x' = x, \tag{8.12}$$

where ζ is a constant spinor that anticommutes with ψ.

We retain the character of the transformations of x_μ and ψ in the transformations of the Poincaré group, and replace the transformations (8.12) by the transformations

$$\psi \to \psi' = \psi + \zeta, \quad \psi^+ \to \psi^{+\prime} = \psi^+ + \zeta^+,$$
$$x_\mu \to x'_\mu = x_\mu = \frac{\alpha}{2i}\left(\zeta^+ \sigma_\mu \psi - \psi^+ \sigma_\mu \zeta\right). \tag{8.13}$$

The resultant structure with ten commuting and four anticommuting parameters has the structure of a group[166] and is the only possible generalization of (8.12) without introducing additional group parameters.

The constant a in the transformations (8.13) is arbitrary and has the dimension of length raised to the fourth power. We postulate that the equations for the neutrino with allowance for the interaction are invariant under the transformations (8.13). We assume also that the terms of the interaction contain the minimum number, compatible with the invariance requirement, of field derivatives.

To construct the phenomenological action integral under the foregoing assumptions, it suffices to use the following differential forms, which are invariant under the transformations (8.13),

$$\omega_\mu = dx_\mu + \frac{\alpha}{2i}\left(\psi^+ \sigma_\mu d\psi - d\psi^+ \sigma_\mu \psi\right). \tag{8.14}$$

The action integral invariantunder transformations (8.13) and the transformations of the Poincaré group is given by

$$S = \frac{1}{\alpha}\int \omega_0 \wedge \omega_1 \wedge \omega_2 \wedge \omega_3, \tag{8.15}$$

where $\wedge$ denotes an external product. The expression (8.15) corresponds to a certain four-dimensional volume in the group-parameter space.

If the 4-volumes are defined by specifying the function $\psi = \psi(x)$, the action integral (8.15) can be written in the following more usual form

$$S = \frac{1}{\alpha}\int |W|\, d^4x, \tag{8.16}$$

where $|W|$ is the determinant of the matrix W,

$$W_{\mu\nu} = \delta_{\mu\nu} + \alpha T_{\mu\nu},$$
$$T_{\mu\nu} = \frac{1}{2i}\left(\psi^+ \sigma_\mu \partial_\nu \psi - \partial_\nu \psi^+ \sigma_\mu \psi\right). \tag{8.17}$$

[166] Lie groups with commuting and anticommuting parameters were recenltly considered by Berezin and Katz [2].

It follows from (8.16) and (8.17) that the action integral as a function of the tensor T takes the form

$$S = \int d^4x \left[\frac{1}{\alpha} + T_{\mu\mu} + \frac{\alpha}{2} \left(T_{\mu\mu} T_{\nu\nu} - T_{\mu\nu} T_{\nu\mu} \right) \right.$$
$$\left. + \frac{\alpha^2}{3!} \sum_P (-1)^P T_{\mu\mu} T_{\nu\nu} T_{\rho\rho} + \frac{\alpha^3}{4!} \sum_P (-1)^P T_{\mu\mu} T_{\nu\nu} T_{\rho\rho} T_{\sigma\sigma} \right], \quad (8.18)$$

where the summation over P corresponds to a sum over all the permutations of the second indices in the products of the tensors T.

The term with $T_{\mu\mu}$ corresponds to the kinetic terms, and the terms with products of two, three, and four tensors T describe interactions in which four, six, and eight fields, respectively, participate. The degrees of the field derivatives in the interaction terms are determined by the number of factors T.

The neutrino interaction with other fields can be determined in a manner that is invariant with respect to the transformations (8.3).

Thus, for example, the action integral for a Dirac particle ϕ is given by

$$S = \int d^4x \left[R_{\mu\mu} + \alpha \left(R_{\mu\mu} T_{\nu\nu} - R_{\mu\nu} T_{\nu\mu} \right) + \frac{\alpha^2}{2} \sum_P (-1)^P R_{\mu\mu} T_{\nu\nu} T_{\rho\rho} \right.$$
$$\left. + \frac{\alpha^3}{3!} \sum_P (-1)^P R_{\mu\mu} T_{\nu\nu} T_{\rho\rho} T_{\sigma\sigma} + m\bar{\phi}\phi |W| \right] \quad (8.19)$$

where

$$R_{\mu\nu} = \frac{1}{2i} \left(\bar{\phi} \gamma_\mu \partial_\nu \phi - \partial_\nu \bar{\phi} \gamma_\mu \phi \right) \quad (8.20)$$

and the tensor T and the determinant $|W|$ are defined in (8.7).

Weak interactions can be included in the scheme under consideration by introducing gauge fields for the approximate unitary symmetry group of the neutrino and other leptons. The electromagnetic interaction for charged leptons is introduced simultaneously. The weak and electromagnetic interactions are turned on simultaneously by the well-known mechanism of spontaneous breaking of the unitary-group symmetry [3]. In the zero lepton mass limit, the unitary symmetry is exact. To obtain the action integral for the leptons in the unitary symmetry limit, it suffices to regard in (8.3), (8.4), and (8.7) the spinor products as invariant products of unitary multiplets. It is also possible to add to formulas (8.3), (8.4), and (8.7) the terms for leptons with opposite chirality. The unitary groups for states with different chirality need not necessarily coincide in this case.

The gauge fields can be introduced into the so-generalized action integral in a manner covariant with respect to the transformations (8.3).

The gravitational interaction can be introduced into the scheme in analogous fashion, by introducing gauge fields corresponding to the Poincaré group. We note that if we introduce also gauge fields corresponding to the transformations (8.3), then, as a consequence of the Higgs effect [4], a massive gauge field with spin 3/2 arises, and the Goldstone particles with spin 1/2 vanish.

References

1. J. Goldstone, Nuovo Cim. **19**, 154 (1961).
2. F.A. Berezin and G.I. Katz, *Matematicheskii sbornik*, **82**, 343 (1970).
3. S. Weinberg, Phys. Rev. Lett. **19**, 1264 (1967).
4. P.W. Higgs, Phys. Rev. **145**, 1150 (1966).

HIGGS EFFECT FOR GOLDSTONE PARTICLES WITH SPIN 1/2

D. V. VOLKOV AND V. A. SOROKA
Physico-Technical Institute, Ukrainian Academy of Sciences

Submitted 3 September 1973
ZhETF Pis. Red. **18**, No. 8, 529–532 (20 October 1973)
Eng. JETP Lett. **18**, *312–314, (1973)*

Abstract: We consider the introduction of gauge fields and the Higgs effect for Goldstone particles with spin 1/2.

Higgs and others [1–3] have noted that when interactions are turned on between Goldstone particles and gauge fields, the Goldstone particles vanish and those gauge fields whose quantum numbers coincide with the quantum numbers of the Goldstone particles acquire mass (the Higgs effect).

We consider in this article a variant of the Higgs effect, in which the "absorption" of the Goldstone particles is effected by a "foreign" gauge field, i.e., by a gauge fields whose quantum numbers differ from those of the Goldstone particles.

We consider as the symmetry group G the direct product of a Poincare group with a certain group of internal symmetry, supplemented by the following transformations:

$$\begin{aligned} \psi &\to \psi' = \psi + \xi, \\ x_\mu &\to x'_\mu = x_\mu - \frac{a}{2i}\left(\xi^\dagger \sigma_\mu \psi - \psi^\dagger \sigma_\mu \xi\right). \end{aligned} \tag{8.21}$$

We use the definitions of [5].

The spontaneous violation of the considered symmetry group under the assumption that the direct product of the Poincare group and of the internal-symmetry group leaves the vacuum invariant, leads to the appearance of a Goldstone particle with spin 1/2 [4, 5].

Let us consider the gauge transformations of the group G with parameters ℓ, u, t, and ξ, which depend on x_μ and gauge fields that are coefficient functions of the differentials ∂x_μ and $\partial\psi$ of the differential form $A(d)$ with the following transformation law:

$$A'(d) = G(t,\xi;\ell,u)A(d)G^{-1}(t,\xi;\ell,u) + \frac{1}{f}G(t,\xi;\ell,u)G^{-1}(t,\xi;\ell,u) \quad (8.22)$$

where $A(d) = A'(d)Z_i$ and Z_i are the generators of group G.

The differential forms

$$\begin{aligned}\tilde{\omega}(d) = H^{-1}(\ell,u)\omega(d)H(\ell,u) &= G^{-1}(t,\xi;\ell,u)\partial G(t,\xi;\ell,u)\\ &+ fG^{-1}(t,\xi;\ell,u)A(d)G(t,\xi;\ell,u)\end{aligned} \quad (8.23)$$

which are invariant to the transformation (8.22) and correspond to the translation generators in the transformations (8.21), take the following form as functions of the gauge fields:

$$\begin{aligned}\omega_\mu(d) &= Dx_\mu + \frac{a}{2i}\left[\psi^\dagger\sigma_\mu\left(D\psi + 2f\Phi(d)\right) - \left(D\psi + 2f\Phi(d)\right)^\dagger\sigma_\mu\psi\right] + fW_\mu(d),\\ \omega_a^\alpha(d) &= \left(D\psi + 2f\Phi(d)\right)_a^\alpha, \quad \omega^{a\dot{\alpha}}(d) = \omega_a^\alpha(d)^\dagger,\\ Dx_\mu &= \partial x_\mu + 2\Omega_{\mu\nu}(d)x^\nu,\\ D\psi &= \partial\psi + f\Omega^{\mu\nu}(d)I_{\mu\nu}\psi - ifV^\alpha(d)I_a\psi,\end{aligned} \quad (8.24)$$

where $I_\mu = (1/4)(\delta_\mu\sigma_\nu - \delta_\nu\sigma_\mu)$, I_a re the generators of the internal-symmetry group, and the gauge fields $\Omega_{\mu\nu}(d), W(d), V^a(d)$, and $\Phi_a^\alpha(d)$ are the coefficients of the expansion of the form $A(d)$ in terms of generators of the considered group.

The forms $\omega(d)$, which correspond to the generators of the Lorentz group, as well as the internal-symmetry group, depend on the parameters ℓ and u. There is no such dependence in the second-order differential forms

$$\begin{aligned}R^{\mu\nu}(\delta,d) &= \delta\Omega^{\mu\nu}(d) - d\Omega^{\mu\nu}(\delta) + 2f\left[\Omega_\lambda^\mu(\delta)\Omega^{\lambda\nu}(d) - \Omega_\lambda^\nu(\delta)\Omega^{\lambda\mu}(d)\right];\\ F^a(\delta,d) &= \delta V^a(d) - \partial V^a(\delta) + fC_{bc}^a V^b(\delta)V^c(d),\end{aligned} \quad (8.25)$$

where C_{bc}^a are the structure constants of the internal symmetry group.

The invariant action integral is made up in the form of outer products of fourth order, which are invariant with respect to the Lorentz group and the internal-symmetry group, of the forms (8.24) and (8.25) (or in the general case in the form of a homogeneous first-order function of such products (see [5, 6])).

The simplest permissible invariant combinations are

$$i\omega_{\dot\alpha}^{\dot\beta}(d_0) \wedge \omega_{\dot\beta}^{\gamma}(d_1) \wedge \omega_{\gamma}^{\dot\delta}(d_1) \wedge \omega_{\dot\delta}^{\alpha}(d_1), \tag{8.26}$$

$$D_0 \wedge \omega_a^{\alpha}(d_1) \wedge \omega_{\alpha\dot\beta}(d_2) \wedge \omega^{a\dot\beta}(d_3) - D_0 \wedge \omega^{a\dot\alpha}(d_1) \wedge \omega_{\dot\alpha\beta}(d_2) \wedge \omega_{\alpha}^{\beta}(d_3), \tag{8.27}$$

$$i\left[R_{\dot\alpha}^{\dot\beta}(d_0,d_1) \wedge \omega_{\dot\beta\gamma}(d_2) \wedge \omega^{\gamma\dot\alpha}(d_3) - R_{\beta}^{\alpha}(d_0,d_1) \wedge \omega^{\beta\dot\gamma}(d_2) \wedge \omega_{\dot\gamma\alpha}(d_3)\right], \tag{8.28}$$

$$\frac{\left[F^a(d_0,d_1) \wedge \omega_{\dot\alpha}^{\dot\beta}(d_2) \wedge \omega_{\gamma}^{\dot\delta}(d_3)\right]\left[F^a(d_0,d_1) \wedge \omega_{\dot\beta}^{\alpha}(d_2) \wedge \omega_{\dot\delta}^{\gamma}(d_3)\right]}{i\omega_{\dot\alpha}^{\dot\beta}(d_0) \wedge \omega_{\dot\beta}^{\gamma}(d_1) \wedge \omega_{\gamma}^{\dot\delta}(d_2) \wedge \omega_{\dot\delta}^{\alpha}(d_3)} \tag{8.29}$$

where $\wedge$ is the outer-product symbol.

Expressions (8.27)–(8.29) contain kinetic terms for gauge fields with spins 3/2, 2, and 1, respectively. Expression (8.26) contains a kinetic term for Goldstone particles.

Al the field variables in the products (8.26)–(8.29) can be parametrized in the form of functions of x_μ can be parametrized in the form of functions of x_μ and the redefined gauge fields corresponding to the coefficients of the differentials.

To consider the Higgs effect it suffices to examine the structure of forms (8.24), (8.25). In the form $\omega_a^{\alpha}(d)$ the gauge form $\Phi(d)$ for spin 3/2 enters together with $D\psi$, so that we can define as a new gauge form

$$f\Phi'(d) = D\psi + f\Phi(d). \tag{8.30}$$

In this case, however, the Goldstone field ψ in the form $\omega_\mu(d)$is not eliminated completely, so that there is no Higgs effect for this field. Moreover, by imposing the invariant condition[167]

$$\omega^{\alpha}(d) = 0 \tag{8.31}$$

we can exclude from consideration the gauge field with spin 3/2.

The gauge form $W_\mu(d)$ enters i n the form $\omega_\mu(d)$ together with other terms that contain, in particular, kinetic terms of Goldstone particles. A change to a new gauge form

$$fW'(d) = \omega_\mu(d) \tag{8.32}$$

[167]The possibility of using conditions such as (8.31) to eliminate some of the considered fields was pointed out to us by V. I. Ogievetsky.

eliminates the variables connected with the Goldstone fields. This circumstance corresponds to the Higgs effect in which the Goldstone particles vanish as a result of a redefinition of the metric tensor.

Owing to redefinition of the gauge field $W_\mu(d)$, the invariants (8.27)–(8.29) correspond to interacting Yang-Mills fields, to a field with spin 3/2, and to gravitational field with a cosmological term (8.26).

A Goldstone effect field with spin 1/2 can be retained only by violating the gauge group. Therefore the program proposed in [4], of including the weak and electromagnetic interactions in the scheme of Goldstone particles, can be carried only with allowance for this violation.

References

1. P. W. Higgs, Phys. Rev. **145**, 1156–1163 (1966).
2. A. A. Migdal and A. M. Polyakov, Zh. Eksp. Teor. Fiz. **51**, 135 (1966) [Sov. Phys. -JETP **24**, 91 (1967)].
3. T. W. Kibble, Phys. Rev. **155**, 1557 (1967).
4. D. V. Volkov and V. P. Akulov, ZhETF Pis.Red. **16**, 621 [JETP Lett. **16**, 438 (1972)].
5. D. V. Volkov and V. P. Akulov, Preprint ITF-73-51R, Kiev. 1973.
6. D. V. Volkov, Fiz. Elem. Chastits At. Yad. **4**, 3 [Sov. J. Part. Nuclei **4**, 1, (1973)].

INDEX